Q_1, Q_2, Q_3 quartiles

$D_1, D_2, \ldots, D_9$ deciles

$P_1, P_2, \ldots, P_{99}$ percentiles

x value of a single score

f frequency with which a value occurs

Σ (capital sigma) summation

n number of scores in a sample

$n!$ factorial

N number of scores in a finite population; also used as the size of all samples combined

$\bar{x}$ mean of the scores in a sample

μ (mu) mean of all scores in a population

s standard deviation of a set of sample values

σ (lower case sigma) standard deviation of all values in a population

s^2 variance of a set of sample values

σ^2 variance of all values in a population

z standard score

$z_{\alpha/2}$ critical value of z

t t distribution

$t_{\alpha/2}$ critical value of t

df number of degrees of freedom

F F distribution

χ^2 chi-square distribution

χ_R^2 right-tailed critical value of chi-square

χ_L^2 left-tailed critical value of chi-square

p probability of an event or the population proportion

q probability or proportion equal to $1 - p$

$\hat{p}$ sample proportion

$\hat{q}$ sample proportion equal to $1 - \hat{p}$

$\bar{p}$ proportion obtained by pooling two samples

$\bar{q}$ proportion or probability equal to $1 - \bar{p}$

$P(A)$ probability of event A

$P(A|B)$ probability of event A assuming event B has occurred

$_nP_r$ number of permutations of n items selected r at a time

$_nC_r$ number of combinations of n items selected r at a time

Fourth Edition
Elementary Statistics

Mario F. Triola

Dutchess Community College
Poughkeepsie, New York

The Benjamin/Cummings Publishing Company, Inc.

Redwood City, California · Fort Collins, Colorado
Menlo Park, California · Reading, Massachusetts · New York
Don Mills, Ontario · Wokingham, U.K. · Amsterdam · Bonn
Sydney · Singapore · Tokyo · Madrid · San Juan

To Marc and Scott

Sponsoring Editor: Lisa Moller
Production Editor: Sharon Montooth
Production Service: Stacey C. Sawyer, San Francisco
Book design: Nancy Benedict
Artist: Mary Burkhardt
Composition: Jonathan Peck Typographers, Ltd.
Cover design: Albert Burkhardt

Library of Congress Cataloging-in-Publication Data

Triola, Mario F.
　　Elementary statistics/Mario F. Triola
　　　　p.　　cm.
　　Includes index.
　　1. Statistics　I. Title.
QA276.12.T76 1989
519.5--dc19
ISBN 0-8053-0271-9　　　　　　　　　　　　　　　88-31460
　B C D E F G H I J – D O – 8 9 3 2 0 9　　　　　CIP

The Benjamin/Cummings Publishing Company, Inc.
390 Bridge Parkway
Redwood City, California 94065

Preface

Why Study Statistics?

Modern times have created modern problems, and many of those problems involve data. Surveys, polls, product testing, quality control, standardized testing, and conservation of resources are but a few of the many different fields that require an intelligent analysis of data. In particular, the working world now demands employees that are better prepared to cope with increased uses of statistics. The *New York Times* recently reported that "when Motorola Inc. introduced a system of quality control at its plant in Arcade, N.Y., it found that many employees lacked the mathematical skills needed to understand the new statistics-based approach." As employees, employers, and as citizens, we must learn at least the elementary concepts that constitute the field of statistics. This book is designed to be an interesting and readable introduction to those concepts.

Audience

This book is an introduction to elementary statistics for students majoring in any field except mathematics. A strong mathematics background is not necessary, but students should have completed a high school algebra course. Although underlying theory is included, this book does not stress the mathematical rigor more suitable for mathematics majors.

In this book, strong emphasis is placed on interesting, clear, and readable writing. Because the many examples and exercises cover a wide variety of different applications, this book is appropriate for many disciplines. The previous editions have been used successfully by majors in psychology, sociology, business, computer science, data processing, biology, education, engineering technology, liberal arts, humanities, history, social science, nursing, health, economics, ecology, agriculture, and many others.

Changes in the Fourth Edition

This fourth edition of *Elementary Statistics* includes all the basic features of previous editions. In response to extensive reader surveys, almost every section has been modified to some extent. Some of the new features are identified below.

Beginning-of-Chapter Features

- List of **chapter sections** along with brief descriptions of their contents
- Statement of a **chapter problem**
- **Overview** of the chapter, including statement of chapter objectives

End-of-Chapter Features

- **Review** of the chapter
- Summary list of important **formulas**
- **Vocabulary list** of important terms introduced in the chapter (Glossary is in Appendix D)
- **Review Exercises**
- **Computer Project**
- **Case Study Activity**
- **Data Project** (a new feature) that uses real data sets

Major Content Changes

- New section in Chapter 9: **Multiple Regression**
- **Odds** are included in Section 3-5
- Section 6-2 (Testing a Claim About a Mean) has been reorganized
- New Appendix B includes two **real data sets**
- **Minitab** displays included in many sections
- New Appendix C on **Minitab**
- Many more examples and exercises use identifiable **real data**

Exercises

The book now has more than **1600 exercises**; an increase of 10%. Approximately **45% of the exercises are new**. Many exercises refer to identifiable **real-world data**. Also, a major effort has been made to group exercises so that there is usually an even-numbered exercise similar to each odd-numbered exercise.

As in previous editions, the exercises are arranged in order of increasing difficulty. The exercises are divided into groups A and B, with B types involving more difficult concepts or a stronger mathematical background. In some cases, the B exercises introduce new concepts, such as skewness, the probaiblity β of a type II error, or quality control charts.

Unlike previous editions, the review exercises no longer follow the same order of the chapter sections. In response to user requests, the order of the review exercises is now random. Students should find this more helpful when they use review exercises in preparing for tests.

The exercise sets include a wide variety of real situations from many diverse fields.

Other Features

- The **flowcharts** help clarify the more complicated procedures
- Appendix D contains an expanded **glossary** of important terms
- Appendix F contains **answers** to all the odd-numbered exercises
- A **symbol table** is included on the front inside cover for quick and easy reference to key symbols
- Copies of Tables A-2 and A-3 are included in the rear inside cover for quick and easy reference
- A detachable **formula card** is enclosed for use throughout the course
- There are now more than 90 of the very popular **margin essays**, including 23 new ones. These short essays illustrate uses of statistics in very real and practical applications. The following is a sample of some of the topics covered.

  Advertising: How commercials are regulated

  Biology: Were Mendel's experimental data manipulated?

  Business: How airlines use sampling to save money by determining revenues from split-ticket sales

  Criminology: How valid are crime statistics?

  Ecology: Use of the capture-recapture method to estimate population sizes

  Economics: How unemployment figures are obtained

  Education: How the use of older norms results in misleading scores on standardized tests

Entertainment: How the new Nielsen T.V. rating system works

Medicine: How the Salk vaccine was tested

Psychology: How to measure a seemingly qualitative characteristic such as disobedience

Sociology: Code of ethics for survey research

Sports: How statistics can be used for better baseball strategy

Computers

This text can be used without any reference to computers. For those
who choose to supplement the course with computers, we have
included **computer projects** at the end of each chapter in Chapters 2
through 11.

We also have two different levels of software available.

- **STATDISK** is an easy-to-use statistical software package that
does not require any previous computer experience. Developed
as a supplement specifically for this textbook, STATDISK is available for the IBM PC and the Apple IIe systems. This software is
provided at no cost to colleges who adopt this text.

- **STATDISK Manual/Workbook** includes instructions on the use
of the STATDISK software package. It also includes experiments
to be conducted by students.

The STATDISK software and the STATDISK Manual/Workbook
have been designed so that instructors can assign computer experiments without using classroom time that may be quite limited.
STATDISK includes a wide variety of programs that can be used
throughout the course, and the experiments do more than number
crunch or duplicate text exercises. They include concepts that can
be discovered through computer use. This text includes several
sample displays that result from the use of STATDISK.

For those who wish to use **Minitab**, we have included Minitab
displays throughout the text. Appendix C summarizes key components of Minitab.

- **Minitab Manual/Workbook**, designed specifically for this text,
includes instructions and examples of Minitab use. It also
includes experiments to be conducted by students.

- **Student Edition of Minitab**, available from Addison-Wesley · Benjamin/Cummings for the price of a textbook. Includes
program software with data sets developed by Minitab, Inc. and
a comprehensive user's manual with tutorials and a reference
section written by Robert L. Schaefer of Miami University,
Oxford, Ohio, and Richard B. Anderson.

Acknowledgments

I extend my sincere thanks for the suggestions made by David Bernklau of Long Island University and the following reviewers of this fourth edition.

Frank Gunnip
Oakland Community College

Mary Parker
Austin Community College

Mike Karelius
American River College

David Stout
University of West Florida

Milton Loyer, Messiah College

I also wish to thank the following individuals who have given me valuable input for the fourth edition.

John Anderson
Mary Baker
James Beaird
Chuck Beals
Sherry Blackman
Allan Bluman
Marjorie Bradford
Elizabeth Bryan
Brian Buhrman
Chris Burditt
Hanna Burger
C. Allan Burns
Dennis Callas
Paul Campbell
Ronald Capletto
Carol Carpenter
Roger Champagne
William Chatfield
Don Cohen
Susan Cribelli
Pablo Echeverria
Carol Edgington
Nancy Ellis
Carlton Evans
Dale Everson
Michael Flaherty
Eugene Franks
John Giovino
Sheldon Gordon

Basil Goungetas
Hoke Griffin
Mel Griffin
Norris Griffith
Lewis Hall
O. Randall Harman
Raymond Hebert
H. Joseph Heffelfinger
Joan Iacocca
Francis Jones
J. Kapoor
Gary King
Kenneth Korbin
John Kurtzke
Theodore Lai
John Lawry
James Leslie
Donna Linksz
J. E. Lockley
William Maneer
George Marrah
Katherine McLain
Michelle Meoli
Eileen Murphy
Jim Murphy
John Murphy
Desmond Navares
Eloise Olson
E. E. Peters

Bruce Phillips
Chester Piascik
James Potts
Robert Pumford
Peter Ross
S. Kara Ryan
Sohindar Sachdeu
Robert Sackett
Larry Scott
Jean Sells
Robert Shaddy
Donald Shirey
Errol Simpson
Lawrence Somer
Edward Spitznagel
Suzanne Stock
Gary Taka
Chris Vertullo
Roslyn Vinnik
Donald Walters
Sam Weaver
Roger Willig
Janet Wilson
John Wilson
R. Ward Wilson
Lynn Yankowski
Arlene Yee
Elmar Zemgalis
Michael Zwilling

I would like to extend special thanks to Donald K. Mason of Elmhurst College for his valuable assistance in checking solutions and providing useful suggestions. I also wish to extend my thanks to Stacey Sawyer,

Lisa Moller, Sharon Montooth, Sally Elliott, and the entire Benjamin/ Cummings staff. Finally, I thank Ginny, Marc, and Scott for their support, encouragement, and assistance.

Supplements

- **STATDISK** software for the IBM PC
- **STATDISK** software for the Apple IIe
- **STATDISK Manual/Workbook**
- **Minitab Manual/Workbook**
- **Student Edition of Minitab** (software and manual)
- **Student Solutions Manual** by Donald K. Mason (provides detailed, worked-out solutions to odd-numbered exercises)
- **Instructor's Resource Guide** (includes all answers, printed test bank, data sets, transparency masters)
- Computer test generator for the IBM PC is available from Benjamin/Cummings

To the Student

I strongly recommend the use of a calculator. You should have one that can be used for finding square roots. Such calculators usually have a key labeled $\sqrt{x}$. Also, it should use algebraic logic instead of chain logic. You can identify the type of logic by pressing the buttons

$$2 + 3 \times 4 =$$

If the result is 14, the calculator uses algebraic logic. If the result is 20, the calculator uses chain logic and it is not very suitable for a statistics course. Some inexpensive calculators can directly compute the mean, standard deviation, correlation coefficient, and the slope and intercept values of a regression line. Such keys are usually identified as Mean or $\bar{x}$; S. Dev, SD, or σ_{n-1}; Corr or r; Slope; and Intcp.

I also recommend that you read the overview carefully when you begin a chapter. Read the next section quickly to get a general idea of the material; reread it carefully. Try the exercises. If you encounter difficulty, return to the section and work some of the examples in the text so that you can compare your solution to the one given.

When working on assignments, first attempt the earlier odd-numbered exercises. Check your answers with Appendix F and verify that you are correct before moving on to the other exercises. Keep in mind that neat and well-organized written assignments tend to produce better results. When you finish a chapter, check the review section to make sure that you didn't miss any major topics. Before taking tests, do the review exercises at the end of the chapters. In addition to helping you review, this will help you cope with a variety of different problems.

You might consider purchasing the *Student Solutions Manual* for this text. Written by Donald K. Mason of Elmhurst College, it gives detailed solutions to many of the odd-numbered text exercises.

M. F. T.
LaGrange, New York
January, 1989

Contents

1 **Chapter One**
Introduction to Statistics 2

Chapter Problem 3
1-1 Overview 4
1-2 Background 5
1-3 Uses and Abuses of Statistics 8
1-4 The Nature of Data 14
Vocabulary List 20
Exercises A 20
Exercises B 24
Case Study Activity 25
Data Project 25

2 **Chapter Two**
Descriptive Statistics 26

Chapter Problem 27
2-1 Overview 28
2-2 Summarizing Data 30
Exercises A 37
Exercises B 40
2-3 Pictures of Data 41
Pie Charts 41
Histograms 42
Frequency Polygons 45
Ogives 46
Stem-and-Leaf Plots 47
Miscellaneous Graphics 50
Exercises A 50
Exercises B 56

2-4 Averages 57
Mean 58
Median 60
Mode 62
Midrange 63
Weighted Mean 65
Exercises A 69
Exercises B 74
2-5 Dispersion Statistics 77
Range 78
Standard Deviation 80
Variance 81
Exercises A 89
Exercises B 94
2-6 Measures of Position 96
Exercises A 104
Exercises B 106
Vocabulary List 107
Review 108
Important Formulas 109
Review Exercises 109
Computer Project 112
Case Study Activity 112
Data Project 113

3 **Chapter Three**
Probability 114

Chapter Problem 115
3-1 Overview 116

3-2 **Fundamentals 120**
   Rounding Off Probabilities 126
   Exercises A 128
   Exercises B 131
3-3 **Addition Rule 132**
   Exercises A 140
   Exercises B 143
3-4 **Multiplication Rule 144**
   Conditional Probabilities 146
   Exercises A 157
   Exercises B 160
3-5 **Complements and Odds 161**
   Complementary Events 161
   Odds 164
   Exercises A 168
   Exercises B 171
3-6 **Counting (Optional) 172**
   Exercises A 182
   Exercises B 184
   Vocabulary List 185
   Review 186
   Important Formulas 187
   Review Exercises 188
   Computer Project 193
   Case Study Activity 194
   Data Project 195

4 **Chapter Four**
Probability
Distributions 196

   Chapter Problem 197
4-1 **Overview 198**
4-2 **Random Variables 198**
   Discrete Random Variables 201
   Exercises A 206
   Exercises B 208
4-3 **Mean, Variance, and Expectation 209**
   Expected Value 212
   Exercises A 214
   Exercises B 217
4-4 **Binomial Experiments 218**
   Exercises A 230
   Exercises B 234

4-5 **Mean and Standard Deviation for the**
   Binomial Distribution 235
   Exercises A 240
   Exercises B 242
4-6 **Distribution Shapes 243**
   Exercises A 252
   Exercises B 254
   Vocabulary List 255
   Review 255
   Important Formulas 256
   Review Exercises 257
   Computer Project 260
   Case Study Activity 260
   Data Project 261

5 **Chapter Five**
Normal Probability
Distributions 262

   Chapter Problem 263
5-1 **Overview 264**
5-2 **The Standard Normal Distribution 266**
   Exercises A 273
   Exercises B 274
5-3 **Nonstandard Normal Distributions 276**
   Exercises A 281
   Exercises B 284
5-4 **Finding Scores When Given**
   Probabilities 286
   Exercises A 290
   Exercises B 292
5-5 **Normal as Approximation to**
   Binomial 293
   Exercises A 303
   Exercises B 305
5-6 **The Central Limit Theorem 305**
   Exercises A 317
   Exercises B 321
   Vocabulary List 322
   Review 322
   Important Formulas 323
   Review Exercises 324
   Computer Project 326
   Case Study Activity 327
   Data Project 327

6 **Chapter Six**
Testing Hypotheses **328**

Chapter Problem 329
6-1 Overview 330
6-2 Testing a Claim About a Mean 331
       Null and Alternative
         Hypotheses 333
       Type I and Type II Errors 336
       Conclusions 337
       Left-Tailed, Right-Tailed,
         Two-Tailed 339
       Assumptions 345
     Exercises A 346
     Exercises B 349
6-3 *P*-Values 350
     Exercises A 355
     Exercises B 356
6-4 *t* Test 357
       Important Properties of the Stu-
         dent *t* Distribution 361
       P-Values 365
     Exercises A 367
     Exercises B 371
6-5 Tests of Proportions 372
       P-Values 376
     Exercises A 377
     Exercises B 379
6-6 Tests of Variances 380
     Exercises A 386
     Exercises B 389
     Vocabulary List 390
     Review 390
     Important Formulas 391
     Review Exercises 392
     Computer Project 394
     Case Study Activity 394
     Data Project 395

7 **Chapter Seven**
Estimates and Sample
Sizes **396**

Chapter Problem 397
7-1 Overview 398

7-2 Estimates and Sample Sizes of
       Means 398
       Determining Sample Size 405
       Small Sample Cases 406
     Exercises A 409
     Exercises B 413
7-3 Estimates and Sample Sizes of
       Proportions 414
       Sample Size 417
     Exercises A 421
     Exercises B 423
7-4 Estimates and Sample Sizes of
       Variances (Optional) 424
     Exercises A 430
     Exercises B 432
     Vocabulary List 433
     Review 433
     Important Formulas 434
     Review Exercises 435
     Computer Project 437
     Case Study Activity 438
     Data Project 439

8 **Chapter Eight**
Tests Comparing Two
Parameters **440**

Chapter Problem 441
8-1 Overview 442
8-2 Tests Comparing Two Variances 442
     Exercises A 450
     Exercises B 453
8-3 Tests Comparing Two Means 454
     Exercises A 471
     Exercises B 478
8-4 Tests Comparing Two Proportions 480
     Exercises A 487
     Exercises B 490
     Vocabulary List 491
     Review 491
     Important Formulas 494
     Review Exercises 494
     Computer Project 498
     Case Study Activity 499
     Data Project 499

9 Chapter Nine
Correlation and Regression 500

Chapter Problem 501
9-1 Overview 502
9-2 Correlation 503
  Method 1 508
  Method 2 508
  Common Errors 512
  Properties of r 516
  Exercises A 516
  Exercises B 521
9-3 Regression 522
  Common Errors 529
  Exercises A 533
  Exercises B 536
9-4 Variation 536
  Exercises A 545
  Exercises B 546
9-5 Multiple Regression 547
  Exercises A 552
  Exercises B 555
  Vocabulary List 556
  Review 556
  Important Formulas 557
  Review Exercises 558
  Computer Project 562
  Case Study Activity 562
  Data Project 563

10 Chapter Ten
Chi-Square and Analysis of Variance 564

Chapter Problem 565
10-1 Overview 566
10-2 Multinomial Experiments 567
  Exercises A 574
  Exercises B 578
10-3 Contingency Tables 579
  Exercises A 586
  Exercises B 592

10-4 Analysis of Variance 593
  Equal Sample Sizes 594
  Unequal Sample Sizes 598
  Exercises A 605
  Exercises B 610
  Vocabulary List 611
  Review 611
  Important Formulas 612
  Review Exercises 613
  Computer Project 616
  Case Study Activity 616
  Data Project 616

11 Chapter Eleven
Nonparametric Statistics 618

Chapter Problem 619
11-1 Overview 620
  Advantages of Nonparametric
    Methods 620
  Disadvantages of Nonparametric
    Methods 620
11-2 Sign Test 623
  Claims Involving Nominal
    Data 627
  Claims About a Median 629
  Exercises A 631
  Exercises B 635
11-3 Wilcoxon Signed-Rank Test for Two
    Dependent Samples 636
  Exercises A 640
  Exercises B 644
11-4 Wilcoxon Rank-Sum Test for Two
    Independent Samples 645
  Exercises A 649
  Exercises B 653
11-5 Kruskal-Wallis Test 654
  Exercises A 658
  Exercises B 662
11-6 Rank Correlation 662
  Case I: Perfect Positive
    Correlation 666
  Case II: Perfect Negative
    Correlation 666
  Case III: No Correlation 666

Exercises A   671
Exercises B   676

11-7   Runs Test for Randomness   677
          Randomness Above and Below the
               Mean or Median   685
          Exercises A   686
          Exercises B   691
          Vocabulary List   692
          Review   692
          Important Formulas   693
          Review Exercises   695
          Computer Project   699
          Case Study Activity   699
          Data Project   699

12 Chapter Twelve
Design, Sampling, and Report Writing   700

Identifying Objectives   701
12-1   Designing the Experiment   702
12-2   Sampling and Collecting Data   703
          Random Sampling   703
          Stratified Sampling   704
          Systematic Sampling   704
          Cluster Sampling   705
          Convenience Sampling   705
          Importance of Sampling   706
**12-3   Analyzing Data and Drawing
          Conclusions   706**
12-4   Writing the Report   707
12-5   A Do-It-Yourself Project   708
          Exercises A   709
          Exercises B   710
          Vocabulary List   711

Appendix A: Tables   712

Appendix B: Data Sets   735

Appendix C: Minitab   741

Appendix D: Glossary   748

Appendix E: Bibliography   753

**Appendix F: Answers to
          Odd-Numbered Exercises   755**

Index   782

Essays

CHAPTER 1

Airline Companies Save by Sampling   7
Capture-Recapture   8
And We Thought We Were Doing So Well   10
The Gender Gap in Wages   11
The *Literary Digest* Poll   13
How Do You Measure Disobedience?   18

CHAPTER 2

Authors of the *Federalist* Papers Identified   30
How Valid Are Crime Statistics?   42
Just an Average Guy   58
Mean and Median Incomes of Veterinarians   61
The Class Size Paradox   65
How Valid Are Unemployment Figures?   69
The Census   79
People Meters   86
Index Numbers   98

CHAPTER 3

Coincidences?   117
Using Coins to Survey Sensitive Issues   120
Should You Guess on SATs?   122
Just How Probable Is Probable?   127
Probability and Prosecution   134
Monkeys Are Not Likely to Type *Hamlet*   136
Not Too Likely   138
Convicted by Probability   145
Multiplication Rule Presents Difficulty to FAA   149

Redundancy   150
Nuclear Power Plant Has an Unplanned
   "Event"   153
Bayes' Theorem   155
You Bet?   162
Long-Range Weather Forecasting   165
The Number Crunch   172
Safety in Numbers   175
Probability and Promotion Contests   177
Voltaire Beats the Lottery   181

CHAPTER 4

Is Parachuting Safe?   199
How Not to Pick Lottery Numbers   203
Prophets for Profits   213
Composite Sampling   225
Who Is Shakespeare?   238
Clusters of Disease   245
Captured Tank Serial Numbers Reveal Population
   Size 247

CHAPTER 5

Reliability and Validity   265
Bullhead City Gets Hotter   267
Pollster Lou Harris   270
How Large Was Shakespeare's Vocabulary?   294
Magazine Study Results Reflect Readership   297
Juxtaposition   306
Consumer Price Index as a Measure of
   Inflation   314

CHAPTER 6

Drug Approval Requires Strict Procedure   331
Why Professional Articles Are Rejected   337
Beware of *P*-Value Misuse   352
Product Testing Is Big Business   367
Misleading Statistics   373
Have Atomic Tests Caused Cancer?   375
The Year Was (Safe)(Unsafe)   376

CHAPTER 7

Excerpts from a Department of Transportation
   Circular   405
Sample Size Too Small   415
How One Telephone Survey Was Conducted   417
The Wisdom of Hindsight   419
Large Sample Size Isn't Good Enough   420
Commercials, Commercials, Commercials   426

CHAPTER 8

The Power of Your Vote   445
Survey Solicits Contributions   470
Survey Medium Can Affect Results   481
Polio Experiment   482
Better Results with Smaller Class Size   483
More Police, Fewer Crimes   485
Exits Polls on Their Way Out?   492

CHAPTER 9

Student Ratings for Teachers   508

What SAT Scores Measure   529
Unusual Economic Indicators   538
Rising to New Heights   548

CHAPTER 10

Dangerous to Your Health   567
Did Mendel Fudge His Data?   569
Cheating Success   572
If You Can Read This . . .   581
Zip Codes Reveal Much   597
Statistics and Baseball Strategy   598

CHAPTER 11

Seat Belts Save Lives   624
Air Is Healthier Than Tobacco   627
Technology Clouds Television Ratings   640
Class Attendance *Does* Help   646
Does Television Watching Cause Violence?   663
Study Criticized as Misleading   668
The All-Male Jury Finds Dr. Spock . . .   678
Uncle Sam Wants You, if You're Randomly
   Selected   680

CHAPTER 12

A Professional Speaks About Sampling Error   704
Polls   705
Invisible Ink Deceives Subscribers   707
Ethics in Experiments   708
Code of Ethics for Survey Research   709

Fourth Edition
Elementary Statistics

Chapter One

CHAPTER CONTENTS

1-1   **Overview**

1-2   **Background**
The **beginning** of statistics and its general **nature** are presented.

1-3   **Uses and Abuses of Statistics**
Examples of **beneficial uses** of statistics are presented, along with some of the common ways that statistics are used to **deceive**.

1-4   **The Nature of Data**
Different ways of **arranging data** are discussed. The four **levels of measurement** (nominal, ordinal, interval, ratio) are also defined.

1 Introduction to Statistics

CHAPTER PROBLEM

Figure 1-1(b) below and Figure 1-2(b) on the top of page 4 are typical of the many different ways in which data are represented. In both figures, Bureau of Labor Statistics data have been used to illustrate discrepancies between the salaries of men and women. However, both figures have been drawn so that the discrepancies are exaggerated. For each figure, can you identify the trick used to create the impression that the differences are more extreme than they really are? We will learn about these and other deceptive techniques in this chapter. (These figures will be discussed further in Section 1-3.)

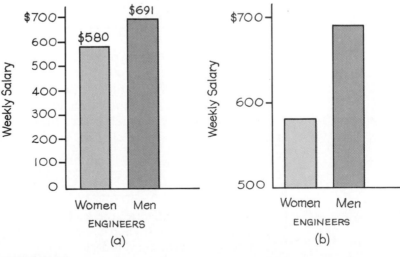

FIGURE 1-1

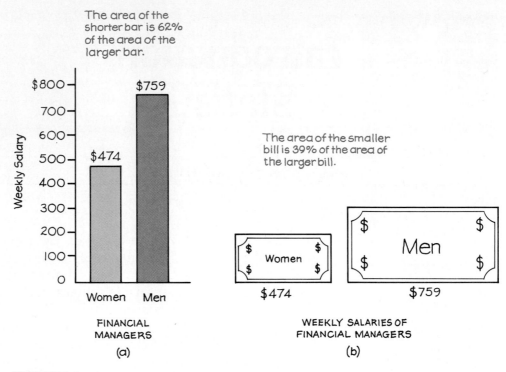

The area of the shorter bar is 62% of the area of the larger bar.

The area of the smaller bill is 39% of the area of the larger bill.

FINANCIAL MANAGERS

(a)

WEEKLY SALARIES OF FINANCIAL MANAGERS

(b)

FIGURE 1-2

1-1 OVERVIEW

The word *statistics* has two basic meanings. We sometimes use this word when we refer to actual numbers derived from data, such as driver fatality rates, consumer price indices, or baseball attendance figures. The second meaning refers to statistics as a subject.

Statistics involves much more than the simple collection, tabulation, and summarizing of data. In this introductory book we will

---

DEFINITION

Statistics is a collection of methods for planning experiments, obtaining data, and then analyzing, interpreting, and drawing conclusions based on the data.

learn how to develop inferences that go beyond the original data. We will learn how to form more general and more meaningful conclusions.

In Chapter 1, we describe the general nature of statistics and present a small sample of beneficial uses as well as some common abuses. Throughout the book we continue to give many real examples of ways that the theories and methods of statistics have been used for our betterment, as well as some of the ways that statistics have been abused. Abuses sometimes stem from ignorance or honest errors, and they are sometimes the result of intentional deception. We will provide a few examples of how statistics have been used for deceptive purposes. These examples will help you to become more aware and more critical when you are presented with statistical claims. As you learn more about the acceptable methods of statistics, you will be better prepared to challenge misleading statements involving statistics.

1-2 BACKGROUND

In the seventeenth century, a successful store owner named John Graunt (1620–1674) had enough spare time to pursue outside interests. His curiosity led him to study and analyze a weekly church publication, called "Bills of Mortality," which listed births, christenings, and deaths and their causes. Based on these studies, Graunt published his observations and conclusions in a work with the catchy title of "Natural and Political Observations Made upon the Bills of Mortality." This 1662 publication was the first real interpretation of social and biological phenomena based on a mass of raw data, and many people consider this to be the birth of statistics.

Graunt made observations about the differences between the birth and mortality rates of men and women. He noted a surprising consistency among events that seem to occur by chance. These and other early observations led to conclusions or interpretations that were invaluable in planning, evaluating, controlling, predicting, changing, or simply understanding some facet of the world we live in.

We say that statistics can be used to predict, but it is very important to understand that we cannot predict with absolute certainty. In fact, statistical conclusions involve an element of uncertainty that can (and often does) lead to incorrect conclusions. It is possible to get 10 consecutive heads when an ordinary coin is tossed 10 times. Yet a statistical analysis of that experiment would lead to the incorrect conclusion that the coin is biased. That conclusion is not, however, certain. It is only a "likely" conclusion, which reflects the very low chance of getting 10 heads in 10 tosses.

In general, mathematics tends to be **deductive** in nature, meaning that acceptable conclusions are deduced with certainty from previous conclusions or assumptions. Statistics is often **inductive** in nature, because inferences are basically generalizations that may or may not correspond to reality. In most branches of mathematics we *prove* results, but in statistics many of our conclusions are associated with different degrees of "likelihood."

In statistics, we commonly use the terms *population* and *sample*.

DEFINITION

A **population** is the complete and entire collection of elements (scores, people, measurements, and so on) to be studied.

DEFINITION

A **sample** is a subset of a population.

Closely related to the concepts of population and sample are the concepts of *parameter* and *statistic*.

population
↕
parameter

DEFINITION

A **parameter** is a numerical measurement describing some characteristic of a *population*.

---

sample
↕
statistic

DEFINITION

A **statistic** is a numerical measurement describing some characteristic of a *sample*.

Of the 315 members of a recent graduating class at a high school in Hyde Park, N.Y., 205 went on to college. Since 205 is 65% of 315, we can say that 65% went on to college. That 65% is a *parameter* (not a statistic) since it is based on the entire population of graduates at this high school. If we could somehow rationalize that this high school

Airline Companies Save by Sampling

In the past, airline companies used an extensive and expensive accounting system to appropriate correctly the revenues from tickets that involved two or more companies. Now, instead of accounting for every ticket involving more than one company, they use a sampling method whereby a small percentage of these split tickets is randomly selected and used as a basis for appropriating all such revenues. The error created by this approach can cause some companies to receive slightly less than their fair share, but these losses are more than offset by the clerical savings accrued by circumventing the 100% accounting method. This concept saves companies millions of dollars each year.

is representative of all high schools in New York State so that we can treat these graduates as a *sample* drawn from a larger population, then the 65% becomes a *statistic*.

Statisticians draw conclusions about an entire population based on the observed data in a sample. Thus they infer a general conclusion from known particular cases in the sample. In most branches of mathematics the procedure is reversed. That is, we first prove the generalized result and then apply it to the particular case. Geometers first prove that, for the population of all triangles, all possess the property that the sum of their respective angles is 180°. They then apply that established result to specific triangles, and they can be certain that the general property will always hold.

As we proceed with our study of statistics, you will learn how to extract pertinent data from samples and how to infer conclusions based on the results of those samples. You will also learn how to assess the reliability of conclusions. Yet you should always realize that, while the tools of statistics can enable you to infer information about a population, you can never predict the behavior of any one individual.

A unique aspect of statistics is its obvious applicability to real and relevant situations. In many branches of mathematics we deal with abstractions that may initially appear to have little or no direct use in the real world, but the elementary concepts of statistics do have direct and practical applications. A wide variety of these applications will be found throughout this book.

Capture-Recapture

Ecologists are often concerned about endangered species and need to estimate the size of some populations, such as the number of blue whales. One common method is to capture a sample of the species, somehow mark each captured member, and then free them. Later, another sample is captured and the ratio of marked subjects, coupled with the size of the first sample, can be used to estimate the population size. For example, suppose you capture 50 fish in a lake, tag them, then release them. A week later you capture 100 fish in the same lake and find that 20 had been tagged. You can now estimate that 20% of the population was tagged, and since you tagged 50 fish, you know that $0.20N = 50$ or $N = 250$. The population size is estimated to be 250 fish. This *capture-recapture* method was used with other methods to estimate the blue whale population, and the result was alarming: the population was as small as 1000. That conclusion led the International Whaling Commission to ban the killing of blue whales so that their extinction could be prevented.

1-3 USES AND ABUSES OF STATISTICS

Short essays that use real-world examples to illustrate the uses and abuses of statistics appear throughout the book. Among the uses are many applications prevalent today in the fields of business, economics, psychology, biology, computer science, military intelligence, English, physics, chemistry, medicine, sociology, political science, agriculture, and education. Statistical theory applied to these diverse fields often results in changes that benefit humanity. Social reforms are sometimes initiated as a result of statistical analyses of factors such as crime rates and poverty levels. Large-scale population planning can result from projections devised by statisticians. Manufacturers can provide better products at lower costs through the effective use of statistics in quality control. Epidemics and diseases can be controlled and anticipated through application of standard statistical techniques. Endangered species of fish and other wildlife can be protected through regulations and laws that are decided on in part by statistical conclusions. Educators may discard innovative teaching techniques if statistical analyses show that traditional techniques are more effective. By pointing to lower fatality rates, legislators can better justify laws such as those governing air pollution, auto inspections, seat belt use, and drunk driving. Amounts and locations of oil, natural gas, and coal can be better determined. Retired employees can benefit from financially stable pension plans. Farmers can benefit from the development of better feed and fertilizer mixtures.

Students choose a statistics course for many different reasons. Some students have no choice and must take a required statistics course. Increasing numbers of other students voluntarily elect to take a statistics course because they recognize its value and application to whatever field they plan to pursue.

Apart from job-motivated or discipline-related reasons, the study of statistics can help you become more critical in your analyses of information so that you are less susceptible to misleading or deceptive claims. You use external data to make decisions, form conclusions, and build your own warehouse of knowledge. If you want to build a sound knowledge base, make intelligent decisions, and form worthwhile opinions, you must be careful to filter out the incoming information that is erroneous or deceptive. As an educated and responsible member of society, you should sharpen your ability to recognize distorted statistical data; in addition, you should learn to interpret undistorted data intelligently.

About a century ago, Benjamin Disraeli said that "there are three kinds of lies: lies, damned lies, and statistics." It has also been said that "figures don't lie; liars figure." Some people have been suspected of using statistics like a drunk uses a lamppost: "more for support than for illumination." Darrell Huff has said that "a well-wrapped statistic is better than Hitler's big lie; it misleads, yet it cannot be pinned on you." According to Sir Josiah Stamp, "The Government is very keen on amassing statistics. They collect them, add them, raise them to the nth power, take the cube root and prepare wonderful diagrams. But you must never forget that every one of these figures comes in the first instance from the village watchman, who just puts down what he damn well pleases."

The preceding statements refer to abuses of statistics in which data are presented in ways that may be misleading. Some abusers of statistics are simply ignorant or careless, whereas others have personal objectives and are willing to suppress unfavorable data while emphasizing supportive data. Here are a few examples of the many ways that data can be distorted.

The term *average* refers to several different statistical measures that will be discussed and defined in Chapter 2. To most people, the

And We Thought We Were Doing So Well

Each year, millions of children take standardized achievement tests so that their learning progress can be measured. Dr. John Cannell, a West Virginia physician, observed that his state's high rate of illiteracy was inconsistent with a state report claiming that West Virginia students were performing at a level *above* the national average. Subsequent investigations showed that, among other problems, old norms were used as a basis for comparison. By comparing the new test results to results obtained years ago, all 50 states had results that were above "the national average." Dr. Cannell charged that "the testing industry wants to sell lots of tests, and the school superintendents desperately need help in improving scores."

average is the sum of all values divided by the number of values, but this is only one type of average. For example, given the 10 annual salaries of $25,000, $25,000, $25,000, $26,000 $28,000, $30,000, $32,000, $34,000, $35,000, and $55,000, we can correctly claim that the "average" is either $31,500 (since $31,500 is the sum of the 10 values divided by 10), or $29,000 (since half of the values are above $29,000 and half are below), or $25,000 (since it is the value that occurs most often), or $40,000 (since this value is exactly midway between the lowest and highest salaries listed). These averages will be defined and discussed in Chapter 2 and there is no need to remember them now, but this example does illustrate that a reported average can be very misleading, since it can be any one of several different numbers. There are no objective criteria that can be used to determine the specific average that is most representative, and the user of statistics is free to select the average that best supports a favored position.

A union contract negotiator can choose the lowest average in an attempt to emphasize the need for a salary increase, while the management negotiator can choose the highest average to emphasize the well-being of the employees. In actual negotiations both sides are usually adept at exposing such ploys, but the typical citizen often accepts the validity of an average without really knowing which specific average is being presented and without knowing how different the picture would look if another average were used. The educated and thinking

The Gender Gap in Wages

Many articles note that, on average, full-time female workers earn only about 70¢ for each $1 earned by full-time male workers. This discrepancy is attributed to such factors as the following: Women tend to have less seniority; they tend to enter and leave the labor market more often; they tend to choose lower-paying jobs more often; and they tend to have less education. But do these factors account for the wage discrepancy, or is sex discrimination another real factor? Researchers at the Institute for Social Research at the University of Michigan analyzed the effects of various key factors and found that about a third of the discrepancy can be explained by differences in education, seniority, work interruptions, and job choices. The other two-thirds of the discrepancy remains "unexplained" by such labor factors.

citizen is not so susceptible to potentially deceptive information; the educated and thinking citizen analyzes and criticizes statistical data so that meaningless or illusory claims are not part of the base on which his or her decisions are made and opinions formulated. If you read that the average annual salary of an American family is $27,735 (the latest available figure from the Bureau of the Census), you should attempt to determine which of the averages that figure represents. If no additional information is available, you should realize that the given figure may be very misleading. Conversely, as you present statistics on data you have accumulated, you should attempt to provide descriptions identifying the true nature of the data.

Many visual devices—such as bar graphs and pie charts—that can be used to exaggerate or deemphasize the true nature of data are also discussed in Chapter 2. In Figure 1-1 we use data from the Bureau of Labor Statistics to show two bar graphs that depict the *same data*, but part (b) is designed to exaggerate the difference between the salaries of women and men. While both bar graphs represent the same set of scores, they tend to produce very different subjective impressions. Too many of us look at a graph superficially and develop intuitive impressions based on the pattern we see. Instead, we should scrutinize the graph and search for distortions of the type illustrated in Figure 1-1. We should analyze the *numerical* information given in the graph instead of being misled by its general shape.

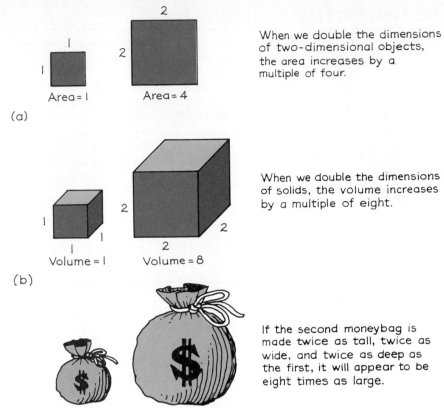

When we double the dimensions of two-dimensional objects, the area increases by a multiple of four.

(a)

When we double the dimensions of solids, the volume increases by a multiple of eight.

(b)

If the second moneybag is made twice as tall, twice as wide, and twice as deep as the first, it will appear to be eight times as large.

(c)

FIGURE 1-3

Drawings of objects may also be misleading. Some objects commonly used to depict data include moneybags, stacks of coins, army tanks (for military expenditures), cows (for dairy production), and houses (for home construction). When we double the dimensions of a two-dimensional figure, the area is quadrupled. In Figure 1-2(a) we use Bureau of Labor Statistics data to show that women financial managers earn about 62% of the salaries earned by their male counterparts. In Figure 1-2(b) we present the *same* data, but we make both dimensions of the smaller bill to be 62% of those for the larger bill. The impact created by Figure 1-2(b) will be greater than that of Figure 1-2(a). In Figure 1-3 we further illustrate how drawings can be used to mislead.

Let's suppose that school taxes in a small community doubled from one year to the next. By depicting the amounts of taxes as bags of

The *Literary Digest* Poll

A classic case of a biased sample is the *Literary Digest* poll conducted during the 1936 presidential race between Alf Landon and Franklin D. Roosevelt. This magazine's poll resulted in a prediction that Landon would defeat Roosevelt. Instead, Roosevelt won by a landslide. What went wrong? George Gallup referred to the flaws of this poll "in which postcard ballots were sent to people who had car registrations and were listed in telephone books, which biased the sample toward people with higher income levels" (recently reported by the Associated Press).

Maurice Bryson says that the real problem with the *Literary Digest* poll is that it depended on *voluntary* responses. According to Bryson, "Ten million sample ballots were mailed to prospective voters, but only 2.3 million were returned. As everyone ought to know, such samples are practically always biased." (See Bryson's "The *Literary Digest* Poll: Making of a Statistical Myth," *The American Statistician*, Vol. 30, No. 4.) He observes that the people who voluntarily respond are those who feel strongly about the issue. Bryson also states that "voluntary response to mailed questionnaires is perhaps the most common method of social-science data collection encountered by statisticians, and perhaps also the worst."

money and by doubling the dimensions of the bag representing the second year, we can easily create the impression that taxes more than doubled (as in Figure 1-3). Another variety of statistical "lying" is often inspired by small sample results. The toothpaste preferences of only 10 dentists should not be used as a basis for a generalized claim such as "Caressed toothpaste is recommended by 7 out of 10 dentists." Even if the sample is large, it must be unbiased and representative of the population from which it comes.

Sometimes the numbers themselves can be deceptive. A mean annual salary of $27,735.29 sounds precise and tends to instill a high degree of confidence in its accuracy. The figure of $27,700 doesn't convey that same sense of precision and accuracy. A statistic that is very precise with many decimal places is not necessarily accurate, though.

Another source of statistical deception involves numbers that are ultimately guesses, such as the crowd count at a political rally, the dog population of Chicago, and the amount of money bet illegally. When the Pope recently visited Miami, officials estimated the crowd size to be 250,000, but the *Miami Herald* used aerial photos and grids to come up with a better estimate of 150,000. As another example, the Associated Press ran an article in which the roach population of New York City was estimated to be 1 billion. That claim was made by a spokesperson for the Bliss Exterminator Company. The head of New York City's bureau of Pest Control would not confirm that figure. He said, "I haven't the slightest idea if that's right. We do strictly rats here."

Continental Airlines ran full-page ads boasting better service. In referring to lost baggage, these ads claimed that this is "an area where we've already improved 100% in the last six months." Do you really believe that they no longer lose any baggage at all? That's what the 100% improvement figure actually means.

"Ninety percent of all our cars sold in this country in the last 10 years are still on the road." Millions of consumers heard that commercial message and got the impression that those cars must be well built in order to persist through those long years of driving. What the auto manufacturer failed to mention was that 90% of the cars it sold in this country were sold within the last three years. The claim was technically correct, but it was very misleading.

The preceding examples comprise a small sampling of the ways in which statistics can be used deceptively. Entire books have been devoted to this subject, including Darrell Huff's *How to Lie with Statistics* and Robert Reichard's *The Figure Finaglers*. Understanding these practices will be extremely helpful in evaluating the statistical data found in everyday situations.

It is often expensive to get good results that will be essentially free of bias. According to *USA Today*, Alan Kay hired three different firms to conduct surveys about arms control. He used a firm that worked for Democrats, one that worked for Republicans, and one that didn't have links to a single political party. While he was able to obtain an accurate overall picture, it cost him $1 million.

1-4 THE NATURE OF DATA

When thinking about collections of data, there is a common tendency to think of them as lists of numbers, such as the cholesterol levels of diseased rats (how disgusting) or the weights of freshly broiled Maine lobsters (that's better). Yet data may be nonnumerical, and even numerical data can belong to different categories with different characteristics. For example, a pollster may compile data consisting of nonnumeric data such as sex, race, and religion of voters in a sample.

Numeric data, instead of being in an unordered list, might be data matched in pairs (discussed in Chapters 8, 9, 11), as in the following two tables.

*Pretraining weights (kg)	99	62	74	59	70
Posttraining weights (kg)	94	62	66	58	70

*Pretraining weights (kg)	99	62	74	59	70
Pretraining heart rate (beats/min)	174	180	171	177	168

Another common arrangement for summarizing sample data is the contingency table (discussed in Section 10-3), such as the one that follows.

	_____ Grade _____				
	A	B	C	D	F
Math	494	689	642	415	610
Business	1895	2048	1675	537	784
History	320	631	585	215	183

Based on data from Dutchess Community College.

In this table the numbers are frequencies (counts) of sample results.

Clearly, the nature of the data can affect the nature of the relevant problem and the method used for analysis. With the paired weight data, a fundamental concern would be whether there is a difference in the pretraining and posttraining weights. Any analysis of these data should attempt to determine whether posttraining weights are significantly less than the pretraining weights. With the paired weight/heart rate data, the fundamental concern would be whether some relationship exists between those two factors. This requires a different method of analysis. With the contingency table, the fundamental concern would be whether the grade distribution is independent of the course. This requires yet another method of analysis. As we consider the topics of later chapters, we will see that the structure of the data does affect our choice of the method we will use. In addition to the structure of the data, our choice of analytical method is also affected by the nature of the data. It is common to classify data according to one of the following four **levels of measurement**.

---

*Based on data from the *Journal of Applied Physiology*, Vol. 62, No. 1.

1. Nominal
2. Ordinal
3. Interval
4. Ratio

The **nominal level of measurement** is characterized by data that consist of names, labels, or categories only. The data cannot be arranged in an ordering scheme.

If we associate *nominal* with "name only," the meaning becomes easy to remember. An example of nominal data is the collection of "yes, no, undecided" responses to a survey question. Data at this nominal level of measurement cannot be arranged according to some ordering scheme. That is, there is no criterion by which values can be identified as greater than or less than other values.

EXAMPLE

The following are other examples of sample data at the nominal level of measurement.

1. The sample consisting of 12 Democrats, 15 Republicans, and 9 Independents.
2. The sample consisting of 14 students from New York, 17 from California, 8 from Connecticut, and 7 from Florida.

It should be obvious that the preceding data cannot be used for calculations because the categories lack any ordering or numerical significance. We cannot, for example, "average" 12 Democrats, 15 Republicans, and 9 Independents. Numbers are sometimes assigned to the different categories, especially when the data are processed by computer. We might find that Democrats are assigned 0, Republicans are assigned 1, and Independents are assigned 2. Even though we now have number labels, those numbers lack any real computational significance. The average of twelve 0s, fifteen 1s, and nine 2s might be 0.9, but that is a meaningless statistic. (0.9 does not represent a liberal Republican!)

FORMULAS

$\bar{x} = \dfrac{\Sigma x}{n}$ Mean

$\bar{x} = \dfrac{\Sigma f \cdot x}{\Sigma f}$ Mean (freq. table)

$s = \sqrt{\dfrac{\Sigma(x - \bar{x})^2}{n - 1}}$ St. dev.

$s = \sqrt{\dfrac{n(\Sigma x^2) - (\Sigma x)^2}{n(n - 1)}}$ St. dev. (shortcut)

$s = \sqrt{\dfrac{n[\Sigma(f \cdot x^2)] - [\Sigma(f \cdot x)]^2}{n(n - 1)}}$ St. dev. (freq. table)

Variance $= s^2$

$P(A \text{ or } B) = P(A) + P(B)$ if A, B are mutually exclusive

$P(A \text{ or } B) = P(A) + P(B) - P(A \text{ and } B)$ if A, B are not mutually exclusive

$P(A \text{ and } B) = P(A) \cdot P(B)$ if A, B are independent

$P(A \text{ and } B) = P(A) \cdot P(B|A)$ if A, B are dependent

$P(\bar{A}) = 1 - P(A)$ Rule of complements

$_nP_r = \dfrac{n!}{(n - r)!}$ Permutations

$_nC_r = \dfrac{n!}{(n - r)! \, r!}$ Combinations

$\mu = \Sigma x \cdot P(x)$ Mean (prob. dist.)
$\sigma = \sqrt{[\Sigma x^2 \cdot P(x)] - \mu^2}$ St. dev. (prob. dist.)

$P(x) = \dfrac{n!}{(n - x)! \, x!} \cdot p^x \cdot q^{n-x}$ Binomial prob.

$\mu = n \cdot p$ Mean (binomial)
$\sigma^2 = n \cdot p \cdot q$ Variance (binomial)
$\sigma = \sqrt{n \cdot p \cdot q}$ St. dev. (binomial)

$z = \dfrac{x - \bar{x}}{s}$ or $\dfrac{x - \mu}{\sigma}$ Standard score

$\mu_{\bar{x}} = \mu$ Central limit theorem

$\sigma_{\bar{x}} = \dfrac{\sigma}{\sqrt{n}}$ St. error

$\sigma_{\bar{x}} = \dfrac{\sigma}{\sqrt{n}} \sqrt{\dfrac{N - n}{N - 1}}$ St. error if $n > 0.05N$

Test Statistics

$z = \dfrac{\bar{x} - \mu}{\sigma/\sqrt{n}}$ Mean—one population (σ known or $n > 30$)

$t = \dfrac{\bar{x} - \mu}{s/\sqrt{n}}$ Mean—one population (σ unknown and $n \le 30$)

$z = \dfrac{\hat{p} - p}{\sqrt{\dfrac{pq}{n}}}$ Proportion—one population

$\chi^2 = \dfrac{(n - 1)s^2}{\sigma^2}$ St. dev. or variance—one population

$F = \dfrac{s_1^2}{s_2^2}$ St. dev. or variance—two populations (where $s_1^2 \ge s_2^2$)

$t = \dfrac{\bar{d} - \mu_d}{s_d/\sqrt{n}}$ Two means—dependent

$z = \dfrac{(\bar{x}_1 - \bar{x}_2) - (\mu_1 - \mu_2)}{\sqrt{\dfrac{\sigma_1^2}{n_1} + \dfrac{\sigma_2^2}{n_2}}}$ Two means—independent (σ_1, σ_2 known or $n_1 > 30$ and $n_2 > 30$)

$t = \dfrac{(\bar{x}_1 - \bar{x}_2) - (\mu_1 - \mu_2)}{\sqrt{\dfrac{s_1^2}{n_1} + \dfrac{s_2^2}{n_2}}}$
Two means—independent (reject $\sigma_1^2 = \sigma_2^2$ and $n_1 \le 30$ or $n_2 \le 30$)

$t = \dfrac{(\bar{x}_1 - \bar{x}_2) - (\mu_1 - \mu_2)}{\sqrt{\dfrac{1}{n_1} + \dfrac{1}{n_2}} \sqrt{\dfrac{(n_1 - 1)s_1^2 + (n_2 - 1)s_2^2}{n_1 + n_2 - 2}}}$
Two means—independent (fail to reject $\sigma_1^2 = \sigma_2^2$ and $n_1 \le 30$ or $n_2 \le 30$)

$z = \dfrac{(\hat{p}_1 - \hat{p}_2) - (p_1 - p_2)}{\sqrt{\bar{p}\bar{q}\left(\dfrac{1}{n_1} + \dfrac{1}{n_2}\right)}}$ Two proportions

$\chi^2 = \Sigma \dfrac{(O - E)^2}{E}$ Multinomial ($df = k - 1$)

$\chi^2 = \Sigma \dfrac{(O - E)^2}{E}$ Contingency table $[df = (r - 1)(c - 1)]$

where $E = \dfrac{(\text{row total})(\text{column total})}{(\text{grand total})}$

FORMULAS

Confidence Intervals

$\bar{x} - E < \mu < \bar{x} + E$ Mean

where $E = z_{\alpha/2}\dfrac{\sigma}{\sqrt{n}}$ (σ known or $n > 30$)

or $E = t_{\alpha/2}\dfrac{s}{\sqrt{n}}$ (σ unknown and $n \le 30$)

$\hat{p} - E < p < \hat{p} + E$ Proportion

where $E = z_{\alpha/2}\sqrt{\dfrac{\hat{p}\hat{q}}{n}}$

$\dfrac{(n-1)s^2}{\chi_R^2} < \sigma^2 < \dfrac{(n-1)s^2}{\chi_L^2}$ Variance

Sample Size

$n = \left[\dfrac{z_{\alpha/2}\sigma}{E}\right]^2$ Mean

$n = \dfrac{[z_{\alpha/2}]^2 \cdot 0.25}{E^2}$ Proportion

$n = \dfrac{[z_{\alpha/2}]^2\hat{p}\hat{q}}{E^2}$ Proportion ($\hat{p}$ and $\hat{q}$ are known)

Linear Correlation/Regression

Correlation $r = \dfrac{n\Sigma xy - (\Sigma x)(\Sigma y)}{\sqrt{n(\Sigma x^2) - (\Sigma x)^2}\sqrt{n(\Sigma y^2) - (\Sigma y)^2}}$

or $r = \dfrac{ms_x}{s_y}$

$m = \dfrac{n\Sigma xy - (\Sigma x)(\Sigma y)}{n(\Sigma x^2) - (\Sigma x)^2}$ or $m = r\dfrac{s_y}{s_x}$ Slope

$b = \dfrac{(\Sigma y)(\Sigma x^2) - (\Sigma x)(\Sigma xy)}{n(\Sigma x^2) - (\Sigma x)^2}$ or $b = \bar{y} - m\bar{x}$

$y' = mx + b$ Eq. of regression line

$r^2 = \dfrac{\text{explained variation}}{\text{total variation}}$

$s_e = \sqrt{\dfrac{\Sigma(y - y')^2}{n - 2}}$ or $\sqrt{\dfrac{\Sigma y^2 - b\Sigma y - m\Sigma xy}{n - 2}}$

$y' - E < y < y' + E$

where $E = t_{\alpha/2}s_e\sqrt{1 + \dfrac{1}{n} + \dfrac{n(x_0 - \bar{x})^2}{n(\Sigma x^2) - (\Sigma x)^2}}$

Analysis of Variance

$F = \dfrac{ns_{\bar{x}}^2}{s_p^2}$ k samples each of size n; num. $df = k - 1$ den. $df = k(n - 1)$

$F = \dfrac{\left[\dfrac{\Sigma n_i(\bar{x} - \bar{\bar{x}})^2}{k - 1}\right]}{\left[\dfrac{\Sigma(n_i - 1)s_i^2}{\Sigma(n_i - 1)}\right]}$ num. $df = k - 1$ den. $df = N - k$ or

$F = \dfrac{MS(\text{treat.})}{MS(\text{error})}$ num. $df = k - 1$ den. $df = N - k$

where N = tot. number of values
c_i = number of values in i^{th} sample
k = number of samples

$MS(\text{treat.}) = \dfrac{SS(\text{treat.})}{df(\text{treat.})}$ $SS(\text{total}) = \Sigma x^2 - \dfrac{(\Sigma x)^2}{N}$

$MS(\text{error}) = \dfrac{SS(\text{error})}{df(\text{error})}$ $df(\text{error}) = N - k$

$df(\text{treat.}) = k - 1$

$SS(\text{treat.}) = \left(\Sigma\dfrac{c_i^2}{n_i}\right) - \dfrac{(\Sigma c_i)^2}{N}$

$SS(\text{error}) = SS(\text{total}) - SS(\text{treat.})$

Nonparametric

$z = \dfrac{(x + 0.5) - (n/2)}{\sqrt{n}/2}$ Sign test for $n > 25$

$z = \dfrac{T - n(n + 1)/4}{\sqrt{\dfrac{n(n + 1)(2n + 1)}{24}}}$ Wilcoxon signed ranks (two dependent samples for $n > 30$)

$z = \dfrac{R - \mu_R}{\sigma_R} = \dfrac{R - \dfrac{n_1(n_1 + n_2 + 1)}{2}}{\sqrt{\dfrac{n_1 n_2(n_1 + n_2 + 1)}{12}}}$

Wilcoxon rank-sum (two independent samples)

$H = \dfrac{12}{n(n + 1)}\left(\dfrac{R_1^2}{n_1} + \dfrac{R_2^2}{n_2} + \cdots + \dfrac{R_k^2}{n_k}\right) - 3(n + 1)$

Kruskal-Wallis (chi-square $df = k - 1$)

$r_s = 1 - \dfrac{6\Sigma d^2}{n(n^2 - 1)}$ Rank correlation

$\left(\text{crit. value for } n > 30: \dfrac{\pm z}{\sqrt{n - 1}}\right)$

$z = \dfrac{G - \mu_G}{\sigma_G} = \dfrac{G - \dfrac{2n_1 n_2}{n_1 + n_2} + 1}{\sqrt{\dfrac{(2n_1 n_2)(2n_1 n_2 - n_1 - n_2)}{(n_1 + n_2)^2(n_1 + n_2 - 1)}}}$

Runs test for $n > 20$

The **ordinal level of measurement** involves data that may be arranged in some order, but differences between data values either cannot be determined or are meaningless.

EXAMPLE

The following are examples of data at the ordinal level of measurement.

1. In a sample of 36 batteries, 12 were rated "good," 16 were rated "better," and 8 were rated "best."
2. In a class of 19 students, 5 required remediation, 10 were average, and 4 were gifted.
3. In a high school graduating class of 463 students, Sally ranked 12th, Allyn ranked 27th, and Mike ranked 28th.

In part (1) of the preceding example, we cannot determine a specific measured difference between "good" and "better." In part (3), we can determine a difference between the rankings of 12 and 27, but the resulting value of 15 doesn't really mean anything. That is, the difference of 15 between ranks of 12 and 27 isn't necessarily the same as the difference of 15 between ranks of 30 and 45. This ordinal level provides information about relative comparisons, but the degrees of differences are not available. We know that a low-income worker earns less than a middle-income worker, but we don't know how much less. Again, data at this level should not be used for calculations.

The **interval level of measurement** is like the ordinal level, with the additional property that we can determine meaningful amounts of differences between data. Data at this level may lack an inherent zero starting point. At this level, differences are meaningful, but ratios are not.

Temperature readings of 25° F and 50° F are examples of data at this measurement level. Those values are ordered and we can determine their difference (often called the *distance* between the two values). However, there is no inherent zero point. The value of 0° F might seem like a starting point, but it is arbitrary and not inherent. The value of

How Do You Measure Disobedience?

The data you collect are at least as important as the statistical methodology you employ. However, it is often difficult to collect usable or relevant data. How do you collect data that relate to a characteristic that doesn't appear to be measurable, such as the level of disobedience in people? Stanley Milgram was a social psychologist who devised a clever experiment that did just that. A researcher instructed a volunteer to operate a control board that gave increasingly painful "electrical shocks" to a third person. Actually, no electric shocks were given and the third person was an actor who feigned increasing levels of pain and anguish. The volunteer began with 15 volts and was instructed to increase the shocks by increments of 15 volts up to a maximum of 450 volts. The disobedience level was the point at which the volunteer refused to follow the researcher's instructions to increase the voltage. Despite the actor's screams of pain and a feigned heart attack, two-thirds of the volunteers continued to obey the researcher. Milgram was surprised by such a high level of obedience. Many people question the ethics of experiments such as this one, however.

0° F does not indicate "no heat," and it is incorrect to say that 50° F is twice as hot as 25° F. For the same reasons, temperature readings on the Celsius scale are also at the interval level of measurement. (Temperature readings on the Kelvin scale are at the ratio level of measurement; that scale has an absolute zero.)

EXAMPLE

The following are other examples of data at the interval level of measurement.

1. Years in which IBM stock split
2. Body temperatures (in degrees Celsius) of hospital patients

---

The **ratio level of measurement** is actually the interval level modified to include the inherent zero starting point. For values at this level, differences and ratios are meaningful.

EXAMPLE

Examples of data at the ratio level are as follows:

1. Heights of pine trees around Lake Tahoe
2. Volumes of helium in balloons
3. Times (in minutes) of runners in a marathon

Ratio
↑
Interval
↑
Ordinal
↑
Nominal

Values in each of these data collections can be arranged in order, differences can be computed, and there is an inherent zero starting point. *This level is called the ratio level because the starting point makes ratios meaningful.* Since a tree 50 feet high *is twice* as tall as a tree 25 feet high and 50° F *is not twice* as hot as 25° F, heights are at the ratio level while Fahrenheit temperatures are at the interval level.

Among the four levels of measurement, the nominal is considered the lowest. It is followed by the ordinal level, the interval level, and then the ratio level, which is highest.

Levels of Measurement		
Level	**Summary**	**Example**
Nominal	Categories only. Data cannot be arranged in an ordering scheme.	Voter distribution: 45 Democrats / 80 Republicans / 90 Independents } Categories only.
Ordinal	Categories are ordered, but differences cannot be determined or they are meaningless.	Voter distribution: 45 low-income voters / 80 middle-income voters / 90 upper-income voters } An order is determined by "low, middle, upper."
Interval	Differences between values can be found, but there may be no inherent starting point. Ratios are meaningless.	Temperatures of steel rods: 45° F / 80° F / 90° F } 90° F is not twice as hot as 45° F.
Ratio	Like interval, but with an inherent starting point. Ratios are meaningful.	Lengths of steel rods: 45 cm / 80 cm / 90 cm } 90 cm is twice as long as 45 cm.

A general and important guideline is that **the statistics based on one level of measurement should not be used for a lower level (but can be used for a higher level)**. We can, for example, calculate the average for data at the interval or ratio level, but not at the lower ordinal or nominal levels. An implication of this guideline is that data obtained from using a Likert scale, such as the one that follows, should

Superior	Good	Average	Poor	Very Poor
1	2	3	4	5

not be used for such calculations since these data are only at the ordinal level. This guideline is sometimes ignored and Likert scale results are often treated as interval- or ratio-level data, even though they are not. But serious errors may result from violating this guideline. If, for data processing requirements, we assign the numbers 0, 1, and 2 to Democrats, Republicans, and Independents (respectively) and proceed to calculate the average, we are creating a meaningless statistic that can lead to incorrect conclusions.

In Chapter 2 we introduce basic important ways of dealing with data sets that are primarily at the interval or ratio levels of measurement.

VOCABULARY LIST

Define and give an example of each term.

statistics (as a	sample	ordinal
discipline)	parameter	interval
population	statistic	ratio
	nominal	

EXERCISES A

1-1   *USA Today* conducted a poll of 800 divorced people who were asked if they wanted to marry again. It was reported that "overall, 58% of divorced people say they don't want to get married again." Do the 800 respondents constitute a sample or a population? If they are a sample, what is the population?

1-2   For one semester at Dutchess Community College it was reported that among 3243 total credit hours for mathematics courses, 18.82% were assigned the grade of F. Does this collection of 3243 credit hours constitute a sample or a population? Explain.

1-3   The Labor Department reported that the median weekly pay of women is about 70% of that for men. One reason for this is discrimination based on sex. Cite a second reason that might help to explain the discrepancy between the salaries of men and women.

1-4   A study by Dr. Ralph Frerichs (UCLA) showed that family incomes correspond to the risk of dying because of heart disease. Higher family income levels corresponded to lower heart disease death rates.
      a. Does this imply that more earned money *causes* the risk of dying of heart disease to be lower?
      b. Identify a factor that could explain the correspondence.

1-5   You plan to conduct a telephone survey of 500 people in your region. What would be wrong with using the telephone directory as the population from which your sample is drawn?

1-6   You plan to conduct a poll of students at your college. What is wrong with polling every 50th student who is leaving the cafeteria?

1-7   In each of the following, determine which of the four levels of measurement (nominal, ordinal, interval, ratio) is most appropriate.
      a. A car is described as subcompact, compact, intermediate, or full-size.
      b. The weights of a sample of machine parts
      c. The colors of a sample of cars involved in alcohol-related crashes
      d. The years in which Republicans won presidential elections
      e. Zip codes

1-8   In each of the following, determine which of the four levels of measurement (nominal, ordinal, interval, ratio) is most appropriate.
      a. Social security numbers
      b. Total annual incomes for a sample of families
      c. Final course averages of A, B, C, D, F
      d. Body temperatures (in degrees Fahrenheit) of a sample of bears captured in Wyoming
      e. At the end of a course an instructor is rated as superior, above average, average, below average, or poor.

1-9   A college conducts a survey of its alumni in an attempt to determine their typical annual salary. Would alumni with very low salaries be likely to respond? How would this affect the result? Identify one other factor that might affect the result.

1-10 One study actually showed that smokers tend to get lower grades in college than nonsmokers. Does this mean that smoking causes lower grades? What other explanation is possible?

1-11 A report by the Nuclear Regulatory Commission noted that a particular nuclear reactor was being operated at "below-average standards." What is the approximate percentage of nuclear power plants that operate at below-average standards? Is a below-average standard necessarily equivalent to a dangerous or undesirable level?

1-12 An employee earning $400 per week was given a 20% cut in pay as part of her company's attempt to reduce labor costs. After a few weeks, this employee's dissatisfaction grew and her threat to resign caused her manager to offer her a 20% raise. The employee accepted this offer since she assumed that a 20% raise would make up for the 20% cut in pay.

   a. What was the employee's weekly salary after she received the 20% cut in pay?

   b. Use the salary figure from part (a) to find a 20% increase and determine the weekly salary after the raise.

   c. Did the 20% cut followed by the 20% raise get the employee back to the original salary of $400 per week?

1-13 The first edition of a textbook contains 1000 exercises. For the second edition, the author removed 100 of the original exercises and added 300 new exercises. Which of the following statements about the second edition are correct?

   a. There are 1200 exercises.

   b. There are 33% more exercises.

   c. There are 20% more exercises.

   d. 25% of the exercises are new.

1-14 What differences are there between the following two statements, and which one do you believe is more accurate?

   a. Drunk drivers cause more than half of all fatal car crashes.

   b. Of all fatal motor vehicle crashes, more than 50% involve alcohol.

1-15 Census takers have found that in obtaining the ages of people, they would get more people of age 50 than of age 49 or of age 51. Can you explain how this might occur?

1-16 A news report states that the police seized forged record albums that had a value of $1 million. How do you suppose the police computed the value of the forged albums, and in what other ways can that value be estimated? Why might the police be inclined to exaggerate the value of the albums?

1-17 A newspaper article reports that "this morning's demonstration was attended by 8755 students." Comment.

1-18 In a typical year, about 46,000 deaths result from motor vehicle accidents, according to data from the National Safety Council.
   a. How many deaths would result from motor vehicle accidents in a typical day?
   b. How many deaths would result from motor vehicle accidents in a typical four-day period?
   c. For the four days of the Memorial Day weekend (Friday through Monday), assume that driving increases by 25% and that there are 630 deaths resulting from motor vehicle accidents. Does it appear that driving is more dangerous over the Memorial Day weekend?

1-19 The Bureau of the Census reports an average family size of 3.21 persons and an average household size of 2.67 persons. Explain the discrepancy.

1-20 A district attorney claims that "organized crime paid $89,541 in bribes to local officials last year." Criticize that claim.

1-21 In a final examination for a statistics course, one student received a grade of 50, and another student received a grade of 100.
   a. If we consider these numbers to represent only the points earned on the exam, then the score of 100 is twice that of 50. What is the corresponding level of measurement?
   b. If we consider these numbers to represent the amount of the subject learned in the course, it is wrong to conclude that the one student knows twice as much as the other. What is the level of measurement in this case?

1-22 Many people question what IQ scores actually measure. Assuming that IQ scores measure intelligence, is a person with an IQ score of 150 twice as intelligent as another person with an IQ score of 75?
   a. What does a yes answer imply about the level of measurement corresponding to IQ scores?
   b. What does a negative answer imply about the level of that data?

1-23 If a recipe requires cooking something at "300° F for three hours," but you decide to cook it at 900° F for one hour, the result will be different. Explain.

1-24 The years 1990, 1988, 1972, 1963, and 1984 form a collection of data at the interval level of measurement. Explain.

1-25 In an advertising supplement inserted in *Time*, the increases in expenditures for pollution abatement were shown in a graph similar to Figure 1-4 on the next page. What is wrong with that figure?

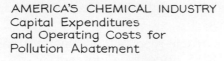

AMERICA'S CHEMICAL INDUSTRY
Capital Expenditures
and Operating Costs for
Pollution Abatement

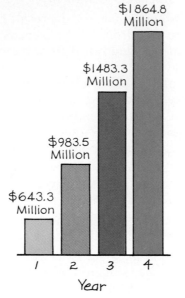

FIGURE 1-4

1-26   Babe Ruth's lifetime batting average is 0.342 (from 2873 hits in 8399 times at bat). Is this a baseball *statistic* or a baseball *parameter*?

1-27   The Minolta Corporation conducted a survey of 703 business owners with fewer than 500 employees. The owners were asked to identify reasons why new businesses fail. Describe the sample and the population.

1-28   A survey questions the sex of the respondent, and results indicate 30 men and 20 women. If a male response is coded as 1 and a female response is coded as 0, the average of the 50 scores is calculated as 0.6. Of what use is that value?

EXERCISES B

1-29   A researcher at the Sloan-Kettering Cancer Research Center was once criticized for falsifying data. Among his data were figures obtained from 6 groups of animals, with 20 animals in each group. These values were given for the percentage of successes in each group: 53%, 58%, 63%, 46%, 48%, 67%. What's wrong?

1-30   If an employee is given a cut in pay of x percent, find an expression for the percent raise that would return the salary to the original amount.

1-31   A *New York Times* editorial criticized a chart caption that described a dental rinse as one that "reduces plaque on teeth by over 300%." What does it mean to reduce plaque by over 300%?

1-32   If you must travel from New York to California, is it safer to go by car or plane? Here are some recent data from the National Safety Council.

	Passenger Miles (billions)	Total Deaths
Car	2467	22,974
Plane	278	201

CASE STUDY ACTIVITY

Collect an example from a current newspaper or magazine in which data have been presented in a potentially deceptive manner. Identify the source from which the example was taken, explain briefly the way in which the data might be deceptive, and suggest how the data might be presented more fairly.

DATA PROJECT

1. Refer to the student data in Appendix B. Identify any pulse rates that are unrealistic. In calculating an average, what should you do with these values?

2. Refer to the student data in Appendix B. For the column of random digits, the survey respondents were asked to enter a three-digit random number. Find the total number of times each digit from 0 through 9 occurs. Do these results appear to be the result of a random selection? Using the column of data consisting of the last three digits of social security numbers, find the total number of times each digit from 0 through 9 occurs. Do these results appear to be the result of a random selection?

Chapter Two

CHAPTER CONTENTS

2-1   **Overview**

2-2   **Summarizing Data**
The construction of **frequency tables** is described.

2-3   **Pictures of Data**
Techniques are given for presenting data using **pie charts**, **histograms**, **frequency polygons**, **ogives**, and **stem-and-leaf plots**.

2-4   **Averages**
These measures of central tendency are defined: **mean**, **median**, **mode**, **midrange**, and **weighted mean**.

2-5   **Dispersion Statistics**
These measures of dispersion are defined: **range**, **standard deviation**, **mean deviation**, and **variance**.

2-6   **Measures of Position**
For comparison purposes, the **standard score** (or z **score**) is defined. Also defined are **percentiles**, **quartiles**, and **deciles**.

2 Descriptive Statistics

CHAPTER PROBLEM

You've just landed a job in beautiful Dutchess County, located in upstate New York. You plan to buy a home, but you don't have a good sense of what homes cost in that area. You know that the cost of comparable homes can vary by large amounts in different locations. A home on the beach at Malibu will cost much more than a similar home in a housing development. Because buying a home is such an expensive and important decision, you decide to do some prior investigation. Table 2-1 lists the actual selling prices of 150 randomly selected homes that were recently sold in Dutchess County.

A visual examination of those selling prices may provide some insight, but it is generally difficult to draw meaningful conclusions from a collection of raw data that are simply listed in no particular order. We need to look further into the data and do something with them so that we can increase our understanding. As one example, we might add the 150 selling prices and then divide that total by 150. The result will be an average that helps us to better understand the data. There are several other things we might do. The major objective of this chapter is to develop a variety of methods that will give us more insight into data sets such as the one listed in Table 2-1 on the following page.

TABLE 2-1 Selling Prices (in Dollars) of 150 Dutchess County Homes					
179,000	126,500	134,500	125,000	142,000	164,000
146,000	129,000	141,900	135,000	118,500	160,000
89,900	169,900	127,500	162,500	152,000	122,500
220,000	141,000	80,500	152,000	231,750	180,000
185,000	265,000	135,000	203,000	141,000	159,000
182,000	208,000	96,000	156,000	185,500	275,000
144,900	155,000	110,000	154,000	151,500	141,000
119,000	108,500	126,500	302,000	130,000	140,000
123,500	153,500	194,900	165,000	179,900	194,500
127,500	170,000	160,000	135,000	117,000	235,000
223,000	163,500	78,000	187,000	133,000	125,000
116,000	135,000	194,500	99,500	152,500	141,900
139,900	117,500	150,000	177,000	136,000	158,000
211,900	165,000	183,000	85,000	126,500	162,000
169,000	175,000	267,000	150,000	115,000	126,500
215,000	190,000	190,000	113,500	116,300	190,000
145,000	269,900	135,500	190,000	98,000	137,900
108,000	120,500	128,500	142,500	72,000	124,900
134,000	205,406	217,000	94,000	189,900	168,500
133,000	180,000	139,500	210,000	126,500	285,000
195,000	97,000	117,000	150,000	180,500	160,000
181,500	124,000	125,900	165,000	122,000	132,000
145,900	156,000	136,000	142,000	140,000	144,900
133,000	196,800	121,900	126,000	164,900	172,000
100,000	129,900	110,000	131,000	107,000	165,900

2-1 OVERVIEW

In analyzing a data set, we should first determine whether we know all values for a complete population, or whether we know only the values for some sample drawn from a larger population. That determination will affect both the methods we use and the conclusions we form.

We use methods of **descriptive statistics** to summarize or *describe* the important characteristics of a known set of data. If the 150 values given in Table 2-1 represent the selling prices of *all* homes sold in Dutchess County, then we have known population data. We might then proceed to improve our understanding of this known population data by computing some average or by constructing a graph. In contrast to descriptive statistics, **inferential statistics** goes beyond mere description. We use inferential statistics when we use sample data to make *inferences* about a population.

Suppose we compute an average of the 150 scores in Table 2-1 and obtain a value of $153,775. If exactly 150 homes are sold in Dutchess County and our list is complete, that average of $153,775 is a parameter that describes and summarizes known population data. But those values are the actual selling prices of 150 homes that were randomly selected from a larger population. Treating those 150 values as a sample drawn from a larger population, we might conclude that the average selling price for all Dutchess County homes sold is $153,775. In so doing, we have made an *inference* that goes beyond the known data.

This chapter deals with the basic concepts of descriptive statistics, Chapter 3 includes an introduction to probability theory, and the subsequent chapters deal mostly with inferential statistics. Descriptive statistics and inferential statistics are the two basic divisions of the subject of statistics.

We use the tools of descriptive statistics in order to understand an otherwise unintelligible collection of data. The following three characteristics of data are extremely important, and they can give us much insight:

1. Representative score, such as an average
2. Measure of scattering or variation
3. Nature of the distribution, such as bell shaped

We can learn something about the nature of the distribution by organizing the data and constructing graphs, as in Sections 2-2 and 2-3. In Section 2-4 we will learn how to obtain representative, or average, scores. We will measure the extent of scattering, or variation, among data as we use the tools found in Section 2-5. Finally, in Section 2-6 we will learn about measures of position so that we can better analyze or compare various scores. As we proceed through this chapter, we will refer to the 150 scores given in Table 2-1, and our insight into that data set will be increased as we reveal its characteristics.

There is one last point that should be made in this overview. When collecting data, we must be extremely careful about the methods we use (common sense is often a critical requirement). If we plan our data collection with care and thoughtfulness, we can often learn much by using simple methods. If our data collection is quick, easy, and without much thought, we may well end up with something that is misleading or worthless. As we consider the methods of this chapter, remember that they will give us misleading results if the sample data are not representative of the population.

Authors of the *Federalist* Papers Identified

In 1787–1788 Alexander Hamilton, John Jay, and James Madison anonymously published the famous *Federalist* papers in an attempt to convince New Yorkers that they should ratify the Constitution. The identity of most of the papers' authors became known, but the authorship of 12 of the papers was contested. Through statistical analysis of the frequencies of various words, we can now conclude that James Madison is the *likely* author of these 12 papers. For many of these disputed papers, the evidence in favor of Madison's authorship is overwhelming to the degree that we can be almost certain of being correct.

2-2 SUMMARIZING DATA

Referring to the 150 home selling prices given in Table 2-1, we see that simply looking at those numbers will not lead to any specific conclusions because the human mind usually cannot assimilate, organize, and condense that much data. In general, any large collection of raw data will remain unintelligible until it is organized and summarized. We now proceed to consider an important method for organizing data.

A **frequency table** is an excellent device for making large collections of data much more intelligible. A frequency table is so named because it lists categories of scores along with their corresponding frequencies. The *frequency* for a category or class is the number of original scores that fall into that class.

This might seem complicated, but the construction of a frequency table is really a simple process. Although construction may be time-consuming and monotonous, it is not very difficult. For extremely large collections of scores, the data can be entered in a computer that is programmed to construct the appropriate frequency table automatically.

The following standard definitions formalize and identify some of the basic terminology associated with frequency tables. These definitions may seem difficult, so it will be helpful to examine Table 2-2, which illustrates them.

TABLE 2-2	
Score	Frequency
1–5	
6–10	
11–15	
16–20	
21–25	
26–30	

This frequency table has six classes.

The **lower class limits** are 1, 6, 11, 16, 21, 26.

The **upper class limits** are 5, 10, 15, 20, 25, 30.

The **class boundaries** are 0.5, 5.5, 10.5, 15.5, 20.5, 25.5, 30.5.

The **class marks** are 3, 8, 13, 18, 23, 28.

The **class width** is 5.

DEFINITIONS

Lower class limits are the smallest numbers that can actually belong to the different classes.

Upper class limits are the largest numbers that can actually belong to the different classes.

The **class boundaries** are obtained by increasing the upper class limits and decreasing the lower class limits by the same amount so that there are are no gaps between consecutive classes. The amount to be added or subtracted is one-half the difference between the upper limit of one class and the lower limit of the following class.

The **class marks** are the midpoints of the classes.

The **class width** is the difference between two consecutive lower class limits (or class boundaries).

The process of actually constructing a frequency table involves these key steps:

1. **Decide on the number of classes your frequency table will contain.** Guideline: The number of classes should be between 5 and 20. The actual number of classes may be affected by convenience or other subjective factors.

2. **Find the class width** by dividing the number of classes into the range. (The range is the difference between the highest and lowest scores.) Round the result *up* to a convenient number. This rounding up (not off) is not only convenient, but it also guarantees that all

of the data will be included in the frequency table. (If the number of classes divides into the range evenly with no remainder, you may need to add another class in order for all of the data to be included.)

$$\text{class width} = \text{round } up \text{ of } \frac{\text{range}}{\text{number of classes}}$$

3. **Select as a starting point either the lowest score or a value slightly less than the lowest score.** This starting point is the lower class limit of the first class.
4. **Add the class width to the starting point to get the second lower class limit.** Add the class width to the second lower class limit to get the third, and so on.
5. **List the lower class limits in a vertical column and enter the upper class limits**, which can be easily identified at this stage.
6. **Represent each score by a tally** in the appropriate class.
7. **Replace the tally marks in each class with the total frequency count** for that class.

EXAMPLE

Using the preceding steps, we construct the frequency table that summarizes the data in Table 2-1.

Step 1: We begin by selecting 10 as the number of desired classes. (Many statisticians recommend that we should generally aim for about 10 classes.)

Step 2: With a minimum of $72,000 and a maximum of $302,000 the range is $302,000 − $72,000 = $230,000, so that

$$\text{class width} = \text{round } up \text{ of } \frac{230,000}{10}$$

$$= \text{round } up \text{ of } 23,000$$

$$= \$25,000 \text{ (rounded for its convenience)}$$

Step 3: Starting with $72,000 would not be good since the resulting class limits would be inconvenient. Instead, we start with $50,000 since it is a more convenient number.

continued ▸

Example, continued

Step 4: Add the class width of $25,000 to the lower limit of $50,000 to get the next lower limit of $75,000. Continuing, the other lower class limits are $100,000, $125,000, and so on.

Step 5: These lower class limits suggest these upper class limits.

$ 50,000	$ 74,999
75,000	99,999
100,000	124,999

Step 6: The tally marks are shown in the middle of Table 2-3.

Step 7: The frequency counts are shown in the extreme right column of Table 2-3.

Note that the resulting frequency table, as shown in Table 2-3, actually has 11 classes instead of the 10 we began with. This is a result of selecting class limits that are convenient. We could have forced the table to have exactly 10 classes, but we would get messy limits such as $72,000–$94,999.

TABLE 2-3		
Selling Price (dollars)	Tally Marks	Frequency
50,000–74,999	I	1
75,000–99,999	⌗⌗ IIII	9
100,000–124,999	⌗⌗ ⌗⌗ ⌗⌗ ⌗⌗ II	22
125,000–149,999	⌗⌗ ⌗⌗ ⌗⌗ ⌗⌗ ⌗⌗ ⌗⌗ ⌗⌗ ⌗⌗ ⌗⌗ II	47
150,000–174,999	⌗⌗ ⌗⌗ ⌗⌗ ⌗⌗ ⌗⌗ ⌗⌗ I	31
175,000–199,999	⌗⌗ ⌗⌗ ⌗⌗ ⌗⌗ III	23
200,000–224,999	⌗⌗ IIII	9
225,000–249,999	II	2
250,000–274,999	III	3
275,000–299,999	II	2
300,000–324,999	I	1

Table 2-3 provides a tremendous advantage by making more intelligible the otherwise unintelligible list of 150 selling prices of homes. Yet this advantage is not gained without some loss. In constructing frequency tables, we may lose the accuracy of the raw data. To see how this loss occurs, consider the first class of $50,000–$74,999. Table 2-3 shows that there is one score in that class, but there is no way to determine from the table exactly what that score is. We cannot reconstruct the original 150 selling prices from Table 2-3. The exact values have been compromised for the sake of comprehension.

Summarizing data generally involves a compromise between accuracy and simplicity. A frequency table with too few classes is simple but not accurate. A frequency table with too many classes is more accurate but not as easy to understand. The best arrangement is arrived at subjectively, usually in accordance with the common formats used in particular applications. Some of these difficulties will be overcome in the next section when stem-and-leaf plots are discussed.

EXAMPLE

A large collection of college grade-point averages ranges from 0.00 to 3.96 and we want to develop a frequency table with 10 classes. Find the upper and lower limits of each class.

Solution

The difference between the highest and lowest scores is 3.96. Dividing by the number of classes (10), we get 0.396, which we round up to 0.40. (Rounding 0.396 up to 1 would cause all the data to fall in the first 4 classes, with the last 6 classes all empty.) With a class width of 0.4, we get these lower class limits: 0.00, 0.40, 0.80, 1.20, 1.60, 2.00, 2.40, 2.80, 3.20, 3.60. From these lower class limits we can see that the upper class limits must be 0.39, 0.79, 1.19, . . . , 3.59, 3.99. The first class of 0.00–0.39 has a class mark of 0.195; the other class marks can be found by halving the sum of the upper and lower limits. The first class of 0.00–0.39 has an upper class boundary of 0.395 (the value midway between 0.39 and 0.40). Since the upper class boundary of 0.395 is 0.005 above the upper limit, it follows that the lower class boundary should be 0.005 below the lower class limit, so the value of −0.005 is determined.

Those lower class limits	suggest these upper class limits.		
0.00–0.39			−0.005–0.395
0.40–0.79			0.395–0.795
0.80–1.19	←Those		0.795–1.195
1.20–1.59	class		1.195–1.595
1.60–1.99	limits		1.595–1.995
2.00–2.39	determine		1.995–2.395
2.40–2.79	these →		2.395–2.795
2.80–3.19	class		2.795–3.195
3.20–3.59	boundaries.		3.195–3.595
3.60–3.99			3.595–3.995

When frequency tables are being constructed, the following guidelines should be followed.

1. **The classes must be mutually exclusive.** That is, each score must belong to exactly one class.
2. **Include all classes**, even if the frequency might be zero.
3. **All classes should have the same width**, although it is sometimes impossible to avoid open-ended intervals such as "65 years or older."
4. **Try to select convenient numbers for class limits.**
5. **The number of classes should be between 5 and 20.**

Common sense is often helpful in constructing frequency tables that are easier to use and understand. For the grade-point averages of the preceding example, we know that the highest and lowest possible values are 4.00 and 0.00, respectively. Also, we may know of special cutoff values like 2.00 (for probation) or 3.00 (for dean's list). Based on this additional information we may be wise to adjust the frequency table so that it is easier to read and comprehend. We might, for example, use the following class limits:

0.00–0.49
0.50–0.99
1.00–1.49
1.50–1.99
2.00–2.49
2.50–2.99
3.00–3.49
3.50–3.99

Realizing that these adjusted class limits do not allow a perfect grade-point average of 4.00, we might also replace the last class of 3.50–3.99 with "3.50 and higher." We no longer have 10 classes and the class widths may not be uniform throughout, but the resulting frequency table is much more readable and usable.

A variation of the standard frequency table is used when cumulative totals are desired. The cumulative frequency for a class is the sum of the frequencies for that class and all previous classes. Table 2-4 is an example of a **cumulative frequency table**, and it corresponds to the same 150 selling prices presented in Table 2-3. A comparison of the frequency column of Table 2-3 and the cumulative frequency column of Table 2-4 reveals that the latter values can be obtained from the former by starting at the top and adding on the successive values. For example, the cumulative frequency of 10 from Table 2-4 represents the sum 1 and 9 from Table 2-3.

TABLE 2-4	
Selling Price	Cumulative Frequency
Less than $75,000	1
Less than $100,000	10
Less than $125,000	32
Less than $150,000	79
Less than $175,000	110
Less than $200,000	133
Less than $225,000	142
Less than $250,000	144
Less than $275,000	147
Less than $300,000	149
Less than $325,000	150

In the next section, we will explore various graphic ways to depict data so that they are easy to understand. Frequency tables will be necessary for some of the graphs, and those graphs are often necessary for considering the way the scores are distributed. Frequency tables become important prerequisites for later, more useful concepts.

2-2 EXERCISES A

2-1 What upper class limits correspond to lower class limits of 3, 6, 9, 12, 15?

2-2 What upper class limits correspond to lower class limits of 0, 10, 20, 30, 40, 50?

2-3 What upper class limits correspond to lower class limits of 0, 2.5, 5.0, 7.5, 10.0?

2-4 What upper class limits correspond to lower class limits of 50, 65, 80, 95, 110, 125?

2-5 For a certain frequency table, the lower limit of the first class is 10. The class width is 20 and there are five classes. Identify the other four lower class limits.

2-6 A frequency table has six classes and the class width is 6. If the lower limit of the first class is 8, identify the other five lower class limits.

2-7 In a frequency table with eight classes, the lower limit of the first class is 7.5, and the class width is 2.5. Identify the other lower class limits.

2-8 In a frequency table with six classes, the lower limit of the first class is 0, and the class width is 0.2. Find the other lower class limits.

In Exercises 2-9 through 2-12, find a suitable value for the class width, given the information provided about the data set.

2-9 Minimum is 50, maximum is 99, 10 classes.

2-10 Minimum is 50, maximum is 91, 6 classes.

2-11 Minimum is 0.67, maximum is 3.24, 11 classes.

2-12 Minimum is 120, maximum is 239, 8 classes.

In Exercises 2-13 through 2-16, identify the class width for the frequency table identified.

2-13 Table 2-5   **2-14** Table 2-6

TABLE 2-5	
IQ	Frequency
80–87	16
88–95	37
96–103	50
104–111	29
112–119	14

TABLE 2-6	
Time (hours)	Frequency
0.0–7.5	16
7.6–15.1	18
15.2–22.7	17
22.8–30.3	15
30.4–37.9	19

2-15   Table 2-7   **2-16**   Table 2-8

TABLE 2-7	
Weight (kg)	Frequency
16.2–21.1	16
21.2–26.1	15
26.2–31.1	12
31.2–36.1	8
36.2–41.1	3

TABLE 2-8	
Sales (dollars)	Frequency
0–21	2
22–43	5
44–65	8
66–87	12
88–109	14
110–131	20

In Exercises 2-17 through 2-20, identify the class marks for the frequency table identified.

2-17   Table 2-5   **2-18**   Table 2-6
2-19   Table 2-7   **2-20**   Table 2-8

In Exercises 2-21 through 2-24, identify the class boundaries for the frequency table identified.

2-21   Table 2-5   **2-22**   Table 2-6
2-23   Table 2-7   **2-24**   Table 2-8

In Exercises 2-25 through 2-28, construct the cumulative frequency table corresponding to the frequency table indicated.

2-25   Table 2-5   **2-26**   Table 2-6
2-27   Table 2-7   **2-28**   Table 2-8

In Exercises 2-29 through 2-32, modify the class limits of the frequency table identified so that the new limits make the table easier to read. Omit the frequencies since they cannot be determined. It may be necessary to change the number of classes.

2-29   Table 2-5   **2-30**   Table 2-6
2-31   Table 2-7   **2-32**   Table 2-8

In Exercises 2-33 through 2-36, use the given information to find the upper and lower limits of the first class.

2-33   Assume that you have a collection of scores representing the weights (rounded up to the nearest pound) of adult males. Also assume that the lowest and highest weights are 110 and 309, respectively. (You wish to construct a frequency table with 10 classes.)

2-34 You have been studying the jail sentences of those convicted of driving while intoxicated. You have a list of sentences ranging from 0 days in jail to 95 days in jail, and you intend to publish a frequency table with 16 classes.

2-35 A scientist is investigating the time (in seconds) required for a certain chemical reaction to occur. The experiment is repeated 200 times and the results vary from 17.3 s to 42.7 s. (You wish to construct a frequency table with 12 classes.)

2-36 A track star has run the mile in 150 different events, and his times ranged from 238.7 s to 289.8 s. (You wish to construct a frequency table with 14 classes.)

2-37 When 40 people were surveyed at the Galleria Mall in Wappingers Falls, N.Y., they reported the distances they drove to the mall, and the results (in miles) are given below. Construct a frequency table with eight classes.

2	8	1	5	9	5	14	10	31	20
15	4	10	6	5	5	1	8	12	10
25	40	31	24	20	20	3	9	15	15
25	8	1	1	16	23	18	25	21	12

2-38 A student working part time for a moving company in Dutchess County, N. Y., collected the following load weights (in pounds) for 50 consecutive customers. Construct a frequency table with eight classes.

8,090	17,810	3,670	10,100	17,330
8,800	7,770	12,430	13,260	10,220
13,490	13,520	4,510	14,760	4,480
7,540	10,630	10,330	10,510	12,700
7,200	12,010	11,450	9,140	7,280
9,110	12,350	14,800	26,580	15,970
11,860	8,450	10,780	5,030	11,430
11,600	7,470	14,310	13,410	7,450
3,250	6,400	8,160	9,310	9,900
6,170	16,200	8,770	6,820	6,390

2-39 Listed below are the actual energy consumption amounts as reported on the electric bills for one residence. Each amount is in kilowatt-hours and represents a two-month period. Construct a frequency table with 10 classes.

728	774	859	882	791	731	838	862	880	831
759	774	832	816	860	856	787	715	752	778
829	792	908	714	839	752	834	818	835	751
837									

2-40 Listed below are the daily sales totals (in dollars) for 44 days at a large retail outlet in Orange County, Calif. Construct a frequency table with 10 classes.

24,145	66,644	52,250	59,708
33,650	29,099	68,945	32,097
37,196	34,423	31,987	36,802
39,897	37,809	35,018	38,366
46,351	41,160	38,275	44,249
65,798	50,857	43,923	61,923
25,597	66,892	55,254	33,310
34,132	30,284	31,995	37,052
37,229	34,804	35,420	38,812
39,921	37,858	38,295	46,285
47,563	42,906	43,952	64,842

2-2 EXERCISES B

2-41 Compare frequency Tables 2-5, 2-6, 2-7, and 2-8. Are there any obvious differences in the *distribution* of the scores? Could such differences be readily observed by examining the raw scores?

2-42 Listed below are two sets of scores that are supposed to be heights (in inches) of randomly selected adult males. One of the sets consists of real heights actually obtained from randomly selected adult males, but the other set consists of numbers that were fabricated. Construct a frequency table for each set of scores. By examining the two frequency tables, identify the set of data that you believe to be false.

70	73	70	72	70	73	70	72
71	73	71	67	71	66	74	76
68	72	67	72	68	75	67	68
71	73	72	70	71	77	66	69
72	68	71	71	72	67	77	75
71	73	69	73	66	76	76	77
71	66	77	67	73	74	69	67

2-43 A table in a recent issue of *USA Today* included the following data that were based on U.S. Census Bureau observations. Refer to the five guidelines for constructing frequency tables. Which of them are not followed?

Age	U.S. Population (millions)
Under 5	18.1
5–13	30.3
14–17	14.8
18–24	28.0
25–34	43.0
35–44	33.1
45–54	22.8
55–64	22.2
65–84	9.1
85/older	2.8

2-44   In constructing a frequency table, Sturges' guideline suggests that the ideal number of classes can be approximated by finding the value of $1 + (\log n)/(\log 2)$, where n is the number of scores. Use this guideline to find the ideal number of classes corresponding to a data collection with the number of scores equal to (a) 50; (b) 100; (c) 150; (d) 500; (e) 1000; (f) 50,000.

2-3 PICTURES OF DATA

In the preceding section, we saw that frequency tables transform a disorganized collection of raw scores into an organized and understandable summary. In this section we consider ways of representing data in pictorial form. The obvious objective is to promote understanding of the data. We attempt to show that one graphic illustration can be a suitable replacement for hundreds of data.

Pie Charts

It is a simple and obvious characteristic of human physiology that people can comprehend one picture better than a more abstract collection of words or data. Consider this statement: Among the 19,257 murders in a recent year, 11,381 (or 59.1%) were committed with firearms, 3957 (or 20.5%) were committed with knives, 1310 (or 6.8%) were committed with personal weapons (hands, feet, and so on), 1099 (or 5.7%) were committed with blunt objects, and the remaining 1510 (or 7.8%) were committed with a variety of other weapons (based on FBI data in the *Uniform Crime Reports*). After reading the last sentence, how much did you comprehend? What impressions did you develop?

How Valid Are Crime Statistics?

Police departments are often judged by the number of arrests because that statistic is readily available and clearly understood. However, this encourages police to concentrate on easily solved minor crimes (such as marijuana smoking) at the expense of more serious crimes, which are difficult to solve. A study of Washington, D.C., police records revealed that police can also manipulate statistics another way. The study showed that more than 1000 thefts in excess of $50 were intentionally valued at less than $50 so that they would be classified as petty larceny and not major crime.

Now examine the **pie chart** in Figure 2-1. Which does a better job of making the data understandable? The abstraction of numbers is overcome by the concrete reality of the slices of the pie.

Pie charts are especially useful in cases involving data categorized according to attributes rather than numbers (data at the nominal level of measurement). The weapons—firearms, knives, blunt objects, and so on—are examples of attributes that are not themselves numbers.

The construction of a pie chart (sometimes called a *circle graph*) is not very difficult. It simply involves the "slicing up" of the pie into the proper proportions. If firearms represent 59.1% of the total, then the wedge representing firearms should be 59.1% of the total. (The central angle should be $0.591 \times 360° = 213°$.)

Histograms

A **histogram** is another common graphic way of presenting data. Histograms are especially important because they show the distribution of the data, and the distribution of data is an extremely important characteristic.

We generally construct a histogram to represent a set of scores after we have completed a frequency table. The standard format for a

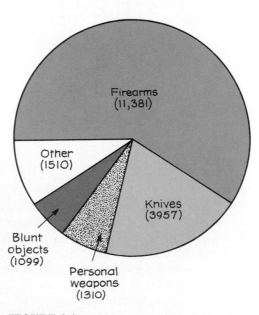

FIGURE 2-1
Murder weapons in the United States

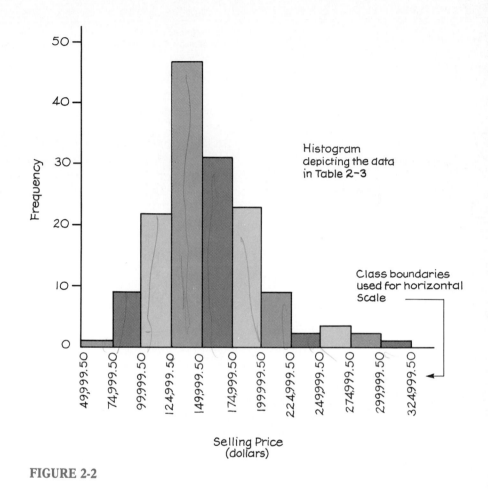

FIGURE 2-2

histogram usually involves a vertical scale that delineates frequencies and a horizontal scale that delineates values of the data being represented. We use shaded bars to represent the individual classes of the frequency table. Each bar extends from its lower class boundary to its upper class boundary so that we can mark the class boundaries on the horizontal scale. (However, improved readability is often achieved by using class limits or class marks instead of class boundaries.) As an example, Figure 2-2 is the histogram that corresponds directly to Table 2-3 in the previous section.

Before constructing a histogram from a completed frequency table, we must give some consideration to the scales used on the vertical and horizontal axes. We can begin by determining the maximum frequency, and the maximum frequency should suggest a suitable scale for the

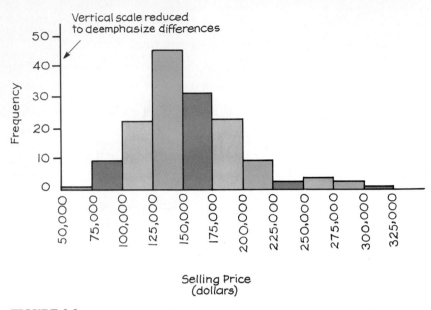

FIGURE 2-3

vertical axis. In Table 2-3 the maximum frequency of 47 will correspond to the tallest bar on the histogram, so 47 (or the next highest convenient number) should be located at the top of the vertical scale, with 0 at the bottom. In Figure 2-2, we designed the vertical scale to run from 0 to 50. The horizontal scale should be designed to accommodate all the classes of the frequency table. Ideally, we should try to follow the rule of thumb that the vertical height of the histogram should be about $\frac{3}{4}$ of the total width. Consider the desired length of the horizontal axis along with the number of classes that must be incorporated. Both axes should be clearly labeled.

The histogram of Figure 2-3 shows the same data as Table 2-3 and Figure 2-2, but with class limits used on the horizontal scale for improved readability. The vertical axis is scaled down to create an impression of smaller differences. The flexibility in manipulating the scales for the vertical and horizontal axes can serve as a great device for deception. A clever person can exaggerate small differences or deemphasize large differences to suit his or her own purpose. However, conscientious and critical students of statistics become aware of such misleading presentations and carefully examine histograms by analyzing the objective numerical data, instead of simply glancing at the overall picture.

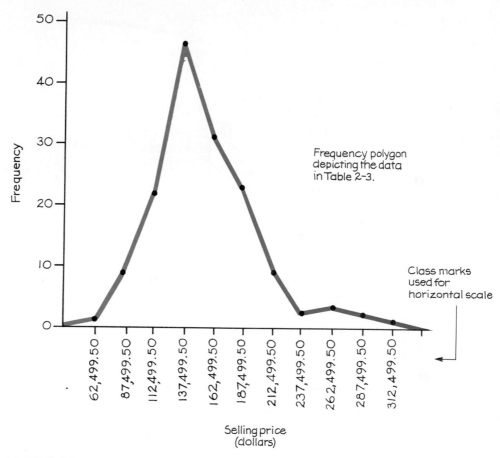

FIGURE 2-4

Frequency Polygons

A **frequency polygon** is a variation of the histogram in which the vertical bars are replaced by dots that are connected to form a line graph. When delineating the horizontal scale of a frequency polygon, we generally use class marks, so that the dots can be located directly above them. In Table 2-3 the class marks are 62,499.50, 87,499.50, . . . , 312,499.50. In Figure 2-4 we identify those values along the horizontal axis, plot the points directly above them, connect those points with straight-line segments, and then extend the resulting line graph at the beginning and end so that it starts and ends with a frequency of 0.

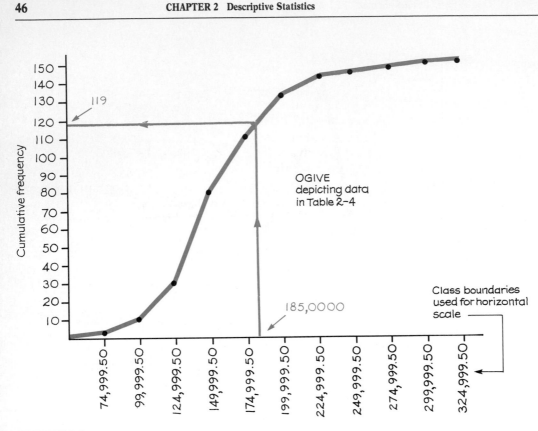

FIGURE 2-5

Ogives

Another common pictorial display is the **ogive** (pronounced "oh-jive"), or **cumulative frequency polygon**. This differs from an ordinary frequency polygon in that the frequencies are cumulative. That is, we add each class frequency to the total of all previous class frequencies, as in the cumulative frequency table described in Section 2-2. The vertical scale is again used to delineate frequencies, but we must now adjust the scale so that the total of all individual frequencies will fit. The horizontal scale should depict the class boundaries. See Figure 2-5. Ogives are useful when, instead of seeking the frequency for a given class interval, we want to know how many of our scores are above or below some level. For example, in doing an analysis of selling prices of homes, a buyer might be more interested in the homes that cost less than $185,000 than in the homes that cost between any two particular values. Figure 2-5 shows that approximately 119 (or 79%) of the sampled homes fall below $185,000.

Stem-and-Leaf Plots

Histograms, frequency polygons, and ogives are extremely useful for graphically displaying the distribution of data. By using those graphic devices, we are able to learn something about the data that is not apparent while the data remain in a list of values. This additional insight is clearly an advantage. However, in constructing histograms, frequency polygons, or ogives we also suffer the disadvantage of distorted data. When we transform raw data into a histogram, for example, we lose some information in the process. Generally we cannot reconstruct the original data set from the histogram, and this shows that some distortion has occurred. We will now introduce another device that enables us to see the distribution of data without losing information in the process. The following concept is part of a relatively new branch of statistics often referred to as **exploratory data analysis (EDA)**. See *Exploratory Data Analysis* by John W. Tukey for reference to other concepts in this field, which is gaining wider acceptance.

In a **stem-and-leaf plot** we sort data according to a pattern that reveals the underlying distribution. The pattern involves separating a number into two parts, usually the first digit and the other digits. The *stem* consists of the leftmost digits and *leaves* consist of the rightmost digits. The method is illustrated in the following example.

EXAMPLE

A librarian records the number of daily microfilm uses and compiles the sample data that follow. Construct a stem-and-leaf plot for this data.

10	11	15	23	27	28	38	38	39	39
40	41	44	45	46	46	52	57	58	65

Solution

We note that the numbers have first digits of 1, 2, 3, 4, 5, or 6 and we let those values become the stem. We then construct a vertical line and list the "leaves" as shown below. The first row represents the numbers 10, 11, 15. In the second row we have 23, 27, 28, and so on.

continued ▶

Solution, continued

Stem	Leaves
1	015
2	378
3	8899
4	014566
5	278
6	5

By turning the page on its side we can see a distribution of these data, which, in this case, roughly approximates a bell shape. We have also retained all the information in the original list. We could reconstruct the original list of values from the stem-and-leaf plot. In the next example we illustrate the construction of a stem-and-leaf plot for data with three significant digits.

EXAMPLE

An aeronautical research team is investigating the stall speed of an ultralight aircraft, and the following sample values are obtained (in knots). Construct a stem-and-leaf plot and the graph suggested by that plot.

21.7	24.0	22.4	22.4	24.3	22.3	22.6	25.2
24.1	21.8	23.2	23.9	23.5	23.2	23.9	23.8

Solution

Note that unlike the previous example, these data are not in order, so two sweeps will be necessary. For the first sweep, we record the data reading across one row at a time to get the following result.

Stem	Leaves
21.	78
22.	4436
23.	295298
24.	031
25.	2

continued ▶

Solution, continued

For the second sweep we arrange each row of leaves in order from low to high as follows.

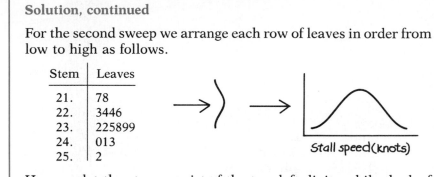

Stem	Leaves
21.	78
22.	3446
23.	225899
24.	013
25.	2

Stall speed (knots)

Here we let the stem consist of the two left digits while the leaf is the digit farthest to the right. Again, we can see the shape of the distribution by turning the page on its side and observing the columnns of digits above the line.

For some data sets, simplified stem-and-leaf plots can be constructed by first rounding the values to two or three significant digits. If necessary, stem-and-leaf plots can be condensed by combining adjacent rows. They can also be stretched out by subdividing rows into those with the digits 0 through 4 and those with digits 5 through 9. Note that the following two stem-and-leaf plots represent the same set of data, but the plot on the right has been stretched out to include more rows. Also note that such changes may affect the apparent shape of the distribution.

Stem	Leaves			
51	6899		51		(last digits of 0 through 4)
52	0347	expand	51	6899	(last digits of 5 through 9)
53	3788	⟶	52	034	(last digits of 0 through 4)

51		(last digits of 0 through 4)
51	6899	(last digits of 5 through 9)
52	034	(last digits of 0 through 4)
52	7	(last digits of 5 through 9)
53	3	(last digits of 0 through 4)
53	788	(last digits of 5 through 9)

In the preceding example we expanded the number of rows. We can also condense a stem-and-leaf plot by combining adjacent rows as shown on the next page.

50	01				
51	4		50–51	01*4	
52	56		52–53	56*368	
53	368		54–55	2457*3499	
54	2457	contract	56–57	0127*358	
55	3499	⟶	58–59	1269*17	
56	0127				
57	358			↑	↑
58	1269				
59	17			(581)	(591)

In the condensed plot, we separated digits in the "leaves" associated with the numbers in each stem by an asterisk. Every row in the condensed plot must include exactly one asterisk so that the shape of the plot is not distorted.

Another useful feature of stem-and-leaf plots is that their construction provides a fast and easy procedure for ranking data (putting data in order). Data must be ranked for a variety of statistical procedures, such as the Wilcoxon rank-sum test in Chapter 11 and the median in the next section.

Miscellaneous Graphics

Numerous pictorial displays other than the ones just described can be used to represent data dramatically and effectively. Some examples are soldiers, tanks, airplanes, stacks of coins, and moneybags. This list is a very small sample of the diverse graphic methods used to convey the nature of statistical data. In fact, there is almost no limit to the variety of different ways that data can be illustrated. However, pie charts, histograms, frequency polygons, and ogives are among the most common devices used. These graphic representations often convey the *distribution* of the data, and an understanding of the distribution is often critically important to the statistician. In the following section we consider ways of measuring other characteristics of data.

2-3 EXERCISES A

2-45   Given the following frequency table, identify the values that should be entered along the horizontal scale. Assume that the frequency table will be used to construct

a. a histogram
b. a frequency polygon
c. an ogive

Distance (mi)	Frequency
1–8	15
9–16	12
17–24	7
25–32	5
33–40	1

Data based on distances traveled by 40 people at the Galleria Mall in Wappingers Falls, N.Y.

2-46   Construct a histogram that corresponds to the frequency table in Exercise 2-45.

2-47   Construct a frequency polygon that corresponds to the frequency table in Exercise 2-45.

2-48   Construct an ogive that corresponds to the frequency table in Exercise 2-45. Use the ogive to estimate the proportion of people who traveled less than 20 mi.

2-49   Given the following frequency table, identify the values that should be entered along the horizontal scale. Assume that the frequency table will be used to construct
a. a histogram
b. a frequency polygon
c. an ogive

Load (lb)	Frequency
0–3,999	2
4,000–7,999	13
8,000–11,999	19
12,000–15,999	12
16,000–19,999	3
20,000–23,999	0
24,000–27,999	1

Data based on weights of loads for 50 clients of a moving company in Dutchess County, N.Y.

2-50   Construct a histogram that corresponds to the frequency table in Exercise 2-49.

2-51   Construct a frequency polygon that corresponds to the frequency table in Exercise 2-49.

2-52   Construct an ogive that corresponds to the frequency table in Exercise 2-49. Use the ogive to estimate the proportion of loads that were under 10,000 lb.

2-53   Given the following frequency table, identify the values that should be entered along the horizontal scale. Assume that the frequency table will be used to construct
a. a histogram
b. a frequency polygon
c. an ogive

Energy (kwh)	Frequency
700–719	2
720–739	2
740–759	4
760–779	4
780–799	2
800–819	2
820–839	8
840–859	2
860–879	2
880–899	2
900–919	1

Data based on energy consumption reported on electric bill for 31 two-month periods.

2-54   Construct a histogram that corresponds to the frequency table in Exercise 2-53.

2-55   Construct a frequency polygon that corresponds to the frequency table in Exercise 2-53.

2-56   Construct an ogive that corresponds to the frequency table in Exercise 2-53. Use the ogive to estimate the proportion of energy levels that *exceed* 830 kwh.

2-57   Given the following frequency table, identify the values that should be entered along the horizontal scale. Assume that the frequency table will be used to construct
a. a histogram
b. a frequency polygon
c. an ogive

Sales (dollars)	Frequency
20,000–24,999	1
25,000–29,999	2
30,000–34,999	9
35,000–39,999	14
40,000–44,999	5
45,000–49,999	3
50,000–54,999	2
55,000–59,999	2
60,000–64,999	2
65,000–69,999	4

Data based on daily sales for 44 days at a large retail outlet in Orange County, Calif.

2-58   Construct a histogram that corresponds to the frequency table in Exercise 2-57.

2-59   Construct a frequency polygon that corresponds to the frequency table in Exercise 2-57.

2-60   Construct an ogive that corresponds to the frequency table in Exercise 2-57.

In Exercises 2-61 through 2-64, list the original numbers in the data set represented by the given stem-and-leaf plots.

2-61
Stem	Leaves
2	00358

2-62
Stem	Leaves
1	001112278
2	3444569
3	013358

2-63
Stem	Leaves
40	6678
41	09999
42	13466
43	088

2-64
Stem	Leaves
68	45 45 47 86
69	33 38 89
70	52 59 93
71	27

In Exercises 2-65 through 2-68, construct the stem-and-leaf plots for the given data sets.

2-65   High temperatures (in degrees Fahrenheit) for the 31 days in July (recorded at Dutchess Community College):

80	68	84	86	85	77	64	81	93	94
97	93	89	82	76	75	83	90	83	84
92	94	90	92	91	84	81	84	79	80
80									

2-66   Pulse rates (number of beats per minute) of 20 male statistics students:

82	74	77	62	78	58	85	74	66	71
58	80	65	60	54	75	71	74	73	82

2-67   Lot sizes (in acres) of 20 homes sold in Dutchess County:

0.75	0.70	0.65	0.75	0.50
0.79	0.50	0.50	0.68	0.65
0.75	0.50	0.50	0.65	0.58
0.80	0.91	0.78	0.68	0.92

2-68   Takeoff distances (in feet) required for a light aircraft:

717	716	736	772	740	741	735	735	710	753	757	747
756	715	718	720	726	721	760	769	771	738	721

2-69   Construct the frequency table corresponding to the following stem-and-leaf plot. The data represent sales in an auto parts store. All values have been rounded to the nearest dollar.

Stem	Leaves
0	00011222233333344555677788999
1	0011122333334455555667
2	00001111333333446678
3	0011111223334445567899
4	0011222333334555667789
5	01334678999
6	11355779
7	
8	0245
9	3
10	26

2-70   a. Using the frequency table from Exercise 2-69, construct the corresponding histogram.

   b. Compare the histogram from part (a) to the shape of the stem-and-leaf plot given in Exercise 2-69.

FIGURE 2-6
Histogram

2-71   Use the histogram in Figure 2-6.
       a. Construct the corresponding frequency table. Use the eleven class
          marks of 15, 16, 17, . . . , 25.
       b. Construct the corresponding ogive.
       c. Refer to the ogive to estimate how many police trainees scored
          below 21.
       d. If 17 is a passing score, estimate the percentage of trainees who
          failed this test.

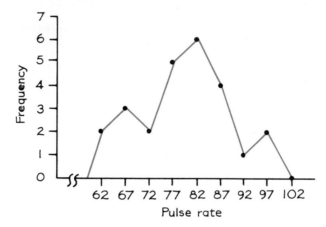

FIGURE 2-7
Frequency polygon

2-72   Use the frequency polygon in Figure 2-7.
       a. Construct the corresponding frequency table.
       b. Construct the corresponding ogive.
       c. Estimate the number of subjects who had pulse rates between 75
          and 84.
       d. Estimate the percentage of subjects with pulse rates below 90.

2-3 EXERCISES B

2-73   Given below are frequency tables for the first 100 digits in the decimal representation of π and the first 100 digits in the decimal representation of 22/7.
a. Construct histograms representing the frequency tables and note any differences.
b. The numbers π and 22/7 are both real numbers, but how are they fundamentally different?

π	
x	f
0	8
1	8
2	12
3	11
4	10
5	8
6	9
7	8
8	12
9	14

22/7	
x	f
0	0
1	17
2	17
3	1
4	17
5	16
6	0
7	16
8	16
9	0

2-74   Use the following data (based on data from the U.S. Commerce Department) to construct a pie chart.

Imports into the United States (in billions of dollars)	
Canada	68.3
Japan	81.9
West Germany	25.1
Taiwan	20.0
United Kingdom	15.4
Other	176.3

2-75   Using a collection of sample data, a frequency table is constructed with 10 classes, then the corresponding histogram is constructed. How is the histogram affected if the number of classes is doubled, but the same vertical scale is used?

2-76   Given the graphs of the following frequency polygons (see next page), construct the graphs of the corresponding ogives.

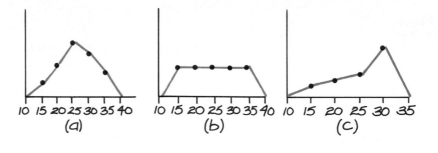

2-4 AVERAGES

In this chapter we describe some of the fundamental methods and tools generally categorized as descriptive statistics. We began with frequency tables that are used to summarize raw data (Section 2-2) and then considered a variety of graphic devices that represent the data in pictorial form (Section 2-3). However, summary and graphic methods cannot be used easily if we are limited to verbal exchanges. In addition they are very difficult to use when we want to make statistical inferences.

Suppose, for example, that we wish to compare the reaction times of two different groups of people. We could collect the two sets of raw data, organize them into two frequency tables, and then construct the appropriate histograms. Next we could compare the relative shapes and positions of the two histograms, but we would be evaluating the strengths of their similarities subjectively. Our conclusions and inferences would be much more reliable if we could develop more objective procedures. Fortunately, we can.

The more objective procedures do require numerical measurements that describe certain characteristics of the data. In this and the following section we are concerned with some of these quantitative measurements. This section deals mainly with **averages**, which are also called **measures of central tendency** since they are measurements that are intended to capture the value at the center of the scores.

An important and basic point to remember is that there are different ways to compute an average. We have already stated that, given a list of scores and instructions to find the average, most people will obligingly proceed first to total the scores and then to divide that total by the number of scores. However, this particular computation is only one of several procedures that are classified under the umbrella description of an average. The most commonly used averages are mean, median, mode, and midrange.

Just an Average Guy

The "average" American male is named Robert. He is 30 years old, 5 ft 9$\frac{1}{2}$ in. tall, weighs 173 lb, wears a size 40 suit, and has a 9$\frac{1}{2}$ shoe size and a 34-in. waist. Each year he eats 12.3 lb of pasta, 26 lb of bananas, 4 lb of potato chips, 18 lb of ice cream, and 79 lb of beef. Each year he also watches television for 2555 h and gets 585 pieces of mail. After eating his share of potato chips, reading some of his mail, and watching some television, he ends the day with 7 h and 43 min of sleep. The next day begins with a 21-min commute to a job at which he will work for 6 h and 7 min.

Mean

The **arithmetic mean** is the most important of all numerical descriptive measurements, and it corresponds to what most people call an *average*.

DEFINITION

The **arithmetic mean** of a list of scores is obtained by adding the scores and dividing the total by the number of scores. This particular average will be employed frequently in the remainder of this text, and it will be referred to simply as the **mean**.

See Figure 2-8.

EXAMPLE

Find the mean of the following quiz scores:

2, 3, 6, 7, 7, 8, 9, 9, 9, 10

Solution

First add the scores.

$$2 + 3 + 6 + 7 + 7 + 8 + 9 + 9 + 9 + 10 = 70$$

Then divide the total by the number of scores present.

$$\frac{70}{10} = 7$$

The mean score is therefore 7.

In many cases, we can provide a formula for computing some measurement. In statistics, these formulas often involve the Greek letter Σ (capital sigma), which is intended to denote the summation of a set of values. Σx means *sum the values that* x *can assume*. If we are working with a specific set of scores, the letter x is used as a variable that can assume the value of any one of those scores. Thus Σx means the sum of all of the scores. In the preceding example, the variable x assumes the values of 2, 3, 6, 7, 7, 8, 9, 9, 9, and 10, so Σx is

$$2 + 3 + 6 + 7 + 7 + 8 + 9 + 9 + 9 + 10, \text{ or } 70$$

In future formulas, we may encounter expressions such as Σx^2. Standard usage of the symbol Σ requires that the values to be added must

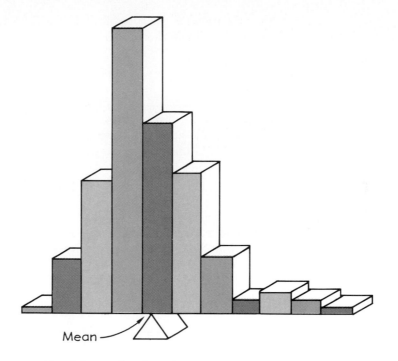

Mean

If a fulcrum is placed at the position of the mean, it will balance the histogram.

FIGURE 2-8

be in the form of the variable following Σ. Thus Σx^2 indicates that the individual scores must be squared *before* they are summed. Referring to the preceding scores, Σx^2 is

$$2^2 + 3^2 + 6^2 + 7^2 + 7^2 + 8^2 + 9^2 + 9^2 + 9^2 + 10^2$$
$$= 4 + 9 + 36 + 49 + 49 + 64 + 81 + 81 + 81 + 100$$
$$= 554$$

NOTATION
Σ denotes **summation** of a set of values.
x is the **variable** usually used to represent the individual raw scores.
n represents the **number** of scores being considered.
$\bar{x}$ denotes the **mean** of a set of **sample** scores.
μ denotes the **mean** of all scores in some **population**.

Since the mean is the sum of all scores (Σx) divided by the number of scores (n), we have the following formula.

FORMULA 2-1   mean $= \dfrac{\Sigma x}{n}$

The result can be denoted by $\bar{x}$ if the available scores are samples from a larger population; if all scores of the population are available, then we can denote the computed mean by μ (mu). Applying the formula to the preceding quiz scores, we get

$$\text{mean} = \frac{\Sigma x}{n} = \frac{70}{10} = 7$$

According to the definition of mean, 7 is the central value of the ten quiz scores given. Other definitions of averages involve different perceptions of how the center is determined. The median reflects another approach.

Median

DEFINITION

The **median** of a set of scores is the middle value when the scores are arranged in order of increasing magnitude.

After first arranging the original scores in increasing (or decreasing) order, the median will be either of the following:

1. If the number of scores is *odd*, the median is the number that is exactly in the middle of the list.
2. If the number of scores is *even*, the median is found by computing the mean of the two middle numbers.

EXAMPLE

Find the median of the scores 7, 2, 3, 7, 6, 9, 10, 8, 9, 9, 10.

Solution

Begin by arranging the scores in increasing order.

continued ▶

Mean and Median Incomes of Veterinarians

In a report of incomes earned by veterinarians, there is a reference to a collection of annual incomes with a mean of $245,287 and a median of $186,447. The report states that "this difference between the mean and the median highlights the fact that a few very high incomes offset many lower incomes to skew the mean upward" (from the *Journal of the American Veterinary Medical Association*, Vol. 19, No. 12). This example illustrates the importance of both the mean and the median for collections of scores that are skewed (lopsided) instead of being symmetrical.

Solution, continued

2, 3, 6, 7, 7, 8, 9, 9, 9, 10, 10

With these eleven scores, the number 8 is located in the exact middle, so 8 is the median.

EXAMPLE

Find the median of the scores 7, 2, 3, 7, 6, 9, 10, 8, 9, 9.

Solution

Again, begin by arranging the scores in increasing order.

2, 3, 6, 7, 7, 8, 9, 9, 9, 10

With these ten scores, no single score is at the exact middle. Instead, the two scores of 7 and 8 share the middle. We therefore find the mean of those two scores.

continued ▶

Solution, continued

$$\frac{7 + 8}{2} = \frac{15}{2} = 7.5$$

The median is 7.5.

Mode

DEFINITION

The **mode** is obtained from a collection of scores by selecting the score that occurs most frequently. In those cases where no score is repeated, we stipulate that there is no mode. In those cases where two scores both occur with the same greatest frequency, we say that each one is a mode and we refer to the data set as being **bimodal**. If more than two scores occur with the same greatest frequency, each is a mode and the data set is said to be **multimodal**.

EXAMPLE

The scores 1, 2, 2, 2, 3, 4, 5, 6, 7, 9 have a mode of 2.

EXAMPLE

The scores 2, 3, 6, 7, 8, 9, 10 have no mode since no score is repeated.

EXAMPLE

The scores 1, 2, 2, 2, 3, 4, 5, 6, 6, 6, 7, 9 have modes of 2 and 6, since 2 and 6 both occur with the same highest frequency.

EXAMPLE

A town meeting is attended by 40 Republicans, 25 Democrats, and 20 Independents. While we cannot numerically average these party affiliations, we can report that the mode is Republican, since that party had the highest frequency.

Midrange

DEFINITION

The **midrange** is that average obtained by adding the highest score to the lowest score and then dividing the result by 2.

$$\text{midrange} = \frac{\text{highest score} + \text{lowest score}}{2}$$

EXAMPLE

Find the midrange of the scores 2, 3, 6, 7, 7, 8, 9, 9, 9, 10.

$$\text{midrange} = \frac{10 + 2}{2} = 6$$

EXAMPLE

For the 150 home selling prices listed in Table 2-1, find the values of the (a) mean; (b) median; (c) mode; (d) midrange.

Solution

a. The sum of the 150 selling prices is $23,066,256, so that

$$\bar{x} = \frac{\$23,066,256}{150}$$

$$= \$153,775$$

continued ▶

Solution, continued

b. After arranging the scores in increasing order, we find that the 75th and 76th scores are both \$144,900, so the median is \$144,900. (The scores can easily be arranged in increasing order by constructing a stem-and-leaf plot or by using a computer program such as STATDISK or Minitab.)

c. The most frequent (five times) selling price is \$126,500, so it is the mode. No other value occurs more than four times.

d. The midrange is found as follows.

$$\text{midrange} = \frac{\text{highest score} + \text{lowest score}}{2}$$

$$= \frac{302{,}000 + 72{,}000}{2}$$

$$= \$187{,}000$$

We now summarize these results.

mean	\$153,775
median	\$144,900
mode	\$126,500
midrange	\$187,000

ROUND-OFF RULE

A simple rule for rounding answers is to carry one more decimal place than was present in the original data. We should round only the final answer and not intermediate values. For example, the mean of 2, 3, 5 is expressed as 3.3. Since the original data were whole numbers, we rounded the answer to the nearest tenth. The mean of 2.1, 3.4, 5.7 is rounded to 3.73.

You should know and understand the preceding four averages (mean, median, mode, and midrange). Other averages (geometric mean, harmonic mean, and quadratic mean) are not used as often, and they will be included only in Exercises B at the end of this section (see Exercises 2-106, 2-107, and 2-108).

The Class Size Paradox

Students might prefer colleges with smaller classes. But there are at least two distinct ways to obtain mean class size, and the results can be dramatically different. In one college of the State University of New York, if we take the numbers of students in the 737 classes, we get a mean of 40 students. But if we were to compile a list of the class sizes for each student and use these numbers, we would get a mean class size of 147. This large discrepancy is due to the fact that there are many students in large classes, while there are few students in small classes.

Suppose a college had only two classes, with 10 students in one and 90 students in the other. One obvious way to compute the mean class size is to compute $(10 + 90)/2 = 50$. But if we ask the 100 students for their class sizes we would get ten 10s and ninety 90s, and those 100 numbers would produce a mean of 82. The college would experience a mean class size of 50, but the students would experience a mean class size of 82.

Without changing the number of classes or faculty, the mean class size experienced by students can be reduced by making all classes close to the same size. This would also improve attendance, if we accept research evidence that attendance is better in smaller classes.

Weighted Mean

The **weighted mean** is useful in many situations where the scores vary in their degree of importance. An obvious example occurs frequently in the determination of a final average for a course that includes four tests plus a final examination. If the respective grades are 70, 80, 75, 85, and 90, the mean of 80 does not reflect the greater importance placed on the final exam. Let's suppose that the instructor counts the respective tests as 15%, 15%, 15%, 15%, and 40%. The weighted mean then becomes

$$\frac{(70 \times 15) + (80 \times 15) + (75 \times 15) + (85 \times 15) + (90 \times 40)}{100}$$

$$= \frac{1050 + 1200 + 1125 + 1275 + 3600}{100}$$

$$= \frac{8250}{100} = 82.5$$

This computation suggests a general procedure for determining a weighted mean. Given a list of scores $x_1, x_2, x_3, \ldots, x_n$ and a corresponding list of weights $w_1, w_2, w_3, \ldots, w_n$, the weighted mean is obtained by computing

FORMULA 2-2   weighted mean $= \dfrac{\Sigma(w \cdot x)}{\Sigma w}$ where w = weight

That is, first multiply each score by its corresponding weight; then find the total of the resulting products, thereby evaluating Σwx. Finally, add the values of the weights to find Σw and divide the latter value into the former.

Formula 2-2 can be modified so that we can approximate the mean from a frequency table. The home selling price data from Table 2-1 have been entered in Table 2-9, where we use the class marks as representative scores and the frequencies as weights. Then the formula for the weighted mean leads directly to Formula 2-3, which can be used to approximate the mean of a set of scores in a frequency table.

FORMULA 2-3 $\bar{x} = \dfrac{\Sigma(f \cdot x)}{\Sigma f}$ where x = class mark

f = frequency

TABLE 2-9			
Selling Price (dollars)	Frequency f	Class Mark x	$f \cdot x$
50,000–74,999	1	62,499.5	62,499.5
75,000–99,999	9	87,499.5	787,495.5
100,000–124,999	22	112,499.5	2,474,989.0
125,000–149,999	47	137,499.5	6,462,476.5
150,000–174,999	31	162,499.5	5,037,484.5
175,000–199,999	23	187,499.5	4,312,488.5
200,000–224,999	9	212,499.5	1,912,495.5
225,000–249,999	2	237,499.5	474,999.0
250,000–274,999	3	262,499.5	787,498.5
275,000–299,999	2	287,499.5	574,999.0
300,000–324,999	1	312,499.5	312,499.5
Total	150		23,199,925

$\uparrow$
Σf

$\uparrow$
$\Sigma(f \cdot x)$

Formula 2-3 is really a variation of Formula 2-1, $\bar{x} = \Sigma x/n$. When data are summarized in a frequency table, Σf is the total number of scores, so that $\Sigma f = n$. Also, $\Sigma(f \cdot x)$ is simply a quick way of adding up all of the scores. Formula 2-3 doesn't really involve a fundamentally different concept; it is simply a variation of Formula 2-1.

We can now compute the weighted mean.

$$\bar{x} = \frac{\Sigma(f \cdot x)}{\Sigma f} = \frac{23{,}199{,}925}{150} = \$154{,}666$$

When we used the original collection of scores to calculate the mean directly, we obtained a mean of $153,775, so the value of the weighted mean obtained from the frequency table is quite accurate. The procedure we use is justified by the fact that a class like $50,000–$74,999 can be represented by its class mark of $62,499.5 and the frequency number indicates that the representative score of $62,499.5 occurs one time. In essence, we are treating Table 2-3 as if it contained one score of $62,499.5, nine values of $87,499.5, and so on.

We have stressed that the four basic measures of central tendency (mean, median, mode, midrange) can all be called averages and that the freedom to select a particular average can be a source of deception. Consider our 150 home selling prices that led to these results.

mean	$153,775
median	$144,900
mode	$126,500
midrange	$187,000

Technically, someone could use any of the four preceding figures as the average home selling price. A real estate salesperson might want to encourage a sale by emphasizing that a particular home is a bargain compared to the high average of $187,000. A wise buyer (such as you, after you've taken this course) might try to get a lower price by emphasizing the low average of $126,500. We can see that ignorance of these concepts can be very costly.

Even when we have no particular self-interest to promote, the selection of the best (most representative) average is not always easy. The different averages have different advantages and disadvantages and there are no objective criteria that determine the most representative average for all data sets. Some of the important advantages and disadvantages are summarized in Table 2-10.

In this section we considered measures designed to identify a central or representative value. In the next section we consider another important characteristic when we use measures of dispersion to describe the variation within sets of data.

TABLE 2-10

Average	Definition	How Common?	Existence	Takes Every Score into Account?	Affected by Extreme Scores?	Miscellaneous Comments
Mean	$\bar{x} = \dfrac{\Sigma x}{n}$	most familiar "average"	always exists	yes	yes	used throughout this book
Median	Middle score	commonly used	always exists	no	no	often a good choice if there are some extreme scores
Mode	most frequent score	sometimes used	might not exist; may be more than one mode	no	no	appropriate for data at the nominal level
Midrange	$\dfrac{\text{high} + \text{low}}{2}$	rarely used	always exists	no	yes	very sensitive to extreme values

General comments:

- For a data collection that is approximately symmetric, the mean, median, mode, and midrange tend to be about the same.
- For a data collection that is obviously asymmetric, it would be good to report both the mean and median.
- The mean is relatively *reliable*. That is, when samples are drawn from the same population, the sample means tend to be more consistent than the other averages.

How Valid Are Unemployment Figures?

The actions of presidents, economists, budget directors, corporation heads, and many other key decision makers are often based on the monthly unemployment figures. Yet these figures have been criticized as being exaggerated or understated. Julius Shickin, a commissioner of the Bureau of Labor Statistics, says that "if you're thinking of unemployment in terms of economic potential, then the answer is no, the figure doesn't overstate the problem. But if you're thinking in terms of hardships then the answer is yes, it does. It all depends upon your value judgments."

Critics point out that, in developing unemployment figures, families are asked whether anyone out of work and aged 16 years or older has actually looked for work sometime within the last four weeks. However, some people have been out of work so long that they have abandoned efforts to find a job, while others do ask about work, but they do it in very casual ways. At the other end of the scale is a chemical engineer who is counted as employed even though she is merely driving a cab 35 hours a week while she looks for meaningful employment. Such people distort the unemployment figure, which is arrived at through a survey involving about 50,000 families each month. The survey data are sent to Washington, processed by the Census Bureau, and then given to the Bureau of Labor Statistics. This procedure, in effect since 1940, costs about $5 million per year.

2-4 EXERCISES A

In Exercises 2-77 through 2-92, find the (a) mean, (b) median, (c) mode, (d) midrange for the given sample data.

2-77 Statistics are sometimes used to compare or identify authors of different works. The lengths of the first 20 words in *The Cat in the Hat* by Dr. Seuss are listed below.

3	3	3	3	5	2	3	3	3	2
4	2	2	3	2	3	5	3	4	4

2-78 The lengths of the first 20 words in the foreword written by Tennessee Williams in *Cat on a Hot Tin Roof* are listed below.

2	6	2	2	1	4	4	2	4	2
3	8	4	2	2	7	7	2	3	11

2-79   In Section 1 of a statistics class, 10 test scores were randomly selected, and the following results were obtained:

74, 73, 77, 77, 71, 68, 65, 77, 67, 66

2-80   In Section 2 of a statistics class, test scores were randomly selected, and the following results were obtained:

42, 100, 77, 54, 93, 85, 67, 77, 62, 58

2-81   Leukemic mice were placed in cages at the Sloan-Kettering Cancer Research Center and the numbers listed below were recorded for the numbers of days they survived without treatment.

10, 9, 8, 8, 9, 9, 7, 15, 9, 9

2-82   The numbers of rooms for 15 homes recently sold in Dutchess, County, N.Y., are listed below.

8, 8, 8, 5, 9, 8, 7, 6, 6, 7, 7, 7, 7, 9, 9

2-83   In "An Analysis of Factors that Contribute to the Efficacy of Hypnotic Analgesia" (by Price and Barber, *Journal of Abnormal Psychology*, Vol. 96, No. 1), the "before" readings (in centimeters) on a mean visual analog scale are given for 16 subjects. They are listed below.

8.8	6.6	8.4	6.5	8.4	7.0	9.0	10.3
8.7	11.3	8.1	5.2	6.3	11.6	6.2	10.9

2-84   In "Determining Statistical Characteristics of a Vehicle Emissions Audit Procedure" (by Lorenzen, *Technometrics*, Vol. 22, No. 4), carbon monoxide emissions data (in g/m) are listed for vehicles. The following values are included.

5.01	14.67	8.60	4.42	4.95	7.24
7.51	12.30	14.59	7.98	11.53	4.10

2-85   In "A Simplified Method for Determination of Residual Lung Volumes" (by Wilmore, *Journal of Applied Psychology*, Vol. 27, No. 1), the oxygen dilution was measured (in liters) for subjects and the following sample values were obtained.

1.361	1.013	1.140	1.649	1.278
1.148	1.824	1.551	1.041	

2-86   The numbers of calories of "light" beer were recorded for 12-oz samples of different brands, and the results are given below.

106, 99, 101, 103, 108, 107, 107, 107, 106

2-87   The reaction times (in seconds) of a group of adult men were found to be 0.74, 0.71, 0.41, 0.82, 0.74, 0.85, 0.99, 0.71, 0.57, 0.85, 0.57, 0.55.

2-88 In a test of hearing, subjects estimate the loudness (in decibels) of sound, and their results are as follows.

68, 72, 70, 71, 68, 68, 75, 62, 80, 73, 68

2-89 When 40 people were surveyed at the Galleria Mall in Wappingers Falls, N.Y., they reported the distances they drove to the mall, and the values (in miles) are given below.

2	8	1	5	9	5	14	10	31	20
15	4	10	6	5	5	1	8	12	10
25	40	31	24	20	20	3	9	15	15
25	8	1	1	16	23	18	25	21	12

2-90 A student working part time for a moving company in Dutchess County, N.Y., collected the following load weights (in pounds) for 50 consecutive customers.

8,090	3,250	12,350	4,510	8,770	5,030	12,700
8,800	6,170	8,450	10,330	10,100	13,410	7,280
13,490	17,810	7,470	11,450	13,260	9,310	15,970
7,540	7,770	6,400	14,800	14,760	6,820	11,430
7,200	13,520	16,200	10,780	10,510	17,330	7,450
9,110	10,630	3,670	14,310	9,140	10,220	9,900
11,860	12,010	12,430	8,160	26,580	4,480	6,390
11,600						

2-91 Listed below are the actual energy consumption amounts as reported on the electric bills for one residence. Each amount is in kilowatt-hours and represents a two-month period.

728	774	859	882	791	731	838	862	880	831
759	774	832	816	860	856	787	715	752	778
829	792	908	714	839	752	834	818	835	751
837									

2-92 Listed below are the daily sales totals (in dollars) for 44 days at a large retail outlet in Orange County, Calif.

24,145	39,921	30,284	43,923	38,366
33,650	47,563	34,804	55,254	44,249
37,196	66,644	37,858	31,995	61,923
39,897	29,099	42,906	35,420	33,310
46,351	34,423	52,250	38,295	37,052
65,798	37,809	68,945	43,952	38,812
25,597	41,160	31,987	59,708	46,285
34,132	50,857	35,018	32,097	64,842
37,229	66,892	38,275	36,802	

In Exercises 2-93 through 2-100, use the given frequency table. (a) Identify the class mark for each class interval. (b) Find the mean using the class marks and frequencies.

2-93

Distance (in thousands of miles)	Frequency
0–39	17
40–79	41
80–119	80
120–159	49
160–199	4

Distances traveled by buses before the first major motor failure (based on data from "Large Sample Simultaneous Confidence Intervals for the Multinomial Probabilities Based on Transformations of the Cell Frequencies" by Bailey, *Technometrics*, Vol. 22, No. 4)

2-94

Age (years)	Frequency
0–9	44
10–19	42
20–29	33
30–39	30
40–49	27
50–59	22
60–69	18
70–79	9

Ages of randomly selected Dutchess County residents (based on data from the *Cornell Community and Resource Development Series*)

2-95

Size of Family (persons)	Frequency
2	51
3	31
4	27
5	12
6	4
7	1
8	1

Family size for 127 randomly selected families (based on data from the Bureau of the Census)

2-96

Time (years)	Number
4	147
5	81
6	27
7	15
7.5–11.5	30

Time required to earn bachelor's degree (based on data from the National Center for Education Statistics)

2-97

Index	Frequency
0.10–0.11	9
0.12–0.13	16
0.14–0.15	12
0.16–0.17	3
0.18–0.19	1

Frequency table of the blood alcohol concentrations of arrested drivers.

2-98

Number	Frequency
1–5	19
6–10	24
11–15	18
16–20	21
21–25	23
26–30	20
31–35	16
36–40	15

Frequency table of lottery numbers selected in a 26-week period

2-99

Family Income (dollars)	Number
0–2,499	3
2,500–7,499	11
7,500–12,499	16
12,500–14,999	9
15,000–19,999	19
20,000–24,999	19
25,000–34,999	36
35,000–49,999	37

Incomes of families (restricted to families with incomes below $50,000 and based on data from the Bureau of the Census)

2-100

SAT (math)	Frequency
200–249	2
250–299	8
300–349	28
350–399	27
400–449	33
450–499	62
500–549	26
550–599	41
600–649	27
650–699	28
700–749	14
750–800	6

SAT math scores of 302 students at Roy C. Ketcham High School in upstate New York

2-4 EXERCISES B

2-101   a. Find the mean, median, mode, and midrange for the following barometric pressure readings (in millibars):

1023.6, 1024.2, 1026.7, 1033.5, 1040.3

b. Subtract 1000 from each score in part (a) and then find the mean, median, mode, and midrange.

c. Is there any relationship between the answer set to part (a) and the answer set to part (b)? If so, what is it?

d. In general, if a constant k is added to (or subtracted from) some set of scores, what happens to the mean, median, mode, and midrange?

e. Find the mean, median, mode, and midrange for the following wavelengths (in kilometers):

0.0236, 0.0242, 0.0267, 0.0335, 0.0403

f. Multiply each score from part (e) by 10,000 and then find the mean, median, mode, and midrange.

g. Is there any relationship between the answer set to part (e) and the answer set to part (f)? If so, what is it?

h. In general, if every score in a data set is multiplied by some constant k, what happens to the mean, median, mode, and midrange?

i. Do these results suggest a way of simplifying the computations for certain types of inconvenient numbers?

2-102   a. A student receives quiz grades of 70, 65, and 90. The same student earns an 85 on the final examination. If each quiz constitutes 20% of the final grade while the final makes up 40%, find the weighted mean.

b. A student earns the grades in the accompanying table. If grade points are assigned as A = 4, B = 3, C = 2, D = 1, F = 0 and the grade points are weighted according to the number of credit hours, find the weighted mean (grade-point average).

Course	Grade	Credit hours
Math	A	4
English	C	3
Art	B	1
Physical Education	A	2
Biology	C	4

2-103   Compare the averages that comprise the solutions to Exercises 2-79 and 2-80. Do these averages discriminate or differentiate between the two lists of scores? Is there any apparent difference between the two data sets that is not reflected in the averages?

2-104 Using the data from Exercise 2-79, change the first score (74) to 900 and find the mean, median, mode, and midrange of this modified data set. How does the extreme value affect these averages?

2-105 Collapse the frequency table in Exercise 2-100 by using the six class intervals of 200–299, 300–399, and so on. Calculate the mean of the collapsed table and compare the result to that obtained in Exercise 2-100. Which result is likely to be better?

2-106 To obtain the **harmonic mean** of a set of scores, divide the number of scores n by the sum of the reciprocals of all scores.

$$\text{harmonic mean} = \frac{n}{\sum \frac{1}{x}}$$

For example, the harmonic mean of 2, 3, 6, 7, 7, 8 is

$$\frac{6}{\frac{1}{2} + \frac{1}{3} + \frac{1}{6} + \frac{1}{7} + \frac{1}{7} + \frac{1}{8}} = \frac{6}{1.4} = 4.3$$

(Note that 0 cannot be included in the scores.)

a. Find the harmonic mean of 2, 3, 6, 7, 7, 8, 9, 9, 9, 10.
b. A group of students drives from New York to Florida (1200 miles) at a speed of 40 mi/h and returns at a speed of 60 mi/h. What is their average speed for the round trip? (The harmonic mean is used in averaging speeds.)
c. A dispatcher of charter buses calculates the average round trip speed (in miles per hour) for a certain route. The results obtained for 14 different runs are listed below. Based on these values, what is the average speed of a bus assigned to this route?

42.6	41.3	38.2	42.9	43.4	43.7	40.8
34.2	40.1	41.2	40.5	41.7	39.8	39.6

2-107 Given a collection of n scores (all of which are positive), the **geometric mean** is the nth root of their product. For example, to find the geometric mean of 2, 3, 6, 7, 7, 8, 9, 9, 9, 10, first multiply the scores.

$$2 \cdot 3 \cdot 6 \cdot 7 \cdot 7 \cdot 8 \cdot 9 \cdot 9 \cdot 9 \cdot 10 = 102,876,480$$

Then take the tenth root of the product, since there are 10 scores.

$$\sqrt[10]{102,876,480} = 6.3$$

The geometric mean is often used in business and economics for finding average rates of change, average rates of growth, or average ratios. The *average growth factor* for money compounded at annual interest rates of 10%, 8%, 9%, 12%, and 7% can be found by computing the geometric mean of 1.10, 1.08, 1.09, 1.12, and 1.07. Find that average growth factor.

2-108   The **quadratic mean** (or root mean square, or R.M.S.) of a set of scores is obtained by squaring each score, adding the results, dividing by the number of scores n, and then taking the square root of that result.

$$\text{quadratic mean} = \sqrt{\frac{\Sigma x^2}{n}}$$

For example, the quadratic mean of 2, 3, 6, 7, 7, 8, 9, 9, 9, 10 is

$$\sqrt{\frac{4 + 9 + 36 + 49 + 49 + 64 + 81 + 81 + 81 + 100}{10}}$$

$$= \sqrt{\frac{554}{10}}$$

$$= \sqrt{55.4} = 7.4$$

The quadratic mean is usually used in physical applications. In power distribution systems, for example, voltages and currents are usually referred to in terms of their R.M.S. values. Find the R.M.S. of the following power supplies (in volts): 151, 162, 0, 81, −68.

2-109   A comparison of the mean and median can reveal information about **skewness**, as illustrated below. Refer to the data given in Exercises 2-77 through 2-80 and, in each case, identify the data as being skewed to the left, symmetric, or skewed to the right.

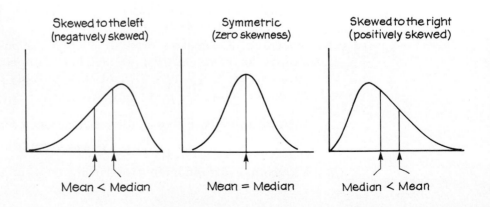

Skewed to the left   Symmetric   Skewed to the right
(negatively skewed)   (zero skewness)   (positively skewed)

Mean < Median   Mean = Median   Median < Mean

2-110 Frequency tables often have open-ended classes such as the one given below. Formula 2-3 cannot be directly applied since we can't determine a class mark for the class of "more than 20." Calculate the mean by assuming that this class is really (a) 21–25; (b) 21–30; (c) 21–40. What can you conclude?

Hours Studying per Week	Frequency
0	5
1–5	96
6–10	57
11–15	25
16–20	11
More than 20	6

Time spent studying by college freshmen (based on data from *The American Freshman* as reported in *USA Today*)

2-111 When data are summarized in a frequency table, the median can be found by first identifying the "median class" (the class that contains the median). Then we assume that the scores in that class are evenly distributed and we can interpolate. This process can be described by

$$(\text{lower limit of median class}) + (\text{class width})\left(\frac{\left(\frac{n+1}{2}\right) - (m+1)}{\text{frequency of median class}}\right)$$

where n is the sum of all class frequencies and m is the sum of the class frequencies that *precede* the median class. Use this procedure to find the median home selling price by referring to Table 2-3.

2-112 To find the 10% **trimmed mean** for a data set, first arrange the data in order. Then delete the bottom 10% of the scores and delete the top 10% of the scores and calculate the mean of the remaining scores. Find the 10% trimmed mean for the data in Exercise 2-90. Also find the 20% trimmed mean for that data set. What advantage does a trimmed mean have over the regular mean?

2-5 DISPERSION STATISTICS

The preceding section was concerned with averages, or measures of central tendency. Those statistics were designed to reflect a certain characteristic of the data from which they came. That is, the averages are supposed to be *central* scores. However, other features of the data may not be reflected at all by the averages. Suppose, for example, that two different groups of 10 students are given identical quizzes, with these results.

Group A	Group B
65	42
66	54
67	58
68	62
71	67
73	77
74	77
77	85
77	93
77	100

Computing the averages, we get

Group A	Group B
mean = 71.5	mean = 71.5
median = 72.0	median = 72.0
mode = 77	mode = 77
midrange = 71.0	midrange = 71.0

From looking at these averages, we can see no difference between the two groups. Yet an intuitive perusal of both groups shows an obvious difference: The scores of group B are much more widely scattered than those of group A. This variability among data is one characteristic to which averages are not sensitive. Consequently, statisticians have tried to design statistics that measure this variability, or dispersion. The three basic **measures of dispersion** are range, variance, and standard deviation.

Range

The **range** is simply the difference between the highest value and the lowest value. For group A, the range of 12 is the difference between 77 and 65. The range of group B is $100 - 42 = 58$. This much larger range suggests greater dispersion. Be sure you avoid confusion between the midrange (an average) and the range (a measure of dispersion). The range is extremely easy to compute, but it's often inferior to other measures of dispersion. The rather extreme example of groups C and D should illustrate this point.

The Census

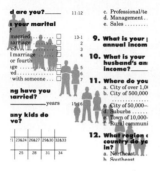

Every 10 years, the U.S. Government undertakes a census intended to obtain information about each American. At stake are over $50 billion in government allocations, representation in Congress, redistricting of state and local governments, and business adjustments to changing populations. The Census Bureau once classified missed people and unanswered questions as unknown, but modern technology and sampling techniques now make possible the imputation of data, which means they make up their own answers. Millions of people are invented and given ages, religions, spouses, jobs, children, incomes, and so on.

There is much political pressure to correct for any undercount, since the error affects some regions and groups much more than others.

Group C	Group D
1	2
20	3
20	4
20	5
20	6
20	14
20	15
20	16
20	17
20	18
Range = 19	Range = 16

The larger range for group C suggests more dispersion than in group D. But the scores of group C are very close together while those of group D are much more scattered. The range may be misleading in this case (and in many other circumstances) because it depends only on the maximum and minimum scores. In general, the value of the range tends to increase as the number of scores increases. Better measures of dispersion have been developed, but the improvement is not achieved without some loss. The loss occurs in the ease of computation. The measures of standard deviation and variance generally have greater value, but they require more difficult calculations.

Standard Deviation

The next measure of dispersion to be considered is the **standard deviation**, which is generally the most important and useful such measure.

FORMULA 2-4 $s = \sqrt{\dfrac{\Sigma(x - \bar{x})^2}{n - 1}}$ sample standard deviation

(Some authors define the standard deviation with a denominator of n, but $n - 1$ provides a better estimate of the standard deviation for a population when the formula is applied to a random sample from that population. Consequently, if the data are an entire population of scores, divide by n, but divide by $n - 1$ when dealing with sample data. For large values of n (such as those greater than 30) it really doesn't make too much difference which choice is made. For example, computations involving 100 scores will result in division by 100 or by 99, which gives values that are essentially the same.) To use Formula 2-4 you should follow this procedure.

PROCEDURE FOR USING FORMULA 2-4

1. Find the mean of the scores ($\bar{x}$).
2. Subtract the mean from each individual score ($x - \bar{x}$).
3. Square each of the differences obtained from step 2. That is, multiply each value by itself. [This produces numbers of the form $(x - \bar{x})^2$.]
4. Add all of the squares obtained from step 3 to get $\Sigma(x - \bar{x})^2$.
5. Divide the preceding total by the number ($n - 1$); that is, 1 less than the total number of scores present.
6. Find the square root of the result.

On students' initial exposure to the standard deviation, there is sometimes a feeling of awe, mystery, fear, or even an urge to practice that old time-honored ritual: withdrawal from the course. These feelings can, of course, be overcome with a little work and a little patience. Computation of the standard deviation involves only the operations of addition, subtraction, multiplication, division, and square root.

Why define a measure of dispersion as in Formula 2-4? When measuring dispersion in a collection of data, one reasonable approach is to begin with the individual amounts by which scores deviate from the mean. For a particular score x, that amount of deviation can be denoted by $x - \bar{x}$. The sum of all such deviations, denoted by

$\Sigma(x - \bar{x})$, might initially seem like a reasonable measure of dispersion, but that sum will *always* equal 0. Scores greater than the mean will cause $(x - \bar{x})$ to be positive, while scores below the mean will cause $(x - \bar{x})$ to be negative values since the mean $\bar{x}$ is at the center of the scores.

How do we prevent this undesirable canceling out of positive and negative values so that the measure of dispersion is not always zero? One alternative is to use absolute values as in $\Sigma|x - \bar{x}|$. This leads to the **mean deviation**

$$\frac{\Sigma|x - \bar{x}|}{n}$$

but this approach tends to be unsuitable for the important methods of statistical inference. Instead, we make all of the terms $(x - \bar{x})$ nonnegative by squaring them. Finally, we take the square root to compensate for the fact that the deviations were all squared.

Variance

If we omit step 6 in the procedure for calculating the standard deviation, we get the **variance**, which is defined by

FORMULA 2-5 $\quad s^2 = \dfrac{\Sigma(x - \bar{x})^2}{n - 1} \quad$ sample variance

By comparing Formulas 2-4 and 2-5, we see that the variance is the square of the standard deviation. The variance will be used later in the book, but we should concentrate on the concept of standard deviation as we try to get a feeling for what this statistic is about. A major obstacle of the variance is that it is not in the same units as the original data. For example, a data set might have a standard deviation of $3.00 and a variance of 9.00 *square* dollars. Since we can't relate well to square dollars, we find it difficult to understand variance.

NOTATION
s denotes the **standard deviation** of a set of **sample** scores.
σ denotes the **standard deviation** of a set of **population** scores. (σ is the lowercase Greek sigma.)
s^2 denotes the **variance** of a set of **sample** scores.
σ^2 denotes the **variance** of a set of **population** scores.
Note: Articles in professional journals and reports often use **SD** for standard deviation and **Var** for variance.

EXAMPLE

Find the standard deviation for these quiz scores: 2, 3, 5, 6, 9, 17.

Solution

Step 1: The mean is obtained by adding the scores (42) and then dividing by the number of scores present (6). The mean is 7.0.

Step 2: Subtracting the mean of 7.0 from each score, we get -5, -4, -2, -1, 2, and 10. (As a quick check, these numbers must always total 0.)

Step 3: Squaring each value obtained from step 2, we get 25, 16, 4, 1, 4, and 100.

Step 4: The sum of all the preceding squares is 150.

Step 5: There are six scores, so we divide 150 by one less than 6. That is, $150 \div 5 = 30$.

Step 6: Find the square root of 30. We get a standard deviation of $\sqrt{30} = 5.5$.

Here is a helpful hint: Organize the computations for standard deviation in the form of a table like Table 2-11.

TABLE 2-11

x	$(x - \bar{x})$	$(x - \bar{x})^2$
2	-5	25
3	-4	16
5	-2	4
6	-1	1
9	2	4
17	10	100
Totals: 42		150

$$\bar{x} = \frac{42}{6} = 7.0$$

$$s = \sqrt{\frac{150}{6-1}} = \sqrt{\frac{150}{5}} = \sqrt{30.0} = 5.5$$

> **ROUND-OFF RULE**
>
> As in Section 2-4, we round off answers by carrying one more decimal place than was present in the original data. We should round only the final answer and not intermediate values. (If we must round intermediate results, we should carry at least twice as many decimal places as will be used in the final answer.)

We will now attempt to make some intuitive sense out of the standard deviation. First, we should clearly understand that the standard deviation is measuring the characteristic of dispersion or variation among scores. Scores close together will yield a small standard deviation, whereas scores spread farther apart will yield a larger standard deviation. See Figure 2-9. From that figure we should see that as the data spread farther apart, the corresponding values of the standard deviation increase.

In attempting to develop a sense for the standard deviation, we might consider the following results from Chebyshev's theorem (see Exercise 2-146):

- At least 75% of all scores will fall within the interval from two standard deviations below the mean to two standard deviations above the mean.

- At least 89% of all scores will fall within three standard deviations of the mean.

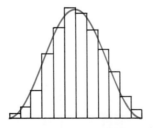

We might also consider the **empirical rule**, which applies to data having a histogram that is approximately bell shaped, as in the figure shown above. Such distributions are important and common. For these bell-shaped distributions, the empirical rule states that

- About 68% of all scores fall within *one* standard deviation of the mean. (*continued on page 85*)

Increasing the spread of the data leads to larger values of variance and standard deviation.

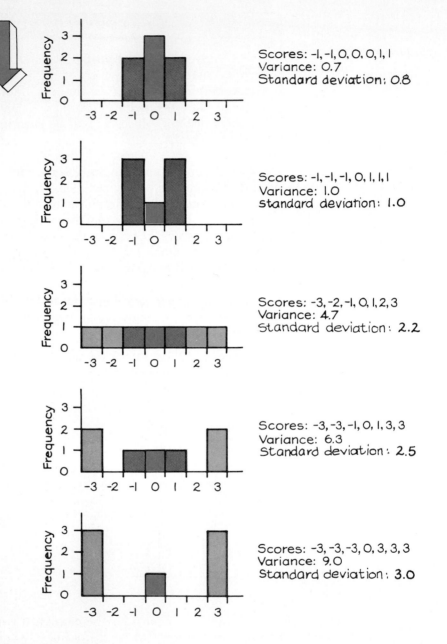

Scores: -1, -1, 0, 0, 0, 1, 1
Variance: 0.7
Standard deviation: 0.8

Scores: -1, -1, -1, 0, 1, 1, 1
Variance: 1.0
standard deviation: 1.0

Scores: -3, -2, -1, 0, 1, 2, 3
Variance: 4.7
Standard deviation: 2.2

Scores: -3, -3, -1, 0, 1, 3, 3
Variance: 6.3
Standard deviation: 2.5

Scores: -3, -3, -3, 0, 3, 3, 3
Variance: 9.0
Standard deviation: 3.0

FIGURE 2-9
*All sets of scores shown here have a mean of zero.
Observe that as the data spread farther apart, the
value of the standard deviation increases.*

- About 95% of all scores fall within *two* standard deviations of the mean.
- About 99% of all scores fall within *three* standard deviations of the mean.

For examples of Chebyshev's theorem and the empirical rule, see the following table.

Given: IQ scores with a mean of 100, a standard deviation of 15, and a distribution that is bell shaped	
Chebyshev's Theorem	Empirical Rule
	About 68% of all scores are between 85 and 115.
At least 75% of all scores are between 70 and 130.	About 95% of all scores are between 70 and 130.
At least 89% of all scores are between 55 and 145.	About 99.7% of all scores are between 55 and 145.

In the table, note that "within two standard deviations of the mean" translates to "within 2(15) of 100" or "within 30 of 100," which is really from 70 to 130.

Take the time to study the results in the preceding table. They show us how the standard deviation can be used to get a sense for how data vary. According to the empirical rule, an IQ score of 147 would be very rare, since it deviates from the mean by more than three standard deviations.

We will now present two additional formulas for standard deviation, but these formulas do not involve a different concept; they are only different versions of Formula 2-4. First, Formula 2-4 can be expressed in the following equivalent form.

FORMULA 2-6 $$s = \sqrt{\frac{n(\Sigma x^2) - (\Sigma x)^2}{n(n-1)}}$$

Formulas 2-4 and 2-6 are equivalent in the sense that they will always produce the same results. Algebra can be used to show that they are equal. Formula 2-6 is called the *shortcut* formula because it tends to be convenient with messy numbers or with large sets of data. Also,

People Meters

Nielsen Media Research recently reported that *60 Minutes* was the number one television show for one particular week. Of all sets in use during the Sunday 7:00 to 8:00 P.M. time slot, 35% were tuned to *60 Minutes* and it was watched by 32,700,000 viewers. How does Nielsen know this? Sampling.

The Nielsen company installs "people meters" in a sample of 2000 households. A people meter is a remote control device that allows each household member to press their assigned number when they watch a show. A meter records the viewers along with the show being watched and transmits the data to a central computer. The central computer then matches the transmitted data with information about the household members so that more detailed information can be derived. The results determine which programs are canceled and the amounts that advertisers are charged for the commercials that we all love so dearly. The people meters are rela- tively new. Before 1988, television ratings were determined by simpler meters that provided data for the number of viewers only. The people meters provide more data, including such factors as age, sex, income, and education.

Formula 2-6 is often used in calculators and computer programs since it requires only three memory registers

$$\left(\text{for } n, \Sigma x, \quad \text{and} \quad \Sigma x^2\right)$$

instead of one memory register for every single score. However, many instructors prefer to use only Formula 2-4 for calculating standard deviations. They argue that Formula 2-4 reinforces the concept that the standard deviation is a type of average deviation while Formula 2-6 obscures that idea. Other instructors have no objections to using Formula 2-6. We have included the shortcut formula so that it is available for those who choose to use it.

For the quiz scores of 2, 3, 5, 6, 9, 17 we could use Formula 2-6 to find the standard deviation as in Table 2-12.

TABLE 2-12	
x	x^2
2	4
3	9
5	25
6	36
9	81
17	289
42	444
↑	↑
Σx	Σx^2

Because there are six scores, $n = 6$. From the table we see that $\Sigma x = 42$ and $\Sigma x^2 = 444$, so that

$$s = \sqrt{\frac{n(\Sigma x^2) - (\Sigma x)^2}{n(n-1)}}$$

$$= \sqrt{\frac{6(444) - (42)^2}{6(6-1)}}$$

$$= \sqrt{\frac{2664 - 1764}{30}}$$

$$= \sqrt{\frac{900}{30}}$$

$$= \sqrt{30}$$

$$= 5.5$$

We can develop a formula for standard deviation when the data are summarized in a frequency table. The result is

FORMULA 2-7

$$s = \sqrt{\frac{n[\Sigma(f \cdot x^2)] - [\Sigma(f \cdot x)]^2}{n(n-1)}} \qquad \begin{array}{l} \text{standard deviation} \\ \text{for frequency table} \end{array}$$

where x = class mark
$\quad f$ = frequency
$\quad n$ = sample size

Table 2-13, on the next page, summarizes the work done in applying Formula 2-7 to a frequency table formed by that table's first two columns.

Measures of dispersion are extremely important in many practical circumstances. Manufacturers interested in producing items of consistent quality are very concerned with statistics such as standard deviations. A producer of car batteries might be pleased to learn that a product has a mean life of four years, but that pleasure would become distress if the standard deviation indicated a very large dispersion, which would correspond to many battery failures long before the mean of four years. Quality control requires consistency, and consistency requires a relatively small standard deviation.

Some exercises in the previous section showed that if a constant is added to (or subtracted from) each value in a set of data, the measures of central tendency change by that same amount. In contrast, the

Score	Frequency f	Class Mark x	$f \cdot x$	$f \cdot x^2$ or $f \cdot x \cdot x$
0–8	10	4	40	160
9–17	5	13	65	845
18–26	15	22	330	7,260
27–35	3	31	93	2,883
36–44	1	40	40	1,600
Total	34		568	12,748

TABLE 2-13

$$s = \sqrt{\frac{n[\Sigma(f \cdot x^2)] - [\Sigma(f \cdot x)]^2}{n(n-1)}}$$

$$= \sqrt{\frac{34(12,748) - (568)^2}{34(34-1)}}$$

$$= \sqrt{\frac{110,808}{1,122}} = \sqrt{98.759} = 9.9$$

measures of dispersion do not change when a constant is added to (or subtracted from) each score. To find the standard deviation for 8001, 8002, . . . , 8009, we could subtract 8000 from each score and work with more manageable numbers without changing the value of the standard deviation, which is a definite advantage. But now suppose a manufacturer learns that the average (mean) readings of a thousand pressure gauges are 10 pounds per square inch (psi) too low. A relatively simple remedy would be to rotate all of the dials to read 10 psi higher. That remedy would correct the mean, but the standard deviation would remain the same. If the gauges are inconsistent and have errors that vary considerably, then the standard deviation of the errors might be too large. If that is the case, each gauge must be individually calibrated and the entire manufacturing process might require an overhaul. Machines might require replacement if they do not meet the necessary tolerance levels. We can see from this example that measures of central tendency and dispersion have properties that lead to very practical and serious implications.

2-5 EXERCISES A

In Exercises 2-113 through 2-128, find the range, variance, and standard deviation for the given data.

2-113 Statistics are sometimes used to compare or identify authors of different works. The lengths of the first 20 words in *The Cat in the Hat* by Dr. Seuss are listed here.

3	3	3	3	5	2	3	3	3	2
4	2	2	3	2	3	5	3	4	4

2-114 The lengths of the first 20 words in the foreword written by Tennessee Williams in *Cat on a Hot Tin Roof* are listed here.

2	6	2	2	1	4	4	2	4	2
3	8	4	2	2	7	7	2	3	11

2-115 In Section 1 of a statistics class, 10 test scores were randomly selected:

74, 73, 77, 77, 71, 68, 65, 77, 67, 66

2-116 In Section 2 of a statistics class, 10 test scores were randomly selected:

42, 100, 77, 54, 93, 85, 67, 77, 62, 58

2-117 Leukemic mice were placed in cages at the Sloan-Kettering Cancer Research Center and the numbers listed below were recorded for the numbers of days they survived without treatment.

10, 9, 8, 8, 9, 9, 7, 15, 9, 9

2-118 The numbers of rooms for 15 homes recently sold in Dutchess, County, N.Y., are listed here.

8, 8, 8, 5, 9, 8, 7, 6, 6, 7, 7, 7, 7, 9, 9

2-119 In "An Analysis of Factors that Contribute to the Efficacy of Hypnotic Analgesia" (by Price and Barber, *Journal of Abnormal Psychology*, Vol. 96, No. 1), the "before" readings (in centimeters) on a mean visual analog scale are given for 16 subjects. They are listed here.

8.8	6.6	8.4	6.5	8.4	7.0	9.0	10.3
8.7	11.3	8.1	5.2	6.3	11.6	6.2	10.9

2-120 In "Determining Statistical Characteristics of a Vehicle Emissions Audit Procedure" (by Lorenzen, *Technometrics*, Vol. 22, No. 4), carbon monoxide emissions data (in g/m) are listed for vehicles. The following values are included.

5.01	14.67	8.60	4.42	4.95	7.24
7.51	12.30	14.59	7.98	11.53	4.10

2-121   In "A Simplified Method for Determination of Residual Lung Volumes" (by Wilmore, *Journal of Applied Psychology*, Vol. 27, No. 1), the oxygen dilution was measured (in liters) for subjects and the following sample values were obtained.

    1.361   1.013   1.140   1.649   1.278
    1.148   1.824   1.551   1.041

2-122   The numbers of calories of "light" beer were recorded for 12-oz samples of different brands:

    106, 99, 101, 103, 108, 107, 107, 107, 106

2-123   The reaction times (in seconds) of a group of adult men were found to be

    0.74, 0.71, 0.41, 0.82, 0.74, 0.85, 0.99, 0.71, 0.57, 0.85, 0.57, 0.55

2-124   In a test of hearing, subjects estimate the loudness (in decibels) of sound, and their results are as follows.

    68, 72, 70, 71, 68, 68, 75, 62, 80, 73, 68

2-125   When 40 people were surveyed at the Galleria Mall in Wappingers Falls, N.Y., they reported the distances they drove to the mall, and the values (in miles) are given here.

2	8	1	5	9	5	14	10	31	20
15	4	10	6	5	5	1	8	12	10
25	40	31	24	20	20	3	9	15	15
25	8	1	1	16	23	18	25	21	12

2-126   A student working part time for a moving company in Dutchess County, N.Y., collected the following load weights (in pounds) for 50 consecutive customers.

8,090	17,810	3,670	10,100	17,330
8,800	7,770	12,430	13,260	10,220
13,490	13,520	4,510	14,760	4,480
7,540	10,630	10,330	10,510	12,700
7,200	12,010	11,450	9,140	7,280
9,110	12,350	14,800	26,580	15,970
11,860	8,450	10,780	5,030	11,430
11,600	7,470	14,310	13,410	7,450
3,250	6,400	8,160	9,310	9,900
6,170	16,200	8,770	6,820	6,390

2-127   Listed here are the actual energy consumption amounts as reported
        on the electric bills for one residence. Each amount is in kilowatt-
        hours and represents a two-month period.

728	816	834
759	714	862
829	791	715
837	860	818
774	839	880
774	731	752
792	856	835
859	752	831
832	838	778
908	787	751
882		

2-128   Listed here are the daily sales totals (in dollars) for 44 days at a large
        retail outlet in Orange County, Calif.

24,145	66,644	52,250	59,708
33,650	29,099	68,945	32,097
37,196	34,423	31,987	36,802
39,897	37,809	35,018	38,366
46,351	41,160	38,275	44,249
65,798	50,857	43,923	61,923
25,597	66,892	55,254	33,310
34,132	30,284	31,995	37,052
37,229	34,804	35,420	38,812
39,921	37,858	38,295	46,285
47,563	42,906	43,952	64,842

2-129   Add 15 to each value given in Exercise 2-117 and then find the range,
        variance, and standard deviation. Compare the results to those of Exer-
        cise 2-117.

2-130   Subtract 5 from each value given in Exercise 2-117 and then find the
        range, variance, and standard deviation. Compare the results to those
        of Exercise 2-117.

2-131   Multiply each value given in Exercise 2-117 by 10 and then find the
        range, variance, and standard deviation. Compare the results to those
        of Exercise 2-117.

2-132   Double each value in Exercise 2-117 and then find the range, variance,
        and standard deviation. Compare the results to those of Exercise
        2-117.

In Exercises 2-133 through 2-140, find the variance and standard deviation for the given data.

2-133

Distance (in thousands of miles)	Frequency
0–39	17
40–79	41
80–119	80
120–159	49
160–199	4

Distances traveled by buses before the first major motor failure (based on data from "Large Sample Simultaneous Confidence Intervals for the Multinomial Probabilities Based on Transformations of the Cell Frequencies" by Bailey, *Technometrics*, Vol. 22, No. 4)

2-134

Age (years)	Frequency
0–9	44
10–19	42
20–29	33
30–39	30
40–49	27
50–59	22
60–69	18
70–79	9

Ages of randomly selected Dutchess County residents (based on data from the *Cornell Community and Resource Development Series*)

2-135

Size of Family (persons)	Frequency
2	51
3	31
4	27
5	12
6	4
7	1
8	1

Family size for 127 randomly selected families (based on data from the Bureau of the Census)

2-136

Time (years)	Number
4	147
5	81
6	27
7	15
7.5–11.5	30

Time required to earn bachelor's degree (based on data from the National Center for Education Statistics)

2-137

Index	Frequency
0.10–0.11	9
0.12–0.13	16
0.14–0.15	12
0.16–0.17	3
0.18–0.19	1

Frequency table of the blood alcohol concentrations of arrested drivers.

2-138

Number	Frequency
1–5	19
6–10	24
11–15	18
16–20	21
21–25	23
26–30	20
31–35	16
36–40	15

Frequency table of lottery numbers selected in a 26-week period

2-139

Family Income (dollars)	Number
0–2,499	3
2,500–7,499	11
7,500–12,499	16
12,500–14,999	9
15,000–19,999	19
20,000–24,999	19
25,000–34,999	36
35,000–49,999	37

Incomes of families (restricted to families with incomes below $50,000 and based on data from the Bureau of the Census)

2-140

SAT (math)	Frequency
200–249	2
250–299	8
300–349	28
350–399	27
400–449	33
450–499	62
500–549	26
550–599	41
600–649	27
650–699	28
700–749	14
750–800	6

SAT math scores of 302 students at Roy C. Ketcham High School in upstate New York

2-141   Which would you expect to have a higher standard deviation: the IQ scores of a class of 25 statistics students or the IQ scores of 25 randomly selected adults? Explain.

2-142   Which would you expect to have a higher standard deviation: the reaction times of 30 sober adults or the reaction times of 30 adults who had recently consumed three martinis each?

2-143   Is it possible for a set of scores to have a standard deviation of zero? If so, how? Is it possible for a set of scores to have a negative standard deviation?

2-144   Test the effect of an *outlier* (an extreme value) by changing the 15 in Exercise 2-117 to 1500. Calculate the range, variance, and standard deviation for the modified set and compare the results to those originally obtained in Exercise 2-117.

2-5 EXERCISES B

2-145   Find the range and standard deviation for each of the following two groups. Which group has less dispersion according to the criterion of the range? Which group has less dispersion according to the criterion of the standard deviation? Which measure of dispersion is "better" in this situation: the range or the standard deviation?

     group C: 1, 20, 20, 20, 20, 20, 20, 20, 20, 20
     group D: 2, 3, 4, 5, 6, 14, 15, 16, 17, 18

2-146   Chebyshev's theorem states that the *proportion* (or fraction) of any set of data lying within K standard deviations of the mean is always *at least* $1 - 1/K^2$, where K is any positive number greater than 1.

     a. Given a mean IQ of 100 and a standard deviation of 15, what does Chebyshev's theorem say about the number of scores within two standard deviations of the mean (that is, 70–130)?

     b. Given a mean of 100 and a standard deviation of 15, what does Chebyshev's theorem say about the number of scores between 55 and 145?

     c. The mean score on the College Entrance Examination Board Scholastic Aptitude Test is 500 and the standard deviation is 100. What does Chebyshev's theorem say about the number of scores between 300 and 700?

     d. Using the data of part (c), what does Chebyshev's theorem say about the number of scores between 200 and 800?

2-147   A large set of sample scores yields a mean and standard deviation of $\bar{x} = 56.0$ and $s = 4.0$, respectively. The distribution of the histogram

is roughly bell shaped. Use the empirical rule to answer the following.
a. What percentage of the scores should fall between 52.0 and 60.0?
b. What percentage of the scores should fall within 8.0 of the mean?
c. About 99.7% of the scores should fall between what two values? (The mean of 56.0 should be midway between those two values.)

2-148   Find the standard deviation of the 150 home selling prices by using (a) the original set of data given in Table 2-1; (b) the frequency table given in Table 2-3. Then compare the results to determine the amount of distortion caused by the frequency table.

2-149   For any population of scores having a standard deviation σ and range R, the relationship $\sigma \leq R/2$ must be true. This relationship might be helpful in determining whether a computed value of σ is at least a reasonable possibility. Verify this relationship for the data given in Exercise 2-117. Use

$$\sigma = \sqrt{\frac{\Sigma(x - \mu)^2}{n}}$$

2-150   If we consider the values $1, 2, 3, \ldots, n$ to be a population, the standard deviation can be calculated by the formula

$$\sigma = \sqrt{\frac{n^2 - 1}{12}}$$

This formula is equivalent to Formula 2-4 modified for division by n instead of $n - 1$, where the values are $1, 2, 3, \ldots, n$.
a. Find the standard deviation of the population $1, 2, 3, \ldots, 100$.
b. Find an expression for calculating the *sample* standard deviation s for the sample values $1, 2, 3, \ldots, n$.

2-151   Find the mean and standard deviation for the scores $18, 19, 20, \ldots,$ 182. Treat those values as a population (see Exercise 2-150).

2-152   Computers commonly use a random number generator, which produces values between 0 and 1. In the long run, all values occur with the same relative frequency. Find the mean and standard deviation for the numbers between 0.00000000 and 0.99999999. (*Hint:* See Exercises 2-131 and 2-150.)

2-153   Given a collection of temperatures in degrees Fahrenheit, the following statistics are calculated:

$$\bar{x} = 40.2 \qquad s = 3.0 \qquad s^2 = 9.0$$

$$\left[C = \frac{5}{9}(F - 32) \right]$$

Find the values of $\bar{x}$, s, and s^2 for the same data set after each temperature has been converted to the Celsius scale.

2-154 a. A set of data yields $\bar{x} = 50.0$ and $s = 8.0$. If we multiply each of the original scores by the constant b and if we then add the constant c to each product, find the values of b and c that will cause the modified data set to yield $\bar{x} = 65.0$ and $s = 12.0$.
       b. If a population has mean μ and standard deviation σ, what must be done to each score in order to have a mean of 0 and a standard deviation of 1?

2-155 The **coefficient of variation**, expressed in percent, is used to describe the standard deviation relative to the mean. It is calculated as

$$\frac{s}{\bar{x}} \cdot 100 \qquad \text{or} \qquad \frac{\sigma}{\mu} \cdot 100$$

       a. Find the coefficient of variation for IQ tests having a mean of 100 and a standard deviation of 15.
       b. Find the coefficient of variation for reading tests having a mean of 250 and a standard deviation of 50.
       c. Find the coefficient of variation for the following sample scores: 2, 2, 2, 3, 5, 8, 12, 19, 22, 30.

2-156 In Exercise 2-109 we introduced the general concept of skewness. Skewness can be measured by **Pearson's index of skewness**:

$$I = \frac{3(\bar{x} - \text{median})}{s}$$

       If $I \geq 1.00$ or $I \leq -1.00$, the data can be considered to be *significantly skewed*. Find Pearson's index of skewness for the data given in Exercise 2-126 and then determine whether or not there is significant skewness.

2-6 MEASURES OF POSITION

There is often a need to compare scores taken from two separate populations with different means and standard deviations. The standard score (or z score) can be used to help make such comparisons.

DEFINITION

The **standard score**, or **z score**, is the number of standard deviations that a given value x is above or below the mean, and it is found by

$$\begin{array}{ccc} \textit{Sample} & & \textit{Population} \\ z = \dfrac{x - \bar{x}}{s} & \text{or} & z = \dfrac{x - \mu}{\sigma} \end{array}$$

Round z to two decimal places.

For example, which is better: a score of 65 on test A or a score of 29 on test B? The class statistics for the two tests are as follows:

Test A	Test B
$\bar{x} = 50$	$\bar{x} = 20$
$s = 10$	$s = 5$

For the score of 65 on test A we get a z score of 1.50, since

$$z = \frac{x - \bar{x}}{s} = \frac{65 - 50}{10} = \frac{15}{10} = 1.50$$

For the score of 29 on test B we get a z score of 1.80, since

$$z = \frac{x - \bar{x}}{s} = \frac{29 - 20}{5} = \frac{9}{5} = 1.80$$

That is, a score of 65 on test A is 1.50 standard deviations above the mean, while a score of 29 on test B is 1.80 standard deviations above the mean. This implies that the 29 on test B is the better score. While 29 is below 65, it has a better *relative* position when considered in the context of the other test results. Later, we will make extensive use of these standard, or z, scores.

The z score provides a useful measurement for making comparisons between different sets of data. Quartiles, deciles, and percentiles are measures of position useful for comparing scores within one set of data or between different sets of data.

Just as the median divides the data into two equal parts, the three **quartiles**, denoted by Q_1, Q_2, and Q_3, divide the ranked scores into four equal parts. Roughly speaking, Q_1 separates the bottom 25% of the ranked scores from the top 75%, Q_2 is the median, and Q_3 separates

INDEX NUMBERS

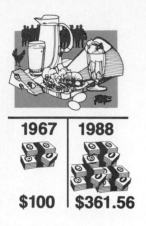

1967 | **1988**

$100 | **$361.56**

The Consumer Price Index (CPI) is one of a family of yardsticks called *index numbers*. An index number allows us to compare the value of some variable relative to its value at some base period. The net effect is that we measure the change of the variable over some time span. We find the value of an index number by evaluating

$$\frac{\text{current value}}{\text{base value}} \times 100$$

The CPI is based on a weighted average of the costs of specific goods and services. Using 1967 as a base year with index 100, the CPI in 1988 was $361.56. This means that a combination of goods and services that cost $100 in 1967 would cost $361.56 in 1988. Many labor contracts have wage adjustments linked to changes in the CPI.

the top 25% from the bottom 75%. To be more precise, at least 25% of the data will be less than or equal to Q_1 and at least 75% will be greater than or equal to Q_1. At least 75% of the data will be less than or equal to Q_3, while at least 25% will be equal to or greater than Q_3. Also, Q_2 is actually the median.

Similarly, there are nine **deciles**, denoted by $D_1, D_2, D_3, \ldots, D_9$, which partition the data into 10 groups with about 10% of the data in each group. There are also 99 **percentiles**, which partition the data into 100 groups with about 1% of the scores in each group. (Quartiles, deciles, and percentiles are examples of fractiles, which partition data into parts that are approximately equal.) A student taking a competitive college entrance examination might learn that he or she scored in the 92nd percentile. This does not mean that the student received a grade of 92% on the test; it indicates roughly that whatever score he or she did achieve was higher than 92% of peers who took a similar test (and also lower than 8% of his or her colleagues). Percentiles are useful for converting meaningless raw scores into meaningful comparative scores. For this reason, percentiles are used extensively in educational testing. A raw score of 750 on a college entrance exam means nothing to most people, but the corresponding percentile of 93% provides useful comparative information.

The process of finding the percentile that corresponds to a particular score x is fairly simple, as indicated in the following definition.

DEFINITION

$$\text{percentile of score } x = \frac{\text{number of scores less than } x}{\text{total number of scores}} \cdot 100$$

EXAMPLE

Table 2-14 lists the 150 home selling prices arranged in order from lowest to highest. Find the percentile corresponding to $100,000.

Solution

From Table 2-14 we see that there are 10 selling prices less than $100,000, so that

$$\text{percentile of } \$100,000 = \frac{10}{150} \cdot 100 = 7 \quad \text{(rounded off)}$$

The selling price of $100,000 is the 7th percentile.

There are several different methods for finding the score corresponding to a particular percentile, but the process we will use is summarized in Figure 2-10.

EXAMPLE

Refer to the 150 home selling prices in Table 2-14 on page 101 and find the score corresponding to the 35th percentile. That is, find the value of P_{35}.

Solution

We refer to Figure 2-10 and observe that the data are already ranked from lowest to highest. We now compute the locator L as follows.

$$L = \left(\frac{k}{100}\right) n = \left(\frac{35}{100}\right) \cdot 150 = 52.5$$

continued ▶

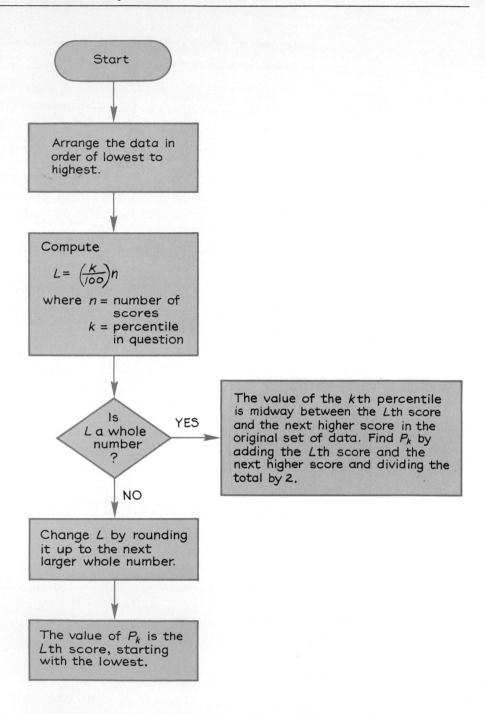

FIGURE 2-10
Procedure for finding the value of the kth percentile P_k

Solution, continued

We answer no when asked if $L = 52.5$ is a whole number, so we are directed to round L *up* (not off) to 53. (In *this* procedure we round L up to the next higher integer, but in most other situations in this book we generally follow the usual process for rounding.) The 35th percentile, denoted by P_{35}, is the 53rd score, starting with the lowest. Beginning with the lowest score of $72,000, we count down the list to find the 53rd score of $134,000, so that $P_{35} = \$134,000$.

TABLE 2-14 *Ranked* Selling Prices (in dollars) of 150 Dutchess County homes					
72,000	120,500	133,000	144,900	164,900	190,000
78,000	121,900	133,000	145,000	165,000	190,000
80,500	122,000	134,000	145,900	165,000	190,000
85,000	122,500	134,500	146,000	165,000	194,500
89,900	123,500	135,000	150,000	165,000	194,500
94,000	124,000	135,000	150,000	165,900	194,900
96,000	124,900	135,000	150,000	168,500	195,000
97,000	125,000	135,000	151,500	169,000	196,800
98,000	125,000	135,500	152,000	169,900	203,000
99,500	125,900	136,000	152,000	170,000	205,406
100,000	126,000	136,000	152,500	172,000	208,000
107,000	126,500	137,900	153,500	175,000	210,000
108,000	126,500	139,500	154,000	177,000	211,900
108,500	126,500	139,900	155,000	179,000	215,000
110,000	126,500	140,000	156,000	179,900	217,000
110,000	126,500	140,000	156,000	180,000	220,000
113,500	127,500	141,000	158,000	180,000	223,000
115,000	127,500	141,000	159,000	180,500	231,750
116,000	128,500	141,000	160,000	181,500	235,000
116,300	129,000	141,900	160,000	182,000	265,000
117,000	129,900	141,900	160,000	183,000	267,000
117,000	130,000	142,000	162,000	185,000	269,900
117,500	131,000	142,000	162,500	185,500	275,000
118,500	132,000	142,500	163,500	187,000	285,000
119,000	133,000	144,900	164,000	189,900	302,000

In the preceding example, we can see that there are 52 scores below $134,000, so the definition on page 99 indicates that the percentile of $134,000 is $(52/150) \cdot 100 = 34.7$, which is approximately 35. As the amount of data increases, these differences become smaller.

In the preceding example we found the 35th percentile. Because of the sample size, the location L first became 52.5, which was rounded to 53 because L was not originally a whole number. In the next example we illustrate a case in which L does begin as a whole number. This condition will cause us to branch to the right in Figure 2-10.

EXAMPLE

Refer to the same home selling prices listed in Table 2-14. Find P_{70}, which denotes the 70th percentile.

Solution

Following the procedure outlined in Figure 2-10 and noting that the data are already ranked from lowest to highest, we compute

$$L = \left(\frac{k}{100}\right) n = \left(\frac{70}{100}\right) \cdot 150 = 105$$

We now answer yes when asked if $L = 105$ is a whole number, and we then see that P_{70} is midway between the 105th and 106th scores. Since the 105th and 106th scores are $165,900 and $168,500, we get

$$P_{70} = \frac{165,900 + 168,500}{2} = \$167,200$$

Once these calculations with percentiles are mastered, similar calculations for quartiles and deciles can be done with the same procedures by noting the following relationships.

Quartiles	Deciles
$Q_1 = P_{25}$	$D_1 = P_{10}$
$Q_2 = P_{50}$	$D_2 = P_{20}$
$Q_3 = P_{75}$	$D_3 = P_{30}$
	$\vdots$
	$D_9 = P_{90}$

Finding Q_3, for example, is equivalent to finding P_{75}.

Other statistics are sometimes defined using quartiles, deciles, or percentiles. For example, the **interquartile range** is a measure of dispersion obtained by evaluating $Q_3 - Q_1$. The **semi-interquartile range** is $(Q_3 - Q_1)/2$, the **midquartile** is $(Q_1 + Q_3)/2$, and the **10–90 percentile range** is defined to be $P_{90} - P_{10}$.

When dealing with large collections of data, more reliable results are obtained with greater ease when statistics software packages are used. See the following STATDISK and Minitab computer displays for the 150 home selling prices listed in Table 2-1. In the STATDISK display, the results correspond to the values entered as thousands of dollars. For example, $179,000 was entered as 179. This is a common technique used to ease the entry of data, but it does affect some results, such as the variance. STATDISK and Minitab also provide histograms.

STATDISK Display

```
   Sample Size = 150        Minimum = 72.0000        Maximum = 302.0000
 Sum of Scores = 23066.2600        Sum of Squares = 3805003.0000

                  MEASURES OF CENTRAL TENDENCY

    Mean = 153.7751        Median = 144.9000        Midrange = 187.0000
 Geom. Mean = 148.6213     Harm. Mean = 143.7444    Quad. Mean = 159.2692

                  MEASURES OF DISPERSION

   Sample St. Dev. = 41.6108          Sample Variance = 1731.4560
 Population St. Dev. = 41.4718      Population Variance = 1719.9130
      Range = 230.0000                Standard Error = 3.3975

              MEASURES OF POSITION - Quartiles

    Q1 = 126.5000           Q2 = 144.9000           Q3 = 179.0000
```

MINITAB DISPLAY (see also Appendix C)

```
MTB > SET C1
DATA> 179000 126500 134500 125000 142000 164000
DATA> 146000 129000 141900 135000 118500 160000
          .
          .
          .
DATA> 133000 196800 121900 126000 164900 172000
DATA> 100000 129900 110000 131000 107000 165900
DATA> ENDOFDATA

MTB > MEAN C1
   MEAN      =        153775
MTB > STDEV C1
   ST.DEV. =          41611
MTB > MEDIAN C1
   MEDIAN =         144900
MTB > SUM C1
   SUM       =      23066258
MTB > SSQ C1
   SSQ       =3.805003E+12
MTB > MAXIMUM C1
   MAXIMUM =         302000
MTB > MINIMUM C1
   MINIMUM =          72000
```

2-6 EXERCISES A

In Exercises 2-157 through 2-168, express all z scores with two decimal places.

2-157   For a set of data, $\bar{x} = 60$ and $s = 10$. Find the z score corresponding to (a) $x = 80$; (b) $x = 40$; (c) $x = 65$.

2-158   For a set of data, $\bar{x} = 100$ and $s = 15$. Find the z score corresponding to (a) $x = 115$; (b) $x = 145$; (c) $x = 85$.

2-159   In an investigation of the number of hours college freshmen spend studying each week, it is found that the mean is 7.06 h and the standard deviation is 5.32 h (based on data from *The American Freshman*). Find the z score corresponding to a freshman who studies 10.00 hours weekly.

2-160   In a study of the time high school students spend working at a job each week, it is found that the mean is 10.7 h and the standard deviation is 11.2 h (based on data from the National Federation of State High School Associations). Find the z score corresponding to a high school student who works 15.0 h each week.

2-161   Using the Burnout Measure, sample subjects are found to have measures with a mean of 2.97 and a standard deviation of 0.60 (based on "Moderating Effect of Social Support on the Stress-Burnout Relationship" by Etzion, *Journal of Applied Psychology*, Vol. 69, No. 4). Find the z score corresponding to a subject with a score of 2.00.

2-162   The heights of six-year-old girls have a mean of 117.8 cm and a standard deviation of 5.52 cm (based on data from the National Health Survey, USDHEW publication 73-1605). Find the z score corresponding to a six-year-old girl who is 106.8 cm tall.

2-163   For a certain population, scores on the Thematic Apperception Test have a mean of 22.83 and a standard deviation of 8.55 (based on "Relationships Between Achievement-Related Motives, Extrinsic Conditions, and Task Performance" by Schroth, *Journal of Social Psychology*, Vol. 127, No. 1). Find the z score corresponding to a member of this population who has a score of 10.00.

2-164   For men aged between 18 and 24 years, serum cholesterol levels (in mg/100 ml) have a mean of 178.1 and a standard deviation of 40.7 (based on data from the National Health Survey, USDHEW publication 78-1652). Find the z score corresponding to a male, aged 18–24 years, who has a serum cholesterol level of 249.3 mg/100 ml.

2-165   Which of the following two scores has the better relative position?
a. A score of 53 on a test for which $\bar{x} = 50$ and $s = 10$
b. A score of 53 on a test for which $\bar{x} = 50$ and $s = 5$

2-166   Two students took different language facility tests. Which of the following results indicates the higher level of language facility?
a. A score of 60 on a test for which $\bar{x} = 70$ and $s = 10$
b. A score of 480 on a test for which $\bar{x} = 500$ and $s = 50$

2-167   Three prospective employees take different tests of communicative ability. Which of the following scores corresponds to the highest level?
a. A score of 60 on a test for which $\bar{x} = 50$ and $s = 5$
b. A score of 230 on a test for which $\bar{x} = 200$ and $s = 10$
c. A score of 540 on a test for which $\bar{x} = 500$ and $s = 15$

2-168   Three students take different tests of neuroticism with the given results. Which is the highest relative score?
a. A score of 3.6 on a test for which $\bar{x} = 4.2$ and $s = 1.2$
b. A score of 72 on a test for which $\bar{x} = 84$ and $s = 10$
c. A score of 255 on a test for which $\bar{x} = 300$ and $s = 30$

In Exercises 2-169 through 2-172, use the 150 ranked home selling prices listed in Table 2-14. Find the percentile corresponding to the given selling price.

2-169 $110,000   **2-170** $125,000   **2-171** $175,000   **2-172** $220,000

In Exercises 2-173 through 2-180, use the 150 ranked home selling prices listed in Table 2-14. Find the indicated percentile, quartile, or decile.

2-173 P_{15} **2-174** P_5 **2-175** P_{80} **2-176** P_{90}

2-177 Q_1 **2-178** Q_2 **2-179** Q_3 **2-180** D_6

In Exercises 2-181 through 2-184, use the following data to find the percentile corresponding to the given value. These numbers are the actual weights (in pounds) of 50 consecutive loads handled by a moving company in Dutchess County, N.Y.

8,090	17,810	3,670	10,100	17,330
8,800	7,770	12,430	13,260	10,220
13,490	13,520	4,510	14,760	4,480
7,540	10,630	10,330	10,510	12,700
7,200	12,010	11,450	9,140	7,280
9,110	12,350	14,800	26,580	15,970
11,860	8,450	10,780	5,030	11,430
11,600	7,470	14,310	13,410	7,450
3,250	6,400	8,160	9,310	9,900
6,170	16,200	8,770	6,820	6,390

2-181 5,030   **2-182** 10,220   **2-183** 12,430   **2-184** 14,760

In Exercises 2-185 through 2-192, use the same load weights given above and find the indicated percentile, quartile, or decile.

2-185 P_{15} **2-186** P_{20} **2-187** P_{80} **2-188** P_{66}

2-189 Q_1 **2-190** Q_3 **2-191** D_3 **2-192** D_9

2-6 EXERCISES B

2-193 Use the ranked home selling prices listed in Table 2-14.
a. Find the interquartile range.
b. Find the midquartile.
c. Find the 10–90 percentile range.
d. Doe $P_{50} = Q_2$? Does P_{50} *always* equal Q_2?
e. Does $Q_2 = (Q_1 + Q_3)/2$?

2-194   Find Q_1, Q_2, and Q_3 for the data summarized in the given stem-and-leaf plot.

$$
\begin{array}{r|l}
5 & 00011122 \\
6 & 001233457899 \\
7 & 4444455678899 \\
8 & 012344567899 \\
9 & 0001177
\end{array}
$$

2-195   Construct a collection of data consisting of 50 scores for which $Q_1 = 20$, $Q_2 = 30$, $Q_3 = 70$.

2-196   Using the scores 2, 5, 8, 9, and 16, first find $\bar{x}$ and s, then replace each score by its corresponding z score. Now find the mean and standard deviation of the five z scores. Will these new values of the mean and standard deviation result from *every* set of z scores?

VOCABULARY LIST

Define and give an example of each term.

descriptive statistics
inferential statistics
frequency table
lower class limits
upper class limits
class boundaries
class marks
class width
cumulative frequency
    table
pie chart
histogram
frequency polygon
ogive
exploratory data
    analysis (EDA)

stem-and-leaf plot
average
measure of central
    tendency
mean
median
mode
bimodal
multimodal
midrange
weighted mean
measure of dispersion
range
standard deviation

mean deviation
variance
empirical rule
standard score
z score
quartiles
deciles
percentiles
interquartile range
semi-interquartile
    range
midquartile
10–90 percentile
    range

REVIEW

Chapter 2 deals mainly with the methods and techniques of descriptive statistics. The main objective of this chapter is to develop the ability to organize, summarize, and illustrate data and to extract from data some meaningful measurements. In Section 2-2 we considered the **frequency table** as an excellent device for summarizing data, while Section 2-3 dealt with graphic illustrations, including **pie charts, histograms, frequency polygons, ogives**, and **stem-and-leaf plots**. These visual illustrations help us to determine the position and distribution of a set of scores. In Section 2-4 we defined the common **averages**, or measures of central tendency. The **mean, median, mode**, and **midrange** represent different ways of characterizing the central value of a collection of data. The **weighted mean** is used to find the average of a set of scores that may vary in importance. In Section 2-5 we presented the usual **measures of dispersion**, including the **range, standard deviation**, and **variance**; these descriptive statistics are designed to measure the variability among a set of scores. The **standard score**, or z **score**, was introduced in Section 2-6 as a way of measuring the number of standard deviations by which a given score differs from the mean. That section also included the common measures of position: **quartiles, percentiles**, and **deciles**.

By now we should be able to organize, present, and describe collections of data composed of single scores. We should be able to compute the key descriptive statistics that will be used in later applications.

Using the concepts developed in this chapter, we now have a better understanding of the 150 home selling prices given in Table 2-1. The histogram allowed us to see the shape of the distribution of those values. We know that the mean is \$153,775, the median is \$144,900, and the standard deviation is \$41,611. The data are summarized in a frequency table and depicted in a histogram, frequency polygon, and ogive. By using these descriptive characteristics, a home buyer could form some meaningful and practical conclusions. Subsequent chapters will consider other ways of using sample statistics.

IMPORTANT FORMULAS	
$$\bar{x} = \frac{\Sigma x}{n}$$	Mean
$$\bar{x} = \frac{\Sigma (f \cdot x)}{\Sigma f}$$	Computing the mean when the data is in a frequency table
$$s = \sqrt{\frac{\Sigma (x - \bar{x})^2}{n - 1}}$$	Standard deviation
$$s^2 = \frac{\Sigma (x - \bar{x})^2}{n - 1}$$	Variance
$$s = \sqrt{\frac{n(\Sigma x^2) - (\Sigma x)^2}{n(n - 1)}}$$	Shortcut formula for standard deviation
$$s = \sqrt{\frac{n[\Sigma (f \cdot x^2)] - [\Sigma (f \cdot x)]^2}{n(n - 1)}}$$	Computing the standard deviation when the data is in a frequency table
$$z = \frac{x - \bar{x}}{s} \quad \text{or} \quad \frac{x - \mu}{\sigma}$$	Standard score or z score

REVIEW EXERCISES

2-197 Based on a *USA Today* poll of 800 divorced people, it was stated that "overall, 58% of divorced people say they don't want to get married again." Does that conclusion involve descriptive statistics or inferential statistics? Explain.

2-198 In one semester at Dutchess Community College, it was reported that among the 3243 students who took mathematics, the failure rate was 18.82%. Does that conclusion involve descriptive statistics or inferential statistics? Explain.

2-199   The values given below are snow depths (in centimeters) measured as part of a study of satellite observations and water resources (based on data in *Space Mathematics* published by NASA). Find the (a) mean; (b) median; (c) mode; (d) midrange; (e) range; (f) variance; (g) standard deviation.

19, 18, 12, 25, 22, 8, 8, 16

2-200   A psychologist gives a subject two different tests designed to measure spatial perception. The subject obtained scores of 66 on the first test and 223 on the other test. The first test is known to have a mean of 75 and a standard deviation of 15, while the second test has a mean of 250 and a standard deviation of 25. Which result is better? Explain.

2-201   The given scores represent the number of cars rejected in one day at an automobile assembly plant. The 50 scores correspond to 50 different randomly selected days. Construct a frequency table with 10 classes.

29	58	80	35	30	23	88	49	35	97
12	73	54	91	45	28	61	61	45	81
83	23	71	63	47	87	36	8	94	26
95	63	86	42	22	44	8	27	20	33
28	91	87	15	67	10	45	67	26	19

2-202   Construct a histogram that corresponds to the frequency table from Exercise 2-201.

2-203   For the data in Exercise 2-201, find (a) Q_1, (b) P_{45}, and (c) the percentile corresponding to the score of 30.

2-204   Use the frequency table from Exercise 2-201 to find the mean and standard deviation for the number of rejects.

2-205   The values given below are the living areas (in square feet) of 12 homes recently sold in Dutchess County, N.Y. Find the (a) mean; (b) median; (c) mode; (d) midrange; (e) range; (f) variance; (g) standard deviation.

| 3060 | 1600 | 2000 | 1300 | 2000 | 1956 |
| 2400 | 1200 | 1632 | 1800 | 1248 | 2025 |

2-206   Construct a stem-and-leaf plot for the data in Exercise 2-205.

2-207   For the data given in Exercise 2-205, find the z score corresponding to (a) 1200; (b) 2400.

2-208   A supplier constructs a frequency table for the number of car stereo units sold daily. Use that table to find the mean and standard deviation.

Number Sold	Frequency
0–3	5
4–7	9
8–11	8
12–15	6
16–19	3

2-209   Using the frequency table given in Exercise 2-208, construct the corresponding histogram.

2-210   Using the frequency table given in Exercise 2-208, construct the corresponding frequency polygon.

2-211   Using the frequency table given in Exercise 2-208, construct the corresponding ogive.

2-212   a. A set of data has a mean of 45.6. What is the mean if 5.0 is added to each score?
        b. A set of data has a standard deviation of 3.0. What is the standard deviation if 5.0 is added to each score?
        c. You just completed a calculation for the variance of a set of scores, and you got an answer of -21.3. What do you conclude?
        d. True or false: If set A has a range of 50 while set B has a range of 100, then the standard deviation for data set A must be less than the standard deviation for data set B.
        e. True or false: In proceeding from left to right, the graph of an ogive can never follow a downward path.

2-213   The following values are the diameters (in micrometers) of virus samples. Find the (a) mean; (b) median; (c) mode; (d) midrange; (e) range; (f) variance; (g) standard deviation.

        175, 183, 168, 191, 181, 183, 170, 174, 184, 181, 182

2-214   For the data given in Exercise 2-213, find the z score corresponding to (a) 175; (b) 182.

2-215   Construct a stem-and-leaf plot for the data in Exercise 2-213.

2-216   The values given here are the lot sizes (in acres) of 12 homes recently sold in Dutchess County, N.Y. Find the (a) mean; (b) median; (c) mode; (d) midrange; (e) range; (f) variance; (g) standard deviation.

0.75	0.26	0.70	0.65	0.75	0.50
0.40	0.33	3.00	0.50	0.25	1.10

COMPUTER PROJECT

a. Many computer systems are supported by software consisting of program packages for various uses. Statistical programs are common among such packages. For example, STATDISK is a statistical package designed as a supplement to this text. Use STATDISK, Minitab, or any other statistics software package to obtain descriptive statistics (mean, variance, standard deviation, and so on) for the 150 home selling prices listed in Table 2-1.

b. Develop a computer program that will take a set of data as input. Output should consist of the mean, variance, standard deviation, and the number of scores. Run the program using the set of 150 home selling prices listed in Table 2-1.

CASE STUDY ACTIVITY

Through observation or experimentation, compile a list of sample data that are at the interval or ratio levels of measurement. Obtain at least 40 values, and try to select data from an interesting or meaningful population.

1. Describe the nature of the data. That is, what do the values represent?

2. What method was used to collect the values?

3. What are some possible reasons why the data might not be representative of the population? That is, what are some possible sources of bias?

4. Find the value of each of the following: sample size, minimum, maximum, mean, median, midrange, range, standard deviation, variance, and the quartiles Q_1 and Q_3.

5. Construct a frequency table and histogram.

DATA PROJECT

1. Refer to Appendix B for the data describing homes recently sold in Dutchess County, N.Y.
   a. For the living areas (in square feet), construct a frequency table and histogram and find the mean, median, mode, midrange, range, standard deviation, variance, and the quartiles Q_1, Q_2, and Q_3.
   b. For the same recent time period during which the 150 homes were sold, *U.S. Housing Markets* reported that the average price of a home in the United States was $121,000. What percentage of homes sold in Dutchess County are above the national average?
   c. Identify a factor that might cause the mean of the 150 selling prices to be different from the mean value of all homes in the region from which the sample was drawn.

2. Refer to Appendix B for the data describing the number of siblings. Each person surveyed reported the total number of brothers and sisters.
   a. Use that data to determine the mean for the total number of children per family, including the person who was surveyed.
   b. Based on Census Bureau data, the "average family size" is 3.21 people, *including the parent(s)*. Explain the discrepancy between this figure and the one obtained from part (a). Why can't we use the data from part (a) to estimate the mean size of a family in the United States?

Chapter Three

CHAPTER CONTENTS

3-1 **Overview**
We identify chapter **objectives** and compare probability and statistics.

3-2 **Fundamentals**
We introduce the **basic definitions** of probability theory.

3-3 **Addition Rule**
We present the rule for finding the probability that one event *or* another event will occur.

3-4 **Multiplication Rule**
We present the rule for finding the probability that two events will *both* occur.

3-5 **Complements and Odds**
We present the rule for finding the probability that an event does *not* occur, and we discuss the odds of various events.

3-6 **Counting (Optional)**
We present techniques for determining the total number of different ways that various events can occur.

3 Probability

CHAPTER PROBLEM

Suppose that a political analyst plans to do a study of voting behavior in a county of 200,000 voters, of which 30% are Republican. The political analyst begins with a simple pilot study by paying a pollster $1200 for in-depth surveys of 12 randomly selected voters. But the pollster returns and reports that each of the 12 voters is a Republican. The political analyst protests and says that the sample is not random. The pollster replies that with such a small sample, it is easy to get 12 Republicans since there are 60,000 of them available. This argument might seem reasonable, but we will be able to analyze it more effectively by the techniques developed in this chapter. We will then learn how to determine whether or not the pollster is correct.

3-1 OVERVIEW

Suppose you discover a way to mark each molecule in an 8-oz glass of water so that each one is recognizable. You then proceed to the nearest ocean beach and pour the water into the first wave. After waiting 20 years for the waters of the world to mix, you board a plane for Venice, Italy. On your arrival, you scoop a glass of water from the first canal you find and examine the water in search of one of the molecules that you dumped in 20 years ago. Can you really expect to find one of the molecules in Venice? As absurd as this problem seems, the answer is actually yes. (If you find this incredible, you are not alone.) In fact, you can expect about 1000 of the original molecules to reappear in the glass.

In a class of 25 students, each is asked to identify the month and day of his or her birth. What are the chances that at least two students will share the same birthday? Again our intuition is misleading; it happens that at least two students will have the same birthday in more than half of all classes with 25 students.

For a worldwide population of 5 billion people, a scientist needs to sample only 1000 (or 0.00002%) in order to correctly estimate the percentage of people with a certain genotype.

The preceding conclusions are based on simple principles of probability, which play a critical role in the theory of statistics. All of us now form simple probability conclusions in our daily lives. Sometimes these determinations are based on fact, while other probability determinations are subjective. In addition to being an important aspect of the study of statistics, probability theory is playing an increasingly important role in a society that must attempt to measure uncertainties. Before firing up a nuclear power plant, we should have some knowledge about the probability of a meltdown. Before arming a nuclear warhead, we should have some knowledge about the probability of an accidental detonation. Before raising the speed limit on our nation's highways, we should have some knowledge of the probability of increased fatalities.

In Chapter 1 we stated that inferential statistics involves the use of sample evidence in formulating inferences or conclusions about a population. These inferential decisions are based on probabilities or likelihoods of events. As an example, suppose that a statistician plans to study the hiring practices of a large company. She finds that of the last 100 employees hired, all are men. Perhaps the company hires men and women at the same rate and the run of 100 men is an extremely rare chance event. Perhaps the hiring practices favor males. What

Coincidences?

Many of us have encountered events that seemed to be remarkable coincidences, and there are some classic cases. John Adams and Thomas Jefferson (the second and third presidents) both died on July 4, 1826 (the 50th anniversary of the signing of the Declaration of Independence). President Lincoln was assassinated in Ford's Theater; President Kennedy was assassinated in a Lincoln car made by the Ford Motor Company. Also, Lincoln and Kennedy were both succeeded by vice presidents named Johnson.

Fourteen years *before* the sinking of the Titanic, Morgan Robertson wrote a novel in which he described the sinking of a passenger ship called the Titan. The fictional Titan and the real Titanic both hit icebergs and they both carried around 3000 passengers. (See Martin Gardner's *The Wreck of the Titanic Foretold?*) Martin Gardner states that "in most cases of startling coincidences, it is impossible to make even a rough estimate of their probability."

should we infer? The statistician, along with most reasonable people, would conclude that men are favored. This decision is based on the very low probability of getting 100 consecutive men by chance alone.

In subsequent chapters we develop methods of statistical inference that rely on this type of thinking. It is therefore important to acquire a basic understanding of probability theory. We want to cultivate some intuitive feeling for what probabilities are, and we want to develop some very basic skills in calculating the probabilities of certain events (see Figure 3-1).

In this chapter we begin by introducing the fundamental concept of mathematical probability and we then proceed to investigate the basic rules of probability: the addition rule, the multiplication rule, and the rule of complements. We also consider techniques of counting the number of ways an event can occur. The primary objective of this chapter is to develop a sound understanding of probability values that will be used in subsequent chapters. A secondary objective is to develop the ability to solve simple probability problems, which are valuable in their own right as they are used to make decisions and better understand our world.

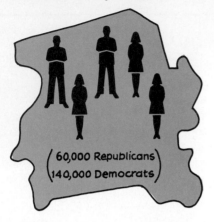

a. Duncan County voters

b. Jackson County voters

60,000 Republicans
140,000 Democrats

? Republicans
? Democrats

(a) **Probability.** With a *known population*, we can see that if one voter is randomly selected, there are 60,000 chances out of 200,000 of picking a Republican. This corresponds to a probability of 60,000/200,000 or 0.3. The population of all voters in the county is known, and we are concerned with the likelihood of obtaining a particular sample (a Republican). We are making a conclusion about a sample based on our knowledge of the population.

(b) **Statistics.** With an *unknown population*, we can obtain a fairly good idea of voter preferences by randomly selecting 500 voters and assuming that this sample is representative of the whole county population. After making the random selections, our sample is known and we can use it to make inferences about the population of all voters in the county. We are making a conclusion about the population based on our knowledge of the sample.

FIGURE 3-1

3-2 FUNDAMENTALS

In considering probability problems, we deal with experiments and events.

DEFINITION
An **experiment** is any process that allows researchers to obtain observations.

DEFINITION

An **event** is any collection of results or outcomes of an experiment.

For example, the random selection of a letter from a computer data bank is an experiment and the result of a vowel is an event. In this case, the event of a vowel will ocur if the letter is *a*, *e*, *i*, *o*, or *u*. We might think of the event of getting a vowel as being a combination of the simpler outcomes corresponding to the individual vowel letters. It is often necessary to consider events that cannot be decomposed into simpler outcomes.

DEFINITION

A **simple event** is an outcome or an event that cannot be broken down any further.

When conducting the experiment of randomly selecting a letter from a data bank, the occurrence of an *e* is a simple event, but the occurrence of a vowel is not a simple event since it can be broken down into the five simple events of *a*, *e*, *i*, *o*, and *u*.

DEFINITION

The **sample space** for an experiment consists of all possible simple events. That is, the sample space consists of all outcomes that cannot be broken down any further.

In the experiment of randomly selecting a letter from a data bank, the sample space consists of the 26 letters of the alphabet.

There is no universal agreement on the actual definition of the probability of an event, but among the various theories and schools of thought, two basic approaches emerge most often. The approaches will be embodied in two rules for finding probabilities. The notation employed will relate P to probability, while capital letters such as A, B, and C will denote specific events. For example, A might represent the event of winning a million-dollar state lottery; $P(A)$ denotes the probability of event A occurring.

Using Coins to Survey Sensitive Issues

It sometimes happens that a pollster wants information about a topic, but respondents are reluctant to be honest because of the sensitive nature of the topic. For example, ask employees if they stole anything from their employer within the past year, and they will be inclined to answer no, whether they stole or not. However, we often need accurate data about such sensitive topics as honesty or sexual practices.

Stanley Warner (York University, Ontario) devised a scheme that leads to more accurate results. As an example, ask employees if they stole within the past year and also ask them to flip a coin. The employees are instructed to answer no if they didn't steal and the coin turns up tails. Otherwise, they should answer yes. The employees are more likely to be honest in their responses because the coin flip helps protect their privacy. Probability theory can then be used to analyze responses so that more accurate results can be obtained.

RULE 1

Empirical (relative frequency) approximation of probability. Conduct (or observe) an experiment a large number of times and count the number of times that event A actually occurs. Then $P(A)$ is *estimated* as follows.

$$P(A) = \frac{\text{number of times } A \text{ occurred}}{\text{number of times experiment was repeated}}$$

RULE 2

Classical approach to probability. Assume that a given experiment has n different simple events, each of which has an **equal chance** of occurring. If event A can occur in s of these n ways, then

$$P(A) = \frac{s}{n} \qquad \text{(requires equally likely outcomes)}$$

The first rule, often referred to as the **empirical**, or **relative frequency**, interpretation of probability, gives us only approximations or

(a) Empirical approach (Rule I)

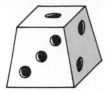

When trying to determine $P(2)$ on a "shaved" die, we must repeat the experiment of rolling it many times and then form the ratio of the number of times 2 occurred to the number of rolls. That ratio is our estimate of $P(2)$.

(b) Classical approach (Rule 2)

With a balanced and fair die, each of the six faces has an equal chance of occurring.

$$P(2) = \frac{\text{number of ways 2 can occur}}{\text{total number of simple events}} = \frac{1}{6}$$

FIGURE 3-2

estimates of $P(A)$. The second rule, often referred to as the **classical** approach, is useful in the analysis of simple experiments that involve equally likely outcomes, such as flipping coins, rolling dice, and so on (see Figure 3-2).

Complicated experiments, such as those involving events that are not equally likely, all require the relative frequency approximation. For example, to determine the probability of a driver having a car accident in a given year, we must study past records in order to arrive at a reasonable approximation. The events of having an accident and not having one are not equally likely, so we cannot use rule 2.

When determining probabilities by using rule 1 (the empirical approach), we obtain an approximation instead of an exact value. As the total number of observations increases, the corresponding approximations tend to get closer to the actual probability. This idea is stated as a theorem commonly referred to as the Law of Large Numbers.

LAW OF LARGE NUMBERS

As an experiment is repeated again and again, the empirical probability (from rule 1) of success tends to approach the actual probability.

Should You Guess on SATs?

Students preparing for the SAT exams are often advised to avoid guessing, but is that good advice? Not necessarily! The correction for guessing is this: "For questions with five answer choices, one-fourth of a point is subtracted for each incorrect response; one-third of a point is subtracted for incorrect responses to questions with four answer choices." While that is a *correction* for guessing, it is not a *penalty*. Principles of probability can be used to show that in the long run, pure random guessing will neither raise nor lower the exam score. Also, students often can eliminate at least one choice and frequently have some intuitive sense for a correct answer. These factors suggest that the best strategy is to guess at every question for which the answer is not known. However, students should not waste too much time on such questions and, if possible, they should try to avoid tricky questions with attractive wrong answers.

The Law of Large Numbers tells us that the empirical approximations from rule 1 tend to get better with more observations. This law reflects a simple notion supported by common sense: In only a few trials, results can vary substantially, but with a very large number of trials, results tend to be more stable and consistent. For example, it would not be unusual to get 4 girls in 4 births, but it would be extremely unusual to get 400 girls in 400 births.

Many experiments involving equally likely outcomes are so complicated that rule 1 is used as a practical alternative to extremely complex computations. Consider, for example, the probability of winning at solitaire. In theory, rule 2 applies, but in reality the better approach requires that we settle for an approximation obtained through rule 1. That is, we should play solitaire many times and record the number of wins along with the number of attempts. The ratio of wins to trials is the approximation we seek. Computers can often be programmed to simulate an event so that rule 1 can be used.

The following examples are intended to illustrate the use of Rules 1 and 2. In some of these examples we use the term *random*.

DEFINITION
In a **random selection** of an element, all elements available for selection have the same chance of being chosen.

This concept of random selection is extremely important in statistics. When making inferences based on samples, we must have a sampling process that is representative, impartial, and unbiased. Also, random selection is different from haphazard selection. Ask people to randomly select one of the 10 digits from 0 through 9, and they tend to favor some digits while ignoring others. Implementation of a random selection process often requires careful and thoughtful planning.

EXAMPLE

On a college entrance examination, each question has five possible answers. If an examinee makes a random guess on the last question, what is the probability that the response is wrong?

Solution

There are five possible outcomes or answers, and there are four ways to answer incorrectly. Random guessing implies that the outcomes are equally likely, so we apply rule 2 to get

$$P(\text{wrong answer}) = \frac{4}{5} \quad \text{or} \quad 0.8$$

EXAMPLE

In a study of 398 cases of deaths from guns in homes, 333 cases were determined to be suicides (based on data from the *New England Journal of Medicine*). If one of these 398 cases is randomly selected, what is the probability it involves a suicide?

Solution

With random selection, the 398 cases are equally likely and rule 2 applies, so that

$$P(\text{suicide}) = \frac{333}{398} = 0.837 \qquad \begin{array}{l} \longleftarrow \text{number of suicides} \\ \longleftarrow \text{total number of cases} \end{array}$$

EXAMPLE

Find the probability that a couple with three children will have exactly two boys. (Assume that boys and girls are equally likely and that the sex of any child is independent of any brothers or sisters.)

Solution

If the couple has exactly two boys in three births, there must be one girl. The possible outcomes are listed and each is assumed to be equally likely.

```
1st 2nd 3rd
```
boy-boy-boy
boy-boy-girl ⟵
boy-girl-boy ⟵ exactly
boy-girl-girl   two
girl-boy-boy ⟵ boys
girl-boy-girl
girl-girl-boy
girl-girl-girl

Of the eight different possible outcomes, three correspond to exactly two boys, so that

$$P(2 \text{ boys in 3 births}) = \frac{3}{8} = 0.375$$

EXAMPLE

Find the probability of a 20-year-old male living to be 30 years of age.

Solution

Here the two outcomes of living and dying are not equally likely, so the relative frequency approximation must be used. This requires that we observe a large number of 20-year-old males and then count those who live to be 30. Suppose that we survey 10,000 20-year-old males and find that 9840 of them lived to be 30 (these are realistic figures based on U.S. Department of Health

continued ▶

Solution, continued

and Human Services data). Then the empirical approximation becomes

$$P(\text{20-year-old male living to 30}) = \frac{9{,}840}{10{,}000} = 0.984$$

This is the basic approach used by insurance companies in the development of mortality tables.

EXAMPLE

If a year is selected at random, find the probability that Thanksgiving Day will be on a (a) Wednesday; (b) Thursday.

Solution

a. Thanksgiving Day is defined so that it will always be on a Thursday. It is therefore impossible for Thanksgiving to be on a Wednesday. When an event is impossible, we say that its probability is 0.

b. It is certain that Thanksgiving will be on a Thursday. When an even is certain to occur, we say that its probability is 1.

The probability of any impossible event is 0.

The probability of any event that is certain to occur is 1.

Since any event imaginable is either impossible, certain, or somewhere in between, it is reasonable to conclude that the mathematical probability of any event is either 0, 1, or a number between 0 and 1 (see Figure 3-3). This property can be expressed as follows:

$$0 \leq P(A) \leq 1 \qquad \text{for any event } A$$

In Figure 3-3, the scale of 0 through 1 is shown on top, whereas the more familiar and common expressions of likelihood are shown on the bottom.

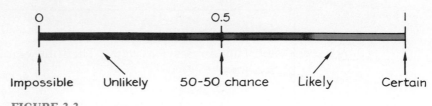

FIGURE 3-3
Possible values for probabilities

Rounding Off Probabilities

Although it is difficult to develop a universal rule for rounding off probabilities, the following guide will apply to most problems in this text.

> Either give the exact fraction representing a probability, or round off final decimal results to three significant digits.

All the digits in a number are significant except for the zeros that are included for proper placement of the decimal point. The probability of 0.00128506 can be rounded to three signifcant digits as 0.00129. The probability of 1/3 can be left as a fraction or rounded in decimal form to 0.333, but not 0.3.

EXAMPLE

An experiment in ESP involves blindfolding the subject and allowing the subject to make one selection from a standard deck of cards. The subject must then identify the correct suit (that is, clubs, diamonds, hearts, or spades). What is the probability of making a correct random guess?

Solution

The standard deck contains 52 cards with 13 of each suit. There are therefore 52 possible outcomes, with 13 outcomes corresponding to whatever suit the subject selects. Thus

$$P(\text{correct guess}) = \frac{13}{52} = \frac{1}{4} = 0.250$$

Just How Probable Is Probable?

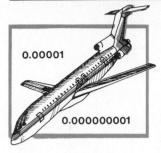

The mathematical definition requires that probability values must be between 0 and 1, with 1 corresponding to an event that is certain to occur and 0 corresponding to an impossible event. But what probability values correspond to terms such as *probable*, *improbable*, or *extremely probable*? In discussing systems design analysis of aircraft, the Federal Aviation Administration (FAA) has defined these terms. The following definitions are taken from an FAA publication.

1. Probable—Probable events may be expected to occur several times during the operational life of each airplane. A probability on the order of 0.00001 or greater (for each hour of flight).

2. Improbable—Improbable events are not expected to occur during the total operational life of a random single airplane of a particular type, but may occur during the total operational life of all airplanes of a particular type. A probability on the order of 0.00001 or less.

3. Extremely improbable—Extremely improbable events are so unlikely that they need not be considered to ever occur, unless engineering judgment would require their consideration. A probability on the order of 1×10^{-9} or less. (Note: 1×10^{-9} = 0.000000001.)

The 1/4 probability of the previous example implies that, on the average, for each four guesses, there will be one correct selection. If our ESP subject does *significantly* better than one correct response in four, then we can conclude that the selections are not made solely on the basis of chance. But what is significant? Thirty correct guesses in 100 trials? Fifty correct guesses in 100 trials? Ninety-nine correct guesses in 100 trials? This decision relates to a key concept of statistical inference and will be explored later in the book.

An important concept of this section is the mathematical expression of probability as a number between 0 and 1. This type of expression is fundamental and common in statistical procedures, and we will use it throughout the remainder of this text. A typical computer output, for example, may involve a *P*-value expression such as "Significance less than 0.001." We will discuss the meaning of *P*-values later, but they are essentially probabilities of the type discussed in this section. We should recognize that a probability of 0.001 (equivalent to 1/1000) corresponds to an event that is very rare in the sense that it occurs an average of only once in a thousand trials.

3-2 EXERCISES A

3-1   Which of the following values *cannot* be probabilities?

$$1.2, \frac{77}{75}, \frac{9}{10}, 0, -\frac{1}{2}, 1, 5, 0.9999, 1.001, \sqrt{2}$$

3-2   a. What is $P(A)$ if event A is certain to occur?
      b. What is $P(A)$ if event A is impossible?
      c. A sample space consists of 14 separate events that are equally likely. What is the probability of each?
      d. According to *Glamour* magazine, if you are a woman born after 1960, your chances of living to age 70 or older are 1 in 1.4. Express this as a probability in decimal form.

3-3   If one person is to be *randomly selected* from a class of 30 students, what is the probability that the tallest person will be chosen?

3-4   Among 80 randomly selected blood donors, 36 were classified as group O (based on data from the Greater New York Blood Program). What is the approximate probability that a person will have group 0 blood?

3-5   On a college entrance exam, each question has five possible answers. If an examinee makes a random guess on the first question, what is the probability that the response is correct?

3-6   If a person is randomly selected, find the probability that his or her birthday is October 18, which is National Statistics Day in Japan. Ignore leap years.

3-7   Among 400 randomly selected drivers in the 20 to 24 age bracket, 136 were in a car accident during the last year (based on data from the National Safety Council). If a driver in that age bracket is randomly selected, what is the approximate probability that he or she will be in a car accident during the next year?

3-8   In a recent national election, there were 12,300,000 citizens in the 18- to 20-year age bracket. Of these 4,400,000 actually voted. Find the empirical probability that a person randomly selected from this group of 18- to 20-year-old citizens did vote in that national election.

3-9   Among eight helicopters sent to rescue American hostages in Iran, three helicopters failed to operate properly. Given the same conditions, what is the empirical probability of failure for a helicopter?

3-10 A jury of four men and eight women has been selected for a sex-discrimination case. If the foreperson of this jury is selected at random, find the probability that a woman will be chosen.

3-11 A computer is used to generate random telephone numbers. Of the numbers generated and in service, 56 are unlisted, and 144 are listed in the telephone directory. If one of these telephone numbers is randomly selected, what is the probability that it is unlisted?

3-12 In a Census Bureau survey of 600 people in the 18 to 25 age bracket, it was found that 237 people smoke. If a person in that age bracket is randomly selected, find the probability that he or she smokes.

3-13 Data from the National Association of Recording Machines show that among 1500 randomly selected record sales, 135 were in the category of country music. If a record sale is randomly selected, what is the approximate probability that it is a country music record?

3-14 Among certain stocks selected in one day, 332 declined and 668 rose. If one of these stocks is selected at random, what is the probability it rose that day?

3-15 In an experiment involving smoke detectors, an alarm was triggered after the subject fell asleep. Of the 95 sleeping subjects, 40 were awakened by the alarm, and 55 did not awake. If one subject is randomly selected, find the probability that he or she is among those who were not awakened by the alarm.

3-16 Data collected by volunteers in the Straphangers Campaign showed that 89 New York City subway cars had broken doors and 286 cars did not. If a car is randomly selected, what is the probability it will have broken doors?

3-17 In a recent presidential election, 44 million voters were in favor of the Republican, and 36 million voters were in favor of the Democratic candidate. Find the probability that a voter selected at random cast a ballot in favor of the Republican.

3-18 In an Environmental Protection Agency survey of cars originally equipped with catalytic converters, it was found that 280 cars still had their converters and 12 cars had them removed. What is the approximate probability of selecting a car with a removed catalytic converter if the selection is random and is limited to cars originally assembled with converters?

3-19 A particular recessive genetic characteristic is present in 235 subjects studied and absent in 755. If one of these subjects is randomly selected, find the probability that the genetic characteristic will be present.

3-20 Among female dieters following the same routine, 16 lost weight, 4 gained weight, and 2 remained the same. If one of these dieters is selected at random, what is the probability that she lost weight?

3-21 Blood groups are determined for a sample of people, and the results are given in the accompanying table (based on data from the Greater New York Blood Program). If one person from this sample group is randomly selected, find the probability that the person has group AB blood.

Blood Group	Frequency
O	90
A	80
B	20
AB	10

3-22   The number of people who receive social security benefits are given in the following table for select states (based on data from the Social Security Administration). If one of these recipients is randomly selected, find the probability of getting a Texan.

State	Number of Social Security Recipients
New York	2,788,649
California	3,284,313
Florida	2,196,141
Maine	198,712
Texas	1,872,383

3-23   The accompanying table summarizes recent driver convictions for select violations in two counties (the data are from the New York State Department of Motor Vehicles).

	Speeding	DWI
Dutchess County	10,589	636
Westchester County	22,551	963

If one of the convictions is randomly selected, find the probability that it is for DWI (driving while intoxicated) in Dutchess County.

3-24   Referring to the data given in Exercise 3-23, find the probability of randomly selecting one of these convictions and getting one for DWI.

3-25   A couple plans to have two children.
   a. List the different outcomes according to the sex of each child. Assume that these outcomes are equally likely.
   b. Find the probability of getting two girls.
   c. Find the probability of getting exactly one child of each sex.

3-26   A couple plans to have four children.
   a. List the 16 different possible outcomes according to the sex of each child. Assume that these outcomes are equally likely.
   b. Find the probability of getting all girls.
   c. Find the probability of getting *at least* one child of each sex.
   d. Find the probability of getting *exactly* two children of each sex.

3-27 A quick quiz consists of three true-false questions and an unprepared student must guess at each one. The guesses will be random.
a. List the different possible solutions.
b. What is the probability of answering the three questions correctly?
c. What is the probability of guessing incorrectly for all three questions?
d. What is the probability of passing by guessing correctly for *at least* two questions?

3-28 Both parents have the brown-blue pair of eye-color genes, and each parent contributes one gene to a child. Assume that if the child has at least one brown gene, that color will dominate and the eyes will be brown. (Actually, the determination of eye color is somewhat more complex.)
a. List the different possible outcomes. Assume that these outcomes are equally likely.
b. What is the probability that a child of these parents will have the blue-blue pair of genes?
c. What is the probability that a child will have brown eyes?

3-2 EXERCISES B

3-29 On a distant planet, one parent from each of three different sexes is necessary for reproduction. Each parent contributes one gene from a triplet of genes that determine the color of a child's antenna. Three parents have the following genes:

red-white-blue
red-white-white
red-red-white

a. List the 27 different possible outcomes. Assume that these outcomes are equally likely.
b. What is the probability that a child receives the blue-red-red combination of genes?
c. The color of a child's antenna is determined by majority. For example, a child receiving the blue-white-white combination would have a white antenna. If one of each color is present, the antenna will be blue. What is the probability that a child has a white antenna?
d. What is the probability that a child has a blue antenna? [See the factors given in part (c).]
e. What is the probability that a child has an antenna that is not white? [See part (c).]

3-30   Someone has reasoned that since we know nothing about the presence of life on Pluto, either life exists there or it does not, so that we have P(life on Pluto) = 0.5. Is this reasoning correct? Explain.

3-31   If two flies land on an orange, find the probability that they lie on the same hemisphere.

3-32   Two points along a straight stick are randomly selected. The stick is then broken at those two points. Find the probability that the three pieces can be arranged to form a triangle.

3-3   ADDITION RULE

In the preceding section we introduced the basic concept of probability and we considered simple experiments and simple events. Many real situations involve compound events.

DEFINITION
Any event combining two or more simple events is called a **compound event**.

Computers are often used to randomly generate telephone numbers of people to be called for a survey or poll. That approach prevents the exclusion of subjects with unlisted numbers (often around 28% of the population). Suppose we consider only the last digit of the randomly generated telephone number. The event of getting an odd last digit is an example of a compound event since it combines the simple events of 1, 3, 5, 7, 9.

In this section we want to develop a rule for finding $P(A \text{ or } B)$, the probability that for a single outcome of an experiment, either event A or event B occurs or both events occur. (Throughout this text we use the inclusive *or*, which means either one, or the other, or both. We will *not* consider the exclusive *or*, which means either one or the other, but not both.)

NOTATION
$P(A \text{ or } B) = P(\text{event } A \text{ occurs or event } B \text{ occurs or they both occur})$

Digits 0 through 9

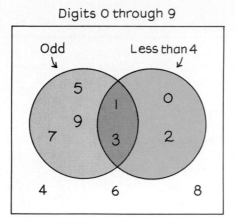

FIGURE 3-4

For example, suppose we stipulate that an experiment consists of using a computer to randomly generate the last digit of a telephone number, and we let events *A* and *B* be defined as follows:

A: the digit is odd
B: the digit is less than 4

The occurrence of *A* or *B* is the occurrence of an odd number or a number less than 4. Similarly, *P(A or B)* is the probability of getting an odd number or a number less than 4. We will now determine the probability of getting *A* or *B* through a direct and simple analysis. We know that there are 10 possible outcomes (0, 1, 2, . . . , 9), so we can identify those that are either odd or less than 4. In Figure 3-4 we use a Venn diagram to categorize the 10 digits. From Figure 3-4 we see that of the 10 possible outcomes, exactly 7 are odd or less than 4. This means that of the 10 simple events that make up the sample space, there are exactly 7 ways that *A* or *B* can occur. Since the 10 simple events are all equally likely, rule 2 from Section 3-2 can be applied, and we get *P(A or B)* = 7/10.

If all problems were as simple and direct as this, we would have no real need for more abstract rules and formulas. But important problems of greater complexity do require formal generalized rules.

Let's use this last example as a basis for constructing one such rule. We know that *P(A)* = *P*(odd number) = 5/10 and *P(B)* = *P*(number less than 4) = 4/10 from Section 3-2. We also know that *P(A or B)* = 7/10 from the above analysis. It would be nice if we could express *P(A or B)* in terms of *P(A)* and *P(B)*, but how can we do that?

Probability and Prosecution

Probability theory is used in forensic science as experts analyze criminal evidence. One particular area of rapid technological advancement is in the analysis of bloodstains. Several years ago, bloodstains could be categorized only according to the four basic blood groups. Current capabilities allow scientists to identify at least a dozen different charac- teristics. It is estimated that the probability of two people having the same set of blood char- acteristics is about 1/1000, and that evi- dence has already been used to help convict murderers.

$P(A$ or $B) = 7/10$ does not seem to suggest any combination of $P(A) = 5/10$ and $P(B) = 4/10$. The problem really reduces to that of counting the number of ways A can occur (5), the number of ways B can occur (4), and getting a total of 7.

The obstacle here is the overlapping of events A and B. There are two ways that A and B will both occur simultaneously. If either 1 or 3 is the outcome, then we have an odd number that is also less than 4. **When combining the number of ways event A can occur with the number of ways B can occur, we must avoid double counting of those outcomes in which A and B happen simultaneously.** If we were to state that

$$P(A \text{ or } B) = P(A) + P(B)$$

in this example, we would be wrong because $7/10 \neq 5/10 + 4/10$. The error arises from the two outcomes of 1 and 3, which were each counted twice. The last equation is corrected by a subtraction to compensate for that double counting.

$$\frac{7}{10} = \frac{5}{10} + \frac{4}{10} - \frac{2}{10}$$

or

$$P(\text{odd or less than 4}) = P(\text{odd}) + P(\text{less than 4}) - P(\text{odd and less than 4})$$

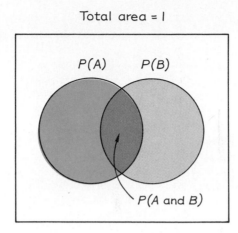

Total area = 1

FIGURE 3-5

The right-hand side of this last equation reflects the five ways of getting an odd number, the four ways of getting a number below 4, and the compensating subtraction of the two ways of getting both. This corrected equation can be generalized in the **addition rule**, which follows.

ADDITION RULE

$$P(A \text{ or } B) = P(A) + P(B) - P(A \text{ and } B)$$

where $P(A \text{ and } B)$ denotes the probability that A and B both occur at the same time as an outcome in the experiment.

See Figure 3-5 for a visual illustration of the addition rule. In this figure we can see that the probability of A or B equals the probability of A (left circle) plus the probability of B (right circle) minus the probability of A and B (football-shaped middle region). This figure shows that the addition of areas of the two circles will cause double counting of the football-shaped middle region. This is the basic concept that underlies the addition rule. Because of this relationship between the addition rule and the Venn diagram shown in Figure 3-5, the notation $P(A \cup B)$ is often used in place of $P(A \text{ or } B)$. Similarly, the notation $P(A \cap B)$ is often used in place of $P(A \text{ and } B)$, so that the addition rule can be expressed as

$$P(A \cup B) = P(A) + P(B) - P(A \cap B)$$

Monkeys Are Not Likely to Type *Hamlet*

A classical assertion holds that a monkey randomly hitting the keys on a typewriter would eventually produce the complete works of Shakespeare if it could continue to type century after century. Dr. William Bennet used the rules of probability to develop a computer simulation that addressed this problem, and he con-

cluded that it would take a monkey about 1,000,000,000,000,000, 000,000,000,000,000, 000,000 years to reproduce Shakespeare's works. In the same spirit, Sir Arthur Eddington wrote this: "There once was a brainy baboon
Who always breathed down a bassoon.
For he said, 'It appears
That in billions of years

I shall certainly hit on a tune!'"

The addition rule is simplified whenever A and B cannot occur simultaneously, so that $P(A \text{ and } B)$ becomes zero.

> **DEFINITION**
>
> Events A and B are **mutually exclusive** if they cannot occur simultaneously.

In the experiment of rolling one die, the event of getting a 2 and the event of getting a 5 are mutually exclusive events, since no outcome can be both a 2 and 5 simultaneously. The following pairs of events are other examples of mutually exclusive pairs in a single experiment. That is, within each of the following pairs of events, it is impossible for both events to occur at the same time.

Pairs of mutually exclusive events

{ Manufacturing a defective electronic component
  Manufacturing a good electronic component }

{ Selecting a voter who is a registered Democrat
  Selecting a voter who is a registered Republican }

{ Testing a subject with an IQ above 100
  Testing a subject with an IQ below 95 }

The following pairs of events are examples that are *not* mutually exclusive in a single trial. That is, within each of the following pairs of events, it is possible that both events occur at the same time.

Pairs of events that are not mutually exclusive

$$\left\{ \begin{array}{l} \text{Selecting a doctor who is a brain surgeon} \\ \text{Selecting a doctor who is a woman} \end{array} \right\}$$

$$\left\{ \begin{array}{l} \text{Selecting a voter who is a registered Democrat} \\ \text{Selecting a voter who is under 30 years of age} \end{array} \right\}$$

$$\left\{ \begin{array}{l} \text{Testing a subject with an IQ above 100} \\ \text{Testing a subject with an IQ above 110} \end{array} \right\}$$

EXAMPLE

A computer randomly generates the last digit of a telephone number. Find the probability that the outcome is an 8 or 9.

Solution

The outcome of 8 and the outcome of 9 are mutually exclusive events. This means that it is impossible for both 8 and 9 to occur together when one digit is selected, so $P(8 \text{ and } 9) = 0$ and the addition rule is applied as follows:

$$P(8 \text{ or } 9) = P(8) + P(9) - P(8 \text{ and } 9)$$

$$= \frac{1}{10} + \frac{1}{10} - 0$$

$$= \frac{2}{10} \quad \text{or} \quad \frac{1}{5}$$

In Figure 3-6 we see that the addition rule can be simplified when the events in question are known to be mutually exclusive. The following examples involve application of the addition rule.

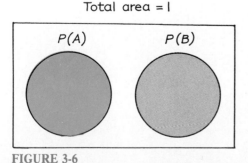

Total area = 1

$P(A)$ $P(B)$

FIGURE 3-6

If events A and B are known to be mutually exclusive, they are completely disjointed and involve no overlapping. When this occurs, the addition rule can be simplified to $P(A \text{ or } B) = P(A) + P(B)$.

Not Too Likely

N. C. Wickramashinghe of University College in Cardiff, Wales, stated that the chance of getting life from a random shuffling of amino acids is about 1 in $10^{40,000}$. He said that this is equivalent to a tornado blowing through a junkyard and creating a jumbo jet in the process.

EXAMPLE

Men were once drafted into the U.S. Army according to the random selection of birthdays. If the 366 different possible birthdays are written on separate slips of paper and mixed in a bowl, find the probability of making one selection and getting a birthday in May or November.

Solution

Let M denote the event of drawing a May date, while N denotes the event of drawing a November date. Clearly, M and N are mutually exclusive because no date is in both May and November. Applying the addition rule, we get

$$P(M \text{ or } N) = P(M) + P(N) = \frac{31}{366} + \frac{30}{366} = \frac{61}{366}$$

The subtraction of $P(M \text{ and } N)$ can be ignored since it is zero.

EXAMPLE

Using the same population of 366 different birthdays, find the probability of making one selection that is the first day of a month or a November date.

Solution

Let F denote the event of selecting a date that is the first of the month. Here F and N are not mutually exclusive because they can occur simultaneously (as on November 1). Applying the addition rule, we get

$$P(F \text{ or } N) = P(F) + P(N) - P(F \text{ and } N)$$

$$= \frac{12}{366} + \frac{30}{366} - \frac{1}{366} \quad \underset{\text{Nov. 1}}{\big\lfloor}$$

$$= \frac{41}{366}$$

EXAMPLE

Data were collected in a study of set belt use. Given the data in Table 3-1, find the probability of randomly selecting one of the 458 subjects and getting someone who wears seat belts or has never attempted suicide. (The table is based on data from the *American Journal of Public Health*, Vol. 67, 1-43-1050.)

TABLE 3-1

		Wear seat belt	Don't wear seat belt	
Have ever attempted suicide?	Yes	9	4	13
	No	233	212	445
		242	216	

(Total: 458 subjects)

Solution

We seek the value of P(wears seat belt or never attempted suicide), which we abbreviate as P(Wears or No). Using the addition rule, we get

$$P(\text{Wears or No}) = P(\text{Wears}) + P(\text{No}) - P(\text{Wears and No})$$

$$= \frac{242}{458} + \frac{445}{458} - \frac{233}{458}$$

$$= \frac{454}{458} = 0.991$$

A commonsense alternative to formal use of the addition rule is to simply count the cells in Table 3-1 that correspond to wearing seat belts or not attempting suicide. See the shaded regions of Table 3-1.

Again, we should be careful to avoid double counting.

Errors made when applying the addition rule often involve double counting. That is, events that are not mutually exclusive are treated as if they were. One positive indication of such an error is a total probability that exceeds 1, but errors involving the addition rule do not always cause the total probability to exceed 1.

3-3 EXERCISES A

3-33 For each pair of events given, determine whether the two events are mutually exclusive for a single experiment.
   a. Selecting a registered voter
       Selecting someone over 65 years of age
   b. Selecting a required course
       Selecting an elective course
   c. Selecting someone born in the United States
       Selecting a citizen
   d. Selecting a voter who favors gun control
       Selecting a conservative
   e. Selecting a person with blond hair (natural or otherwise)
       Selecting a person with brown eyes

3-34 For each pair of events given, determine whether the two events are mutually exclusive for a single experiment.
   a. Selecting a high school graduate
       Selecting someone who is unemployed
   b. Selecting an unmarried person
       Selecting a person with an employed spouse
   c. Selecting a dominant personality type
       Selecting a submissive personality type
   d. Selecting a voter who is a registered Democrat
       Selecting a voter who favors a Republican candidate
   e. Selecting a consumer with an unlisted phone number
       Selecting a consumer who does not drive

3-35 In a state lottery, a three-digit number is randomly selected between 000 and 999 inclusive. Find the exact probability that the number selected is less than 100 or greater than 900.

3-36 Among 200 seats available on one international airliner, 40 are reserved for smokers (including 16 aisle seats) and 160 are reserved for non-smokers (including 64 aisle seats). If a late passenger is randomly assigned a seat, find the probability of getting an aisle seat or one in the smoking section.

3-37 If a birth date is randomly selected from the 366 different possibilities (all equally likely), find the probability that it is a date in November or December.

3-38 If a computer randomly generates the last digit of a telephone number, find the probability that it is odd or greater than 2.

3-39 A labor study involves a sample of 12 mining companies, 18 construction companies, 10 manufacturing companies, and 3 wholesale companies. If a company is randomly selected from this sample group, find the probability of getting a mining or construction company.

3-40 In one local survey, 100 subjects indicated their opinions on a zoning ordinance. Of the 62 favorable responses, there were 40 males. Of the 38 unfavorable responses, there were 15 males. Find the probability of randomly selecting one of these subjects and getting a male or a favorable response.

In Exercises 3-41 through 3-44, refer to the data in Table 3-2, which describes the age distribution for residents of Dutchess County, N.Y. (The table is based on data from the Cornell Community and Resource Development Series.) *In each case, assume that one resident is randomly selected.*

3-41 Find the probability of selecting a resident under 10 years of age or over 69 years of age.

3-42 Find the probability of selecting a resident with an age between 20 and 39.

3-43 Find the probability of selecting a resident with an age under 30 or between 20 and 49.

3-44 Find the probability of selecting a resident with an age under 20 or between 10 and 39.

TABLE 3-2	
Age	Number
0–9	42,284
10–19	40,182
20–29	32,010
30–39	28,798
40–49	25,932
50–59	21,211
60–69	17,253
70 or over	14,625

In Exercises 3-45 through 3-48, use the following data: In a survey of randomly selected adults, 150 were ticketed for speeding and 140 others received parking tickets during the last year. For this group, nobody received both a speeding ticket and a parking ticket. Men received 100 of the speeding tickets and 90 of the parking tickets. (These figures are based on data from a survey by R. H. Bruskin Associates.) In each case assume that one of these 290 individuals is randomly selected.

3-45   Find the probability of selecting a woman.

3-46   Find the probability of selecting a woman or someone ticketed for speeding.

3-47   Find the probability of selecting a man or someone ticketed for speeding.

3-48   Find the probability of selecting a woman or someone ticketed for parking.

In Exercises 3-49 through 3-52, refer to Table 3-3, which is based on data from AT&T and the Automobile Association of America. In each case assume that one of the 758 drivers is randomly selected.

3-49   Find the probability that the selected driver had a car accident in the last year.

3-50   Find the probability that the selected driver uses a cellular phone or had a car accident in the last year.

3-51   Find the probability that the selected driver did not use a cellular phone.

3-52   Find the probability that the selected driver did not use a cellular phone or did not have an accident in the last year.

TABLE 3-3			
		Car accident in last year?	
		Yes	No
Use a cellular phone?	Yes	23	282
	No	46	407

In Exercises 3-53 through 3-60, refer to the accompanying figure, which describes the blood groups and Rh types of 100 people. (The table is based on data from the Greater New York Blood Program.) In each case assume that one of the 100 subjects is randomly selected and find the indicated probability.

3-53 P(group A or group B)

3-54 P(type Rh^+)

3-55 P(group A or type Rh^-)

3-56 P(group O or type Rh^+)

3-57 P(group A or B or AB)

3-58 P(type Rh^-)

3-59 P(group O or type Rh^-)

3-60 P(group A or group O or type Rh^-)

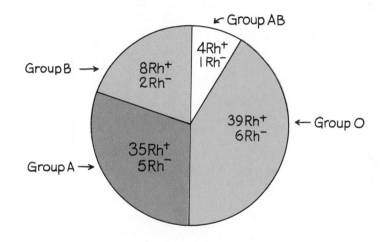

3-3 EXERCISES B

3-61 If $P(A \text{ or } B) = 1/3$, $P(B) = 1/4$, $P(A \text{ and } B) = 1/5$, find $P(A)$.

3-62 Find $P(B)$ if $P(A \text{ or } B) = 0.6$, $P(A) = 0.6$, and A and B are mutually exclusive events.

3-63 If $P(A \text{ or } B) = 79/120$, $P(A) = 11/24$, and $P(B) = 13/40$, find the value of $P(A \text{ and } B)$.

3-64 If events A and B are mutually exclusive, $P(A) = 0.123$ and $P(B) = 0.456$, find
  a. $P(A \text{ and } B)$
  b. $P(A \text{ or } B)$
  c. $P(A \text{ or not } B)$

3-65   a. If $P(A) = 0.4$ and $P(B) = 0.5$, what is known about $P(A$ or $B)$ if A and B are mutually exclusive events?

b. If $P(A) = 0.4$ and $P(B) = 0.5$, what is known about $P(A$ or $B)$ if A and B are not mutually exclusive events?

c. $P(A$ or $B) = 0.8$ while $P(A) = 0.4$. What is known about events A and B?

3-66   Find the largest possible value for

a. $P(A$ or $B)$

b. $P(A) + P(B)$

c. $P(A) + P(B)$ if A, B are mutually exclusive

d. $P(A) + P(B)$ if $A, B,$ and C are mutually exclusive

3-67   How is the addition rule changed if the exclusive *or* is used instead of the inclusive *or*? Recall that the exclusive *or* means either one or the other, but not both.

3-68   Given that $P(A$ or $B) = P(A) + P(B) - P(A$ and $B)$, develop a rule for $P(A$ or B or $C)$.

3-4 MULTIPLICATION RULE

In Section 3-3 we developed a rule for finding the probability that A or B will occur in a given experiment. In the addition rule, we used the term $P(A$ and $B)$, which denoted the probability that A and B both occurred. Determination of a value for $P(A$ and $B)$ is often easy when we are dealing with a single simple experiment, but now we want to develop a more general rule for finding $P(A$ and $B)$, where A occurs in one experiment, while B occurs in another. This is an important distinction: the $P(A$ and $B)$ used in the addition rule is different from the $P(A$ and $B)$ used in this section. In the addition rule, $P(A$ and $B)$ means that A and B both occur simultaneously in a single event, but the $P(A$ and $B)$ we will now consider involves two separate events with the outcome of A followed by the outcome of B.

We begin with a simple example, which will suggest a preliminary multiplication rule. Then we use another example to develop a variation and ultimately obtain a generalized multiplication rule.

Probability theory is used extensively in the analysis and design of tests. Practical considerations often require that standardized tests allow only those answers that can be corrected easily, such as true-false or multiple choice. Let's assume that the first question on a test is a true-false type while the second question is multiple choice with five possible answers (a, b, c, d, e). We will use the following two questions. Try them!

Convicted by Probability

Mrs. Juanita Brooks was robbed in Los Angeles. According to witnesses, the robber was a Caucasian woman with blond hair in a ponytail, who escaped in a yellow car driven by a black male with a mustache and beard. Janet and Malcolm Collins were arrested and convicted after a college mathematics instructor testified that there is only about one chance in 12 million that any couple would have the characteristics described by the witnesses. The following estimated probabilities were presented in court.

P(yellow car) $= \frac{1}{10}$
P(man with mustache) $= \frac{1}{4}$
P(woman with hair in a ponytail) $= \frac{1}{10}$
P(woman with blond hair) $= \frac{1}{3}$

P(black man with beard) $= \frac{1}{10}$
P(interracial couple in car) $= \frac{1}{1000}$

The convictions were reversed by the California Supreme Court, which noted that no evidence was presented to support the stated probabilities and the independence of the characteristics was not established.

---

1. "Jimmy Carter was the first U.S. president who was born in a hospital." (true, false)

2. What foreign city receives the largest number of American tourists each year?
   a. Montreal
   b. London
   c. Paris
   d. Rome
   e. Tijuana

We want to determine the probability of getting both answers correct by making random guesses. We begin by listing the complete sample space of different possible answers.

| T,a | T,b | T,c | T,d | T,e | 1 case is correct |
| F,a | F,b | F,c | F,d | F,e | 10 equally likely cases |

If the answers are random guesses, then the 10 possible outcomes are equally likely. The correct answers are *true* and *e*, so that

$$P(\text{both correct}) = P(\text{true and } e) = \frac{1}{10}$$

Considering the component answers of *true* and *e*, respectively, we see that with random guesses we have $P(\text{true}) = 1/2$ while $P(e) = 1/5$.

Recognizing that 1/10 is the product of 1/2 and 1/5, we observe that $P(\text{true and } e) = P(T) \cdot P(e)$ and we use this observation as a basis for formulating the following rule.

RULE
Multiplication rule (preliminary) $$P(A \text{ and } B) = P(A) \cdot P(B)$$

EXAMPLE
Three floppy disks are produced and one of them is defective. Two disks are randomly selected for testing, but the first is replaced before the second selection is made. Find the probability that both disks are good.
Solution
Letting G represent the event of selecting a good disk, we want $P(G \text{ and } G)$. With $P(G) = 2/3$, we apply the multiplication rule to get $$P(G \text{ and } G) = P(G) \cdot P(G) = \frac{2}{3} \cdot \frac{2}{3} = \frac{4}{9}$$

Conditional Probabilities

Although the preceding solution is mathematically correct, common sense suggests that product testing should be conducted without replacement of the items already tested. There is always the chance that you could test the same item twice. Also, we cannot be sure of the reliability of any generalization based on a specific case, such as our preliminary multiplication rule. We would be wise to test that rule in a variety of cases to see if any errors arise. We will see that the preliminary rule is sometimes inadequate, as in the improved testing procedure in the next example.

EXAMPLE

Let's again assume that we have three floppy disks of which one is defective. We will again randomly select two disks, but we will *not* replace the first selection. We will find the probability of getting two good disks with this improved testing procedure.

Solution

To understand the sample space better, we will represent the defective disk by D and the two good disks by G_1 and G_2. Assuming that the first selection is not replaced, we now illustrate the different possible outcomes in Figure 3-7, which is an example of a **tree diagram**. By examining this list of six equally likely outcomes, we see that only two cases correspond to two good disks so that $P(G \text{ and } G) = 2/6$ or $1/3$. In contrast, our preliminary multiplication rule was used in the preceding example to produce a result of $4/9$. Here, the correct result is $1/3$, so the preliminary multiplication rule does not fit this case.

In this example, the preliminary multiplication rule does not take into account the fact that the first selection is not replaced. $P(G)$ again begins with a value of $2/3$, but after getting a good disk on the first selection, there would be one good disk and one defective disk remaining so that $P(G)$ becomes $1/2$ on the second selection.

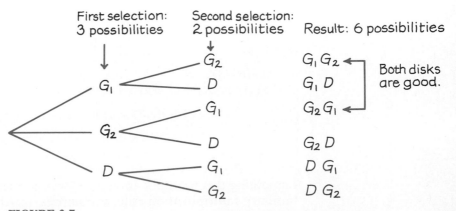

FIGURE 3-7

Here is the key concept of this last example: **Without replacement of the first selection, the second probability is affected by the first result.** There are many other cases where a probability is affected by another event, or even by additional knowledge you may acquire. The probability of getting to class on time may be affected by the probability of your car starting. An estimated probability of a football team winning the Super Bowl may be affected by news of an injured quarterback. Since this dependence of an event on some other event is so important, we provide a special definition.

DEFINITION

Two events A and B are **independent** if the occurrence of one does not affect the probability of the occurrence of the other. (Several events are similarly independent if the occurrence of any does not affect the probabilities of the occurrence of the others.) If A and B are not independent, they are said to be **dependent**.

The preliminary multiplication rule holds if the events A and B are independent; in that case $P(B)$ is not affected by the occurrence of A. But in our last example, where the first disk was not replaced before the second selection, $P(G)$ was affected by the result of the first selection. The correct result of 1/3 could be obtained by examining the listed sample space (Figure 3-7) or by calculating $\frac{2}{3} \cdot \frac{1}{2} = \frac{1}{3}$, where the probability of 1/2 was affected by the first selection. The probability of 1/2 is obtained by noting that after selecting a good disk and removing it, we now have one good disk and one bad disk, so that $P(G) = 1/2$. We have found the probability that the second disk is good given that the first disk was good. We generalize this observation as follows.

RULE

Let $P(B|A)$ represent the probability of B occurring after assuming that A has already occurred. (We can read $B|A$ as "B given A.")

Multiplication rule for dependent events

$$P(A \text{ and } B) = P(A) \cdot P(B|A)$$

Combining this multiplication rule for dependent events with the preliminary multiplication rule, we summarize the key concept of this section as follows.

Multiplication Rule Presents Difficulty to FAA

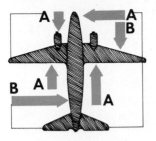

In a publication discussing the analysis of component failures in aircraft, the Federal Aviation Administration (FAA) refers specifically to the multiplication rule as it relates to "system independence and redundancy." The FAA states that "the most often encountered difficulty with quantitative analyses presented to the FAA has been the improper treatment of events which are not mutually independent. The probability of occurrence of two events which are mutually independent may be multiplied to obtain the probability that both events occur using the formula: $P(A \text{ and } B) = P(A) \cdot P(B)$. This multiplication will produce an incorrect solution if A and B are not mutually independent."

MULTIPLICATION RULE

$P(A \text{ and } B) = P(A) \cdot P(B)$ if A and B are **independent**.

$P(A \text{ and } B) = P(A) \cdot P(B|A)$ if A and B are **dependent**.

In the last expression, we can easily solve for $P(B|A)$, and the result suggests the following definition.

DEFINITION

The **conditional probability** of B given A is

$$P(B|A) = \frac{P(A \text{ and } B)}{P(A)}$$

If $P(B|A) = P(B)$, then the occurrence of event A had no effect on the probability of event B. This is often used as a test for independence. Another equivalent test for independence involves checking for equality of $P(A \text{ and } B)$ and $P(A) \cdot P(B)$. If they're equal, events A and B are independent. If $P(A \text{ and } B) \neq P(A) \cdot P(B)$, then A and B are dependent events.

In some cases, the distinction between dependent and independent events is quite obvious, while in other cases it is not. Getting heads on a coin and 6 on a die clearly involves independent events. But

Redundancy

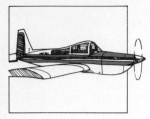

Reliability is often designed into systems through redundancy of critical components. For example, the typical single-engine light aircraft has only one engine, but each cylinder can be fired by either of two separate and independent spark plugs. There are two independent electrical systems so that a failure of one will not lead to engine failure. Aircraft used for instrument flight typically have two separate radios. Such redundancy is an application of the multiplication rule in probability theory. If the probability of failure of one component, such as a spark plug, is 0.001, the probability of failure of dual components, such as both plugs firing the same cylinder, is $0.001 \times 0.001 = 0.000001$. In this case, there is clearly safety in numbers.

consider the data in Table 3-4, which summarizes 60 responses to a survey question. Are the events of being a male and answering yes independent? For this group,

$$P(M \text{ and yes}) \neq P(M) \cdot P(\text{yes})$$

24 respondents are males who answered yes

$$\frac{24}{60} \quad \neq \quad \frac{40}{60} \quad \cdot \quad \frac{38}{60} \quad \begin{array}{l} \text{40 males among 60 respondents} \\ \text{38 yes responses out of 60} \end{array}$$

Since $P(M \text{ and yes}) \neq P(M) \cdot P(\text{yes})$, we conclude that the events of being a male and answering yes are dependent for selections from this group of 60 people. In general, if we can verify or refute the equation $P(A \text{ and } B) = P(A) \cdot P(B)$, then we can establish or disprove the independence of events A and B.

TABLE 3-4			
		Male	Female
Do you favor a no-smoking rule on airliners?	Yes	24	14
	No	16	6

EXAMPLE

Using the data in Table 3-4, find the probability of selecting someone from the group who answered yes, if we already know that the person selected is a male.

Solution

We want to find the probability of getting a yes answer, given that a male (M) was selected. That is, we want $P(\text{yes}|M)$.

First approach: Use Table 3-4 directly. If we know a male was selected, we know that we have 40 people, and 24 of them answered yes, so that

$$P(\text{yes}|M) = \frac{24}{40} = 0.600$$

Second approach:

$$P(\text{yes}|M) = \frac{P(\text{yes and } M)}{P(M)}$$

$$= \frac{24/60}{40/60} = \frac{24}{40} = 0.600$$

Given the condition that a male was selected, the conditional probability $P(\text{yes}|M) = 0.600$.

So far we have discussed two events, but the multiplication rule can be easily extended to several events. In general, **the probability of any sequence of independent events is simply the product of their corresponding probabilities**. The next two examples illustrate this extension of the multiplication rule.

EXAMPLE

A couple plans to have four children. Find the probability that all four children are girls. Assume that boys and girls are equally likely and that the sex of each child is independent of the sex of any other children.

continued ▶

Example, continued

Solution

The probability of getting a girl is $1/2$ and the four births are independent of each other, so

$$P(\text{four girls}) = \frac{1}{2} \cdot \frac{1}{2} \cdot \frac{1}{2} \cdot \frac{1}{2} = \frac{1}{16}$$

EXAMPLE

A pill designed to prevent certain physiological reactions is advertized as 95% effective. Find the probability that the pill will work in all of 15 separate and independent applications.

Solution

Since the 15 applications are independent, we get

$$P(15 \text{ successes}) = 0.95 \cdot 0.95 \cdot 0.95 \cdots 0.95 \ (15 \text{ factors})$$
$$= 0.95^{15}$$
$$= 0.463$$

The preceding two examples illustrate an extension of the multiplication rule for independent events; the next example illustrates a similar extension of the multiplication rule for dependent events.

In this section, we have seen that the independence of events must be considered in the determination of probabilities like $P(A \text{ and } B)$, since the component probabilities may be affected. Replacement of randomly selected items usually involves independent events, while failure to replace selected items usually causes the events to be dependent.

EXAMPLE

A Congressional committee of four men and six women is appointed. If a reporter randomly selects three different committee members for interviews, find the probability that they are all women.

continued ▶

Nuclear Power Plant Has an Unplanned "Event"

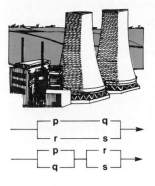

Scientists at MIT once estimated the likelihood of a serious nuclear energy accident to be about one in a million, but an "event" at Three Mile Island and a major accident at Chernobyl have changed that. We now know that the backup and fail-safe systems were not as infallible as proponents of nuclear energy had claimed. The reassessment of the likelihood of a nuclear catastrophe is certain to have a strong impact on future energy policies. Scientists will undoubtedly conduct a careful review of the design of nuclear power plants, including the use of redundant, or backup, components. Compare the two systems at left and determine which is likely to be more reliable. Assume that p, q, r, and s are valves and that each system supplies water necessary for cooling purposes.

Example, continued

Solution

Letting A, B, and C represent the respective events of randomly selecting a woman on the first, second, and third attempts, we see that those events are dependent since successive probabilities are affected by previous results. The probability of getting two different women on the first two selections can be found through direct application of the multiplication rule for dependent events.

$$P(A \text{ and } B) = P(A) \cdot P(B|A)$$

$$= \frac{6}{10} \cdot \frac{5}{9} = \frac{30}{90}$$

If the probability of the first woman is 6/10 and the probability of the second woman is 5/9, then the probability of the third woman must be 4/8. This seems reasonable since, after the first two women are selected, there would be eight remaining members, of which four are women. Therefore

$$P(A \text{ and } B \text{ and } C) = P(A) \cdot P(B|A) \cdot P(C|A \text{ and } B)$$

$$= \frac{6}{10} \cdot \frac{5}{9} \cdot \frac{4}{8}$$

$$= \frac{120}{720} = \frac{1}{6}$$

EXAMPLE

In Section 3-1 we described a pollster who supposedly surveyed 12 randomly selected voters and got 12 Republicans. The population consisted of 200,000 voters, of which 30% (or 60,000) are Republicans. The pollster argued that this could easily happen, but can it?

Solution

We seek the probability of getting 12 Republicans in 12 selections. The events of getting the 12 Republicans are dependent since the sampling was done without replacement. We get

$$P(12 \text{ Republicans}) = \frac{60,000}{200,000} \times \frac{59,999}{199,999} \times \cdots \times \frac{59,989}{199,989}$$

$$= 0.000000531$$

When small samples are drawn from large populations, the common practice is to treat the events as being independent so that the calculations are simplified. Taking advantage of the fact that each component fraction in the above expression is approximately 0.30, we get

$$P(12 \text{ Republicans}) = (0.30) \times (0.30) \times \cdots \times (0.30) \qquad (12 \text{ times})$$

$$= 0.30^{12} = 0.000000531$$

In any event, we see that the probability of getting 12 Republicans with random selections is incredibly small. The selection process used by the pollster can now be criticized with sound justification.

In this last example we noted the common practice of treating events as independent when small samples are drawn from large populations. A common guideline is to assume independence whenever the size of the sample is at most 5% of the size of the population. If a sample of 12 voters is selected from a population of 60,000, we can assume independence since 5% of 60,000 is 3000 and the sample size of 12 is clearly less than that. If the sample size had been 3001, then we should not assume independence.

When using the multiplication rule for finding probabilities in compound events, tree diagrams are sometimes helpful in determining the number of possible outcomes. We used one earlier in Figure 3-7. In a tree diagram we depict schematically the possible outcomes of an experiment as line segments emanating from one starting point. In

Bayes' Theorem

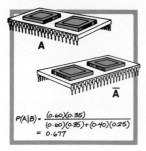

$$P(A|B) = \frac{(0.60)(0.35)}{(0.60)(0.35)+(0.40)(0.25)}$$
$$= 0.677$$

Thomas Bayes (1702–1761) suggested that probabilities should be revised as we acquire additional knowledge about an event. He formulated what is now known as Bayes' theorem. One case of the theorem leads to the following expression. (A more general expression can be found in many advanced texts.)

$$P(A|B) = \frac{P(A) \cdot P(B|A)}{P(A) \cdot P(B|A) + P(\bar{A}) \cdot P(B|\bar{A})}$$

For example, suppose that 60% of a company's computer chips are manufactured in one factory (denoted by A), while 40% are produced in its other factory (denoted by $\bar{A}$). Given a randomly selected chip, the probability it came from the first factory is

$P(A) = 0.60$. Suppose we now learn that the randomly selected chip is defective and the defect rates for the two factories are 35% (for A) and 25% (for $\bar{A}$) respectively. Given the defective chip, we can now find the probability that it came from factory A as follows. (We denote a defective chip by B for "bad.")

$$P(A|B) = \frac{(0.60)(0.35)}{(0.60)(0.35) + (0.40)(0.25)}$$
$$= 0.677$$

Included in its list of varied applications, Bayes' theorem is used in paternity suits to calculate the probability that a defendant is the father of a child.

Figure 3-8 we show the tree diagram that summarizes the possibilities for parents who plan to have three children. Note that the tree diagram in Figure 3-8 presents the eight different possible outcomes as the eight different possible paths begin at the left and end at the right. Assuming that boys and girls are equally likely, we see that the eight paths are equally likely, so each of the eight possible outcomes has a probability of 1/8. Such diagrams are helpful in counting the number of possible outcomes if the number of possibilities is not too large. In cases involving large numbers of choices, the use of tree diagrams is impractical. However, they are useful as visual aids to provide insight into the multiplication rule. Returning to the situation discussed earlier in this section, if we have the true-false question and the multiple-choice question, we can use a tree diagram to summarize the outcomes. See Figure 3-9.

Assuming that both answers are random guesses, all 10 branches are equally likely and the probability of getting the correct pair (T, e) is 1/10. For each response to the first question there are 5 responses for the second. The total number of outcomes is therefore 5 taken 2 times, or 10. The tree diagram in Figure 3-9 illustrates the reason for the use of multiplication.

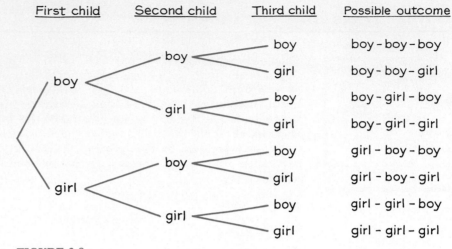

FIGURE 3-8

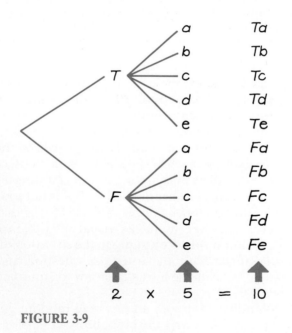

FIGURE 3-9

3-4 EXERCISES A

3-69 For each of the following pairs of events, classify the two events as independent or dependent.
   a. Randomly selecting a Massachusetts voter who is a Democrat
      Randomly selecting a California voter who is a Democrat
   b. Making a correct guess on the first question of a multiple choice quiz
      Making a correct guess on the second question of the same multiple choice quiz
   c. Randomly selecting a defective component from a bin of 15 good and 5 defective components
      Randomly selecting a second component that is defective (assume that the same bin is used and the first selection was not replaced)
   d. Events A and B, where $P(A) = 0.40$, $P(B) = 0.60$, and $P(A \text{ and } B) = 0.20$
   e. Events A and B, where $P(A) = 0.2$, $P(B) = 0.3$, and $P(A \text{ and } B) = 0.06$

3-70 For each of the following pairs of events, classify the two events as independent or dependent.
   a. Finding your kitchen light inoperable
      Finding your battery-operated flashlight inoperable
   b. Finding your kitchen light inoperable
      Find your microwave oven inoperable
   c. Getting an unfavorable reaction when a certain drug is administered to a mouse
      Getting an unfavorable reaction when the same drug is administered to another mouse
   d. Events A and B, where $P(A) = 0.90$, $P(B) = 0.80$, and $P(A \text{ and } B) = 0.72$
   e. Events A and B, where $P(A) = 0.36$, $P(B) = 0.50$, and $P(A \text{ and } B) = 0.15$

3-71 According to the U.S. Census Bureau, 62% of Americans over the age of 18 are married. Find the probability of getting two married people (not necessarily married to each other) when two different Americans over the age of 18 are randomly selected.

3-72 The first two answers to a true-false quiz are both true. If random guesses are made for those two questions, find the probability that both answers are correct.

3-73 According to *Popular Science*, 57% of all chain saw injuries involve arms or hands. If four different chain saw injury cases are randomly selected, find the probability that they all involve arms or hands.

3-74   On a TV program (ABC's *Nightline*) on smoking, it was reported that there is a 60% success rate for those trying to stop through hypnosis. Find the probability that for eight randomly selected smokers who undergo hypnosis, they all successfully stop smoking.

3-75   A Roper poll showed that 18% of adults regularly engage in swimming. If three adults are randomly selected, find the probability that they all regularly engage in swimming.

3-76   According to a *U.S. News and World Report* article, an analyst for Paine Webber estimated that 35% of the computer chips manufactured by Intel are acceptable. Find the probability of getting all good chips in a batch of six.

3-77   A batch of computer chips includes two that are defective and four that are acceptable. If two of these chips are randomly selected without replacement, find the probability that they are both defective.

3-78   Six defective fuses are present in a bin of 80 fuses. The entire bin is approved for shipping if no defects show up when three randomly selected fuses are tested.
a. Find the probability of approval if the selected fuses are replaced.
b. Find the probability of approval if the selected fuses are not replaced.
c. Comparing the results of parts (a) and (b), which procedure is more likely to reveal a defective fuse? Which procedure do you think is better?

3-79   One couple attracted media attention when their three children, who were born in different years, were all born on July 4. Find the probability that three randomly selected people were all born on July 4.

3-80   If two people are randomly selected, find the probability that the second person has the same birthday as the first.

3-81   Four firms using the same auditor independently and randomly select a month in which to conduct their annual audits. What is the probability that all four months are different?

3-82   An experiment begins with four female mice and two male mice in the same cage. One mouse is randomly selected each day and put in a separate cage. Find the probability that the first three removals are all females.

3-83   A computer simulation involves the random generation of digits from 1 through 5. If two digits are generated, find the probability that they are both odd in each case.
a. The first selected digit cannot be selected again.
b. The first selected digit is available for the second selection.

3-84   A container holds 12 eggs, of which 5 are fertile.

    a. Find the probability of randomly selecting 3 eggs that are all fertile if each egg is replaced before the next selection is made.

    b. Find the probability of randomly selecting 3 eggs that are all fertile if the eggs selected are not replaced.

3-85 Special thermometers are used in a scientific experiment and there is a 50% chance that any such thermometer is in error by more than 2°. If 10 thermometers are used, find the probability that they are all in error by more than 2°.

3-86 A circuit requiring a 500-ohm resistance is designed with five 100-ohm resistors arranged in series. The proper resistance is achieved if all five resistors function correctly. There is a 0.992 probability that any individual resistor will not fail. What is the probability that all five of the resistors will work correctly to provide the 500-ohm resistance?

3-87 If a death is selected at random, there is a 0.0478 probability that it was caused by an accident, according to data from *Statistical Abstract of the United States*. Find the probability that five randomly selected deaths were all accidental.

3-88 Using U.S. Census Bureau data, we know that 60% of those who are eligible to vote actually do vote. If a pollster surveys 10 people eligible to vote, what is the probability that they all vote?

3-89 Of 2 million components produced by a manufacturer in one year, 5,000 are defective. If two of these components are randomly selected and tested, find the probability that they are both good in each of the following cases.

    a. The first selected component is replaced.

    b. The first selected component is not replaced.

3-90 An approved jury list contains 20 women and 20 men. Find the probability of randomly selecting 12 of these people and getting an all-male jury.

3-91 A manager can identify employee theft by checking samples of shipments. Among 36 employees, two are stealing. If the manager checks on four different randomly selected employees, find the probability that neither of the thieves will be identified.

3-92 A delivery company has 12 trucks, and when 3 are inspected, it is found that all 3 have faulty brakes. The owner claims that the other trucks all have good brakes and it was just chance that led to selecting the trucks that happened to have faulty brakes. Find the probability of that event, assuming that the manager's claim is correct.

3-93 According to *Glamour* magazine, in general your chances of having a supernatural experience are 1 in 17. But if you live in California, your chances of having a supernatural experience are 1 in 11. *(continued)*

a. If two different Americans are randomly selected, find the probability that they both have had a supernatural experience.

b. If two Californians are randomly selected, find the probability that they both have had a supernatural experience.

3-94   The probability of a hang glider participant dying in a given year is 0.008 (based on the book *Acceptable Risks* by Imperato and Mitchell). If 10 hang glider participants are randomly selected, find the probability that they all survive a year.

3-95   In a Riverhead, N.Y., case, nine different rape victims listened to voice recordings of five different men. All nine women identified the same voice as that of the rapist. If the voice identifications had been made by random guesses, find the probability that all nine women would select the same person.

3-96   An employer must hire 4 people from a pool of 62 men and 73 women. If the selections are made randomly, what is the probability that all four are of the same sex?

3-4 EXERCISES B

3-97   Use a calculator to compute the probability that of 25 people, no two share the same birthday.

3-98   Use a calculator to compute the probability that of 50 people, no two share the same birthday.

3-99   A poll was taken on the campus of a large university in order to determine student attitudes about a variety of issues. Fifty students (40 males and 10 females) were polled on their involvement in campus activities, and the results showed 20 responses of yes and 30 responses of no. This data is summarized in the following table.

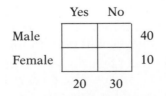

a. If one of the 50 students is randomly selected, find the probability of getting a male.

b. If one of the 50 students is randomly selected, find the probability of getting a female.

c. If one of the 50 students is randomly selected, find the probability of getting a student who answered yes.

    d. If one of the 50 students is randomly selected, find the probability of getting a student who answered no.

    e. Assuming that the sex of the respondent has no effect on the response, find the probability of randomly selecting one of the 50 polled students and getting a male who answered yes.

    f. Using the probability from part (e) and the fact that there are 50 respondents, what would you expect to be the number of males who answered yes?

    g. If the number from part (f) is entered in the appropriate box in the chart, are the numbers in the other boxes then determined? If so, find them.

3-100 Do Exercise 3-99 after changing the data to correspond to the table given below.

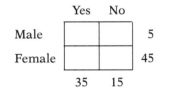

3-5 COMPLEMENTS AND ODDS

Complementary Events

In this section we begin by defining complementary events and present one last rule of probabilities that relates to these events. We then examine the relationship between probabilities and odds.

DEFINITION
The **complement** of event A, denoted by $\bar{A}$, consists of all outcomes not in A.

    The complement of event A is the event A does *not* occur. If A represents an outcome of answering a test question correctly, then the complementary event $\bar{A}$ represents a wrong answer. Sometimes the determination of the complement of an event is more difficult. For

You Bet?

Only 40% of the money bet in the New York State lottery is returned in prizes, so the "house" has a 60% advantage. The house advantage at race-tracks is usually about 15%. In casinos, the house advantage is 5.26% for roulette, 5.9% for blackjack, 1.4% for craps, and 3% to 22% for slot machines. Some professional gamblers can systematically beat the house at blackjack if they use sophisti-cated card-counting techniques. The basic idea is to keep a record of cards that have been played so that the player will know when the deck has a dispro-portionately large number of high cards. When this occurs, the casino is more likely to lose and the player begins to place large bets. Many casinos eject card counters or shuffle the decks more frequently.

example, if A is the event of getting at least one boy in seven births, the complement of A is the event of getting all girls. (It's *not* the event of getting at least one girl.)

The definition of complementary events implies that they must be mutually exclusive, since it is impossible for an event and its opposite to occur at the same time. Also, we can be absolutely certain that either A does or does not occur. That is, either A or $\bar{A}$ must occur. These observations enable us to apply the addition rule for mutually exclusive events as follows:

$$P(A \text{ or } \bar{A}) = P(A) + P(\bar{A}) = 1$$

We justify $P(A \text{ or } \bar{A}) = P(A) + P(\bar{A})$ by noting that A and $\bar{A}$ are mutually exclusive, and we justify the total of 1 by our absolute cer-tainty that A either does or does not occur. This result of the addition rule leads to the following three *equivalent* forms.

RULE OF COMPLEMENTARY EVENTS
$P(A) + P(\bar{A}) = 1$
$P(\bar{A}) = 1 - P(A)$
$P(A) = 1 - P(\bar{A})$

Total area = 1

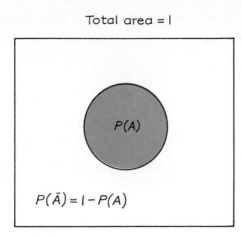

FIGURE 3-10

The first form comes directly from our original result, while the second (see Figure 3-10) and third variations involve very simple equation manipulations. A major advantage of the rule of complementary events is that it can sometimes be used to significantly reduce the workload required to solve certain problems. As an example, let's consider a very ambitious couple planning to have seven children. We want to determine the probability of getting at least one girl among those seven children. The direct solution to this problem is messy, but a simple indirect approach is made possible by our rule of complementary events. Let's denote the event of getting at least one girl in seven by G. We will begin by finding $P(\bar{G})$, the probability of *not* getting at least one girl (which is equivalent to getting seven boys). Now $P(\bar{G})$ is relatively easy to compute if we make two reasonable assumptions:

1. $P(\text{boy}) = 1/2$.
2. The sexes of successive babies are independent of those of any younger children.

(Neither of these two assumptions is exactly correct, but they can be used with extremely good results.)

Applying the multiplication rule for independent events, we get

$$P(\bar{G}) = P(7 \text{ consecutive boys}) = \frac{1}{2} \cdot \frac{1}{2} \cdot \frac{1}{2} \cdot \frac{1}{2} \cdot \frac{1}{2} \cdot \frac{1}{2} \cdot \frac{1}{2} = \frac{1}{128}$$

However, we are seeking $P(G)$ so that the rule of complementary events can be applied.

$$P(G) = 1 - P(\bar{G})$$

$$= 1 - \frac{1}{128}$$

$$= \frac{127}{128}$$

As complex as this solution may appear, it is trivial in comparison to the alternate solutions that involve a direct approach.

EXAMPLE

A conference is attended by 40 doctors and 10 psychologists. If three different participants are randomly selected for a panel discussion, find the probability that at least one is a psychologist.

Solution

Let A represent the event of selecting at least one psychologist when three participants are randomly selected. Then $\bar{A}$ becomes the event of not getting at least one psychologist. That is, $\bar{A}$ signifies that no psychologists are selected, so all three participants are doctors. We use the multiplication rule to get

$$P(\bar{A}) = P(\text{all 3 are doctors})$$

$$= \frac{40}{50} \cdot \frac{39}{49} \cdot \frac{38}{48}$$

$$= 0.504$$

We now use the rule of complementary events to find $P(A)$.

$$P(A) = 1 - P(\bar{A})$$

$$= 1 - 0.504$$

$$= 0.496$$

The complement of "at least one psychologist" is "no psychologists."

Odds

Having discussed complementary events, we should now be familiar with the notation $P(\bar{A})$. We will use that notation as we proceed to consider odds and their relationship to probabilities.

Long-Range Weather Forecasting

A high school teacher in Millbrook, New York, created a controversy by conducting a classroom experiment in which long-range weather forecasts were made by throwing darts. The dart method resulted in 7 correct predictions among 20 winter storms, while a nearby private forecasting service correctly predicted only 6 of the storms. The teacher claimed that while short-range weather forecasts are usually quite accurate, long-range forecasts are not.

Many expressions of likelihood use odds instead of probability values between 0 and 1. The use of odds makes it easier to deal with the money exchanges that occur with betting at racetracks and casinos. However, odds are extremely awkward in mathematical and scientific calculations. We will now examine procedures for converting between probabilities and odds. We begin by defining the odds against an event and the odds in favor of an event.

DEFINITION

The **odds against event A** occurring are the ratio $P(\bar{A})/P(A)$, usually expressed in the form of $a{:}b$ where a and b are integers having no common factors.

The **odds in favor of event A** occurring are the ratio $P(A)/P(\bar{A})$, usually expressed as the ratio of two integers having no common factors.

From these two definitions we see that the odds against an event are the reciprocal of the odds in favor. If the odds against an event are 3:2, the odds in favor are 2:3.

EXAMPLE

If $P(A) = 2/5$, find (a) the odds against A; (b) the odds in favor of event A.

Solution

a. Using the definition, we find the odds against event A as follows.

$$\text{odds against } A = \frac{P(\bar{A})}{P(A)} = \frac{3/5}{2/5} = \frac{3}{2}$$

which we express as 3:2.

b. The odds in favor of event A are found as follows.

$$\text{odds in favor of event } A = \frac{P(A)}{P(\bar{A})} = \frac{2/5}{3/5} = \frac{2}{3}$$

expressed as 2:3.

If odds are given and are not specified as being in favor or against, they are probably odds against the event occurring. If we learn only that a horse has odds of 50:1, we can safely assume that those odds are *against* the horse winning. Don't bet the ranch on such a horse.

The preceding two definitions allow us to convert probabilities into odds. To convert from odds to probabilities, we convert the given ratio into two fractions whose sum is 1. Given the odds *a:b* against event *A*, we know that

$$\frac{P(\bar{A})}{P(A)} = \frac{a}{b} = \frac{\dfrac{a}{a+b}}{\dfrac{b}{a+b}} \quad \left.\begin{array}{l} \leftarrow \text{divide numerator} \\ \text{and denominator} \\ \leftarrow \text{by } a+b \end{array}\right]$$

The fractions $a/(a+b)$ and $b/(a+b)$ have a sum of 1 and they are in the ratio of *a:b*, so that $P(\bar{A}) = a/(a+b)$ and $P(A) = b/(a+b)$. We summarize the procedures in the following table.

Converting Between Odds and Probabilities	
Odds → Probabilities	Probabilities → Odds
Given: Odds *against* event *A* are *a:b* (or odds *in favor* of event *A* are *b:a*) Then $P(\bar{A}) = \dfrac{a}{a+b}$ $P(A) = \dfrac{b}{a+b}$	Given: $P(A)$ or $P(\bar{A})$ (express as the ratio of two integers having no common factors) Then odds *against* event $A = \dfrac{P(\bar{A})}{P(A)}$, odds *in favor* of event $A = \dfrac{P(A)}{P(\bar{A})}$

EXAMPLE

The odds against a woman having twins are 92:1 (from the *World Book of Odds* by Neft, Cohen, and Deutsch). Find the probability of a woman having twins.

continued ▶

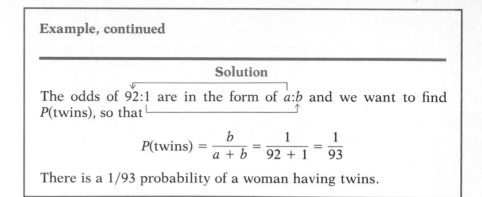

Example, continued

Solution

The odds of 92:1 are in the form of a:b and we want to find P(twins), so that

$$P(\text{twins}) = \frac{b}{a + b} = \frac{1}{92 + 1} = \frac{1}{93}$$

There is a 1/93 probability of a woman having twins.

We have stated that the use of odds makes it easier to deal with the money exchanges that typically occur with bets. We will now show how this is done. For bets, the odds against an event represent the ratio of the net profit to the amount bet.

Odds against event A = (net profit):(amount bet)

EXAMPLE

A gambler bets $2 on a horse with odds of 10:1. How much does the gambler win if the horse wins the race?

Solution

With 10:1 odds against winning, we get

$$10:1 = (\text{net profit}):(\text{amount bet})$$

This shows that for each dollar bet, there will be a net profit of $10. Since $2 is bet, the winning horse will provide a $20 net profit. The gambler would collect a total of $22, which includes the original $2 bet and the net profit of $20.

The preceding example illustrates a major advantage of odds. With odds, it is easy to see the relationship between the bet and the winnings. But suppose a circuit has 10:1 odds against a failure. What are the odds against two such separate and independent circuits both failing? The best way to solve this problem is to first convert the 10:1 odds to the corresponding probability of failure for a circuit (1/11). We now use the multiplication rule for independent events to get

$$\frac{1}{11} \times \frac{1}{11} = \frac{1}{121}$$

as the probability of both circuits failing. The probability of 1/121 can now be converted to odds and the result is 120:1. There is no easy corresponding multiplication rule for odds that allows us to multiply 10:1 by 10:1 to get a product of 120:1. The moral is this: when performing calculations involving likelihoods of events, use probability values (between 0 and 1), not odds.

3-5 EXERCISES A

In Exercises 3-101 through 3-104, use words to describe the complement of the given event.

3-101   A manufactured system is defective.

3-102   Among five mice used in an experiment, none survived.

3-103   Among five true-false questions, at least one answer is "false."

3-104   Among three components, there is at least one failure.

In Exercises 3-105 through 3-112, determine the probability of the given event and the probability of the complementary event.

3-105   A defective disk drive is randomly selected from 50 available disk drives, of which five are defective. (Only one selection is made.)

3-106   A letter is randomly selected from the alphabet and the result is a vowel (*a, e, i, o, u*).

3-107   When a multiple choice test question is answered by guessing, the response is correct. (Assume five possible answers, of which one is correct.)

3-108   When a computer generates a random number from 1 through 5 (integers only), the result is odd.

3-109   A baby is born and that baby is a girl.

3-110   Two babies are born and they are both girls.

3-111   In a class of 10 men and 15 women, one student is randomly called and that student is a woman.

3-112   In a class of 10 men and 15 women, two different students are called and they are both women.

In Exercises 3-113 through 3-116, find (a) the odds against event A *and (b) the odds in favor of event* A. *In each case, the probability of event* A *is given.*

3-113   2/7   3-114   3/8   3-115   0.4   3-116   0.8

In Exercises 3-117 through 3-120, find the probability of event A *given the described odds.*

3-117   The odds against event A are 8:1.

3-118   The odds against event A are 9:4.

3-119   The odds in favor of event A are 3:7.

3-120   The odds in favor of event A are 9:8.

3-121   According to data from the U.S. Department of Labor, 10% of United States employees are aged 55 years or over. If two employees are randomly selected, find the probability that at least one of them is 55 or over.

3-122   In a recent national election, 52% of the voters were women, according to a *New York Times*/CBS News poll. For a random selection of four voters, find the probability of getting at least one woman.

3-123   In a survey of college freshmen by the Cooperative Institutional Research Institute, 76% stated that getting a better job was very important in their decision to attend college. If three different freshmen are randomly selected, find the probability that at least one indicated a better job as being very important in the decision to go to college.

3-124   A true-false test of four questions is given and an unprepared student must make random guesses in answering each question. What is the probability of at least one correct response?

3-125   A certain method of contraception is found to be 95% effective. What is the probability of at least one pregnancy on the part of 10 different women using this method of contraception?

3-126   An employee needs to call any one of five colleagues. Assume that the five colleagues are random selections from a population in which 28% have unlisted numbers. Find the probability that at least one of the five fellow workers will have a listed number.

3-127   A circuit is designed so that a critical function is properly performed if at least one of four identical and independent components does not fail. The probability of failure for any one of these four components is 0.081. Find the probability that the critical function will be properly performed.

3-128   Three programmers are given a problem, which they work on independently. The probabilities of completing a working program by the end of the day are 1/5, 2/5, and 1/4, respectively. Find the probability that a working program will be available by the end of the day.

3-129   In a 10-horse race, if all horses had the same chance of winning, what would be fair odds with no advantage to the track?

3-130   Find the odds against correctly guessing the day of the week on which July 4, 1776, fell.

3-131   Find the odds against correctly guessing the first number in a security code that consists of a sequence of three numbers, each between 1 and 50 inclusive.

3-132   Find the odds against correctly guessing someone's birthday if you know only that they were born in 1973.

3-133   If you bet that when two dice are rolled, the outcome is "snake eyes" (1 on each of two dice), a casino will give you odds of 30:1. What would be the fair odds if the casino did not have an advantage?

3-134   If you bet that when two dice are rolled, the total is 11, a casino will give you odds of 15:1. What would be the fair odds if the casino did not have an advantage?

3-135   For a casino's "Big Six Wheel," if you bet $1 you are given odds of 1:1. There are 54 different slots on the wheel and 23 of them are winners for the $1 bet. What would be fair odds if the casino did not have an advantage?

3-136   The American roulette wheel has 38 different slots numbered 0, 00, and 1 through 36. If you bet on any individual number, the casino gives you odds of 35:1. What would be fair odds if the casino did not have an advantage?

3-137   Under certain conditions, there is a 0.015 probability that a randomly selected driver will fail a breathalyzer test for sobriety. If the police test 100 randomly selected drivers, what is the probability of finding at least one driver who fails the sobriety test?

3-138   It is found that for medical students given selected case studies, there is a 0.850 probability of a correct diagnosis. For three students each given one case, what is the probability of at least one incorrect diagnosis?

3-139   A $2 bet results in a net profit of $18. Find the odds against winning and find the probability of winning.

3-140   A $5 bet results in a net profit of $11. Find the odds against winning and find the probability of winning.

3-141   A nuclear weapon will not be misfired if at least one of five separate and independent fail-safe mechanisms functions properly. The estimated likelihoods of failure for these fail-safe devices are 0.105, 0.200, 0.001, 0.115, and 0.340, respectively. What is the probability that a misfire will not occur?

3-142   A critical component in a circuit will work properly only if three other components all work properly. The probabilities of a failure for the three other components are 0.010, 0.005, and 0.012. Find the probability that at least one of these three components will fail.

3-143 Find the odds against at least one correct answer when five true-false questions are all answered with random guesses.

3-144 Find the odds against at least one correct answer when three multiple-choice questions are answered with random guesses. Assume that each question has five possible answers of which one is correct.

3-5 EXERCISES B

3-145 a. If $P(A) = 1$ and A and B are complementary events, what is known about event B?
b. If $P(A) = 0$ and A and B are complementary events, what is known about event B?
c. If $P(A)$ is at least 0.7 and A and B are complementary events, what is known about event B?

3-146 a. Develop a formula for the probability of not getting either A or B on a single trial. That is, find an expression for $P(\overline{A \text{ or } B})$.
b. Develop a formula for the probability of not getting A or not getting B on a single trial. That is, find an expression for $P(\bar{A} \text{ or } \bar{B})$.
c. Compare the results from parts (a) and (b). Are they the same or are they different?

3-147 The odds against an event are 7:3. What are the odds against this event occurring in all of three separate and independent trials?

3-148 The odds on the 11 horses in a certain race are given as 8:1, 40:1, 99:1, 95:1, 40:1, 6:1, 8:1, 9:1, 72:1, 25:1, and 9:1. Convert these odds to probabilities and find the total. What do you conclude about the total?

3-149 All pairs of complementary events must be mutually exclusive pairs. Must all mutually exclusive pairs of events also be complementary? Support your response with specific examples.

3-150 Find the probability that, of 25 randomly selected people, at least 2 share the same birthday.

3-151 Find the probability that, of 50 randomly selected people, at least 2 share the same birthday.

3-152 A company has been manufacturing resistors at a cost of 50¢ each, and there is a probability of 0.900 that any such resistor will work. An employee suggests that another process will generate resistors in groups of four at a cost of 10¢ for each resistor. There is a 50% failure rate for the resistors produced by this second method. Is it better to produce four resistors at 10¢ each, or the one resistor at a cost of 50¢? Explain.

The Number Crunch

Every so often tele-
phone companies split
regions with one area
code into regions with
two or more area codes
because the increased
number of telephones
in the area have nearly
exhausted the possible
numbers that can be
listed under a single
code. A seven-digit
telephone number can-
not begin with a 0 or
1, but if we allow all
other possibilities, we
get $8 \times 10 \times 10 \times 10 \times 10 \times 10 \times 10 =$
8,000,000 different pos-
sible numbers! Even
so, after surviving for
80 years with the sin-
gle area code of 212,
New York City was
recently partitioned
into the two area codes
of 212 and 718. Los
Angeles, Houston, and
San Diego have also
endured split area
codes.

3-6 COUNTING (Optional)

In calculating probabilities, we often obtain the total number of different possible outcomes by counting them. However, it sometimes
happens that the number of different possible outcomes is extremely
large so that the usual counting procedure becomes impractical. In
this section we introduce techniques that can be used to replace the
standard counting process with fast and efficient procedures. In addition to their use in probability problems, these counting techniques
have also grown in importance because of their applicability to problems relating to computers and programming.

Suppose that we use a computer to randomly select one of the two
Rh types (positive, negative) and one of the four blood groups A, O, B,
or AB. Let's assume that we want the probability of getting a positive
Rh type and group A blood. We can represent this probability as
$P(\text{positive and } A)$ and apply the multiplication rule to get

$$P(\text{positive and } A) = \frac{1}{2} \cdot \frac{1}{4} = \frac{1}{8}$$

We could also arrive at the same probability by examining the tree
diagram in Figure 3-11.

In Figure 3-11 we see that there are eight branches that, by the
random computer selection method, are all equally likely. Since only
one branch corresponds to $P(\text{positive and } A)$ we get a probability of
1/8. Apart from the calculation of the probability, this solution does

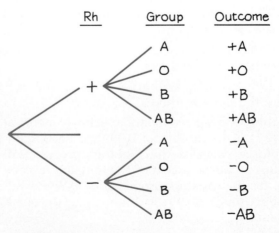

FIGURE 3-11

reveal another principle, which is a generalization of the following specific observation: with *two* Rh types and *four* blood groups, there are *eight* different possibilities for the compound event of selecting a factor and type. We now state this generalized principle.

FUNDAMENTAL COUNTING RULE

For a sequence of two events in which the first event can occur *m* ways and the second event can occur *n* ways, the events together can occur a total of $m \cdot n$ ways.

EXAMPLE

The first question on a standard test is true-false, while the second question is multiple choice with possible answers of *a, b, c, d, e*. How many different possible answer sequences are there for these two questions?

Solution

The first question can be answered two ways and the second question can be answered five ways, so the total number of different possible answer sequences is given by $2 \cdot 5 = 10$.

The **fundamental counting rule** given above easily extends to situations involving more than two events, as illustrated in the following example.

EXAMPLE

In computer design, if a byte is defined to be a sequence of eight bits, and each bit must be a 0 or 1, how many different bytes are possible?

Solution

Since each bit can occur in two ways (0 or 1) and we have a sequence of eight bits, the total number of different possibilities is given by $2 \cdot 2 \cdot 2 \cdot 2 \cdot 2 \cdot 2 \cdot 2 \cdot 2 = 256$.

The next rule uses the factorial symbol !, which denotes the product of decreasing whole numbers as illustrated below. Many calculators have a factorial key.

$$5! = 5 \cdot 4 \cdot 3 \cdot 2 \cdot 1 = 120$$

$$4! = 4 \cdot 3 \cdot 2 \cdot 1 = 24$$

$$3! = 3 \cdot 2 \cdot 1 = 6$$

$$2! = 2 \cdot 1 = 2$$

$$1! = 1$$

$$0! = 1 \text{ (by definition)}$$

$$n! = n(n - 1)(n - 2) \cdots 1$$

Using the factorial symbol, we now present the factorial rule.

FACTORIAL RULE

n different items can be arranged in order $n!$ different ways.

This **factorial rule** reflects the fact that the first item may be selected n different ways, the second item may be selected $n - 1$ ways, and so on.

EXAMPLE

Routing problems are extremely important in many real applications. AT&T wants to route telephone calls through the shortest networks. Federal Express wants to find the shortest routes for its deliveries. Suppose a computer salesperson must visit three separate cities denoted by A, B, C. How many routes are possible?

Solution

Using the factorial rule, we see that the three different cities (A, B, C) can be arranged in $3! = 6$ different ways. In Figure 3-12, we can see exactly why there are 6 different routes.

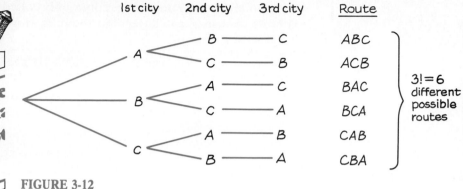

FIGURE 3-12

Safety in Numbers

Some hotels have abandoned the traditional room key in favor of an electronic key made of paper and aluminum foil. A central computer changes the access code to a room as soon as a guest checks out. A typical electronic key has 32 different positions that are either punched or left untouched. This configuration allows for 2^{32} or 4,294,967,296 different possible codes, so it is impractical to develop a complete set of keys or try to make an illegal entry by trial and error.

EXAMPLE

A presidential candidate plans to visit the capitol of each of the 50 states. How many different routes are possible?

Solution

The 50 state capitols can be arranged 50! ways, so that the number of different routes is 50! or 30,414 *followed by 60 zeros*; that is an incredibly large number! Now we can see why the symbol ! is used for factorials!

The preceding example is a variation of a classical problem called the traveling salesman problem. It is especially interesting because the large number of possibilities precludes a direct computation for each route, even if computers are used. The time for the fastest computer to directly calculate the shortest possible route is about

1,000,000,000,000,000,000,000,000,000,000,000,000,000 *centuries*!

(A Bell Laboratories mathematician claims that the shortest distance for the 48 contiguous states is 10,628 miles. See *Discover*, July 1985.) Clearly, those who use computers to solve problems must be able to recognize when the number of possibilities is so large.

In the factorial counting rule, we determine the number of different possible ways we can arrange a number of items in some type of ordered sequence. Sometimes we don't want to include all of the items available. When we refer to arrangements, we imply that *order* is taken into account. Arrangements are commonly called permutations, which explains the use of the letter P in the following rule.

PERMUTATIONS RULE

The number of **permutations** (or arrangements) of r items selected from n available items is

$$_nP_r = \frac{n!}{(n-r)!}$$

It must be emphasized that in applying the preceding **permutations rule**, we must have a total of n items available, we must select r of the n items, and we must consider rearrangements of the same items to be different. In the following example we are asked to find the total number of different arrangements that are possible. That clearly suggests use of the permutations rule.

EXAMPLE

If an editor must arrange five articles in a magazine and there are eight articles available, how many different arrangements are possible?

Solution

Here we want the number of arrangements of $r = 5$ items selected from $n = 8$ available articles so that the number of different possible arrangements is

$$_8P_5 = \frac{8!}{(8-5)!} = \frac{8!}{3!} = 6,720$$

This permutation rule can be thought of as an extension of the fundamental counting rule. We can solve the preceding problem by using the fundamental counting rule in the following way. With 8 articles available and with space for only 5 articles, we know that there are 8 choices for the first article, 7 choices for the second article, 6 choices for the third article, 5 choices for the fourth article, and 4 choices for the fifth article. The number of different possible arrangements is therefore $8 \cdot 7 \cdot 6 \cdot 5 \cdot 4 = 6720$, but $8 \cdot 7 \cdot 6 \cdot 5 \cdot 4$ is actually $8! \div 3!$. In general, whenever we select r items from n available items,

Probability and Promotion Contests

$21,000,000

We have all encountered contests designed to promote a product. Some of those contests didn't go exactly as the sponsors had hoped. The Beatrice Company ran one such contest that involved matching numbers on scratch cards with scores from Monday night football games. Because the game cards had too few permutations, Frank Maggio was able to identify patterns. He recruited help and was able to collect around 4000 winning cards worth $21 million. The contest was canceled and lawsuits were filed.

The Pepsi people ran another contest in which letters were printed on bottle caps. Contestants were asked to spell out their names, but a large number of people with short names like Ng forced the cancelation of this contest.

the number of different possible arrangements is $n! \div (n - r)!$, and this is expressed in the permutations rule.

When we intend to select r items from n available items but do not take order into account, we are really concerned with possible combinations rather than permutations. That is, when different orderings of the same items are to be counted separately, we have a permutation problem, but when different orderings are *not* to be counted separately, we have a combination problem and may apply the following rule.

COMBINATIONS RULE

The number of **combinations** of r items selected from n available items is

$$_nC_r = \frac{n!}{(n - r)! \, r!}$$

[other notations for $_nC_r$ are $\binom{n}{r}$ and $C(n, r)$.]

The combinations rule makes sense when we think of it as a modification of the permutations rule. Note that

$$_nC_r = \frac{n!}{(n - r)! \, r!} = \frac{n!}{(n - r)!} \cdot \frac{1}{r!} = {_nP_r} \cdot \frac{1}{r!} = \frac{_nP_r}{r!}$$

Realizing that the **combinations rule** disregards the order of the r selected items while the permutations rule counts all of the different orderings separately, we see that $_nC_r$ will be a fraction of $_nP_r$. For any particular selection of r items, the factorial rule shows that there are $r!$ different arrangements. We know that any particular selection of r items would be counted as only one combination, but it would yield $r!$ different arrangements. As a result, the total number of combinations will be $1/r!$ of the total number of permutations, and that result is expressed in the combinations rule.

EXAMPLE

If 12 jurors are to be selected and there are 15 candidates available, how many different combinations of jurors are possible?

Solution

We want to select $r = 12$ jurors from $n = 15$ available candidates. We are not concerned with the order in which the jurors are selected; we are concerned only with the different possible combinations. Applying the last rule we get

$$_{15}C_{12} = \frac{15!}{(15 - 12)!\ 12!} = \frac{15!}{3!\ 12!}$$

$$= \frac{15 \cdot 14 \cdot 13 \cdot \cancel{12} \cdot \cancel{11} \cdot \cancel{10} \cdot \cancel{9} \cdot \cancel{8} \cdot \cancel{7} \cdot \cancel{6} \cdot \cancel{5} \cdot \cancel{4} \cdot \cancel{3} \cdot \cancel{2} \cdot \cancel{1}}{(3 \cdot 2 \cdot 1)\cancel{12} \cdot \cancel{11} \cdot \cancel{10} \cdot \cancel{9} \cdot \cancel{8} \cdot \cancel{7} \cdot \cancel{6} \cdot \cancel{5} \cdot \cancel{4} \cdot \cancel{3} \cdot \cancel{2} \cdot \cancel{1})}$$

$$= \frac{15 \cdot 14 \cdot 13}{3 \cdot 2 \cdot 1} = 455$$

We stated that the counting techniques of this section are sometimes used in probability problems. The following examples illustrate such applications.

EXAMPLE

In the New York State lottery, first prize is won if a player selects the correct six-number combination when six different numbers from 1 through 44 are drawn. If a player selects one particular six-number combination, find the probability of winning.

continued ▶

Example, continued

Solution

Since 6 different numbers are selected from 44 different possibilities, the total number of combinations is

$$_{44}C_6 = \frac{44!}{(44 - 6)! \, 6!} = \frac{44!}{38! \, 6!} = 7,059,052$$

With only one combination selected, the player's probability of winning is only $1/7,059,052$.

EXAMPLE

A home security device with ten buttons is disarmed when three different buttons are pushed in the proper sequence. (No button can be pushed twice.) If the correct code is forgotten, what is the probability of disarming this device by randomly pushing three of the buttons?

Solution

The number of different possible three-button sequences is

$$_{10}P_3 = \frac{10!}{(10 - 3)!} = 720$$

The probability of randomly selecting the correct three-button sequence is therefore $1/720$.

EXAMPLE

A dispatcher sends a delivery truck to eight different locations. If the order in which the deliveries are made is randomly determined, find the probability that the resulting route is the shortest possible route.

Solution

With eight locations there are 8! or 40,320 different possible

continued ▶

Solution, continued

routes. Among those 40,320 different possibilities, only two routes will be shortest (actually the same route in two different directions). Therefore, there is a probability of only 2/40,320 or 1/20,160 that the selected route will be the shortest possible route.

In this last example, application of the appropriate counting technique made the solution easily obtainable. If we had to determine the number of routes directly by listing them, we would labor for over 11 hours while working at the rapid rate of one route per second! Clearly, these counting techniques are extremely valuable.

Since there is often confusion when choosing between the permutations rule and the combinations rule, we provide the following example, which is intended to emphasize the difference between them.

EXAMPLE

Five students (Al, Bob, Carol, Donna, and Ed) have volunteered for service to the student government.

a. If three of the students are to be selected for a special *committee*, how many different committees are possible?

b. If three of the students are to be nominated for the offices of president, vice president, and secretary, how many different *slates* are possible?

Solution

a. When forming the committee, order of selection is irrelevant. The committee of Al, Bob, and Ed is the same as that of Bob, Al, and Ed. Therefore, we want the number of combinations of 5 students when 3 are selected. We get

$$_5C_3 = \frac{5!}{(5-3)!\,3!} = 10 \qquad \begin{array}{l}\text{Use combinations} \\ \text{when order is} \\ \text{irrelevant.}\end{array}$$

There are 10 different possible committees.

b. When forming slates of candidates, the order is relevant. The slate of Al for president, Bob for vice president, and Ed for

continued ▶

Voltaire Beats the Lottery

In 1729, the philosopher Voltaire became rich when he successfully implemented a scheme for beating a lottery run by the City of Paris. The Paris government ran this lottery to repay municipal bonds. Since the bonds had lost some value, the government added substantial amounts of money with the net effect that the value of the lottery prize was greater than the cost of all tickets. Voltaire organized a group that bought up all tickets in the monthly lottery. For more than a year, this group won the monthly lottery and acquired a substantial net gain of about 7,500,000 francs.

More recently, a bettor in the New York State lottery tried to take advantage of an exceptionally large prize that grew from a lack of winners in previous drawings. Realizing that there were 12,271,512 combinations of numbers costing 50¢ each, this bettor wanted to write a check for $6,135,756 that would guarantee a winning number since it would cover every combination. The director of the lottery refused to accept the check and explained that the nature of the lottery would have been changed.

Solution, continued

secretary is different from the slate of Bob, Al, and Ed for president, vice president, and secretary, respectively. Here we want the number of permutations of 5 students when 3 are selected. We get

$$_5P_3 = \frac{5!}{(5-3)!} = 60 \qquad \text{Use permutations when order is relevant.}$$

There are 60 different possible slates.

The concepts and rules of probability theory presented in this chapter consist of elementary and fundamental principles. A more complete study of probability is not necessary at this time since our main objective is to study the elements of statistics, and we have already covered the probability theory that we will need. We hope that this chapter generates some interest in probability for its own sake. The importance of probability is continuing to grow as it is used by more and more scientists, economists, politicians, biologists, insurance specialists, executives, and other professionals.

3-6 EXERCISES A

In Exercises 3-153 through 3-168, evaluate the given expressions.

3-153 $7!$ 3-154 $9!$ 3-155 $\dfrac{70!}{68!}$ 3-156 $\dfrac{92!}{89!}$

3-157 $(9-3)!$ 3-158 $(20-12)!$ 3-159 $_6P_2$ 3-160 $_6C_2$

3-161 $_{10}C_3$ 3-162 $_{10}P_3$ 3-163 $_{52}C_2$ 3-164 $_{52}P_2$

3-165 $_nP_n$ 3-166 $_nC_n$ 3-167 $_nC_0$ 3-168 $_nP_0$

3-169   Data are grouped according to sex (female, male) and according to income level (low, middle, high). How many different possible categories are there?

3-170   How many different ways can five cars be arranged on a carrier truck with room for five vehicles?

3-171   A computer operator must select 4 jobs from among 10 available jobs waiting to be completed. How many different arrangements are possible?

3-172   A computer operator must select 4 jobs from 10 available jobs waiting to be completed. How many different combinations are possible?

3-173   An IRS agent must audit 12 returns from a collection of 22 flagged returns. How many different combinations are possible?

3-174   A health inspector has time to visit 7 of the 20 restaurants on a list. How many different routes are possible?

3-175   Using a word processor, a pollster develops a survey of 10 questions. The pollster decides to rearrange the order of the questions so that any lead-in effect will be minimized. How many different versions of the survey are required if all possible arrangements are included?

3-176   How many different seven-digit telephone numbers are possible if the first digit cannot be 0 or 1?

3-177   An airline mail route must include stops at seven cities.
a. How many different routes are possible?
b. If the route is randomly selected, what is the probability that the cities will be arranged in alphabetical order?

3-178   A six-member FBI investigative team is to be formed from a list of 30 agents.
a. How many different possible combinations can be formed?
b. If the selections are random, what is the probability of getting the six agents with the most experience?

3-179   How many different social security numbers are possible? Each social security number is a sequence of nine digits.

3-180 A pollster must randomly select 3 of 12 available people. How many different groups of 3 are possible?

3-181 A union must elect 4 officers from 16 available candidates. How many different slates are possible if one candidate is nominated for each office?

3-182 a. How many different zip codes are possible if each code is a sequence of 5 digits?
b. If a computer randomly generates 5 digits, what is the probability it will produce your zip code?

3-183 A typical combination lock is opened with the correct sequence of three numbers between 0 and 49 inclusive. How many different sequences are possible? Are these sequences combinations or are they actually permutations?

3-184 One phase of an automobile assembly requires the attachment of eight different parts, and they can be attached in any order. The manager decides to find the most efficient arrangement by trying all possibilities. How many different arrangements are possible?

3-185 A space shuttle crew has available 10 main dishes, 8 vegetable dishes, 13 desserts, and 3 appetizers. If the first meal includes two desserts and one item from each of the other categories, how many different combinations are possible?

3-186 In Denys Parson's *Directory of Tunes and Musical Themes*, melodies for more than 14,000 songs are listed according to the following scheme: The first note of every song is represented by an asterisk (*), and successive notes are represented by *R* (for repeat the previous note), *U* (for a note that goes up), or *D* (for a note that goes down). Beethoven's "Fifth Symphony" begins as *RRD*. Classical melodies are represented through the first 16 notes. Using this scheme, how many different classical melodies are possible?

3-187 A television program director has 14 shows available for Monday night and 5 shows must be chosen.
a. How many different possible combinations are there?
b. If 650 different combinations are judged to be incompatible, find the probability of randomly selecting 5 shows that are compatible.

3-188 A telephone company employee must collect the coins at 40 different locations.
a. How many different routes are possible?
b. If two of the routes are the shortest, find the probability of randomly selecting a route and getting one of the two shortest routes.

3-189 A representative of an environmental protection agency plans to sample water at 10 different ponds randomly selected from 25 available ponds. (*continued*)

a. How many different combinations are possible?

b. What is the probability of randomly selecting the 10 ponds with the lowest pollution levels?

3-190   A multiple-choice test consists of 10 questions with choices a, b, c, d, e.

a. How many different answer keys are possible?

b. If all 10 answers are random guesses, what is the probability of getting a perfect score?

3-191   The Bureau of Fisheries once asked Bell Laboratories for help in finding the shortest route for getting samples from locations in the Gulf of Mexico. How many different routes are possible if samples must be taken from 24 locations?

3-192   A manager must choose 5 secretaries from among 12 applicants and assign them to different stations.

a. How many different arrangements are possible?

b. If the selections are random, what is the probability of getting the 5 youngest secretaries selected in order of age?

3-6 EXERCISES B

3-193   We say that a sample is *random* if all possible samples of the same size have the same probability of being selected.

a. If a random sample is to be drawn from a population of size 60, find the probability of selecting any one individual sample consisting of 5 members of the population.

b. Write a general expression for the probability of selecting a particular random sample of size n from a population of size N.

3-194   By constructing a tree diagram (such as Figure 3-12), we can see the permutations rule work.

a. Use the permutations rule to calculate the number of different possible arrangements that can be formed when three letters are selected from A, B, C, D, E.

b. Construct a tree diagram (similar to Figure 3-12) showing all possibilities when three letters are selected from A, B, C, D, E.

c. Relate the results of parts (a) and (b).

3-195   The number of permutations of n items when x of them are identical to each other and the remaining $n - x$ are identical to each other is given by

$$\frac{n!}{(n - x)! \, x!}$$

If a sequence of 10 trials results in only successes and failures, how many ways can 3 successes and 7 failures be arranged?

3-196 If a couple with 10 grandchildren is randomly selected, what is the probability that the 10 grandchildren consist of 2 boys and 8 girls? (See Exercise 3-195.)

3-197 Most calculators or computers cannot directly calculate 70! or higher. When n is large, $n!$ can be *approximated* by

$$n! = 10^K \quad \text{where } K = (n + 0.5)\log n + 0.39908993 - 0.43429448n$$

Evaluate 50! using the factorial key on a calculator and also by using the approximation given here.

3-198 The Bureau of Fisheries once asked Bell Laboratories for help in finding the shortest route for getting samples from 300 locations in the Gulf of Mexico. There are 300! different possible routes. If 300! is evaluated, how many digits are used in the result? (See Exercise 3-197.)

3-199 Five managers gather for a meeting. If each manager shakes hands with every other manager exactly once, what is the total number of handshakes?

3-200 a. How many different ways can three people be seated at a round table? (Assume that if everyone moves to the right, the seating arrangement is the same.)
b. How many different ways can four people be seated at a round table?
c. How many different ways can n people be seated at a round table?

VOCABULARY LIST

Define and give an example of each term.

experiment	compound event	tree diagram
event	addition rule	complement of an
simple event	mutually exclusive	event
sample space	events	odds against
empirical	multiplication rule	odds in favor
approximation of	tree diagram	fundamental
probability	independent events	counting rule
classical approach to	dependent events	factorial rule
probability	conditional	permutations rule
random selection	probability	combinations rule

REVIEW

This chapter introduced the basic concept of **probability**. We began with two rules for finding probabilities. Rule 1 represents the *empirical* approach, whereby the probability of an event is approximated by actually conducting or observing the experiment in question.

RULE 1
$P(A) = \dfrac{\text{number of times } A \text{ occurred}}{\text{number of times experiment was repeated}}$

Rule 2 is called the **classical** approach, and it applies only if all of the outcomes are equally likely.

RULE 2
$P(A) = \dfrac{s}{n} = \dfrac{\text{number of ways } A \text{ can occur}}{\text{total number of different outcomes}}$

We noted that the probability of any impossible event is 0, while the probability of any certain event is 1. Also, for any event A,

$$0 \le P(A) \le 1$$

In Section 3-3, we considered the **addition rule** for finding the probability that A **or** B will occur. In evaluating $P(A \text{ or } B)$, it is important to consider whether the events are *mutually exclusive*—that is, whether they can both occur at the same time (see Figure 3-13).

In Section 3-4 we considered the **multiplication rule** for finding the probability that A **and** B will occur. In evaluating $P(A \text{ and } B)$, it is important to consider whether the events are *independent*—that is whether the occurrence of one event affects the probability of the other event (see Figure 3-14).

In Section 3-5 we considered **complements and odds**. Using the addition rule, we were able to develop the **rule of complementary events**: $P(A) + P(\bar{A}) = 1$. We saw that this rule can sometimes be used to simplify probability problems. We also defined the **odds against event A** as $P(\bar{A})/P(A)$; the **odds in favor of event A** are $P(A)/P(\bar{A})$. Both ratios are normally expressed in the form of $a{:}b$, where a and b are

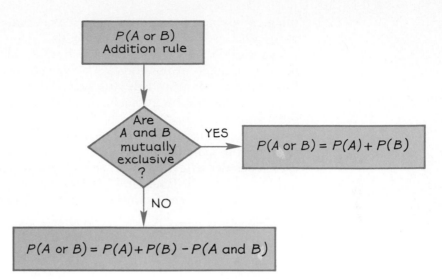

FIGURE 3-13
Finding the probability that for a single trial, either event A or B (or both) will occur

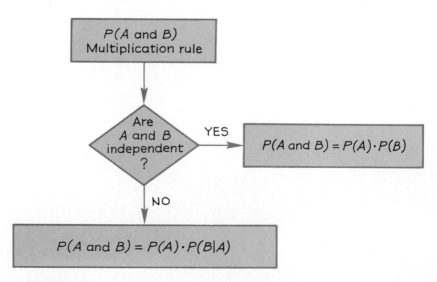

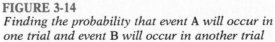

FIGURE 3-14
Finding the probability that event A will occur in one trial and event B will occur in another trial

integers having no common factors. We studied procedures for converting between probabilities and odds.

In Section 3-6 we considered techniques for determining the total number of different possibilities for various events. We presented the **fundamental counting rule**, the **factorial rule**, the **permutations** formula, and the **combinations** formula, all summarized below with the other important formulas from this chapter.

Most of the material that follows this chapter deals with statistical inferences based on probabilities. As an example of the basic approach used, consider a test of someone's claim that a quarter is fair. If we flip the quarter 10 times and get 10 consecutive heads, we can make one of two inferences from these sample results:

1. The coin is actually fair and the string of 10 consecutive heads is a fluke.
2. The coin is not fair.

The statistician's decision is based on the **probability** of getting 10 consecutive heads which, in this case, is so small (1/1024) that the inference of unfairness is the better choice. The purpose of this example is to emphasize the important role played by probability in the standard methods of statistical inference.

IMPORTANT FORMULAS

$0 \leq P(A) \leq 1$ for any event A

$P(A \text{ or } B) = P(A) + P(B)$ if A, B are mutually exclusive

$P(A \text{ or } B) = P(A) + P(B) - P(A \text{ and } B)$ if A, B are not mutually exclusive

$P(A \text{ and } B) = P(A) \cdot P(B)$ if A, B are independent

$P(A \text{ and } B) = P(A) \cdot P(B|A)$ if A, B are dependent

$P(\bar{A}) = 1 - P(A)$

$P(A) = 1 - P(\bar{A})$

$P(A) + P(\bar{A}) = 1$

odds against event $A = \dfrac{P(\bar{A})}{P(A)}$ ← } usually expressed as $a{:}b$, where a, b are integers with no common factors

odds in favor of A $= \dfrac{P(A)}{P(\bar{A})}$ ←

continued ▶

> **Important Formulas, continued**
>
> $m \cdot n$ = the total number of ways two events can occur, if the first can occur m ways while the second can occur n ways
>
> $n!$ = the number of ways n different items can be arranged
>
> $$_nP_r = \frac{n!}{(n - r)!}$$ the number of *permutations* (arrangements) when r items are selected from n available items
>
> $$_nC_r = \frac{n!}{(n - r)!\, r!}$$ the number of *combinations* when r items are selected from n available items

REVIEW EXERCISES

3-201   a. Find $P(A \text{ and } B)$ if $P(A) = 0.2$, $P(B) = 0.4$, and A and B are both independent.
b. Find $P(A \text{ or } B)$ if $P(A) = 0.2$, $P(B) = 0.4$, and A and B are mutually exclusive.
c. Find $P(\bar{A})$ if $P(A) = 0.2$.
d. Find the odds against A if $P(A) = 0.2$.
e. Find $P(C)$ if the odds against event C are 12:1.

3-202   In a televised *60 Minutes* (CBS) report, it was stated that 85% of high school students graduate. If three high school students are randomly selected, what is the probability they all graduate?

3-203   The following table summarizes the ratings for 6,665 films. (The table is based on data from the Motion Picture Association of America.) If one of these movies is randomly selected, find the probability that it has a rating of R.

Rating	Number
G	873
PG	2505
R	2945
X	342

3-204   Using the same data from Exercise 3-203, find the probability of randomly selecting one of the movies and getting one with a rating of G or PG.

3-205   In a televised *Nightline* (ABC) report, it was stated that 3/4 of smokers want to quit but don't.
  a. If you randomly select a smoker, find the odds against getting someone who wants to quit but doesn't.
  b. If two smokers are randomly selected, find the probability that they both want to quit but don't.

3-206   a. What is the probability of an event that is known to be impossible?
  b. What is the probability of an event that will definitely occur?
  c. What is $P(A)$ if $P(\bar{A}) = 0.22$?
  d. If events A and B are mutually exclusive, must they be complementary events?
  e. If events A and B are complementary, must they be mutually exclusive?

3-207   If seven different customers arrive for service in random order, what is the probability that they will be in alphabetical order?

3-208   A psychologist conducts experiments in which a monkey must select the correct interconnecting parts. When 5 parts are selected from 12 available parts, how many different combinations are there?

3-209   A survey is made in a neighborhood of 65 Democrats and 15 Republicans. Of the Democrats 35 are women, while 5 of the Republicans are women. If one subject from this group is randomly selected, find the probability of getting each outcome.
  a. A female or a Democrat
  b. A female Democrat
  c. A Democrat or a Republican

3-210   If a couple plans to have four children, find the probability that they are all of the same sex. Assume that boys and girls are equally likely and that the sexes are independent.

3-211   A research project involves one patient with group A blood, one with group B blood, one with group O blood, and one with group AB blood. If one of these patients is randomly selected, what is the probability of getting the patient with group O blood?

3-212   A medical study involves the subjects summarized in the accompanying table. If one subject is randomly selected, find the probability of getting someone with blood group A or B.

	Blood group			
	A	O	B	AB
Healthy	7	20	3	9
Ill	3	10	2	6

3-213 Using the medical data from Exercise 3-212, if one subject is randomly selected, find the probability of getting someone who is healthy or has group O blood.

3-214 Using the medical data from Exercise 3-212, if one subject is randomly selected, find the probability of getting someone who is ill or has group A blood.

3-215 In one of New York State's lottery games, the probability of winning is 1/6,135,756. Find the odds against winning.

3-216 Eight men and seven women have applied for a temporary job. If three different applicants are randomly selected from this group, find the probability of each event.
   a. All three are women.
   b. There is at least one woman.

3-217 Evaluate the following.
   a. $8!$ b. $_8P_6$ c. $_{10}C_8$ d. $_{80}C_{78}$

3-218 Of 120 auto ignition circuits, there are 18 defects. If two circuits are randomly selected, find the probability that they are both defective in each case.
   a. The first selection is replaced before the second selection is made.
   b. The first selection is not replaced.

3-219 A hearing is attended by 12 people in favor of a construction moratorium and 8 people opposed. If one person is randomly selected to be the first speaker, what is the probability that he or she is opposed?

3-220 According to Census Bureau data, 52% of women aged 18 to 24 years do not live at home with their parents. If we randomly select five different women in that age bracket, what is the probability that none of them live at home with their parents?

3-221 A question on a history test requires that five events be arranged in the proper chronological order. If a random arrangement is selected, what is the probability that it will be correct?

3-222 In the televised NBC White Paper *Divorce*, it was reported that 85% of divorced women are not awarded any alimony. If four divorced women are randomly selected, what is the probability that at least one of them is not awarded alimony?

3-223 In a test of sensory perception, a subject must select the three brightest colors and arrange them in the order of decreasing brightness. If eight colors are available, how many different arrangements are possible when three colors are selected?

3-224 One method has a 60% rate of success in helping people to discontinue cigarette smoking. If eight randomly selected smokers use this method, what is the probability that at least one of them will give up smoking?

3-225 In 60 patients given a vaccine, 35 unfavorable reactions occurred. If the records of two different patients are randomly selected, find the probability that they both had unfavorable reactions.

3-226 A city council decides to appoint 6 members to a zoning committee. If the appointments are to be made from a list of 14 available candidates, how many different ways can the zoning committee be formed?

3-227 It has been found that a particular psychological disorder is improved through group therapy with a success rate of 40%. If ten subjects with this disorder are randomly selected and treated with group therapy, what is the probability that at least one of them will improve?

3-228 If a couple plans to have five children, find the probability that they are all girls. Assume that boys and girls have the same chance of being born and that the sex of any child is not affected by the sex of any other children.

3-229 When buying a home computer system, a programmer can choose any one of six central processing units, any one of four disk drives, any one of eight printers, and any one of four modems. How many different configurations are possible?

3-230 In a certain state, 30% of the voters are Republican. If four voters are randomly selected for an exit poll, what is the probability of getting at least one Republican?

3-231 In one segment of a radio show, a disk jockey must select a program consisting of 4 songs arranged in some order. If 12 songs are available from the rotation, how many different programs are possible?

3-232 In attempting to gain access to a computer data bank, a computer is programmed to automatically dial every phone number with the prefix 478 followed by four digits. How many such telephone numbers are possible?

3-233 The odds against event *A* are 7:1. Find the odds against event *A* occurring in two separate and independent trials.

3-234 Fifty-five percent of enlisted personnel are married, according to a Department of Defense survey of 89,000 enlisted military personnel. If a crew of six enlisted military personnel is formed by random selection, what is the probability that none of them are married?

3-235 In assigning flight numbers, a dispatcher can choose any one of six cities for departures and any one of four other cities for arrivals. How many different routes are possible?

3-236 According to Census Bureau data, among mothers who begin working, 30.6% have less than a high school education. If one working mother is randomly selected, what is the probability that she has at least a high school education?

COMPUTER PROJECT

In many situations, probability problems can be solved by writing a computer program that simulates the relevant circumstances. In determining the probability of winning the game of solitaire, for example, it is easier to program a computer to play solitaire than it is to develop calculations using the rules of probability. The computer can then play solitaire 1000 times and the probability of winning is estimated to be the number of wins divided by 1000. Such simulations often involve the computer's ability to generate random numbers. The subroutine RND is available in many BASIC implementations. One common use causes an entry of RND(X) to return a value between 0 and 1, such as 0.23560387. That result can be manipulated to produce desired values.

```
10 RANDOMIZE
20 LET R = INT (100 * RND(X))
30 IF R > 37 THEN 20
40 PRINT R
```

In the short program above, we take a value like 0.23560387, multiply by 100 to get 23.560387, and then take only the integer part to get 23. If the result is greater than 37, we go back and try again. Those three lines produce randomly selected values from the list 0, 1, 2, . . . , 37. This short program simulates the spinning of a roulette wheel if we stipulate that 37 corresponds to 00.

a. Enter those lines and run the program 20 times. If you were betting on the number 7, how many times did you win?

b. In roulette, if you bet $1 on "odd," you win $1 if the result is an odd number. (Remember that in our program, 37 should not win since it represents 00.) Modify the program so that you begin with $10 and you bet $1 on each spin until you either reach $20 or go broke. The final output should indicate whether you reached $20 or went broke first.

c. A gambler is in Las Vegas with only $10 but must have $20 for the return trip home, or else he must walk. Use computer simulations to determine which of the following two strategies is more likely to get him the $20 he needs for the return trip.
   i. Bet the entire $10 on "odd" for one spin of the roulette wheel.
   ii. Bet on odd, $1 at a time, until reaching $20 or going broke.

CASE STUDY ACTIVITY

In Section 3-1 we stated that at least 2 students will have the same birthday in more than half of the classes with 25 students. Exercise 3-150 requires the theoretical solution to that problem, but many such problems can be solved by developing a **simulation**. Let the days of the year be represented by the numbers from 001 through 365. Now refer to a page randomly selected from a telephone book and record the last three digits of 25 telephone numbers, but ignore 000 and any cases above 365. Check to determine whether or not at least two of the dates (represented by the three-digit numbers) are the same. It should take about 30 minutes to repeat this experiment 10 times. Find the estimated probability, which is the number of times matched dates occurred, divided by the number of times the experiment is repeated. (Another approach is to use a computer instead of a telephone book. STATDISK can be used to repeatedly generate 25 numbers between 1 and 365.)

DATA PROJECT

A psychologist plans to conduct a study of people who are tall and live in relatively smaller homes. Refer to the data sets in Appendix B to answer the following.

a. Estimate the probability that a randomly selected adult male is at least 6 ft tall.

b. Estimate the probability that if a recently sold home is randomly selected, its living area is less than 1500 sq ft.

c. Assume that a male is randomly selected from a home that was recently bought. Use the results from parts (a) and (b) to estimate the probability of getting someone at least 6 ft tall who lives in a home with a living area less than 1500 sq ft.

d. In doing the calculations for part (c), identify any assumptions that were made about the events.

e. Estimate the probability that a recently sold home has fewer than six rooms.

f. Estimate the probability that a recently sold home has a living area less than 1500 sq ft *and* has fewer than six rooms. (Assume that these events are independent and use the results from parts (b) and (e).

g. Are the events of having a living area less than 1500 sq ft and having fewer than six rooms really independent or dependent?

h. Inspect the data and find the actual number of homes that have a living area less than 1500 sq ft and also have fewer than six rooms. Use this result to estimate the probability described in part (f) and compare the result to the one obtained there.

Chapter Four

CHAPTER CONTENTS

4-1 **Overview**
We identify chapter **objectives**.

4-2 **Random Variables**
We describe **discrete** and **continuous random variables** and **probability distributions**.

4-3 **Mean, Variance, and Expectation**
Given a probability distribution, we describe methods of determining the **mean**, **standard deviation**, and **variance**, and we define the **expected value** of a probability distribution.

4-4 **Binomial Experiments**
We learn how to calculate probabilities in **binomial experiments**.

4-5 **Mean and Standard Deviation for the Binomial Distribution**
We learn how to calculate the **mean**, **standard deviation**, and **variance** for a binomial distribution.

4-6 **Distribution Shapes**
We investigate the shape or nature of various probability distributions and establish a correspondence between probability and area.

4 Probability Distributions

CHAPTER PROBLEM

The Federal Aviation Administration studied the possibility of allowing twin-engine commercial jets to make transatlantic flights previously open only to jets with at least three engines. The lowered requirement is of great interest to manufacturers of twin-engine jets (such as the Boeing 767). Also, the two-engine jets use about half the fuel of jets with three or four engines. Obviously, the key issue in approving the lowered requirement is the probability of a twin-engine jet making a safe transatlantic crossing. This probability should be compared to that of three- and four-engine jets. Clearly, such a study involves a thorough understanding of the related probabilities, and the contents of this chapter will enable us to develop such an understanding. We will return to this case study as we develop the relevant principles.

4-1 OVERVIEW

In Chapter 2 we discussed the histogram as a device for showing the frequency distribution of a set of data. In Chapter 3 we discussed the basic principles of probability theory. In this chapter we combine those concepts to develop probability distributions that are basically theoretical models of the frequency distributions we produce when we collect sample data. We construct frequency tables and histograms using *observed* real scores, but we construct probability distributions by presenting possible outcomes along with their *probable* frequencies.

Suppose a casino manager suspects cheating at a dice table. The manager can compare the frequency distribution of the actual sample outcomes to a theoretical model that describes the frequency distribution likely to occur with fair dice. In this case the probability distribution serves as a model of a theoretically perfect population frequency distribution. In essence, we can determine what the frequency table and histogram would be like for a pair of fair dice rolled an infinite number of times. With this perception of the population of outcomes, we can then determine the values of important parameters such as the mean, variance, and standard deviation.

The concept of a probability distribution is not limited to casino management. In fact, the remainder of this book and the very core of inferential statistics depend on some knowledge of probability distributions. To analyze the effectiveness of a new drug, for example, we must know something about the probability distribution of the symptoms the drug is intended to correct.

This chapter deals mostly with discrete cases, while subsequent chapters involve continuous cases. We begin by distinguishing between discrete and continuous random variables.

4-2 RANDOM VARIABLES

In Chapter 3 we defined an experiment to be any process that allows us to obtain observations. Some experiments give us observations that are quantitative (such as weights), while others give us observations that are qualitative (such as colors). Qualitative outcomes can often be expressed with numbers. For example, the qualitative performances of Olympic divers are numerically rated by judges. If we can associate each outcome of an experiment with a single number, then we have a type of variable whose values are determined by chance. Such variables are called *random variables*.

Is Parachuting Safe?

About 35 people die each year from injuries sustained while parachuting. In comparison, a typical year includes about 200 scuba diving fatalities, 7000 drownings, 900 bicycle deaths, 800 lightning deaths, and 1150 deaths from bee stings.

Of course, this does not necessarily mean that it's safer to parachute than to ride a bicycle or swim. A fair comparison should include fatality *rates*, not just the number of fatalities. It has been estimated that in para-

chuting, 1 fatality will occur in about 25,000 jumps.

The author, with much trepidation, made two parachute jumps but gave up the sport after missing the spacious drop zone both times.

DEFINITION
A **random variable** has a single numerical value for each outcome of an experiment.

The values of a random variable are the numbers we associate with the different outcomes that make up the sample space for the experiment. As an example, consider the experiment of randomly selecting three voters. We can let the random variable represent the number of selected voters who are college graduates. That random variable can then assume the possible values of 0, 1, 2, and 3.

EXAMPLE
A quiz consists of 10 multiple-choice questions. Let the random variable represent the number of correct answers. This random variable can take on the values of 0, 1, 2, 3, 4, 5, 6, 7, 8, 9, 10.

EXAMPLE

A jet has two independently operating engines. We can test those two engines with performance results that are acceptable (*A*) or defective (*D*). The set of all possible outcomes can be represented by this sample space.

    AA   *AD*   *DA*   *DD*

If the random variable represents the *number* of defective engines, then its possible values are 0, 1, and 2 (for 0 defects, 1 defect, and 2 defects). In this example, note that the outcomes are not numbers, but we can associate a number (0, 1, or 2) with each outcome so that we do have a random variable.

Random variables may be discrete or continuous. We know what a finite number of values is (1, or 2, or 3, and so on), but our definition of a discrete random variable also involves the concept of a **countable** number of values. As an example, suppose that a random variable represents the number of times a die must be rolled before a 6 turns up. This random variable can assume any one of the values 1, 2, 3, ... We now have an infinite number of possibilities, but they correspond to the counting numbers. Consequently this type of infinity is called countable. In contrast, the number of points on a continuous scale is not countable and represents a higher degree of infinity. There is no way to count the points on a continuous scale, but we can count the number of times a die is rolled, even if the rolling seems to continue forever.

DEFINITION

A **discrete random variable** has either a finite number of values or a countable number of values.

For example, suppose that a random variable can assume the value that represents the number of U.S. Senators present for a roll call. That random variable can assume only one of 101 different values (0, 1, 2, . . . , 99, 100) and, since 101 is a finite number, the random variable is discrete. The random variable representing the number of times a die must be rolled before a 6 turns up is also a discrete random variable because the number of rolls is 1, or 2, or 3, ... (The number of rolls can be counted.)

Just as count data are usually associated with discrete random variables, measurement data are usually associated with continuous random variables.

DEFINITION
A **continuous random variable** has infinitely many values, and those values can be associated with points on a continous scale in such a way that there are no gaps or interruptions.

As an example, we stipulate that a random variable can assume the value representing the exact speed of a car at a particular instant. The car might be traveling 42.135724 . . . kilometers per hour. This random variable can assume any value that corresponds to a point on the continuous interval shown in Figure 4-1. (We assume that the car never exceeds 80 kilometers per hour.) We now have an infinite number of possible values, which are not countable since there is a correspondence with the continuous scale of Figure 4-1. This random variable is continuous, not discrete.

FIGURE 4-1

Random variables that represent heights, weights, times, and temperatures are usually continuous, as are those that represent speeds.

In reality, it is usually impossible to deal with exact values of a continuous random variable. Instead, we usually convert continuous values to discrete values by rounding off to a limited number of decimal places. If each speed between 0 kilometers per hour and 80 kilometers per hour is rounded off to the nearest integer value, we reduce the total number of possibilities to the finite number of 81. In this way, continuous random variables are made discrete. This chapter is involved mostly with discrete random variables.

Discrete Random Variables

Inferential statistics is often used to make decisions in a wide variety of different fields. We begin with sample data and attempt to make inferences about the population from which the sample was drawn. If the sample is very large we may be able to develop a good estimate of the population frequency distribution, but samples are often too small for that purpose. The practical approach is to use information about the sample along with general knowledge about population distributions.

Much of this general information is included in this and the following chapters. Without a knowledge of probability distributions, users of statistics would be severely limited in the inferences they could make. We intend to develop the ability to work with discrete and continuous probability distributions, and we begin with the discrete case because it is simpler.

DEFINITION

A **probability distribution** gives the probability for each value of the random variable.

TABLE 4-1

x	$P(x)$
0	$\frac{1}{4}$
1	$\frac{1}{2}$
2	$\frac{1}{4}$

For example, suppose a drug is administered to two patients where the random variable is the number of cures (0, 1, or 2), and assume that the probability of a cure is $\frac{1}{2}$. The probability of zero cures is $\frac{1}{4}$, the probability of one cure is $\frac{1}{2}$, while the probability of two cures is $\frac{1}{4}$. Table 4-1 summarizes the probability distribution for this situation. (The given probabilities can be easily verified by listing the four cases in the sample space, beginning with "cure-cure.")

There are various ways of graphing these probability distributions, but we include only the histogram, which was introduced in Chapter 2.

The horizontal axis delineates the values of the random variable, while the vertical scale represents probabilities. In Figure 4-2, we show

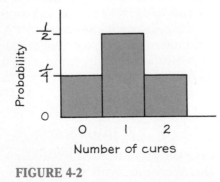

FIGURE 4-2

How Not to Pick Lottery Numbers

In the New York State lottery, a player makes an entry by selecting six different numbers between 1 and 44. After six numbers are randomly selected, any entries with the correct six numbers share in the top prize, which has run as high as $30 million (won by a single person). Since the winning numbers are randomly selected, any combination of six numbers will have the same chance as any other combination.

However, some numbers are better bets than others! The combination of 1, 2, 3, 4, 5, 6 is a poor choice because many people select that particular combination; if those numbers were selected, the top prize would be very small since it would be split among so many people. In a typical week, the first prize total was $2,775,282; if the winning numbers had been 1, 2, 3, 4, 5, 6, there would have been 5930

winners who would have received only $468 each. Since the chances of winning with any six-number combination are the same, it makes sense to pick a combination not selected by many other players. Avoid patterns like 2, 4, 6, 8, 10, 12 or combinations that form a pattern on the entry card. Players would be wise to pick their numbers randomly, not according to some pattern that other players might also follow.

the histogram representing the probability distribution of the last example. Note that along the horizontal axis, the values of 0, 1, and 2 are located at the centers of the rectangles. This implies that the rectangles are each 1 unit wide, so the areas of the three rectangles are $1 \cdot \frac{1}{4}$, $1 \cdot \frac{1}{2}$, $1 \cdot \frac{1}{4}$ $\left(\text{or } \frac{1}{4}, \frac{1}{2}, \text{ and } \frac{1}{4}\right)$.

In general, if we stipulate that each value of the random variable is assigned a width of 1 on the histogram, then the areas of the rectangles will total 1. **We can therefore associate the probability of each numerical outcome with the area of the corresponding rectangle.** This correspondence between probability and area is an important concept that will be used many times in later chapters.

Suppose we have an experiment with an identified discrete random variable. What do we know about any two events that lead to different values of the random variable? They must be mutually exclusive since their simultaneous occurrence would necessarily lead to the same value of the random variable. In the experiment of giving the drug to two patients, for example, the three values of the random variable (0, 1, 2) are mutually exclusive outcomes. (Among two patients, we cannot get exactly 1 cure and 2 cures at the same time.) Knowing that all the values of the random variable will cover all events of the entire sample space, and knowing that events that lead to different values of the random variable are mutually exclusive, we can conclude that the sum of $P(x)$ for all values of x must be 1. Also, $P(x)$ must be between 0 and 1 for any value of x.

REQUIREMENTS FOR $P(x)$ TO BE A PROBABILITY DISTRIBUTION

1. $\Sigma P(x) = 1$ where x assumes all possible values

2. $0 \leq P(x) \leq 1$ for every value of x

These two requirements for probability distributions are actually direct descendents of the corresponding rules of probabilities (discussed in Chapter 3).

EXAMPLE

Does $P(x) = x/5$ (where x can take on the values of 0, 1, 2, 3) determine a probability distribution?

Solution

If a probability distribution is determined, it must conform to the preceding two requirements. But

$$\Sigma P(x) = P(0) + P(1) + P(2) + P(3)$$

$$= \frac{0}{5} + \frac{1}{5} + \frac{2}{5} + \frac{3}{5}$$

$$= \frac{6}{5} \qquad\qquad \Sigma P(x) \neq 1$$

so that the first requirement is not satisfied and a probability distribution is not determined.

EXAMPLE

Does $P(x) = x/10$ (where x can be 0, 1, 2, 3, or 4) determine a probability distribution?

Solution

For the given function we conclude that

$$P(0) = \frac{0}{10} = 0 \qquad P(3) = \frac{3}{10}$$

$$P(1) = \frac{1}{10} \qquad P(4) = \frac{4}{10}$$

$$P(2) = \frac{2}{10} \qquad \left(\frac{0}{10} + \frac{1}{10} + \frac{2}{10} + \frac{3}{10} + \frac{4}{10} = 1 \right.$$

$$\left. \text{so that } \Sigma P(x) = 1. \right)$$

The sum of these probabilities is 1, and each $P(x)$ is between 0 and 1, so both requirements are satisfied. Consequently, a probability distribution is determined. The graph of this probability distribution is shown in Figure 4-3. Note that the sum of the areas of the rectangles is 1, and each rectangle has an area between 0 and 1.

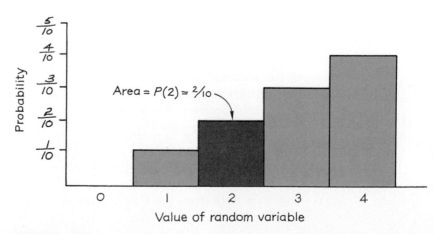

FIGURE 4-3
Value of random variable

We saw in the overview that probability distributions are extremely important in the study of statistics. We have just considered probability distributions of discrete random variables, and later chapters will consider fundamental probability distributions of continuous random variables.

4-2 EXERCISES A

4-1   Of the given variables, which are discrete and which are continuous?
    a. The water temperature of a particular location in a pond
    b. The number of defective vending machines on a particular campus ("all of them" doesn't count)
    c. The total weights of all vending machines on a particular college campus
    d. The mean amount of time required for students to complete a test
    e. The number of trials required for a mouse to find its way through a maze

4-2   Of the given variables, which are discrete and which are continuous?
    a. The mean weight of all students in a particular class
    b. The number of students in a particular class
    c. The number of Democrats that are among the 50 people surveyed in an exit poll
    d. The mean age of the 50 people surveyed in an exit poll
    e. The number of games of chess a particular person plays before winning a game

In Exercises 4-3 through 4-12, determine whether a probability distribution is given. In those cases where P(x) does not determine a probability distribution, identify the requirement that is not satisfied.

4-3   In the accompanying table (based on past results), x represents the number of games required to complete a baseball World Series contest.

x	$P(x)$
4	0.120
5	0.253
6	0.217
7	0.410

4-4   In the accompanying table, x represents the number of children (under 18 years of age) in families. (The table is based on data from the U.S. Census Bureau.)

x	$P(x)$
0	0.48
1	0.21
2	0.19
3	0.08

4-5 In the accompanying table, x represents the number of long-distance personal calls made by an individual in one month.

x	$P(x)$
0	0.32
1	0.08
2	0.12
3	0.09
4	0.07
5	0.06
6	0.04

4-6 In the accompanying table, x represents the number of dots that turn up on a loaded die.

x	$P(x)$
1	1/12
2	1/12
3	1/12
4	1/12
5	1/12
6	1/12

4-7 $P(x) = x$ for $x = 0.1, 0.3, 0.6$

4-8 $P(x) = x$ for $x = 0, 1/2, 1/4$

4-9 $P(x) = x - 0.5$ for $x = 0.5, 0.6, 0.7, 0.8, 0.9$

4-10 $P(x) = x - 2.5$ for $x = 2, 3$

4-11 $P(x) = \dfrac{1}{2(2 - x)!x!}$ for $x = 0, 1, 2$

4-12 $P(x) = \dfrac{3}{4(3 - x)!x!}$ for $x = 0, 1, 2, 3$

In Exercises 4-13 through 4-20, do each of the following:

a. *List the values that the random variable* x *can assume.*

b. *Determine the probability* P(x) *for each value of* x.

c. *Summarize the probability distribution as a table that follows the format of Table 4-1.*

d. *Construct the histogram that represents the probability distribution for the random variable* x. *(See Figure 4-3.)*

e. *Indicate the area of each rectangle in the histogram of part (d) so that there is a correspondence between area and probability.*

4-13 A drug is administered to three patients and the random variable x represents the number of cures that occur. The probabilities corresponding to 0 cures, 1 cure, 2 cures, and 3 cures are found to be 0.125, 0.375, 0.375, and 0.125, respectively.

4-14 A manufacturer produces gauges in batches of three and the random variable x represents the number of defects in a batch. The probabilities corresponding to 0 defects, 1 defect, 2 defects, and 3 defects are found to be 0.70, 0.20, 0.09, and 0.01, respectively.

4-15   A computer simulation involves the repeated generation of a random number. The only possibilities are 0, 1, 2, and they are equally likely.

4-16   A computer is used to select the last digit of telephone numbers to be dialed for a poll. The possible values are 0, 1, 2, . . . , 9, and they are equally likely.

4-17   The random variable x represents the number of girls in a family of three children. (*Hint:* Assuming that boys and girls are equally likely, we get $P(2) = 3/8$ by examing the sample space of *bbb, bbg, bgb, bgg, gbb, gbg, ggb, ggg.*)

4-18   The random variable x represents the number of boys in a family of four children. (See Exercise 4-17.)

4-19   Two dice are rolled. The random variable x represents the total number of dots that turn up. (*Hint:* The sample space consists of 36 cases and $P(2) = 1/36$.)

4-20   A batch of 20 computer chips contains exactly 6 that are defective. Two different chips are randomly selected without replacement, and the random variable x represents the number of defective chips selected. (*Hint:* Use the multiplication rule to first find $P(0)$ and $P(2)$.)

4-2 EXERCISES B

4-21   Let $P(x) = 0.4(0.6)^{x-1}$ where $x = 1, 2, 3, \ldots$. Is $P(x)$ a probability distribution?

4-22   a. Let $P(x) = 1/2^x$ where $x = 1, 2, 3, \ldots$. Is $P(x)$ a probability distribution?
b. Let $P(x) = 1/2x$ where $x = 1, 2, 3, \ldots$. Is $P(x)$ a probability distribution?

4-23   According to an analyst for Paine Webber, Intel's yield for computer chips is around 35%. Assume that we have four randomly selected chips and that the probability of an acceptable chip is 0.35. Let x represent the number of acceptable chips in groups of four and use the multiplication rule and the rule of complements (from Chapter 3) to complete the table so that a probability distribution is determined.

x	$P(x)$
0	
1	0.384
2	0.311
3	
4	

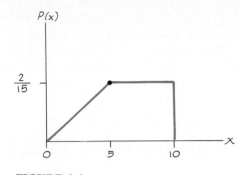

FIGURE 4-4

4-24   Given that x is a continuous random variable with the distribution shown in Figure 4-4, find each of the following.
   a. The total area of the enclosed region
   b. The probability that x is less than 5
   c. The probability that x is less than 7
   d. The probability that x is between 6 and 7
   e. The probability that x is between 3 and 7

4-3   MEAN, VARIANCE, AND EXPECTATION

We know how to find the mean and variance for a given set of scores (see Chapter 2), but suppose we have scores that are "conceptualized" in the sense that we know the probability distribution instead of a sample of observations or measurements. For example, suppose that we want to determine the mean and variance for the numbers of boys that will occur in pairs of independent births. (Assume that boys and girls are equally likely.) We have no specific results to work with, but we do know that the outcomes of 0, 1, and 2 boys have probabilities of 1/4, 1/2, 1/4, respectively. One way to find the mean and variance for the numbers of boys is to pretend that theoretically ideal results actually occurred. We list the different possible outcomes.

Outcome	Number of boys
boy-boy	2
boy-girl	1
girl-boy	1
girl-girl	0

Since they are equally likely, we can pretend that the four results actually did occur. The mean of 2, 1, 1, and 0 is 1, while the variance is 0.5. (The variance is found by applying Formula 2-4. In this use of Formula 2-4 we divide by n instead of $n - 1$ because we assume that we have all scores of the population.) Instead of considering four trials that yield theoretically ideal results, we could pretend that a large number of trials (say 4000) yielded these theoretically ideal results:

		Number of boys	Frequency
boy-boy	(2 boys 1000 times)	2	1000
boy-girl	(1 boy 1000 times)	1	2000
girl-boy	(1 boy 1000 times)	0	1000
girl-girl	(0 boys 1000 times)		

Our random variable represents the number of boys in two births, so the preceding results suggest a list of 4000 scores that consist of 1000 twos, a total of 2000 ones, and 1000 zeros. The mean of those 4000 scores is again 1, while the variance is again 0.5. In this example, it really makes no difference whether we presume 4 theoretically ideal trials or 4000.

Instead of pretending that theoretically ideal results have occurred, we can find the mean of a discrete random variable by using Formula 4-1.

FORMULA 4-1 $$\mu = \Sigma x \cdot P(x)$$

This formula is justified by relating $P(x)$ to its role in describing the relative frequency with which x occurs. Recall that the mean of any list of scores is the sum of those scores divided by the total number of scores n. We usually compute the mean in that order; that is, first sum the scores and then divide the total by n. However, we can obtain the same result by dividing each individual score by n and then summing the quotients. For example, to find the mean of 2, 1, 1, 0 we usually write

$$\frac{2 + 1 + 1 + 0}{4} = 1$$

but we can also compute the mean as

$$\frac{2}{4} + \frac{1}{4} + \frac{1}{4} + \frac{0}{4} = 1$$

Similarly, we can find the mean of the preceding 4000 scores by writing

$$\frac{(2 \cdot 1000) + (1 \cdot 2000) + (0 \cdot 1000)}{4000} = 1$$

but we can also compute that mean as

$$\mu = \left(2 \cdot \frac{1000}{4000}\right) + \left(1 \cdot \frac{2000}{4000}\right) + \left(0 \cdot \frac{1000}{4000}\right) = 1$$

The right side of this last equation corresponds to $\Sigma\, x \cdot P(x)$ in Formula 4-1.

For each specific value of x, $P(x)$ can be considered the relative frequency with which x occurs. $P(x)$ takes into account the repetition of specific x scores when we compute the mean. It also incorporates division by n directly into the summation process.

Similar reasoning enables us to take the variance formula from Chapter 2 ($\sigma^2 = \Sigma\, (x - \mu)^2/N$) and apply it to a random variable of a probability distribution to get

FORMULA 4-2 $\sigma^2 = \Sigma(x - \mu)^2 \cdot P(x)$

Again the use of $P(x)$ takes into account the repetition of specific x scores and simultaneously accomplishes the division by N. This latter formula for variance is usually manipulated into an equivalent form to facilitate computations, as shown in Formula 4-3.

FORMULA 4-3 $\sigma^2 = [\Sigma x^2 \cdot P(x)] - \mu^2$ (shortcut version of Formula 4-2)

Formula 4-3 is a shortcut version that will always produce the same result as Formula 4-2. Formula 4-3 is usually easier to work with, while Formula 4-2 is easier to understand directly.

To apply Formula 4-3 to a specific case, we square each value of x and multiply that square by the corresponding probability and then add all of those products. We then subtract the square of the mean. The standard deviation σ can be easily obtained by simply taking the square root of the variance, as shown in Formula 4-4. If the variance is found to be 0.5, the standard deviation is $\sqrt{0.5}$, or about 0.7.

FORMULA 4-4 $\sigma = \sqrt{[\Sigma x^2 \cdot P(x)] - \mu^2}$

EXAMPLE

Use Formula 4-1, 4-3, and 4-4 to find the mean, variance, and standard deviation of the random variable that represents the number of boys in two independent births. (Assume that a girl or a boy is equally likely to occur.)

continued ▶

Example, continued

Solution

In the following table, the two leftmost columns summarize the probability distribution. The three rightmost columns are created for the purposes of the computations that are required.

x	$P(x)$	$x \cdot P(x)$	x^2	$x^2 \cdot P(x)$
0	$\frac{1}{4}$	0	0	0
1	$\frac{1}{2}$	$\frac{1}{2}$	1	$\frac{1}{2}$
2	$\frac{1}{4}$	$\frac{2}{4}$	4	1
Total		1		1.5

$$\uparrow \qquad\qquad\qquad \uparrow$$
$$\Sigma x \cdot P(x) = 1 \qquad \Sigma x^2 \cdot P(x) = 1.5$$

From the table we see that $\mu = \Sigma x \cdot P(x) = 1$ and

$$\sigma^2 = [\Sigma x^2 \cdot P(x)] - \mu^2 = (1.5) - 1^2 = 0.5$$

The standard deviation is the square root of the variance, so that

$$\sigma = \sqrt{0.5} = 0.7$$

An important advantage of these techniques is that a probability distribution is actually a model of a theoretically perfect population frequency distribution. Since the probability distribution allows us to perceive the population, we are able to determine the values of important parameters such as the mean, variance, and standard deviation. This in turn allows us to make the inferences that are necessary for decision making in a multitude of different professions.

Expected Value

We will now consider an application of the mean of a discrete random variable. First, it should be stressed that when we calculate the mean of a discrete random variable, we are not getting the value that we expect will occur most often. In fact, we often obtain a value that cannot occur in any one trial (such as 1.5 girls in 3 births). The mean of a discrete random variable is the theoretical mean outcome for infinitely many trials. We can think of that mean as the expected value in the sense that it is the average (mean) value that we would expect

Prophets for Profits

For the unreasonable price of $40 per year, subscribers can receive a weekly publication "designed to be of help in picking a winning set of Lotto or N.Y. State Lottery numbers." Apparently blind to the fact that the lottery numbers are randomly selected each week, this booklet recommends numbers to bet. Some numbers are recommended as "hot" because they have been coming up often, others are recommended as "due" because they haven't been coming up often. Other choices involve horoscopes, dreams, and numbers that have "appeared or talked to" a seeress.

Jean Simpson wrote the book *Hot Lotto Numbers* that supposedly helps readers use astrology, numerology, and a deck of cards to pick winning lottery numbers. In reality, books such as this are of no help in predicting winning numbers.

to get if the trials could continue indefinitely. The uses of expected value (also called expectation or mathematical expectation) are extensive and varied, and they play a very important role in an area of application called *decision theory*.

DEFINITION

The **expected value** of a discrete random variable is $E = \Sigma\, x \cdot P(x)$

From Formula 4-1 we see that $E = \mu$. That is, the mean of a discrete random variable and its expected value are the same.

As an example, consider the numbers game started many years ago by members of organized crime groups and now run legally by many organized governments. You can bet that the three-digit number of your choice will be the winning number selected. Suppose that you bet $1 on the number 327. What is your expected value? For this bet there are two simple outcomes: You win or you lose. Since you have the number 327 and there are 1000 possibilities (from 000 to 999), your probability of winning is 1/1000 and your probability of losing is 999/1000. The typical winning payoff is 499 to 1, meaning that for each $1 bet, you would be given $500 and your net return is therefore $499. Table 4-2 summarizes this situation.

| TABLE 4-2 |
| The Numbers Game |

Event	x	$P(x)$	$x \cdot P(x)$
Win	$499	$\dfrac{1}{1000}$	$\dfrac{\$499}{1000}$
Lose	−$1	$\dfrac{999}{1000}$	$-\dfrac{\$999}{1000}$
Total			$\dfrac{-\$500}{1000} = -50¢$

From Table 4-2 we can see that when we bet $1 in the numbers game, our expected value is

$$E = \Sigma x \cdot P(x) = -50¢ \qquad \text{(from Table 4-2)}$$

This means that in the long run, for each $1 bet, we can expect to lose an average of 50¢. The house "take" is 50%, which is very greedy when compared to the house take in casino games such as slot machines (13%), roulette (5.25%), or craps (1.4%).

In Chapter 10 we will use the concept of expected value to compare actual survey results to expected results; the degree of similarity or disparity will allow us to form some meaningful conclusions.

4-3 EXERCISES A

In Exercises 4-25 through 4-32, find the mean, variance, and standard deviation of the random variable x.

4-25

x	$P(x)$
0	0.2
1	0.5
2	0.3

4-26

x	$P(x)$
0	0.2
1	0.7
2	0.1

4-27

x	$P(x)$
1	0.15
2	0.45
3	0.35
4	0.05

4-28

x	$P(x)$
5	0.20
6	0.10
7	0.45
8	0.25

4-29

x	$P(x)$
5	1/4
10	1/4
20	1/4
50	1/4

4-30

x	$P(x)$
5	1/20
10	3/20
20	7/20
50	9/20

4-31

x	$P(x)$
2	4/30
4	6/30
6	10/30
8	6/30
10	4/30

4-32

x	$P(x)$
2	1/5
4	1/5
6	1/5
8	1/5
10	1/5

4-33 Using Census Bureau data, the accompanying table is constructed. It describes the probability distribution for males aged 18 to 24 who live at home with their parents. The random variable x represents the number of such males when four males aged 18 to 24 are randomly selected. Find the mean, standard deviation, and variance for the random variable x.

x	$P(x)$
0	0.026
1	0.154
2	0.346
3	0.346
4	0.130

4-34 If five randomly selected people are tested for group O blood, the accompanying table (based on data from the Greater New York Blood Program) describes the probability distribution. In that table, x represents the number of people with group O blood. Find the mean, standard deviation, and variance for the number of people with group O blood.

x	$P(x)$
0	0.050
1	0.206
2	0.337
3	0.276
4	0.113
5	0.018

4-35 According to an analyst for Paine Webber, Intel's yield for computer chips is around 35%. For a batch of six randomly selected computer chips, the accompanying table describes the probability distribution for the number of defects. Find the mean, standard deviation, and variance for the number of defects.

x	$P(x)$
0	0.002
1	0.020
2	0.095
3	0.235
4	0.328
5	0.244
6	0.075

4-36   According to the Hertz Corporation, 69% of all workers commute in their own cars. For eight randomly selected workers the accompanying table describes the probability distribution for the number of workers who commute in their own cars. Find the mean, variance, and standard deviation.

x	$P(x)$
0	0.000
1	0.002
2	0.012
3	0.053
4	0.147
5	0.261
6	0.290
7	0.185
8	0.051

4-37   On any given weekday, the probabilities of 0, 1, or 2 accidents on a certain highway are 0.650, 0.300, and 0.050 respectively. Find the mean, variance, and standard deviation for the number of accidents in a day.

4-38   A commuter airline company finds that for a certain flight, the probabilities of 0, 1, 2, or 3 vacant seats are 0.705, 0.115, 0.090, and 0.090 respectively. Find the mean, variance, and standard deviation for the number of vacant seats.

4-39   Based on past results found in the *Information Please Almanac*, we estimate that there is a 0.120 probability that a baseball World Series contest will last four games, a 0.253 probability that it will last five games, a 0.217 probability of six games, and a 0.410 probability of seven games. Find the mean, standard deviation, and variance for the numbers of games that World Series contests last.

4-40   A computer store finds that the probabilities of selling 0, 1, 2, 3, or 4 microcomputers in one day are 0.245, 0.370, 0.210, 0.095, and 0.080 respectively. Find the mean, variance, and standard deviation for the number of microcomputer sales in one day.

4-41   If you have a 1/4 probability of gaining $500 and a 3/4 probability of gaining $200, what is your expected value?

4-42   If you have a 1/10 probability of gaining $1000 and a 9/10 probability of losing $300, what is your expected value?

4-43   If you have a 1/10 probability of gaining $200, a 3/10 probability of losing $300, and a 6/10 probability of breaking even, what is your expected value?

4-44   A car wash loses $30 on rainy days and gains $120 on days when it does not rain. If the probability of rain is 0.15, what is the expected value?

4-45   A contractor bids on a job to construct a building. There is a 0.7 probability of making a $175,000 profit, and there is a probability of 0.3 that the contractor will break even. What is the expected value?

4-46   A contractor bids on a job to construct a building. There is a 0.7

probability of making a $175,000 profit and a 0.3 probability of losing $250,000. What is the expected value?

4-47 A flight-training school makes a profit of $200 on fair days but loses $150 for each day of bad weather. If the probability of bad weather is 0.4, what is the expected value?

4-48 A popular magazine runs a sweepstakes as an advertising method. The various prizes are listed, along with their approximate numerical odds of winning.

First prize: $25,000   (one chance in 15,660,641)
Second prize: $24,000   (one chance in 5,455,950)
Third prize: $5,000   (one chance in 3,132,128)
Fourth prize $1,000   (one chance in 313,213)
Fifth prize: $25   (one chance in 7,830)

a. Compute the expected value.
b. If the only cost of entering this contest is a stamp, how much does a person win or lose in the long run?

4-3 EXERCISES B

4-49 A couple plans to have five children. Let the random variable be the number of girls that will occur. Assume that a boy or a girl is equally likely to occur and that the sex of any successive child is unaffected by previous brothers or sisters.

x	$P(x)$
0	
1	
2	
3	
4	
5	

a. List the 32 different possible simple events.
b. Enter the probabilities in the table at left, where x represents the number of girls in the five births.
c. Graph the histogram for the probability distribution from (b).
d. Find the mean number of girls that will occur among the five births.
e. Find the variance for the number of girls that will occur.
f. Find the standard deviation for the number of girls that will occur.
g. On the histogram in part (c), identify the location of the mean from part (d). Also identify the location of the value that is exactly one standard deviation above the mean. Then identify the location of the value that is exactly one standard deviation below the mean.

4-50 The variance for the discrete random variable x is 1.25.
a. Find the variance of the random variable $5x$. (Each value of x is multiplied by 5.)
b. Find the variance of the random variable $x/5$.
c. Find the variance of the random variable $x + 5$.
d. Find the variance of the random variable $x - 5$.

4-51   A discrete random variable can assume the values 1, 2, . . . , n and those values are equally likely.

a. Show that $\mu = (n + 1)/2$.   b. Show that $\sigma^2 = (n^2 - 1)/12$.

(*Hint:* $1 + 2 + 3 + \cdots + n = n(n + 1)/2$.

$$1^2 + 2^2 + 3^2 + \cdots + n^2 = n(n + 1)(2n + 1)/6.)$$

4-52   Verify that $\sigma^2 = [\Sigma x^2 \cdot P(x)] - \mu^2$ is equivalent to $\sigma^2 = \Sigma(x - \mu)^2 \cdot P(x)$. (*Hint:* For constant c, $\Sigma cx = c\,\Sigma x$. Also, $\mu = \Sigma x \cdot P(x)$.)

4-4 BINOMIAL EXPERIMENTS

Many real and important applications require that we find the probability that some event will occur x times out of n different trials. In any one trial, either the event occurs or it does not, so we are dealing with an element of "twoness" present in situations such as these:

- Elections: voters choosing between two candidates
- Manufacturing: products being acceptable or defective items
- Education: subjects passing or failing a test
- Medicine: new drugs being effective or ineffective
- Psychology: mental health treatment being effective or ineffective
- Agriculture: crops being profitable or nonprofitable
- Advertising: television viewers recalling or not recalling a sponsor's name

All these situations exhibit the element of twoness that results in a special type of discrete probability distribution called the *binomial distribution*. A binomial distribution is a list of outcomes and probabilities for a binomial experiment.

DEFINITION

A **binomial experiment** is one that meets all the following requirements:

1. The experiment must have a fixed number of trials.
2. The trials must be independent.
3. Each trial must have all outcomes classified into two categories.
4. The probabilities must remain constant for each trial.

This definition indicates that in a binomial experiment, we have a fixed number of independent trials where each outcome has only two classifications. The term independent simply means that the outcome of one trial will not affect the probabilities of outcomes on subsequent trials. The birth of a girl does not affect the sex of the next baby that is born, so these events are independent.

In Section 3-4 we noted that it is a common practice to treat events as independent when small samples are drawn from large populations. One common guideline is to assume independence whenever the size of the sample is at most 5% of the size of the population. This approach is used extensively by professional statisticians in industry and research.

Many experiments have outcomes that fall into more than two categories. To treat them as *binomial* experiments, we must develop a way to classify all outcomes into exactly two categories. For a multiple-choice test question, there may be five possible answers, but the two categories of "right" and "wrong" are suitable for a binomial experiment. The requirement of two categories does not necessarily exclude experiments with more than two outcomes or simple events.

NOTATION

S and F (success and failure) denote the two possible categories of all outcomes; p and q will denote the probabilities of S and F, respectively, so that

$$P(S) = p$$

$$P(F) = 1 - p = q$$

n will denote the fixed number of trials.

x will denote a specific number of successes in n trials so that x can be any whole number between 0 and n, inclusive.

p denotes the probability of success in *one* of the n trials.

q denotes the probability of failure in *one* of the n trials.

$P(x)$ denotes the probability of getting exactly x successes among the n trials.

The word *success* as used here does not necessarily correspond to a good event. Selecting a defective parachute may be classified a success, even though the results of such a selection may be less than pleasant. Either of the two possible categories may be called the success S as long as the corresponding probability is identified as p. The value of q can always be found by subtracting p from 1. If $p = 0.4$, then $q = 1 - 0.4$, or 0.6.

We will soon introduce a formula for computing probabilities in a binomial experiment, but we begin with an example intended to illustrate the reasoning underlying that formula. Try to follow this example carefully so that the ensuing formula will not seem to be a mysterious revelation.

According to a study by the American College Testing Program, there is a 30% dropout rate among freshmen at four-year public colleges. Looking at the bright side, we can also say that 70% of those freshmen continue their studies. Let's assume that five freshmen students are randomly selected from four-year public colleges. Let's also stipulate that a success is a freshman who continues while a failure is a dropout. If we want to find the probability that all five freshmen continue, we can use the multiplication rule (Section 3-4) to compute

P(5 freshmen continue)

$$= P(\text{success}) \cdot P(\text{success}) \cdot P(\text{success}) \cdot P(\text{success}) \cdot P(\text{success})$$

$$= 0.7 \times 0.7 \times 0.7 \times 0.7 \times 0.7$$

$$= 0.16807$$

The probability of 0.7 remains constant for the five trials because this example involves five separate and independent random selections.

Let's again consider the random selection of five freshmen from four-year public colleges, but this time we will determine the probability that exactly two of them continue their studies (and the other three drop out). That is, we want the probability of two successes and three failures. Since $P(\text{success}) = 0.7$, it follows that $P(\text{failure}) = 0.3$. We may be tempted to solve this problem by computing

$$0.7 \times 0.7 \times 0.3 \times 0.3 \times 0.3 = 0.01323$$

This solution is very common and natural, but it is incomplete because it contains an implicit assumption that the *first* two freshmen continued while the last three dropped out. But we are seeking the probability of exactly two successes with no stipulation that the dropouts must be the last three freshmen. If we represent success by S and failure by F, the product of 0.01323 corresponds to the event S-S-F-F-F. However, there are other ways or orders of listing 2 successes and 3 failures. There are actually 10 possible arrangements, and each one has a probability of 0.01323. We list them in the margin at the left.

S-S-F-F-F
F-F-F-S-S
F-F-S-F-S
F-S-F-S-F
S-F-F-F-S
F-F-S-S-F
F-S-S-F-F
S-F-F-S-F
S-F-S-F-F
F-S-F-F-S

There are 10 ways to arrange 2 successes and 3 failures.

Since there are 10 ways of getting exactly 2 successes among the 5 freshmen and each different way has a probability of 0.01323, there is a total probability of $10 \times 0.01323 = 0.1323$ that 2 of the 5 freshmen will continue. We can generalize this specific result by stating that with n independent trials in which $P(S) = p$ and $P(F) = q$, the multiplication rule indicates that the probability of the first x cases being successes while the remaining cases are failures is

$$\underbrace{p \cdot p \cdots p}_{x \text{ times}} \cdot \underbrace{q \cdot q \cdots q}_{(n - x) \text{ times}}$$

or

$$p^x \cdot q^{n-x}$$

But $P(x)$ denotes the probability of x successes among n trials in *any* order, so that $p^x \cdot q^{n-x}$ must be multiplied by the number of ways the x successes and $n - x$ failures can be arranged. The number of ways in which it is possible to arrange x successes and $n - x$ failures is shown in Formula 4-5.

FORMULA 4-5 $\dfrac{n!}{(n - x)!x!}$ Number of outcomes with exactly x successes among n trials

The factorial symbol, introduced in Section 3-6, denotes the product of decreasing factors. Many calculators have a factorial key and some sample factorials are listed here.

$$3! = 3 \cdot 2 \cdot 1 = 6$$
$$2! = 2 \cdot 1 = 2$$
$$1! = 1$$
$$0! = 1 \text{ (by definition)}$$
$$n! = n \cdot (n - 1)(n - 2) \ldots 1$$

The expression given as Formula 4-5 does correspond to $_nC_r$, also introduced in Section 3-6. (Coverage of Section 3-6 is not necessary for this chapter.) We will not derive Formula 4-5, but its role should become clear in the example that follows. Formula 4-5 can be derived by using the binomial theorem for the expansion of $(a + b)^n$.

We can now combine this counting device (Formula 4-5) with the direct application of the multiplication rule for independent events to get a general formula for computing what we will call binomial probabilities.

FORMULA 4-6
Binomial probability formula

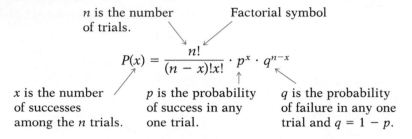

$$P(x) = \frac{n!}{(n - x)!x!} \cdot p^x \cdot q^{n-x}$$

n is the number of trials.

Factorial symbol

x is the number of successes among the n trials.

p is the probability of success in any one trial.

q is the probability of failure in any one trial and $q = 1 - p$.

Let's use this formula to repeat the solution to our original problem by finding the probability that exactly two freshman continue among the five randomly selected from public four-year colleges. We have $P(\text{continue}) = P(S) = p = 0.7$. Also, $n = 5$, $q = 0.3$, and $x = 2$, so that

$$P(2) = \frac{5!}{(5 - 2)!2!} \cdot 0.7^2 \cdot 0.3^{5-2}$$

$$= \frac{120}{6 \cdot 2} \cdot 0.7^2 \cdot 0.3^3$$

$$= 10 \cdot 0.01323 = 0.1323$$

This specific example can help us see the rationale underlying Formula 4-6. The component $n!/(n - x)!x!$ counts the number of ways of getting x successes in n trials. In this example we have 10 ways of getting 2 successes in 5 trials. The product $p^x \cdot q^{n-x}$ gives us the probability of getting exactly x successes among n trials in one particular order. The product of those factors is the total probability representing all ways of getting exactly x successes in n trials. We can think of the binomial probability formula as the product of these components:

The number of outcomes with exactly x successes among n trials

The probability of x successes among n trials for any one particular order

$$P(x) = \frac{n!}{(n - x)!x!} \cdot p^x \cdot q^{n-x}$$

EXAMPLE

According to one study conducted at the University of Texas at Austin, 2/3 of all Americans can do routine computations. If an employer were to hire eight randomly selected Americans, what is the probability that exactly five of them can do routine computations?

Solution

The experiment is binomial because of the following.

1. We have a fixed number of trials (8).
2. The trials are independent since the employees are randomly selected.
3. There are two categories, since each employee either can or cannot do routine computations.
4. The probability of 2/3 remains constant from trial to trial.

We begin by identifying the values of $n, p, q,$ and x so that we can apply the binomial probability formula. We have

$$n = 8 \text{ (number of trials)}$$

$$p = 2/3 \text{ (probability of success)}$$

$$q = 1/3 \text{ (probability of failure)}$$

$$x = 5 \text{ (desired number of successes)}$$

We should check for consistency by verifying that what we call a success, as counted by x, is the same success with probability p. That is, we must be sure that x and p refer to the same concept of success. Using the values for $n, p, q,$ and x in the binomial probability formula, we get

$$P(5) = \frac{8!}{(8-5)!5!} \cdot \left(\frac{2}{3}\right)^5 \cdot \left(\frac{1}{3}\right)^{8-5}$$

$$= \frac{40,320}{(6)(120)} \cdot \frac{32}{243} \cdot \frac{1}{27}$$

$$= 0.273$$

There is a probability of 0.273 that five of the eight employees can do routine computations.

In the last example we found $P(5)$. We could also use the binomial probability formula of find $P(0)$, $P(1)$, $P(2)$, $P(3)$, $P(4)$, $P(6)$, $P(7)$, and $P(8)$ so that the complete probability distribution for this case will be known. The results are shown in Table 4-3, where x denotes the number of employees who can do routine computations. We can depict Table 4-3 in the form of a histogram, as in Figure 4-5. The shape of a probability distribution is often a critically important feature. (We consider the characteristic of distribution shape in Section 4-6.)

TABLE 4-3	
x	$P(x)$
0	0.000
1	0.002
2	0.017
3	0.068
4	0.171
5	0.273
6	0.273
7	0.156
8	0.039

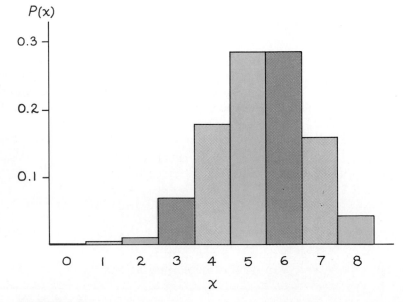

FIGURE 4-5

Composite Sampling

In the 1940s, the United States Army included the Kahn-Wasserman test in its physical examination of inductees. This test, designed to detect the presence of syphilis, required blood samples that were given chemical analyses. Since the testing procedure was time-consuming and expensive, a more efficient one was sought. One researcher noted that if samples from a large number of blood specimens were mixed in pairs and then tested, the total number of chemical analyses would be greatly reduced. Syphilitic inductees could be identified by retesting the few blood samples that were included in the pairs indicating the presence of syphilis. But if the total number of analyses was reduced by pairing blood specimens, why not put them in groups of three or four or more? Probability theory was used to find the most efficient group size, and a general theory was developed for detecting the defective members of any population. Bell Laboratories, for example, used this approach to identify defective capacitors by putting groups of them in a vacuum and testing for leakage. This technique, referred to as *composite sampling*, has also been used to sample bales of wool and to test for pesticides in drinking water.

An alternative to computing with the binomial probability formula involves the use of the table of binomial probabilities (Table A-1 in Appendix A). To use this table, first locate the relevant value of n in the leftmost column and then locate the corresponding value of x that is desired. At this stage, one row of numbers should be isolated. Now align that row with the proper probability of p by using the column across the top. The isolated number represents the desired probability (missing its decimal point at the beginning). A very small probability such as 0.000000345 is indicated by 0+. For example, the table indicates that for $n = 10$ trials, the probability of $x = 2$ successes when $P(S) = p = 0.05$ is 0.075. (A more precise value is 0.0746348, but the table values are approximate.)

EXAMPLE

Find the probability that of five babies, there are exactly four girls. (Assume that a boy or a girl is equally likely to occur.)

Solution

This is a binomial experiment with $n = 5$, $P(S) = p = 0.5$, and $x = 4$. Applying the binomial probability formula, we get

$$P(4) = \frac{5!}{(5 - 4)!4!} \cdot 0.5^4 \cdot 0.5^{5-4}$$

$$= \frac{120}{1 \cdot 24} \cdot 0.0625 \cdot 0.5$$

$$= 0.15625$$

The probability of getting four girls in five births is 0.156. Using Table A-1 we also get 0.156.

Although the values in Table A-1 are approximate and only selected values of p are included, the use of such a table often provides quick and easy results. The next example illustrates situations in which the use of Table A-1 can replace several computations with the binomial probability formula.

EXAMPLE

The operation of a component in a computer system is so critical that three of these components are built in for backup protection. The latest quality control tests indicate that this type of component has a 10% failure rate. (For optimists, that's a 90% success rate.) Assume that this is a binomial experiment and use Table A-1 to find the probabilities of the following events.

a. Among the three components, at least one continues to work.
b. Among the three components, none continue to work.

Solution

For both parts of this problem, we stipulate that a success corresponds to a component that continues to work. With three com-

continued ▶

Solution, continued

ponents and a 10% failure rate, we have

$$n = 3 \qquad p = 0.9 \qquad q = 0.1$$

In the space below we show the portion of Table A-1 that relates to this problem. We have also constructed the table describing the probability distribution.

n	x	p .90		x	$P(x)$
3	0 ...	001		0	0.001
	1 ...	027	$\longrightarrow$	1	0.027
	2 ...	243		2	0.243
	3 ...	729		3	0.729

From Table A-1

We can now proceed to find the indicated probabilities.

a. If at least one component continues to work, we have $x = 1$ or 2 or 3. We get

$$P(\text{at least } 1) = P(1 \text{ or } 2 \text{ or } 3)$$

$$= P(1) + P(2) + P(3) \qquad \text{(addition rule)}$$

$$= 0.027 + 0.243 + 0.729 \qquad \text{(from preceding table)}$$

$$= 0.999$$

The probability of at least one working component is 0.999.

b. The probability of no component working is represented by $P(0)$. From the preceding table we see that $P(0) = 0.001$. There is a 0.001 probability that none of the three components will continue to work.

To *really* appreciate the ease of using Table A-1, let $n = 25$, $p = 0.6$, and find the probability of at least 13 successes using Table A-1 and the binomial probability formula. The use of Table A-1 involves looking up the 13 probabilities (for $x = 13, 14, 15, \ldots, 25$) and adding them. The use of the binomial probability formula involves using that formula 13 times, computing the 13 different probabilities, and then adding them. Given the choice in this case, no sane person would choose the formula. However, we must use the formula when the values of n or p do not allow us to use Table A-1.

We use Table A-1 or the binomial probability formula when we wish to determine probabilities for a binomial experiment.

Many computer statistics packages include an option for generating binomial probabilities. The following sample output from STATDISK illustrates one such program. The user simply enters n and p and the entire probability distribution table is displayed. The column labeled "Cum. Prob." represents cumulative probabilities by adding up the values of $P(x)$ as it goes down the column. The Minitab display shows the two required commands followed by the resulting output.

STATDISK DISPLAY

```
This program uses the BINOMIAL PROBABILITY FORMULA
to find probabilities and cumulative probabilities.

You must enter

  n = the number of trials or sample size (Max. 125)
  P = the probability of success on a single trial

              n = 5

            P = 0.30000

Mean = 1.50000 St. Dev. = 1.02470 Variance = 1.05000

         X      P(X)       Cum.Prob.
         -----------------------------
         0     0.16807      0.16807
         1     0.36015      0.52822
         2     0.30870      0.83692
         3     0.13230      0.96922
         4     0.02835      0.99757
         5     0.00243      1.00000
```

MINITAB DISPLAY

```
MTB > PDF;
SUBC> BINOMIAL n=5 P=.3.

   BINOMIAL WITH N =   5  P = 0.300000
      K              P( X = K )
      0                 0.1681
      1                 0.3601
      2                 0.3087
      3                 0.1323
      4                 0.0284
      5                 0.0024
```

TABLE 4-4 Three Engines	
x	$P(x)$
0	0.9997872491
1	0.0002127358
2	0.0000000151
3	0.0000000000

TABLE 4-5 Two Engines	
x	$P(x)$
0	0.9998581611
1	0.0001418339
2	0.0000000050

At the beginning of this chapter, we briefly discussed the Federal Aviation Administration's consideration of allowing twin-engine jets to make transatlantic flights. (Existing regulations required at least three engines.) A realistic estimate for the probability of an engine failing on a transatlantic flight is 1/14,100. Using that probability and the binomial probability formula, we can develop the probability distributions as summarized in Tables 4-4 and 4-5, where x represents the number of engines that fail on a transatlantic flight.

Using those tables and assuming that a flight will be completed if at least one engine works, we get

P(safe flight with twin-engine jet) = P(0 or 1 engine failures)

$$= 0.9998581611 + 0.0001418339 = 0.9999999950$$

P(safe flight with three-engine jet) = P(0, 1, or 2 engine failures)

$$= 0.9997872491 + 0.0002127358 + 0.0000000151$$
$$= 1.0000000000 \text{ (rounded to 10 decimal places)}$$

The result of 1.0000000000 is actually a rounded-off form of the more precise result of 0.9999999999996433. Comparing the two resulting probabilities, we see that the three-engine jet would be safer, as expected, but the difference doesn't seem to be too significant. All of those leading nines in the results suggest that both configurations are quite safe. The Federal Aviation Administration used this type of reasoning when it changed its regulations to allow transatlantic flights by jets with only two engines. We could analyze the significance of those final probabilities further, but this illustration does show a very useful application of the binomial probability formula.

To keep this section in perspective, remember that the binomial probability formula is only one of many probability formulas that can be used for different situations. It is, however, among the most important and most useful of all discrete probability distributions. In practical cases it is often used in problems such as quality control, voter analysis, medical research, military intelligence, and advertising.

4-4 EXERCISES A

4-53 Which of the following can be treated as binomial experiments?
a. Testing a sample of 5 capacitors (with replacement) from a population of 20 capacitors, of which 40% are defective
b. Testing a sample of 5 capacitors (without replacement) from a population of 20 capacitors of which 40% are defective
c. Tossing an unbiased coin 500 times
d. Tossing a biased coin 500 times
e. Surveying 1700 television viewers to determine whether or not they watched a particular show

4-54 Which of the following can be treated as binomial experiments?
a. Surveying 1200 registered voters to determine their preferences for the person to serve as the next president
b. Surveying 1200 registered voters to determine whether or not they would again vote for the current president
c. Sampling a randomly selected group of 500 prisoners to determine whether or not they have been in prison before
d. Sampling a randomly selected group of 500 prisoners to determine the lengths of their current sentences
e. Testing 500 randomly selected drivers to determine whether or not their blood alcohol content levels are over 0.10%

4-55 In a binomial experiment, a trial is repeated n times. Find the probability of x successes given the probability p of success on a given trial. (Use the given values of n, x, and p and Table A-1.)
a. $n = 10, x = 3, p = 0.5$ b. $n = 10, x = 3, p = 0.4$
c. $n = 7, x = 0, p = 0.1$ d. $n = 7, x = 0, p = 0.99$
e. $n = 7, x = 7, p = 0.01$

4-56 In a binomial experiment, a trial is repeated n times. Find the probability of x successes given the probability p of success on a given trial. (Use the given values of n, x, and p and Table A-1.)
a. $n = 15, x = 5, p = 0.7$ b. $n = 12, x = 11, p = 0.6$
c. $n = 9, x = 6, p = 0.1$ d. $n = 8, x = 5, p = 0.95$
e. $n = 14, x = 14, p = 0.9$

4-57 For each of the following, use Formula 4-5 to find the number of ways you can arrange x successes and $n - x$ failures.
  a. $n = 5, x = 3$ b. $n = 5, x = 0$
  c. $n = 8, x = 7$ d. $n = 8, x = 3$
  e. $n = 8, x = 8$

4-58 For each of the following, use Formula 4-5 to find the number of ways you can arrange x successes and $n - x$ failures.
  a. $n = 6, x = 2$ b. $n = 6, x = 6$
  c. $n = 6, x = 0$ d. $n = 10, x = 3$
  e. $n = 20, x = 18$

4-59 In a binomial experiment, a trial is repeated n times. Find the probability of x successes given the probability p of success on a single trial. Use the given values of n, x, and p and the binomial probability formula. Leave answers in the form of fractions.
  a. $n = 5, x = 3, p = 1/4$
  b. $n = 4, x = 2, p = 1/3$
  c. $n = 5, x = 1, p = 2/3$

4-60 In a binomial experiment, a trial is repeated n times. Find the probability of x successes given the probability p of success on a single trial. Use the given values of n, x, and p and the binomial probability formula. Leave answers in the form of fractions.
  a. $n = 6, x = 2, p = 1/2$
  b. $n = 3, x = 1, p = 3/7$
  c. $n = 4, x = 4, p = 2/3$

In Exercises 4-61 through 4-80, identify the values of n, x, p, *and* q, *and find the value requested.*

4-61 Find the probability of getting exactly four girls in ten births. (Assume that male and female births are equally likely.)

4-62 Find the probability of getting exactly six girls in seven births. (Assume that male and female births are equally likely.)

4-63 A Gallup poll showed that among convenience store shoppers, 60% gave closeness of location as their primary reason for shopping there. Find the probability that among five randomly selected convenience store customers, three of them give closeness of location as their primary reason for choosing that store.

4-64 In a study conducted by *Glamour* magazine, 30% of the teachers who left teaching did so because they were laid off. Assume that we randomly select 10 teachers who left their profession. Find the probability that exactly 4 of them were laid off.

4-65 According to *Discover* magazine, 95% of airline passengers will survive a crash under certain conditions. Given those conditions, what is the probability that exactly 16 of 20 passengers will survive a crash?

4-66   The probability of a computer component being defective is 0.01. Find the probability of getting exactly 2 defective components in a sample of 12.

4-67   According to data from the Greater New York Blood Program, 40% of all individuals have group A blood. If six individuals give blood, find the probability that
a. None of the individuals has group A blood.
b. Exactly three of the individuals have group A blood.
c. At least three of the individuals have group A blood.

4-68   According to a Roper poll, 10% of all adults exercise by jogging. Assume that four adults are randomly selected. Find the probability that
a. There is exactly one jogger.
b. There is at least one jogger.
c. There are at least two joggers.

4-69   According to the Department of Defense, 93% of Air Force recruits have graduated from high school. If 15 Air Force recruits are randomly selected, find the probability that 13 of them have graduated from high school.

4-70   According to FBI data, 44% of those murdered are killed with handguns. If 30 murder cases are randomly selected, find the probability that 12 of the victims were killed with handguns.

4-71   In a study conducted by the Centers for Disease Control, it was found that 15% of New Yorkers average two or more alcoholic drinks per day. If 60 New Yorkers are randomly selected and surveyed, find the probability that 12 of them average two or more alcoholic drinks per day.

4-72   A multiple-choice test has 30 questions, and each one has five possible answers, of which one is correct. If all answers are guesses, find the probability of getting exactly four correct answers.

4-73   A study by the EPA (Environmental Protection Agency) showed that of the cars built with catalytic converters, 4.4% have them removed. If 50 cars built with catalytic converters are randomly selected and checked, find the probability that
a. All cars continue to have their catalytic converters.
b. One car has the catalytic converter removed.
c. Two cars have the catalytic converter removed.

4-74   The National Coffee Association reports that among individuals in the 20 to 29 age bracket, 41% drink coffee. If five people in that age bracket are randomly selected, find the probability that

a. They all drink coffee.

b. There is exactly one coffee drinker.

c. There are two coffee drinkers.

4-75 In Table A-1, the probability corresponding to $n = 3$, $x = 2$, and $p = 0.01$ is shown as 0+. Find the exact probability represented by this 0+.

4-76 In Table A-1, the probability corresponding to $n = 5$, $x = 4$, and $p = 0.10$ is shown as 0+. Find the exact probability represented by this 0+.

4-77 A binomial experiment consists of three trials with a 1/3 probability of success in each trial. Construct the probability distribution table for this experiment. (Use the same format as Table 4-3).

4-78 A binomial experiment consists of four trials with a 1/4 probability of success in each trial. Construct the probability distribution table for this experiment. (Use the same format as Table 4-3.)

4-79 The Company Review 88 offers a nursing review course with the claim that 98% pass the National Council Licensure Examination for Registered Nurses (NCLEX-RN). Among 75 randomly selected examinees who take the course, find the probability that 74 pass the NCLEX-RN.

4-80 At a sobriety checkpoint conducted by police in Beacon, New York, 502 drivers were screened and two were arrested for DWI (driving while intoxicated). Assuming this arrest rate, if 100 drivers are randomly selected and screened, find the probability that exactly one will be arrested for DWI.

4-81 Data from the U.S. Census Bureau show that among those in the 18 to 24 age bracket, 40.8% vote. If 16 individuals from that age bracket are randomly selected, find the probability that fewer than 2 of them vote.

4-82 In a study conducted by UCLA and the American Council on Education, it was found that 33% of college freshmen support increased military spending. If 10 college freshmen are randomly selected, find the probability that fewer than 3 support increased military spending.

4-83 Data from Survey Sampling, Inc., show that in Las Vegas, 46.4% of the telephones have unlisted numbers. If 10 telephone numbers are randomly selected, find the probability that more than 8 have unlisted numbers.

4-84 Among medical school seniors, 15.1% are choosing family practice as their specialty, according to data from the Association of American Medical Colleges. If eight medical school seniors are randomly selected, find the probability that at least three of them choose family practice.

4-4 EXERCISES B

4-85 Suppose that an experiment meets all conditions to be binomial except that the number of trials is not fixed. Then the **geometric distribution**, which gives us the probability of getting the first success on the xth trial, is described by $P(x) = p(1 - p)^{x-1}$, where p is the probability of success on any one trial. Assume that the probability of a defective computer component is 0.2. Find the probability that the first defect is in the seventh component tested.

4-86 The **Poisson distribution** is used as a mathematical model describing the probability distribution for the arrivals of entities requiring service (such as cars arriving at a gas station, planes arriving at an airport, or moviegoers arriving at a theater). The Poisson distribution is defined by the equation

$$P(x) = \frac{\mu^x \cdot e^{-\mu}}{x!}$$

where x represents the number of arrivals during a given time interval (such as 1 hour), μ is the mean number of arrivals during the same time interval, and e is a constant approximately equal to 2.718. Assume that $\mu = 15$ cars per hour for a gas station and find the probability of each event.

a. No arrivals in an hour
b. One arrival in an hour
c. Ten arrivals in an hour
d. Fifteen arrivals in an hour
e. Twenty arrivals in an hour
f. Thirty arrivals in an hour

4-87 If we sample from a small finite population without replacement, the binomial distribution should not be used because the events are not independent. If sampling is done without replacement and the outcomes belong to one of two types, we can use the **hypergeometric distribution**. If a population has A objects of one type, while the remaining B objects are of the other type, and if n samples are drawn without replacement, then the probability of getting x objects of type A and $n - x$ objects of type B is

$$P(x) = \frac{A!}{(A - x)!x!} \cdot \frac{B!}{(B - n + x)!(n - x)!} \div \frac{(A + B)!}{(A + B - n)!n!}$$

Five people are randomly selected (without replacement) from a population of seven men and three women. Find the probability of getting four men and one woman.

4-88   The binomial distribution applies only to cases involving two types of outcomes, whereas the **multinomial distribution** involves more than two categories. Suppose we have three types of mutually exclusive outcomes denoted by A, B, and C. Let $P(A) = p_1$, $P(B) = p_2$, and $P(C) = p_3$. In n independent trials, the probability of x_1 outcomes of type A, x_2 outcomes of type B, and x_3 outcomes of type C is given by

$$\frac{n!}{(x_1!)(x_2!)(x_3!)} \cdot p_1^{x_1} \cdot p_2^{x_2} \cdot p_3^{x_3}$$

a. Extend the result to cover six types of outcomes.
b. A genetics experiment involves six mutually exclusive genotypes identified as $A, B, C, D, E,$ and F, and they are all equally likely. If 20 offspring are tested, find the probability of getting exactly five A's, four B's, three C's, two D's, three E's, and three F's.

4-5 MEAN AND STANDARD DEVIATION FOR THE BINOMIAL DISTRIBUTION

The binomial distribution is a probability distribution, so the mean, variance, and standard deviation for the appropriate random variable can be found from the formulas presented in Section 4-3.

FORMULA 4-1 $\mu = \Sigma x \cdot P(x)$

FORMULA 4-3 $\sigma^2 = [\Sigma x^2 \cdot P(x)] - \mu^2$

FORMULA 4-4 $\sigma = \sqrt{[\Sigma x^2 \cdot P(x)] - \mu^2}$

However, these formulas, which apply to all probability distributions, can be made much simpler for the special case of binomial distributions. Given the binomial probability formula and the above general formulas for μ, σ, and σ^2, we can pursue a series of somewhat complicated algebraic manipulations that ultimately lead to the following desired result.

For a binomial experiment,

FORMULA 4-7 $\mu = n \cdot p$

FORMULA 4-8 $\sigma^2 = n \cdot p \cdot q$

FORMULA 4-9 $\sigma = \sqrt{n \cdot p \cdot q}$

The formula for the mean does make sense intuitively. If we were to analyze 100 births, we would expect to get about 50 girls, and np in this experiment becomes $100 \cdot \frac{1}{2}$, or 50. In general, if we consider p to be the proportion of successes, then the product np will give us the actual number of expected successes among n trials.

The variance and standard deviation are not so easily justified, and we prefer to omit the complicated algebraic manipulations that lead to the second formula. Instead, we will show that these simplified formulas (Formulas 4-7, 4-8, and 4-9) do lead to the same results as the more general formulas (Formulas 4-1, 4-3, and 4-4).

EXAMPLE

According to an analyst from Paine Webber, the Intel Corporation has a 35% yield for the computer chips it produces—35% of the chips are good and 65% are defective. Let's assume that this estimate is correct and that we have randomly selected a sample of four different chips. Find the mean, variance, and standard deviation for the numbers of good chips in such groups of four.

Solution

In this binomial experiment we have $n = 4$ and $P(\text{good chip}) = 0.35$. It follows that $P(\text{defective chip}) = q = 0.65$. We will find the mean and standard deviation by using two methods.

Method 1: Use Formulas 4-7, 4-8, and 4-9, which apply to binomial experiments only.

$$\mu = n \cdot p = 4 \cdot 0.35 = 1.4 \qquad \text{(Formula 4-7)}$$

$$\sigma^2 = n \cdot p \cdot q = 4 \cdot 0.35 \cdot 0.65 = 0.91 \qquad \text{(Formula 4-8)}$$

$$\sigma = \sqrt{n \cdot p \cdot q} = \sqrt{0.91} = 0.95 \qquad \text{(Formula 4-9)}$$

Method 2: Use Formulas 4-1, 4-3, and 4-4, which apply to all discrete probability distributions. (Note: Method 1 provided us with the solutions we sought, but we want to show that these same values will result from the use of the more general formulas from Section 4-3.) We begin by computing the mean using Formula 4-1: $\mu = \Sigma x \cdot P(x)$. The possible values of x are 0, 1, 2, 3, 4, but we also need the values of $P(0)$, $P(1)$, $P(2)$, $P(3)$, and $P(4)$. We use the binomial probability formula to find those values and enter them in Table 4-6.

continued ▶

Solution, continued

		TABLE 4-6		
x	$P(x)$	$x \cdot P(x)$	x^2	$x^2 \cdot P(x)$
0	0.179	0	0	0
1	0.384	0.384	1	0.384
2	0.311	0.622	4	1.244
3	0.111	0.333	9	0.999
4	0.015	0.060	16	0.240
Total		1.399		2.867

We now use results from Table 4-6 to apply the general formulas from Section 4-3 as follows.

$$\mu = \Sigma x \cdot P(x) = 1.4 \qquad \text{(rounded off)}$$

$$\sigma^2 = [\Sigma x^2 \cdot P(x)] - \mu^2$$

$$= 2.867 - 1.4^2$$

$$= 0.91 \qquad \text{(rounded off)}$$

$$\sigma = \sqrt{0.91} = 0.95 \qquad \text{(rounded off)}$$

The two methods produced the same results, except for minor discrepancies due to rounding. There are two important points that we should recognize. First, the simplified binomial formulas (Formulas 4-7, 4-8, and 4-9) do lead to the same results as the more general formulas that apply to all discrete probability distributions. Second, the binomial formulas are much simpler, they provide fewer opportunities for arithmetic errors, and they are generally more conducive to a positive outlook on life. If we know an experiment is binomial, we should use the simplified formulas.

Who Is Shakespeare?

A poll of 1553 randomly selected adult Americans revealed that

- 89% could identify Shakespeare.
- 58% could identify Napoleon.
- 47% could identify Freud.
- 92% could identify Columbus.
- 71% knew what happened in 1776.

Ronald Berman, a past director of the National Endowment for the Humanities, said "I don't worry about them [those who don't know what happened in 1776]. I worry about the people who take polls." He goes on to say that the poll should ask substantive questions, such as: What is the difference between democracy and totalitarianism?

In the following example, we use only the simplified binomial formulas.

EXAMPLE

A test consists of 100 multiple-choice questions with possible answers of *a*, *b*, *c*, *d*, and *e*. For people who know nothing and guess the answer to each question, find the mean and standard deviation for the number of correct answers per person.

Solution

For each person, the number of trials is $n = 100$ and the probability of correctly guessing an answer is $p = 1/5$, so that $q = 4/5$. (We get $p = 1/5$ because there is one correct answer among the five possible answers.) We now proceed to find the mean and standard deviation.

$$\mu = n \cdot p = 100 \cdot \tfrac{1}{5} = 20.0$$

$$\sigma = \sqrt{n \cdot p \cdot q} = \sqrt{100 \cdot \tfrac{1}{5} \cdot \tfrac{4}{5}} = 4.00$$

For the preceding example, the mean number of guesses is 20.0, so that a score of 20.0% on the test is actually an indication of no knowledge. The value of the standard deviation can be used to determine a reasonable range of scores for those who know nothing and guess. For example, using the empirical rule from Chapter 3, we can conclude that about 95% of all scores should be between 12 and 28. Using the same empirical rule, we can conclude that about 99.7% of all scores should be between 8 and 32. If someone gets more than 32 correct answers, it is very unlikely that they are guessing at all questions. A more likely explanation is that they probably know some of the answers.

When *n* is large and *p* is close to 0.5, the binomial distribution tends to resemble the smooth curve that approximates the histograms in Figures 4-6 and 4-7. Note that the data tend to form a bell-shaped curve. In Chapter 5, we will use this property for solving certain applied problems.

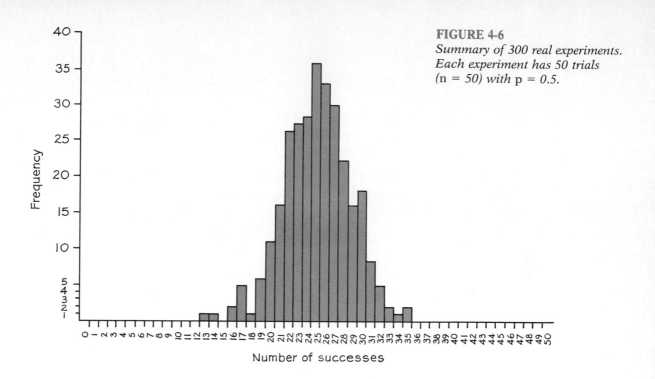

FIGURE 4-6
Summary of 300 real experiments. Each experiment has 50 trials (n = 50) with p = 0.5.

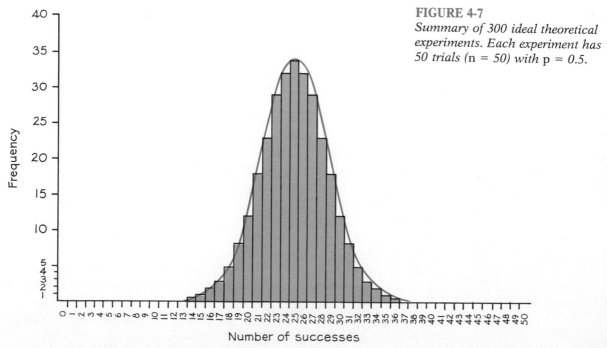

FIGURE 4-7
Summary of 300 ideal theoretical experiments. Each experiment has 50 trials (n = 50) with p = 0.5.

4-5 EXERCISES A

In Exercises 4-89 through 4-100, find the mean μ, variance σ^2, and standard deviation σ for the given values of n and p. Assume that the binomal conditions are satisfied in each case.

4-89 $n = 64, p = 0.5$ 4-95 $n = 534, p = 0.173$

4-90 $n = 100, p = 0.5$ 4-96 $n = 898, p = 0.392$

4-91 $n = 8, p = 0.6$ 4-97 $n = 16, p = 1/5$

4-92 $n = 6, p = 0.3$ 4-98 $n = 27, p = 1/4$

4-93 $n = 36, p = 0.25$ 4-99 $n = 253, p = 2/3$

4-94 $n = 40, p = 0.85$ 4-100 $n = 652, p = 3/8$

In Exercises 4-101 through 4-120 find the indicated values.

4-101   For a true-false test with 50 questions, several students are unprepared and all of their answers are guesses. Find the mean, variance, and standard deviation for the numbers of correct answers for such students.

4-102   For a multiple-choice test with 30 questions, each question has possible answers of $a, b, c,$ and d, one of which is correct. For people who guess at all answers, find the mean, variance, and standard deviation for the number of correct answers.

4-103   Find the mean, variance, and standard deviation for the numbers of girls in families with four children. Assume that boys and girls are equally likely and also assume that the sex of any child is independent of any brothers or sisters.

4-104   Find the mean, variance, and standard deviation for the numbers of girls in families with six children. Assume that boys and girls are equally likely and also assume that the sex of any child is independent of any brothers or sisters.

4-105   In a recent national election, 48% of the voters were men, according to a *New York Times*/CBS News Poll. Assume that this figure is correct and that these voters are randomly selected in groups of 64. Find the mean, variance, and standard deviation for the number of male voters per group.

4-106   According to a survey of adults by the Roper Organization, 64% of adults have money in regular savings accounts. If we plan to conduct a survey with groups of 50 randomly selected adults, find the mean, variance, and standard deviation for the numbers who have regular savings accounts.

4-107   Among the 6665 films rated by the Motion Picture Association of America, 2945 have been rated R. Twenty rated films are randomly selected for a study. Find the mean, variance, and standard deviation for the number of R films in randomly selected groups of 20.

4-108 According to data from the U.S. Bureau of Labor Statistics, 70.4% of women in the 20 to 24 age bracket are working. Find the mean, variance, and standard deviation for the numbers of working women in randomly selected groups of 150 women between 20 and 24 years of age.

4-109 A study conducted by the National Transportation and Safety Board showed that among injured airline passengers, 47% of the injuries were caused by failure of the plane's seat. Two hundred different airline passenger injuries are to be randomly selected for a study. Find the mean, variance, and standard deviation for the number of injuries caused by seat failure in such groups of 200.

4-110 One test of extrasensory perception involves the determination of a color. Fifty blindfolded subjects are asked to identify the one color selected from the possibilities of red, yellow, green, blue, black, and white. Assuming that all 50 subjects make random guesses, find the mean, variance, and standard deviation for the number of correct responses in such groups of 50.

4-111 Of all individual tax returns, 37% include errors made by the taxpayer. If IRS examiners are assigned randomly selected returns in batches of 12, find the mean and standard deviation for the number of erroneous returns per batch.

4-112 A pathologist knows that 14.9% of all deaths are attributable to a myocardial infarction. Find the mean and standard deviation for the number of such deaths that will occur in typical communities of 5000 people.

4-113 In standard English text, the letter *e* occurs with a relative frequency of 0.130. Find the mean and standard deviation for the number of times *e* will be found on standard pages of 2600 letters.

4-114 According to an Environmental Protection Agency study, of cars originally built with catalytic converters, 4.4% have had them removed. One point of a certain highway is passed by 12,600 cars per hour. Assume that all of those cars were originally equipped with catalytic converters. What is the mean and standard deviation for the hourly numbers of cars with their catalytic converters removed?

4-115 In planning for no-smoking areas on an aircraft, it is estimated that 37% of all adults smoke. Find the mean and standard deviation for the number of smokers in groups of 240 adults.

4-116 A survey has shown that 43% of all unregistered voters prefer the Democratic party. A follow-up study is to be conducted with 1200 randomly selected unregistered voters. Find the mean and standard deviation for the numbers of Democrats in groups of 1200.

4-117   Among all military personnel, 13.8% are officers. A study of military personnel involves random selections in groups of 50. Find the mean and standard deviation for the number of officers per group.

4-118   Among Americans aged 20 years or over, 12.5% sleep at least nine hours each night. A study of dreams requires 36 volunteers who are at least 20 years old. Find the mean and standard deviation for the number of such people in groups of 36 who sleep at least nine hours each night.

4-119   A psychologist studying suicides found that 63.8% of the men and 38.0% of the women used firearms. A sample of 200 cases is randomly selected from the population of male suicides. For such groups, find the mean and standard deviation for the number of cases in which firearms are used.

4-120   A manufacturer tested a sample of semiconductor chips and found that 35 were defective and 190 were good. If additional tests are to be conducted with random samples of 160 semiconductor chips, find the mean, variance, and standard deviation for the numbers of defects in such groups of 160.

4-5 EXERCISES B

In Exercises 4-121 through 4-124, consider as unusual anything that differs from the mean by more than twice the standard deviation. That is, unusual values are either less than $\mu - 2\sigma$ or greater than $\mu + 2\sigma$.

4-121   Is it unusual to get 450 girls and 550 boys in 1,000 independent births?

4-122   Is it unusual to find 5 defective transistors in a sample of 20 if the defective rate is 20%?

4-123   A company manufactures an appliance, gives a warranty, and 95% of its appliances do not require repair before the warranty expires. Is it unusual for a buyer of 10 such appliances to require warranty repairs on 2 of the items?

4-124   A candidate is favored by 616 voters in a poll of 1100 randomly selected voters. If this candidate is actually favored by 50% of all voters, find the lowest and highest *usual* numbers of supporters in groups of 1100. Does the poll result of 616 supporters seem to indicate a chance sample fluctuation?

4-6 DISTRIBUTION SHAPES

We have already stated that the *distribution* of data is an extremely important characteristic and may strongly affect the methods we use or the conclusions we draw. As we consider distributions of data in this section, we have two main objectives:

1. We will identify some of the more common distributions and show how such identification often becomes critical.

2. We will introduce a method of working with uniform distributions that are continuous instead of discrete. This method will show how to obtain probabilities by finding areas; it will be a good preparation for a very important concept to be introduced in Chapter 5.

One of the most fascinating aspects associated with a study of statistics is the surprising regularity and predictability of events that seem to happen by chance. In this section we consider some common collections of data, along with their corresponding histograms. Examination of a histogram representing data is extremely helpful in characterizing the way that the data are distributed. An understanding of the distribution may in turn lead to helpful and valuable inferences. Let's begin with a simple example involving the analysis of the last question on an IQ test. Assume that this is a multiple-choice question with five possible answers and that the responses of 100 subjects are summarized in Table 4-7 on the next page. The histogram that corresponds to Table 4-7 is Figure 4-8. You do not have to be an expert statistician to recognize that Figure 4-8 depicts a distribution that is essentially even. A distribution that is evenly spread over the range of possibilities is called **uniform**.

If we determine that the correct response is *b*, we might expect a higher frequency for that correct response. The nature of the distribution should tell us something about the validity or usefulness of that last test question. Because the responses are uniformly distributed, there is good reason to believe that the subjects are guessing. Perhaps there is insufficient time to consider all questions carefully and the last few responses are last-minute guesses, or perhaps this last question is so incredibly difficult that everybody makes random guesses. In any event, the distribution of responses suggests that the test can be improved by changing the last question.

TABLE 4-7	
Response	Frequency
a	20
b	22
c	19
d	18
e	21

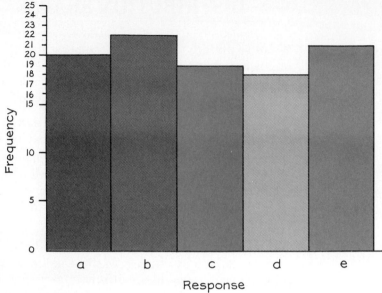

FIGURE 4-8

TABLE 4-8	
Height (inches)	Frequency
61–64	4
65–68	27
69–72	46
73–76	20
77–80	3

Let's consider a second example involving a manufacturer of compact cars who is concerned with the comfort of very tall drivers. The manufacturer seeks information about the distribution of heights and compiles the sample data for car buyers summarized in Table 4-8.

Examination of the sampling distribution leads the manufacturer to conclude that relatively few buyers (about 3%) are more than 76 in. tall (see Figure 4-9). Since the accommodation of these towering torsos would demand expensive design changes, the car manufacturer elects to sacrifice that tall share of the market that equals 3%. Had the distribution been uniform, as in Figure 4-8, then the manufacturer would have proceeded with the necessary design changes, thereby avoiding the sacrifice of about 20% of the potential market.

The key observation at this stage is the essential difference between the distributions of Figures 4-8 and 4-9. Both distributions are very real and arise naturally in various circumstances, yet their inherent differences lead to different inferences. Figure 4-10 presents some of the common shapes of histograms that arise in real data problems. The uneven bars of histograms have been smoothed to form continuous curves, as was done in Figure 4-7.

We must be careful when using histograms to determine the nature of the distribution. In Figure 4-11(a), we show a computer printout of a histogram in which the triglyceride levels of 20 subjects were entered. The last score was incorrectly entered with an extra zero so that 1600 was used instead of 160. In Figure 4-11(b) we show the computer print-

Clusters of Disease

Periodically, much media attention is given to a cluster of cases of a disease in a given community. Typical of this was a New Jersey community's cluster of 13 leukemia cases in a 5-year period. The normal leukemia rate for a comparable community would be only 1 case in 10 years.

Research studies of such clusters can lead to valuable conclusions. For example, such a study led to the discovery that asbestos fibers can be carcinogenic. That conclusion was reached through analysis of a cluster of cancer cases near African asbestos mines.

When we analyze diseases and deaths in an attempt to identify clusters, however, we must be careful to avoid the error of positioning a cluster boundary in a deceiving way. We can sometimes create an artificial cluster that doesn't really exist by locating the outer boundary so that it just barely includes cases of disease or death. That would be somewhat like gerrymandering, even though it might be unintentional.

Recent cases undergoing investigation include a cluster of cancer deaths near the Pilgrim nuclear power plant in Massachusetts and a cluster of three cases of Lou Gehrig's disease among teammates on the San Francisco 49ers football team.

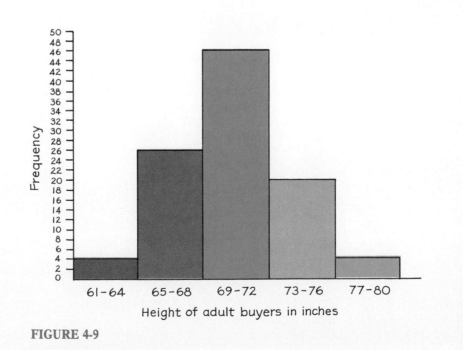

Height of adult buyers in inches

FIGURE 4-9

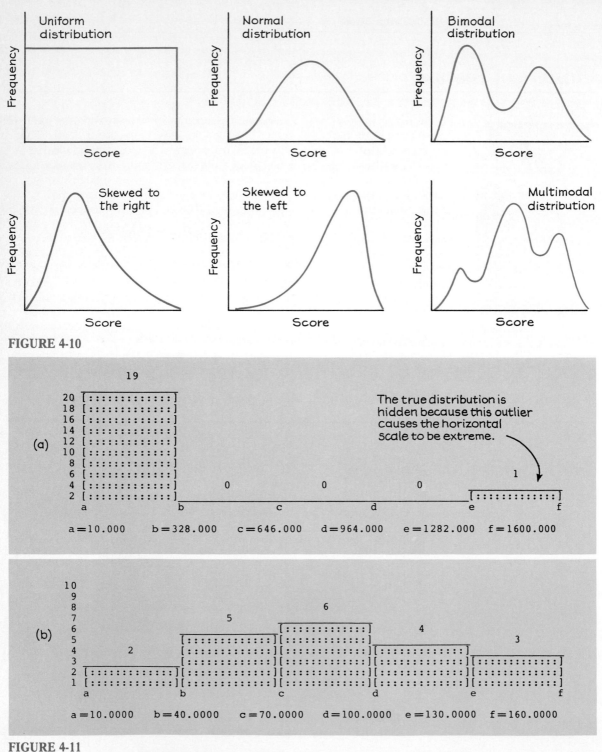

FIGURE 4-10

FIGURE 4-11
The true distribution is hidden because this outlier causes the horizontal scale to be extreme.

Captured Tank Serial Numbers Reveal Population Size

During World War II, allied intelligence specialists wanted to determine the number of tanks Germany was producing. Traditional spy techniques provided unreliable results, but statisticians obtained accurate estimates by analyzing serial numbers on captured tanks. As one example, records show that Germany actually produced 271 tanks in June of 1941. The estimate based on serial numbers was 244, but traditional intelligence methods resulted in the far out estimate of 1550. (See "An Empirical Approach to Economic Intelligence in World War II" by Ruggles and Brodie, *Journal of the American Statistical Association*, Vol. 42, 1947.)

out for the corrected data set. From Figure 4-11(b) we conclude that the data are normally distributed, but that was not at all apparent from Figure 4-11(a), even though only one score was incorrect. In this case, the outlier caused a severe distortion of the histogram. In other cases, outliers may be correct values of data but may continue to disguise the true nature of the distribution through histograms such as the one shown in Figure 4-11(a).

The example of Figure 4-11(a) suggests that we should avoid making important conclusions about the distribution of data without exploring and verifying the values themselves. In addition to the histogram, we might also construct a **box-and-whisker diagram** (sometimes called a **boxplot**), which reveals more information about how the data are spread out. In order to construct such a diagram, we must first determine the minimum, maximum, and median values. We must also obtain the values of the two **hinges**. To find the hinges,

1. Arrange the data in increasing order.
2. Find the median. (With an odd number of scores, it's the middle score; with an even number of scores, it's the mean of the two middle scores.)
3. List the lower half of the data from the minimum score up to and including the median found in step 2. The left hinge is the median of these scores.
4. List the upper half of the data starting with the median and including the scores up to and including the maximum. The right hinge is the median of these scores.
5. Now list the minimum, the left hinge (from step 3), the median (from step 2), the right hinge (from step 4), and the maximum.

This procedure will vary somewhat with different textbooks. The 20 scores depicted in Figure 4-11(a) are arranged in increasing order and listed below.

10 28 43 49 50 60 66 75 83 86 90 93 94 108 121 126 127 131 142 1600

minimum   median (88)   maximum

10 28 43 49 50 (60) 66 75 83 86 88   88 90 93 94 108 (121) 126 127 131 142 1600

The *left hinge* of 60 is found by determining the median of this bottom half of the data.

The *right hinge* of 121 is found by determining the median of this top half of the data.

Note that the hinges are different from the quartiles. For this data set, $Q_1 = 55$ and $Q_3 = 123.5$, while the hinges are 60 and 121. To construct a box-and-whisker diagram, we now begin with a horizontal scale as in Figure 4-12(a). We "box" in the hinges as shown and we use "whiskers" to connect the minimum score to a hinge and the maximum score to a hinge. Figure 4-12(b) shows the box-and-whisker diagram for the data set after the incorrect score of 1600 has been corrected to 160.

Note that by the procedures used here, approximately one-fourth of the values should fall between the low score and the left hinge, the two middle quarters are in the boxes, and approximately one-fourth of the values are between the right hinge and the maximum. The diagram therefore shows how the data are spread out. The uneven spread shown in Figure 4-12(a) is in strong contrast to the even spread shown in Figure 4-12(b). Suspecting that triglyceride levels are normally distributed, we would expect to see a box-and-whisker diagram like the one in Figure 4-12(b), whereas the diagram in Figure 4-12(a) would raise suspicion and lead to further investigation. In Figure 4-13 we show some common distributions along with the corresponding box-and-whisker plots. Minitab can be used to create box-and-whisker diagrams. See the sample Minitab display.

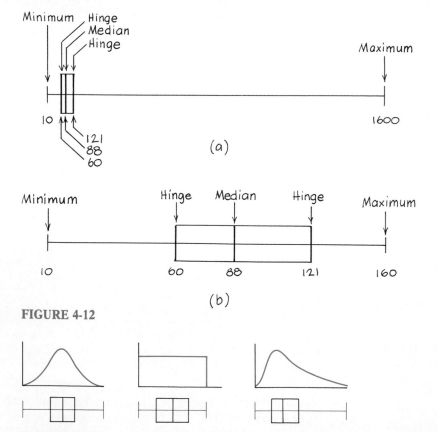

FIGURE 4-12

FIGURE 4-13

MINITAB DISPLAY

```
MTB > SET C1
DATA> 10 28 43 49 50 60 66 75 83 86
DATA> 90 93 94 108 121 126 127 131 142 160
DATA> ENDOFDATA
MTB > BOXPLOT C1


                        ------------------------
       ----------------I            +            I------------
                        ------------------------
     +---------+---------+---------+---------+---------+------C1
     0        30        60        90       120       150
```

Whether we use histograms or box-and-whisker plots, an understanding of the data requires some knowledge of the distribution. Armed with that knowledge and a knowledge of the key parameters (like the mean and standard deviation), we can often formulate useful inferences. The following simple example illustrates an important concept, which will subsequently be applied to more useful and more realistic situations.

Let's assume that we have an electric circuit designed so that the voltage levels are *uniformly* distributed between 5.00 V and 9.00 V, as shown in Figure 4-14. If we assign 1 to the area of the enclosed rectangle in Figure 4-14, then we can establish a natural and usable correspondence between the area and probability. Since the length of that rectangle is $9.00 - 5.00 = 4.00$, we can make its height 0.25 so that the area becomes $4.00 \times 0.25 = 1$. We can now proceed to solve a variety of probability problems by determining the corresponding areas.

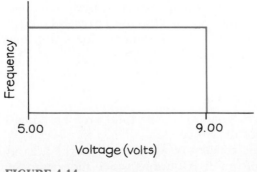

FIGURE 4-14

EXAMPLE

Given the uniform distribution of Figure 4-15, find the probability that a randomly selected voltage level will be between 7.00 V and 8.50 V.

Solution

In Figure 4-15 we show the area corresponding to the voltage levels between 7.00 V and 8.50 V. Note that the vertical scale is marked with 0.25, so that the total area of the complete rectangle is 1. The *area* of the shaded region is $0.25 \times 1.50 = 0.375$, so there is a 0.375 *probability* of randomly selecting a value between 7.00 V and 8.50 V.

EXAMPLE

Given the same uniform distribution shown in Figure 4-16, find the probability that a randomly selected voltage level is greater than 8.20 V.

Solution

Since all voltage levels are between 5.00 V and 9.00 V, "greater than 8.20 V" can be interpreted to mean "between 8.20 V and 9.00 V," and this is the region shaded in Figure 4-16. The dimensions of this shaded rectangle are 0.25 and 0.80 (which is $9.00 - 8.20$). The area of that shaded rectangle is 0.200, so there is a 0.200 probability of randomly selecting a value that is greater than 8.20 V.

In the preceding two examples we assigned 0.25 to the height of the rectangle so that the total area becomes 1. For a general uniform distribution with minimum value a and maximum value b, we can assign the value

$$\frac{1}{b - a}$$

to the height so that the total area becomes 1. Note that $b - a$ is the value of the range.

This concept of assigning 1 to the area under a curve works well for uniform distributions, since the rectangles are easy to work with.

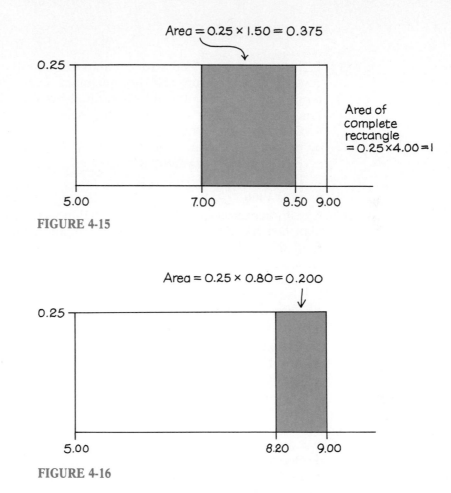

FIGURE 4-15

FIGURE 4-16

Normal distributions similar to the one depicted in Figure 4-17 occur often in reality, but the area computations are much more difficult because of the curve involved. Methods for normal distribution computations are pursued in Chapter 5. The methods we learn here will help to prepare us for the concepts presented in Chapter 5.

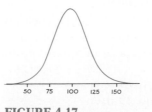

FIGURE 4-17

4-6 EXERCISES A

4-125   Technicians are trained with a computer simulation that randomly generates temperatures for one sector of a chemical reactor. Those temperatures are uniformly distributed between 400° C and 500° C. Find the probability that a randomly selected temperature is
a. Greater than 480° C
b. Between 410° C and 490° C
c. Between 450° C and 600° C
d. Less than 465° C
e. Less than 375° C

4-126   A customer service unit is designed so that waiting times are uniformly distributed with a minimum of 1.20 min and a maximum of 9.20 min. Find the probability that the waiting time for a randomly selected customer is
a. Less than 2.20 min
b. Greater than 5.20 min
c. Between 2.00 min and 3.00 min
d. Between 5.00 min and 9.00 min

4-127   A car wash mechanism is designed so that the times (in seconds) required for one cycle are uniformly distributed between 260 s and 290 s. Find the probability that a randomly selected cycle requires a time that is
a. Greater than 270 s
b. Between 265 s and 287 s
c. Less than 268 s
d. Between 260 s and 280 s
e. Less than the mean time

4-128   While conducting experiments, a marine biologist selects water depths from a uniformly distributed collection of depths that vary between 2.00 m and 7.00 m. Find the probability that a randomly selected depth is
a. Between 2.25 m and 5.00 m
b. Greater than 6.00 m
c. Between 6.40 m and 7.00 m
d. Less than 3.60 m
e. Within 2.00 m of the mean

4-129   A traffic engineer analyzes a model in which vehicle speeds are uniformly distributed with a mean of 35 mi/h and a range of 10 mi/h. Find the probability that the speed of a randomly selected vehicle is
a. Less than 35 mi/h
b. Greater than 32 mi/h
c. Between 33 mi/h and 39 mi/h
d. Less than the median
e. Between 35 mi/h and 45 mi/h

4-130   A medical researcher artificially introduces cholesterol into blood samples so that the resulting mixtures have cholesterol levels uniformly distributed with a mean of 280 and a range of 120. Find the probability that a randomly selected mixture is
a. Between the minimum level and 250
b. Less than 275
c. Greater than 300
d. Between 280 and 350
e. *Not* between 235 and 285

4-131   A machine is designed to pour 50 cm^3 of a drug into a bottle. Because of a flaw, the machine pours amounts that are uniformly distributed with a mean of 51.5 cm^3 and a range of 1.2 cm^3.
a. What is the minimum volume of the drug poured by the machine?
b. If a poured sample is randomly selected, find the probability that it is between 51.0 cm^3 and 52.0 cm^3.
c. If a poured sample is randomly selected, find the probability that it is within 0.2 cm^3 of the mean.
d. If a poured sample is randomly selected, find the probability that it is less than 51.0 cm^3 or greater than 52.0 cm^3.
e. If a poured sample is randomly selected, find the probability that it is less than the mean or greater than 52.0 cm^3.

4-132   A psychologist experiments with learning by giving subjects different times to complete the same task. The times are randomly selected from a uniform distribution with a mean of 24 min and a range of 20 min. Find the probability that a randomly selected time is
a. Between 24 and 44 min
b. Between 15 and 20 min
c. Less than 30 min
d. Within 5 min of the mean
e. Less than 15 min or greater than 30 min

In Exercises 4-133 through 4-140, use the given data to construct box-and-whisker diagrams. Identify the values of the minimum, maximum, median, and hinges.

4-133   Ages of selected full-time undergraduate students (in years):

17.2, 17.9, 18.6, 18.8, 19.3, 19.3, 20.0, 20.1, 23.4, 26.3

4-134   Monthly rental costs of apartments in one region (in dollars):

540, 545, 555, 560, 560, 570, 575, 590, 650, 730

4-135   Blood alcohol contents of drivers given breathalyzer tests:

0.02, 0.04, 0.08, 0.08, 0.09, 0.10, 0.10, 0.12, 0.13, 0.19

4-136   Time intervals (in months) between adjacent births for selected families:

12.9	13.4	18.3	24.7	31.2	31.3	32.0	32.1	33.4
33.8	34.1	36.2	41.7	41.9	52.5			

4-137   Blood pressure levels (in mm of mercury) for patients who have taken 25 mg of the drug captopril.

198	180	142	157	181	183	162	130	170	164
170	173	173	175	195	190	193	157	159	138

4-138   Number of words typed in a 5-min civil service test taken by 25 different applicants.

174	181	219	213	213	207	106	111	143	160
166	350	183	198	193	190	190	185	220	221
229	257	243	281	308					

4-139   Time (in hours of operation) between failures for prototypes of computer printers:

34	22	4	9	27	36	12	40	29	32
35	25	7	9	26	36	45	43	41	2
31	31	30	14	15	18	10	27	38	21

4-140   Time (in seconds) required for long-distance telephone calls:

45	12	21	180	27	33	35	38	41	42
43	65	63	60	60	152	126	121	49	50
56	59	59	59	68	73	85	98	107	

4-6 EXERCISES B

4-141   Given the accompanying triangular distribution, find the probability of randomly selecting a score that is
a. Less than 3.
b. Between 2 and 4.
(*Hint:* The area of any triangle = $\frac{1}{2}$ × base × height.)

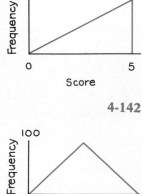

4-142   Given the accompanying distribution, find the probability of randomly selecting a score that is
a. Less than 12
b. Between 5 and 15
c. Between 5 and 19

4-143 Make sketches of histograms that correspond to the given box-and-whisker diagrams.

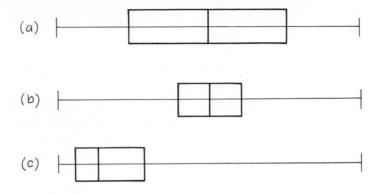

(a)

(b)

(c)

4-144 A population of data is uniformly distributed with a minimum value of a and a maximum value of b. Find the probability that a randomly selected value is within one standard deviation of the mean. (*Hint:* For a continuous uniform distribution with range $b - a$, $\sigma^2 = (b - a)^2/12$.)

VOCABULARY LIST

Define and give an example of each term.

random variable
discrete random
   variable
continuous random
   variable

probability
   distribution
expected value
binomial experiment
binomial probability
   formula

uniform distribution
box-and-whisker
   diagram
hinge

REVIEW

The central concern of this chapter was the concept of a probability distribution. Here we dealt mostly with **discrete** probability distributions, while successive chapters deal with continuous probability distributions.

In an experiment yielding numerical results, the **random variable** can take on those different numerical values. A **probability distribution** consists of all values of a random variable, along with their corresponding probabilities. By constructing a histogram of a probability distribution, we can see a useful correspondence between those probabilities and the areas of the rectangles in the histogram.

Of the infinite number of different probability distributions, special attention is given to the important and useful **binomial probability distribution**, which is characterized by these properties:

1. There is a fixed number of trials (denoted by n).
2. The trials must be independent.
3. Each trial must have outcomes that can be classified in *two* categories.
4. The probabilities involved must remain constant for each trial.

We saw that probabilities for the binomial distribution can be computed by using Table A-1 or by using the binomial probability formula, where n is the number of trials, x is the number of successes, p is the probability of a success, and q is the probability of a failure.

For the special case of the binomial probability distribution, the mean, variance, and standard deviations of the random variable can be easily computed by using the formulas given in the summary below.

We concluded the chapter with a section emphasizing the importance of determining the shape of a distribution and the usefulness of making a correspondence between probability and area in a histogram. We also introduced **box-and-whisker diagrams** as a way of obtaining information about the distribution and spread of data sets.

IMPORTANT FORMULAS

Requirements for a discrete probability distribution:

1. $\Sigma P(x) = 1$ for all possible values of x.
2. $0 \leq P(x) \leq 1$ for any particular value of x.

For *any* discrete probability distribution:

$$\text{mean} \quad \mu = \Sigma x \cdot P(x) \qquad \text{variance} \quad \sigma^2 = \Sigma(x - \mu)^2 \cdot P(x)$$

or

$$\text{variance} \quad \sigma^2 = [\Sigma x^2 \cdot P(x)] - \mu^2$$

$$\text{standard deviation} \quad \sigma = \sqrt{[\Sigma x^2 \cdot P(x)] - \mu^2}$$

Expected value of discrete random variable: $E = \Sigma x \cdot P(x)$

Binomial probability formula: $P(x) = \dfrac{n!}{(n-x)!x!} \cdot p^x \cdot q^{n-x}$

For *binomial* probability distributions:

$$\text{mean} \quad \mu = n \cdot p \qquad \text{variance} \quad \sigma^2 = n \cdot p \cdot q$$

$$\text{standard deviation} \quad \sigma = \sqrt{n \cdot p \cdot q}$$

REVIEW EXERCISES

4-145   Among the voters in a certain region, 40% are Democrats, and five voters are randomly selected.
   a. Find the probability that all five are Democrats.
   b. Find the probability that exactly three of the five are Democrats.
   c. Find the mean number of Democrats that would be selected in such groups of five.
   d. Find the variance of the number of Democrats that would be selected in such groups of five.
   e. Find the standard deviation for the number of Democrats that would be selected in such groups of five.

4-146   A probability distribution $P(x)$ is described by the accompanying table.
   a. Complete the table.
   b. Find the mean of the random variable x.
   c. Find the standard deviation of the random variable x.

x	$P(x)$
0	0.20
1	0.70
5	

4-147   Temperatures (in degrees Fahrenheit) are uniformly distributed with a minimum of 40° F and a maximum of 80° F. If a temperature reading is randomly selected, find the probability that it is between 50° F and 75° F.

4-148   In a study of sleep patterns of adults, it has been found that 23% of the sleep time is spent in the REM (rapid eye movement) stage.
   a. If a sleeping adult is observed at five randomly selected times, find the probability that exactly one of the five observations will be made during a REM stage.
   b. Find the standard deviation for the number of REM stages observed in groups of five observations.

4-149   If $P(x)$ is described by the accompanying table, does $P(x)$ form a probability distribution? Explain.

x	$P(x)$
0	0.4
1	0.4
2	0.4

4-150   Construct the box-and-whisker diagram for the following times (in months) served by convicted felons.

   26, 48, 20, 36, 32, 32, 33, 30, 29, 24

4-151   In a quality control study, it was found that 5% of all auto frames manufactured by one company had at least one defective weld.
   a. Find the probability of getting three frames with defective welds when 20 frames are randomly selected.
   b. If frames are examined in groups of 20, find the mean number of defective frames in each group.
   c. Find the standard deviation for the number of defective frames in groups of 20.

x	$P(x)$
5	0.6
10	
20	0.1

4-152   A probability distribution $P(x)$ is described by the accompanying table.
a. Fill in the missing probability.
b. Find the mean of the random variable x.
c. Find the variance of the random variable x.

4-153   The annual divorce rate is described as follows: "There are 5.3 divorced people per 1000 population."
a. Find the probability that among 10 randomly selected people, exactly 1 person is divorced.
b. Find the standard deviation for the numbers of divorced people in groups of 1000.

4-154   Steel rods are produced so that their lengths are uniformly distributed with a mean of 8.00 ft and a range of 0.88 ft. Find the probability that for a randomly selected rod, the length is
a. Greater than 7.75 ft
b. Between 7.50 ft and 8.40 ft

4-155   Construct the box-and-whisker diagram for the following weights of subjects about to begin a diet program. Label the hinges, the median, and the minimum and maximum.

155, 153, 171, 142, 142, 165, 167, 163

4-156   If $P(x)$ is described by the accompanying table, does $P(x)$ form a probability distribution? Explain.

x	$P(x)$
−1	0.35
0	0.15
1	0.40
2	0.10

4-157   A recessive genetic trait is known to occur in 50% of all offspring born within a certain population. A study involves the random selection of offspring in groups of 16.
a. Find the probability of selecting a group and getting exactly eight offspring with the recessive trait.
b. Find the mean number of offspring having the recessive trait in such groups of 16.
c. Find the variance for the number of offspring having the recessive trait in such groups of 16.

4-158   It has been found that the probabilities of 0, 1, 2, 3, and 4 wrong answers on a quiz are 0.20, 0.35, 0.30, 0.10, and 0.05, respectively.
a. Summarize the corresponding probability distribution.
b. Find the mean of the random variable x.
c. Find the standard deviation of the random variable x.

4-159   A voltage regulator is designed to produce uniformly distributed volt-age levels with a minimum of 7.00 V and a maximum of 9.00 V. If a reading is taken, find the probability that the voltage is below 8.55 V.

4-160   In a study of middle-aged adults (40 to 65 years), it was found that 7.8% suffer from hypertension. A follow-up study begins with the random selection of 20 middle-aged adults.
        a. Find the probability that exactly one-fourth of the selected subjects suffer from hypertension.
        b. Find the mean number of hypertension cases found in such groups of 20.
        c. Find the standard deviation for the numbers of hypertension cases found in groups of 20.

4-161   Does $P(x) = (x + 1)/5$ (for $x = -1, 0, 1, 2$) determine a probability distribution? Explain.

4-162   Construct the box-and-whisker diagram for the following scores achieved by subjects on a personality test.

95	95	91	51	51	103	120	112	74	68	60	59
58	53	83	105	106	106	108	54	57	56	56	96

4-163   An experiment in parapsychology involves the selection of one of the numbers 1, 2, 3, 4, and 5, and they are all equally likely.
        a. Summarize the corresponding probability distribution.
        b. Find the mean of the random variable x.
        c. Find the variance of the random variable x.

4-164   Does $P(x) = x/12$ (for $x = 1, 2, 3, 4$) determine a probability distribution? Explain.

4-165   A solid-state device requires 15 electronic components, two of which cost less than 50¢ each. One component is randomly selected from each of eight such devices.
        a. Find the probability of getting at least one of the components costing less than 50¢.
        b. Find the mean number of components costing under 50¢ in such groups of eight.
        c. Find the standard deviation for the numbers of components costing under 50¢ in such groups of eight.

4-166   Grade stakes used in surveying have lengths that are uniformly distributed between 85 cm and 95 cm. If one such stake is randomly selected, find the probability that its length differs from the mean by more than 2 cm.

4-167   Construct the box-and-whisker diagram for the following daily customer counts achieved by one store:

42	45	48	59	67	72	74	75	75	77
78	78	78	81	82	82	88	94	97	99

4-168   A study of enrollments at one college shows that 40% of all full-time undergraduates are under 20 years of age.
   a. If 10 full-time undergraduates at this college are randomly selected, find the probability that at least half of them are under 20.
   b. Find the mean number of full-time undergraduates at this college that would be found in groups of 10.
   c. Find the standard deviation for the number of full-time undergraduates at this college that would be found in groups of 10.

COMPUTER PROJECT

   a. Develop your own program to compute probabilities using the binomial probability formula. The program should take the values of n, x, and p as input, and output should be the corresponding probability of x successes in n trials of a binomial experiment. Run the program to generate all of the probabilities represented by 0+ in the section of Table A-1 for which $n = 10$.
   b. Use an existing software package that is capable of producing binomial probabilities. With $p = 0.5$, find the largest value of n that will not lead to an error. Also, find the probabilities represented by 0+ in the section of Table A-1 for which $n = 10$.

CASE STUDY ACTIVITY

For families of four children, assume that girls and boys are equally likely. Find the probabilities of 0, 1, 2, 3, and 4 girls, respectively, and summarize the results in the form of a probability distribution table. Calculate the mean and standard deviation for the numbers of girls in families with four children. Now proceed to collect data from 100 simulated families as follows. Refer to a telephone book and record the last four digits of 100 telephone numbers. Letting the digits 0, 1, 2, 3, 4 represent girls and letting 5, 6, 7, 8, 9 represent boys, find the number of girls in each of the 100 "families." Summarize the results in a frequency table and find the mean and standard deviation. Compare the results from the observed data to the theoretical results.

DATA PROJECT

When purchasing desks for classrooms, some consideration is being given to left-handed students. Refer to the data sets in Appendix B to answer the following.

a. Estimate the probability that a randomly selected person is left-handed.

b. The Appendix B data describing left-handedness and right-handedness was collected from college students in statistics classes. Are there any reasons why this group might have a proportion of left-handedness that is different from that for the population of everyone in the United States? If yes, identify the reasons. If no, describe some other population for which the proportion of left-handed people would be significantly different.

c. Using the result from part (a), find the probability of getting exactly 2 left-handed students in a class of 10.

d. Using the result from part (a), find the probability that when 25 students register for a class, the number of left-handed students is less than 4.

e. For classes of 25 students, find the mean, variance, and standard deviation for the number of left-handed students. (Use the result from part (a).)

Chapter Five

CHAPTER CONTENTS

5-1   **Overview**
We identify chapter **objectives** and describe the **normal distribution**.

5-2   **The Standard Normal Distribution**
We define the **standard normal distribution** and describe methods for determining probabilities by using that distribution.

5-3   **Nonstandard Normal Distributions**
We use the z score (or standard score) to work with normal distributions in which the mean is not 0 or the standard deviation is not 1.

5-4   **Finding Scores When Given Probabilities**
With normal distributions, we determine the values of scores that correspond to various given probabilities.

5-5   **Normal as Approximation to Binomial**
We can sometimes use the **normal** distribution to determine probabilities in a **binomial** experiment.

5-6   **The Central Limit Theorem**
The sampling distribution of sample means tends to be a normal distribution with mean μ and standard deviation $\sigma/\sqrt{n}$.

5 Normal Probability Distributions

CHAPTER PROBLEM

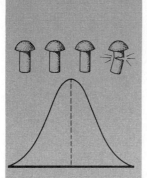

A manufacturer produces shear pins, which are used to protect heavy construction machinery by snapping when the load reaches 3000 lb. When the pin snaps, the blade disengages so that the gears and engine are not damaged. Using the principles covered in Chapters 2 and 4, a quality control analyst has determined that the actual breaking points of the shear pins are normally distributed with a mean of 2860 lb and a standard deviation of 52 lb. A parts buyer complains that many of the pins are defective because they break only when the load is well above 3000 lb. In this chapter we will learn how to find the actual percentage of such defective pins so that our quality control analyst can respond to the buyer. This will be one of many applications of the principles covered in this chapter.

5-1 OVERVIEW

Chapter 4 explained the difference between discrete random variables and continuous random variables and dealt mostly with discrete probability distributions. We concluded Chapter 4 by considering continuous uniform distributions of the type shown in Figure 5-1.

We saw in Section 4-6 that with a uniform distribution, we could solve many probability problems by determining areas like the shaded region in Figure 5-1. In this chapter we apply the same technique to normal distributions. This type of distribution is extremely important and fundamental to statistics because it can be applied to a wide range of real circumstances.

The normal distribution was originally developed as a result of studies dealing with errors that occur in various experiments. We now recognize that many real and natural occurrences, as well as many physical measurements, have frequency distributions that are approximately normal. Blood cholesterol levels, heights of adult women, weights of five-year-old boys, diameters of New York McIntosh apples, and lengths of newborn sharks are all examples of collections of values whose frequency polygons will closely resemble the normal probability distribution (see Figure 5-2).

Scores on standardized tests, such as IQ tests or college entrance examinations, also tend to be normally distributed. We will see in this chapter that sample means tend to be normally distributed, even if the underlying distribution is not. Also, the normal distribution is often

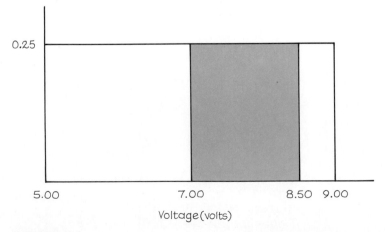

FIGURE 5-1

Reliability and Validity

The reliability of data refers to the consistency with which the same results occur, but the validity of data refers to how well the data measure what they are supposed to measure. The reliability of an IQ test can be judged by comparing scores for the test given on one date to the scores for the same test given on another date. Many critics claim that the popular IQ tests are reliable but not valid. That is, they produce consistent results, but they do not really measure intelligence levels. It is much easier to use statistics to analyze reliability than validity, but validity is the more important characteristic. To test the validity of an intelligence test, we might compare the test scores to another indicator of intelligence, such as academic performance.

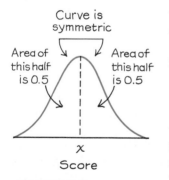

FIGURE 5-2
The normal distribution

used as a good approximation to other types of distributions—both discrete and continuous.

The smooth bell-shaped curve shown in Figure 5-2 depicts the **normal distribution** that can be described by the equation

$$y = \frac{e^{-(x-\mu)^2/2\sigma^2}}{\sigma\sqrt{2\pi}} \tag{5-1}$$

where μ represents the mean score of the entire population, σ is the standard deviation of the population, π is approximately 3.142, and e is approximately 2.718. Equation 5-1 relates the horizontal scale of x values to the vertical scale of y values. We will not use this equation in actual computations since we would need some knowledge of calculus to do so.

In reality, collections of scores may have frequency polygons that approximate the normal distribution illustrated in Figure 5-2, but we cannot expect to find perfect normal distributions that conform to the precise relationship of Equation 5-1. As a realistic example, consider the 150 IQ scores summarized in Table 5-1 and illustrated in Figure 5-3. The histogram of Figure 5-3 closely resembles the smooth bell-shaped curve of Figure 5-2, but it does contain some imperfections. In a theoretically ideal normal distribution, the tails extend infinitely far in both directions as they get closer to the horizontal axis. But this property is usually inconsistent with the limitations of reality. For example, IQ scores cannot be less than zero, so no histogram of IQ scores can extend infinitely far to the left. Similarly, objects cannot

TABLE 5-1	
Frequency Table	
IQ	Frequency
55.0–64.9	1
65.0–74.9	5
75.0–84.9	15
85.0–94.9	31
95.0–104.9	39
105.0–114.9	36
115.0–124.9	15
125.0–134.9	4
135.0–144.9	3
145.0–154.9	1

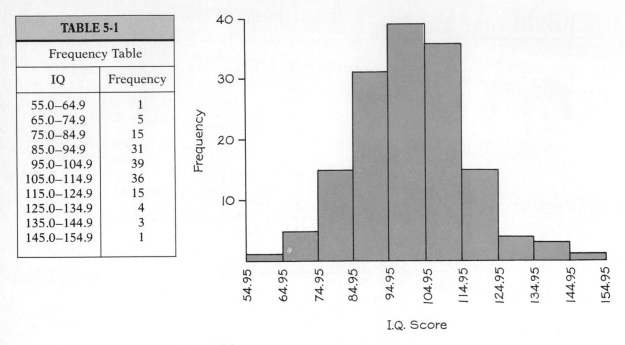

FIGURE 5-3

have negative weights nor can they have negative distances. Limitations such as these require that frequency distributions of real data can only approximate a normal distribution.

This chapter presents the standard methods used to work with normally distributed scores, and it includes applications. In addition to the importance of the normal distribution itself, the methods are important in establishing basic patterns and concepts that will apply to other continuous probability distributions.

5-2 THE STANDARD NORMAL DISTRIBUTION

There are actually many different normal probability distributions, each dependent on only two parameters: the population mean μ and the population standard deviation σ. Figure 5-4 shows three different normal distributions of IQ scores where the differences are due to changes in the mean and standard deviation. A change in the value of

Bullhead City Gets Hotter

A firefighter in Bullhead City, Arizona, was given the responsibility of reporting daily weather statistics to the National Weather Service. One day, a Weather Service representative demanded that the thermometer be moved from the firehouse lawn to a more natural setting. The firefighter selected a dry and dusty area 100 yards away from the cooler lawn. This move resulted in readings about 5° higher, which often made Bullhead City the hottest place in the United States. As Bullhead City gained a newfound prominence in many television weather reports, some residents denounced the notoriety as a handicap to business, while other residents felt that business was helped. Under more standardized conditions, measuring instruments—such as thermometers and scales—tend to produce errors that are normally distributed.

the population mean μ causes the curve to be shifted to the right or left. A change in the value of σ causes a change in the shape of the curve; the basic bell shape remains, but the curve becomes fatter or skinnier, depending on σ. Among the infinite possibilities, one particular normal distribution is of special interest.

DEFINITION

The **standard normal distribution** is a normal probability distribution that has a mean of 0 and a standard deviation of 1.

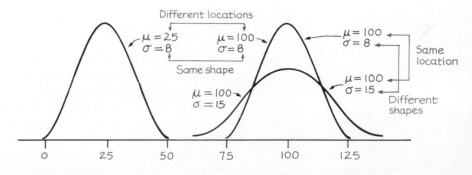

FIGURE 5-4

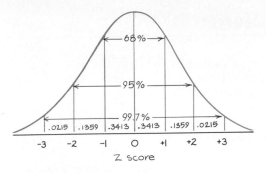

FIGURE 5-5
Standard normal distribution

If we had to perform calculations with Equation 5-1 and we could choose any values for μ and σ, we would soon recognize that $\mu = 0$ and $\sigma = 1$ lead to the simplest form of that equation. Working with this simplified form, mathematicians are able to perform various analyses and computations. Figure 5-5 shows a graph of the standard normal distribution with some of the computed results. For example, the area under the curve and bounded by scores of 0 and 1 is 0.3413. The area under the curve and bounded by scores of −1 and −2 is 0.1359. The sum of the six known areas in Figure 5-5 is 0.9974, but if we include the small areas in the two tails we get a total of 1. (In any probability distribution, the total area under the curve is 1.) Nearly all (99.7%) of the values lie within three standard deviations from the mean.

Figure 5-5 illustrates only six probability values, but a more complete table has been compiled to provide more precise data. Table A-2 (in Appendix A) gives the probability corresponding to the area under the curve bounded on the left by a vertical line above the mean of zero and bounded on the right by a vertical line above any specific positive score denoted by z (see Figure 5-6). Note that when you use Table A-2, the hundredths part of the z score is found across the top row. To find the probability associated with a score between 0 and 1.23, for example, begin with the z score of 1.23 by locating 1.2 in the left column. Then find the value in the adjoining row of probabilities that is directly below 0.03. There is a probability of 0.3907 of randomly selecting a score between 0 and 1.23. It is essential to remember that this table is designed only for the standard normal distribution, which has a mean of 0 and a standard deviation of 1. Nonstandard cases will be considered in the next section.

Since normal distributions originally resulted from studies of experimental errors, the following examples that deal with errors in measurements should be helpful.

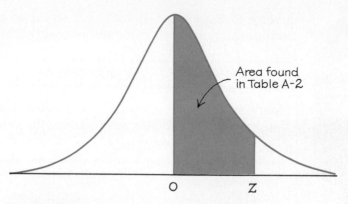

FIGURE 5-6

The standard normal distribution. The area of the shaded region bounded by the mean of zero and the positive number z can be found in Table A-2.

EXAMPLE

A manufacturer of scientific instruments produces thermometers that are supposed to give readings of 0° C at the freezing point of water. Tests on a large sample of these instruments reveal that some readings are too low (denoted by negative numbers) and that some readings are too high (denoted by positive numbers). Assume that the mean reading is 0° C while the standard deviation of the readings is 1.00° C. Also assume that the frequency distribution of errors closely resembles the normal distribution. If one thermometer is randomly selected, find the probability that, at the freezing level of water, the reading is between 0° and +1.58°.

Solution

We are dealing with a standard normal distribution and we are looking for the area of the shaded region in Figure 5-6 with $z = 1.58$. We find from Table A-2 that the shaded area is 0.4429. The probability of randomly selecting a thermometer with an error between 0° and +1.58° is therefore 0.4429. As in Section 4-6, the area corresponds to the probability.

The solutions to this and the following examples are contingent on the values listed in Table A-2. But these values did not appear spontaneously. They were arrived at through calculations that relate directly to Equation 5-1. Table A-2 serves as a convenient means of circumventing difficult computations with that equation.

EXAMPLE

With the thermometers from the preceding example, find the probability of randomly selecting one thermometer that reads (at the freezing point of water) between 0° and −2.43°.

Solution

We are looking for the region shaded in Figure 5-7(a), but Table A-2 is designed to apply only to regions to the right of the mean (zero) as in Figure 5-6. However, by observing that the normal probability distribution possesses symmetry about zero, we see that the shaded regions in parts (a) and (b) of Figure 5-7 have the same area. Referring to Table A-2, we can easily determine that the shaded area of Figure 5-7(b) is 0.4925, so the shaded area of Figure 5-7(a) must also be 0.4925. That is, the probability of randomly selecting a thermometer with an error between 0° and −2.43° is 0.4925.

We incorporate an obvious but useful observation in the following example, but first go back for a minute to Figure 5-6. A vertical line directly above the mean of zero divides the area under the curve into

Pollster Lou Harris

Lou Harris has been in the polling business since 1947, and he has been involved with about 250 political campaigns. While George Gallup believes that pollsters should remain detached and objective, Lou Harris prefers a more personal involvement. He advised John F. Kennedy to openly attack the prejudice against a Catholic becoming president. Kennedy followed that advice and won the nomination.

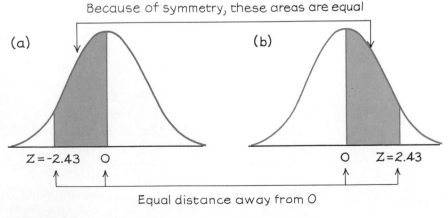

Because of symmetry, these areas are equal

(a)   (b)

Z = −2.43   O   O   Z = 2.43

Equal distance away from O

FIGURE 5-7

two equal parts, each containing an area of 0.5. Since we are dealing with a probability distribution, the total area under the curve must be 1.

EXAMPLE

With these same thermometers, we again make a random selection. Find the probability that the chosen thermometer reads (at the freezing point of water) greater than +1.27°.

Solution

We are again dealing with normally distributed values having a mean of 0° and a standard deviation of 1°. The probability of selecting a thermometer that reads above +1.27° corresponds to the shaded area of Figure 5-8. Table A-2 cannot be used to find that area directly, but we can use the table to find the adjacent area of 0.3980. We can now reason that since the total area to the right of zero is 0.5, the shaded area is 0.5 − 0.3980, or 0.1020. We conclude that there is a 0.1020 probability of randomly selecting one of the thermometers with a reading greater than +1.27°.

We are able to determine the area of the shaded region in Figure 5-8 by an *indirect* application of Table A-2. The following example illustrates yet another indirect use.

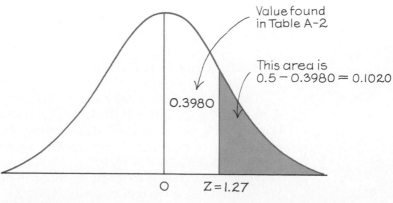

FIGURE 5-8

EXAMPLE

Back to the same thermometers: Assume that one thermometer is randomly selected and find the probability that it reads (at the freezing point of water) between 1.20° and 2.30°.

Solution

The probability of selecting a thermometer that reads between +1.20° and +2.30° corresponds to the shaded area of Figure 5-9. However, Table A-2 is designed to provide only for regions bounded on the left by the vertical line above zero. We can use the table to find the areas of 0.3849 and 0.4893 as shown in this figure. If we denote the area of the shaded region by A, we can see from the figure that

$$0.3849 + A = 0.4893$$

so that

$$A = 0.4893 - 0.3849$$

$$= 0.1044$$

The probability we seek is therefore 0.1044.

NOTATION

$P(a < z < b)$ denotes the probability that the z score is between a and b.

$P(z > a)$ denotes the probability that the z score is greater than a.

$P(z < a)$ denotes the probability that the z score is less than a.

Using this notation, we can express the result of the last example as $P(1.20 < z < 2.30) = 0.1044$. With a continuous probability distribution, such as the normal distribution, $P(z = a) = 0$. That is, the probability of getting any one precise value is 0. From this we can conclude that $P(a \leq z \leq b) = P(a < z < b)$.

Our notation should clearly distinguish between z scores and areas. For example, it is correct to write $P(0 < z < 1.20) = 0.3849$, but it is *wrong* to write $1.20 = 0.3849$.

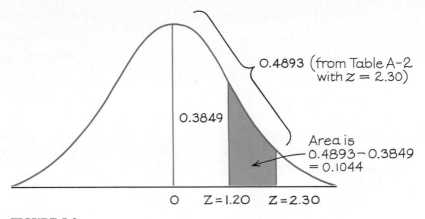

FIGURE 5-9

The examples of this section were contrived so that the mean of 0 and the standard deviation of 1 coincided exactly with the values of the standard normal distribution described in Table A-2. In reality, it would be unusual to find such a nice relationship, since typical normal distributions involve means different from 0 and standard deviations different from 1.

These nonstandard normal distributions introduce another problem. What table of probabilities can be used, since Table A-2 is designed around a mean of 0 and a standard deviation of 1? For example, IQ scores are normally distributed with a mean of 100 and a standard deviation of 15. Scores in this range are far beyond the scope of Table A-2. Section 5-3 examines these nonstandard normal distributions and the methods used in dealing with them.

5-2 EXERCISES A

In Exercises 5-1 through 5-36, assume that the readings on the thermometers are normally distributed with a mean of 0° and a standard deviation of 1.00°. A thermometer is randomly selected and tested. In each case, draw a sketch and find the probability that the reading in degrees is

5-1	Between 0 and 0.25	5-2	Between 0 and 1.00
5-3	Between 0 and 1.50	5-4	Between 0 and 1.96
5-5	Between −1.00 and 0	5-6	Between −0.75 and 0
5-7	Between 0 and −1.75	5-8	Between 0 and −2.33
5-9	Greater than 1.00	5-10	Greater than 0.37

5-11	Greater than 1.83	**5-12**	Greater than 2.05
5-13	Less than -1.00	**5-14**	Less than -2.17
5-15	Less than -0.91	**5-16**	Less than -1.37
5-17	Greater than -1.00	**5-18**	Greater than -0.09
5-19	Less than 3.05	**5-20**	Less than 0.42
5-21	Between -1.00 and 2.00	**5-22**	Between -0.25 and 0.75
5-23	Between -2.00 and 1.50	**5-24**	Between -1.96 and 1.96
5-25	Between 1.00 and 2.00	**5-26**	Between 1.96 and 2.33
5-27	Between 1.28 and 2.58	**5-28**	Between 0.27 and 2.27
5-29	Between -0.83 and -0.51	**5-30**	Between -2.00 and -1.50
5-31	Between -0.25 and -1.35	**5-32**	Between -1.07 and -2.11
5-33	Greater than 0	**5-34**	Less than 0
5-35	Less than -0.50 or greater than 1.50		
5-36	Less than -1.96 or greater than 1.96		

In Exercises 5-37 through 5-48, assume that the readings on the thermometers are normally distributed with a mean of 0° and a standard deviation of 1.00°. Find the indicated probability where z is the reading in degrees.

5-37	$P(z > 2.58)$	**5-38**	$P(-1.36 < z < 1.36)$
5-39	$P(z < 1.28)$	**5-40**	$P(1.25 < z < 1.68)$
5-41	$P(z < -0.57)$	**5-42**	$P(0 < z < 1.68)$
5-43	$P(-2.80 < z < -1.36)$	**5-44**	$P(z > -0.50)$
5-45	$P(-1.09 < z < 0)$	**5-46**	$P(-2.73 < z < 2.51)$
5-47	$P(z < 2.45)$	**5-48**	$P(-0.81 < z < 0.63)$

5-2 EXERCISES B

5-49 Assume that $\mu = 0$ and $\sigma = 1$ for a normally distributed population. Find the percentage of data that are
a. Within 1 standard deviation of the mean
b. Within 1.96 standard deviations of the mean
c. Between $\mu - 3\sigma$ and $\mu + 3\sigma$
d. Between 1 standard deviation below the mean and 2 standard deviations above the mean
e. More than 2 standard deviations away from the mean

5-50   Assume that we have the same normally distributed thermometer readings with a mean of 0° and a standard deviation of 1.00°. If 5% of the thermometers are rejected because they read too high and another 5% are rejected because they read too low, what is the maximum error that will not lead to rejection?

5-51   Assume that we have the same normally distributed thermometer readings with a mean of 0° and a standard deviation of 1.00°. If a buyer establishes a maximum acceptable error and rejects 2% of these thermometers, what are the highest and lowest acceptable readings?

5-52   Assume that z scores are normally distributed with a mean of 0 and a standard deviation of 1.
        a. If $P(0 < z < a) = 0.4778$, find a.   d. If $P(z > d) = 0.8508$, find d.
        b. If $P(-b < z < b) = 0.7814$, find b.   e. If $P(z < e) = 0.0062$, find e.
        c. If $P(z > c) = 0.0329$, find c.

5-53   In Equation 5-1, if we let $\mu = 0$ and $\sigma = 1$ we get

$$y = \frac{e^{-x^2/2}}{\sqrt{2\pi}}$$

which can be approximated by

$$y = \frac{2.7^{-x^2/2}}{2.5}$$

Graph the last equation after finding the y coordinates that correspond to the following x coordinates: $-4, -3, -2, -1, 0, 1, 2, 3,$ and 4. (A calculator capable of dealing with exponents will be helpful.) Attempt to determine the approximate area bounded by the curve, the x-axis, the vertical line passing through 0 on the x-axis, and the vertical line passing through 1 on the x-axis. Compare this result to Table A-2.

5-54   Suppose we begin with a population of normally distributed values with a mean of 0 and a standard deviation of 1.00, and we add 5 to every score. If a value is randomly selected from this modified population, find the probability that it is between 4.00 and 7.00.

5-55   Suppose we begin with a population of normally distributed values with a mean of 0 and a standard deviation of 1.00, and we add the positive number k to every score. If a value is randomly selected from this population, find the probability that it is greater than $k + 1$.

5-56   A population has a mean of 0 and a standard deviation of 1.00 but is uniformly distributed instead of being normally distributed. Such a distribution has minimum and maximum values of $-\sqrt{3}$ and $\sqrt{3}$, respectively. Find the probability of randomly selecting a value between 0 and 1.00 with this uniform distribution, and compare it to the area between 0 and 1.00 for a normal distribution with a mean of 0 and a standard deviation of 1.00.

5-3 NONSTANDARD NORMAL DISTRIBUTIONS

In Section 5-2 we considered only the standard normal distribution, but in this section we will extend the same basic concepts to include nonstandard normal distributions. This inclusion will greatly expand the variety of practical applications we can make since, in reality, most normally distributed populations will have either a nonzero mean and/or a standard deviation different from 1.

We continue to use Table A-2, but we require a way of standardizing these nonstandard cases. This is done by letting

$$z = \frac{x - \mu}{\sigma} \qquad (5\text{-}2)$$

so that z is the number of standard deviations that a particular score x is away from the mean. We call z the **z score** or **standard score** and it is used in Table A-2 (see Figure 5-10).

Suppose, for example, that we are considering a normally distributed collection of IQ scores known to have a mean of 100 and a standard deviation of 15. If we seek the probability of randomly selecting one IQ score that is between 100 and 130, we are concerned with the area shown in Figure 5-11. The difference between 130 and the mean of 100 is 30 IQ points, or exactly 2 standard deviations. The shaded area in Figure 5-11 will therefore correspond to the shaded area of Figure 5-6, where $z = 2$. We get $z = 2$ either by reasoning that 130 is 2 standard deviations above the mean of 100 or by computing

$$z = \frac{x - \mu}{\sigma} = \frac{130 - 100}{15} = \frac{30}{15} = 2$$

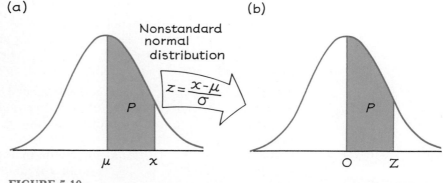

(a)   (b)

FIGURE 5-10

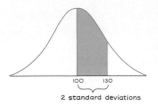

2 standard deviations

FIGURE 5-11

With $z = 2$, Table A-2 indicates that the shaded region we seek has an area of 0.4772, so that the probability of randomly selecting an IQ score between 100 and 130 is 0.4772. Thus Table A-2 can be indirectly applied to any normal probability distribution if we use Equation 5-2 as the algebraic way of recognizing that the z score is actually the number of standard deviations that x is away from the mean. The following examples illustrate that observation. Before considering the next example we should note that IQ and many other types of test scores are discrete whole numbers, while the normal distribution is continuous. We ignore that conflict in this and the following section since the results are minimally affected, but we will introduce a "continuity correction factor" in Section 5-5 when it really becomes necessary. However, we might note for the present that finding the probability of getting an IQ score of *exactly* 130 would require that the discrete value of 130 be represented by the continuous interval from 129.5 to 130.5. From that brief example we can see how the continuity correction factor can be used to deal with discrete data.

EXAMPLE

If IQ scores are normally distributed with a mean of 100 and a standard deviation of 15, find the probability of randomly selecting a subject with an IQ between 100 and 133.

Solution

Referring to Figure 5-12, we seek the probability associated with the shaded region. In order to use Table A-2, the nonstandard data must be standardized by applying Equation 5-2.

$$z = \frac{x - \mu}{\sigma} = \frac{133 - 100}{15} = \frac{33}{15} = 2.20$$

The score of 133 therefore differs from the mean of 100 by 2.20 standard deviations. Corresponding to a z score of 2.20, Table A-2 indicates a probability of 0.4861. There is therefore a probability of 0.4861 of randomly selecting a subject having an IQ between 100 and 133. We could also express this result as $P(100 < x < 133) = 0.4861$ by using the same notation introduced in Section 5.2. Note that in this nonstandard normal distribution, we represent the score in its original units by x, not z. Note also that $P(100 < x < 133) = P(0 < z < 2.20) = 0.4861$.

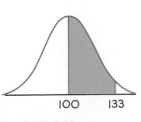

100   133

FIGURE 5-12

When Table A-2 is used in conjunction with Equation 5-2, the nonstandard population mean corresponds to the standard mean of 0. As a result, probabilities extracted directly from Table A-2 must represent regions whose left boundary is the line above the mean.

EXAMPLE

In the beginning of this chapter, we described the production of shear pins with normally distributed breaking points. The mean is 2860 lb, the standard deviation is 52 lb, and a pin is considered defective if its breaking point is above 3000 lb. A buyer complains that many of the pins are defective. Find the percentage of defective pins.

Solution

In Figure 5-13, the shaded region corresponds to the area representing shear pins with a breaking point of 3000 lb or more. We cannot find the area of that shaded region directly, but we can use Equation 5-2 to find the adjacent area immediately to the left of the region we seek.

$$z = \frac{x - \mu}{\sigma} = \frac{3000 - 2860}{52} = \frac{140}{52} = 2.69$$

We now use Table A-2 to get an area of 0.4964 for the region bounded by 2860 and 3000. Since the total area to the right of 2860 is 0.5, the shaded region must be $0.5 - 0.4964$ or 0.0036. The proportion of defective shear pins is 0.0036, which is equivalent to 0.36%. Since only 0.36% of the pins are defective, the buyer is wrong in claiming that many of the pins are defective.

EXAMPLE

The weights of men aged 18 to 74 are normally distributed with a mean of 173 lb and a standard deviation of 30 lb (based on data from the National Health Survey, USDHEW publication 79-1659). Find the percentage of such weights between 190 lb and 225 lb. Among 400 men aged 18 to 74 years, how many are expected to weigh between 190 lb and 225 lb?

continued ▶

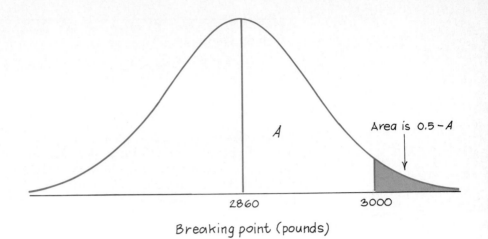

FIGURE 5-13

Example, continued

Solution

Figure 5-14 shows the relevant region B that represents the weights between 190 lb and 225 lb. We cannot find the area of that region directly since the left boundary is not above the mean. Instead, we use Equation 5-2 to find the total area for regions A and B combined and then proceed to subtract the area of region A. To find the area of regions A and B combined, we let $\mu = 173$, $\sigma = 30$, and $x = 225$, so that

$$z = \frac{x - \mu}{\sigma} = \frac{225 - 173}{30} = 1.73$$

From Table A-2 with $z = 1.73$ we get an area of 0.4582. To find the area of region A, we let $\mu = 173$, $\sigma = 30$, and $x = 190$, so that

$$z = \frac{x - \mu}{\sigma} = \frac{190 - 173}{30} = 0.57$$

From Table A-2 with $z = 0.57$ we get an area of 0.2157.

continued ▶

> **Solution, continued**
>
> $$\text{area } B = (\text{areas } A \text{ and } B \text{ combined}) - (\text{area } A)$$
>
> $$= 0.4582 - 0.2157$$
>
> $$= 0.2425$$
>
> We conclude that 24.25% of men who are aged 18 to 74 weigh between 190 lb and 225 lb. If 400 men in that age bracket are randomly selected, we expect that 24.25% of them will weigh between 190 lb and 225 lb. The actual number expected to weigh between those amounts is
>
> $$400 \times 0.2425 = 97.0$$

In this section we extended the concept of Section 5-2 to include more realistic nonstandard normal probability distributions. We noted that the formula $z = (x - \mu)/\sigma$ algebraically represents the number of standard deviations that a particular score x is away from the mean. However, all the examples we have considered so far are of the same general type: a probability (or percentage) determined by using the normal distribution (described in Table A-2) when given the values of the mean, standard deviation, and relevant score(s). In some practical cases, the probability (or percentage) is known and we must determine the relevant score(s). Problems of this type are considered in the following section.

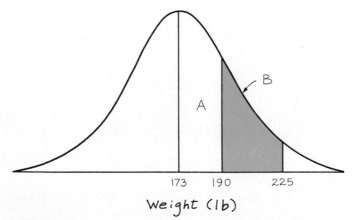

FIGURE 5-14

5-3 EXERCISES A

In Exercises 5-57 through 5-64, assume that IQ scores are normally distributed with a mean of 100 and a standard deviation of 15. An IQ score is randomly selected from this population. Draw a graph and find the indicated probability.

5-57 $P(100 < x < 145)$

5-58 $P(x < 127)$

5-59 $P(x > 140)$

5-60 $P(88 < x < 112)$

5-61 $P(110 < x < 120)$

5-62 $P(120 < x < 130)$

5-63 $P(x < 100)$

5-64 $P(85 < x < 95)$

In Exercises 5-65 through 5-88, answer the given questions. In each case, draw a graph.

5-65 Using the Burnout Measure, subjects have measures that are normally distributed with a mean of 2.97 and a standard deviation of 0.60 (based on "Moderating Effect of Social Support on the Stress-Burnout Relationship," by Etzion, *Journal of Applied Psychology*, Vol. 69, No. 4). Find the probability that a randomly selected subject has a burnout level between 2.97 and 3.87.

5-66 The heights of six-year-old girls are normally distributed with a mean of 117.80 cm and a standard deviation of 5.52 cm (based on data from the National Health Survey, USDHEW publication 73-1605). Find the probability that a randomly selected six-year-old girl has a height between 117.80 cm and 120.56 cm.

5-67 A company manager does a study of the lengths of time clients are kept on hold when calling for information. Those times (in seconds) are found to be normally distributed with a mean of 72.0 s and a standard deviation of 15.0 s. Find the probability that a call will be kept on hold between 60.0 s and 72.0 s.

5-68 For a certain population, scores on the Thematic Apperception Test are normally distributed with a mean of 22.83 and a standard deviation of 8.55 (based on "Relationships Between Achievement-Related Motives, Extrinsic Conditions, and Task Performance," by Schroth, *Journal of Social Psychology*, Vol. 127, No. 1). For a randomly selected subject, find the probability that the score is between 4.02 and 22.83.

5-69 A study shows that Michigan teachers have measures of job dissatisfaction that are normally distributed with a mean of 3.80 and a standard deviation of 0.95 (based on "Stress and Strain from Family Roles and Work-Role Expectations," by Cooke and Rousseau, *Journal of*

Applied Psychology, Vol. 69, No. 2). If subjects with scores above 4.00 are to be given additional tests, what percentage will fall into that category?

5-70   For a certain population, scores on the Miller Analogies Test are normally distributed with a mean of 58.84 and a standard deviation of 15.94 (based on "Equivalencing MAT and GRE Scores Using Simple Linear Transformation and Regression Methods," by Kagan and Stock, *Journal of Experimental Education*, Vol. 49, No. 1). If subjects who score below 27.00 are to be given special training, what is the percentage of subjects who will be given the special training?

5-71   Scores on a college entrance examination are normally distributed with a mean of 500 and a standard deviation of 100. One college gives priority acceptance to subjects scoring above 650. What percentage of subjects are eligible for priority acceptance?

5-72   Using one measure of attractiveness, scores are normally distributed with a mean of 3.93 and a standard deviation of 0.75 (based on "Physical Attractiveness and Self Perception of Mental Disorder," by Burns and Farina, *Journal of Abnormal Psychology*, Vol. 96, No. 2). Find the probability of randomly selecting a subject with a measure of attractiveness that is less than 2.55.

5-73   The serum cholesterol levels in men aged 18 to 74 are normally distributed with a mean of 178.1 and a standard deviation of 40.7. All units are in mg/100 ml and the data are based on the National Health Survey (USDHEW publication 78-1654). If a man aged 18 to 24 is randomly selected, find the probability that his serum cholesterol level is between 100 and 200.

5-74   For a certain group, scores on the Mathematics Usage Test are normally distributed with a mean of 23.9 and a standard deviation of 8.7 (based on "Study of the Measurement Bias of Two Standardized Psychological Tests," by Drasgow, *Journal of Applied Psychology*, Vol. 72, No. 1). If a subject is randomly selected from this group, find the probability of a score between 25.0 and 30.0.

5-75   Scores on an antiaircraft artillery exam are normally distributed with a mean of 99.56 and a standard deviation of 25.84 (based on "Routinization of Mental Training in Organizations: Effects on Performance and Well Being," by Larson, *Journal of Applied Psychology*, Vol. 72, No. 1). For a randomly selected subject, find the probability of a score between 110.00 and 150.00.

5-76   In tests conducted on jet pilots, it has been found that blackout thresholds are normally distributed with a mean of 4.7 G and a standard deviation of 0.8 G. Find the probability of randomly selecting a jet pilot with a blackout threshold that is less than 3.5 G.

5-77 A standard IQ test produces normally distributed results with a mean of 100 and a standard deviation of 15. If an average IQ is defined to be any IQ between 90 and 109, find the probability of randomly selecting an IQ that is average.

5-78 In one college, it has been found that first-semester students have grade-point averages that are normally distributed with a mean of 2.46 and a standard deviation of 0.65. What percentage of students will have grade-point averages between 1.00 and 2.00?

5-79 Scores on the numeric part of the Minnesota Clerical Test are normally distributed with a mean of 119.3 and a standard deviation of 32.4. This test is used for selecting clerical employees. (The data are based on "Modification of the Minnesota Clerical Test to Predict Performance on Video Display Terminals," by Silver and Bennett, *Journal of Applied Psychology*, Vol. 72, No. 1.) If a firm requires scores above 172, find the percentage of subjects who don't qualify.

5-80 In a study of employee stock ownership plans, satisfaction by employees is measured and found to be normally distributed with a mean of 4.89 and a standard deviation of 0.63 (based on "Employee Stock Ownership and Employee Attitudes: A Test of Three Models," by Klein, *Journal of Applied Psychology*, Vol. 72, No. 2). If a subject from this population is randomly selected, find the probability of a job satisfaction score less than 6.78.

5-81 A traffic study conducted at one point on an interstate highway shows that vehicle speeds (in mi/h) are normally distributed with a mean of 61.3 and a standard deviation of 3.3. If a vehicle is randomly checked, what is the probability that its speed is between 55.0 and 60.0?

5-82 A consumer product testing team analyzes the energy consumed by color television sets and finds that the consumption levels (in kwh) are normally distributed with a mean of 320 and a standard deviation of 7.5. For a randomly selected set, find the probability that the level is between 325 and 335.

5-83 For males born in upstate New York, the gestation times are normally distributed with a mean of 39.4 weeks and a standard deviation of 2.43 weeks (based on data from the New York State Department of Health, Monograph No. 11). For a randomly selected male born in upstate New York, find the probability that the gestation time differs from the mean by more than 3.00 weeks.

5-84 In a study of the coliform contamination in streams, an environmentalist finds that one region has normally distributed coliform levels (number of cells per 100 ml) with a mean of 122 and a standard deviation of 14. For a randomly selected sample, find the probability that the coliform level differs from the mean by more than 25.

5-85   For a certain group of students, scores on an algebra placement test are normally distributed with a mean of 18.4 and a standard deviation of 5.1. (This is based on data from "Factors Affecting Achievement in the First Course in Calculus," by Edge and Friedberg, *Journal of Experimental Education*, Vol. 52, No. 3.) If 50 different students are randomly selected from this population, how many of them are expected to score above 16.0?

5-86   The weights of women aged 18 to 24 are normally distributed with a mean of 132 lb and a standard deviation of 27.4 lb (based on data from the National Health Survey, USDHEW publication 79-1659). If 150 women 18 to 24 years old are randomly selected, how many of them are expected to weigh between 100 lb and 150 lb?

5-87   The systolic blood pressures of adults are normally distributed with a mean of 129.8 and a standard deviation of 21.9. (Units are in mm of Hg and the data are based on the National Health Survey, USDHEW publication 78-1648.) If 500 adults are randomly selected for a medical research project, how many of them are expected to have systolic blood pressures above 180.0?

5-88   For Southern girls aged 6 to 11, arithmetic scores on the Wide Range Achievement Test are normally distributed with a mean of 27.0 and a standard deviation of 5.28. (The figures are based on data from the National Health Survey, USDHEW publication 72-1011.) If 750 Southern girls aged 6 to 11 are randomly selected, how many of them are expected to score between 30.0 and 40.0?

5-3 EXERCISES B

5-89   Assume that the following scores are representative of a normally distributed population.
   a. Find the mean $\bar{x}$ of this sample.
   b. Find the standard deviation s of this sample.
   c. Find the percentage of these sample scores that are between 51 and 54, inclusive.
   d. Find the percentage of *population* scores between 51 and 54. Use the sample values of $\bar{x}$ and s as estimates of μ and σ.

50	51	52	51	47	57
51	50	50	51	50	47
51	48	48	50	49	52
49	50	53	47	53	52
51	48	51	54	51	50

Age	f
15–19	220
20–22	173
23–24	55
25–29	98
30–34	62
35–44	85
45–59	36
60–65	4

5-90 The accompanying frequency table summarizes the age distribution for a number of students randomly selected from the population of students at a large university. Although the data do not appear to be normally distributed, assume that they are and find $\bar{x}$ and s. Then use those statistics as estimates of μ and σ in order to find the proportion of students over 21 years of age. How does the result compare to the sample statistics summarized in the table?

5-91 A population has a mean of 100 and a standard deviation of 15 but is uniformly distributed. Such a distribution has a minimum of 74 and a maximum of 126. Find the probability of randomly selecting a value between 80 and 110 with this uniform distribution, and compare it to the area between 80 and 110 for a normal distribution with the same mean of 100 and standard deviation of 15.

5-92 The normal distribution is used in the **control charts** that are an important tool for quality control in industry. In Figure 5-15 we show an example of a control chart reflecting a production process that has gone out of control. Given a production process designed for $\mu = 50.0$ and $\sigma = 4.0$, construct the control chart for the production values given below for 32 consecutive work days. (The values are listed by rows so that 51.2 is the first day, 50.8 is the second day, and so on.) What is happening to production?

51.2	50.8	52.0	49.8	49.7	53.4	46.2
49.3	55.0	48.3	50.4	42.6	58.3	58.0
47.6	43.5	42.9	45.6	52.7	59.4	55.3
41.0	40.8	38.0	56.5	60.4	62.5	57.3
37.8	48.6	62.8	60.7			

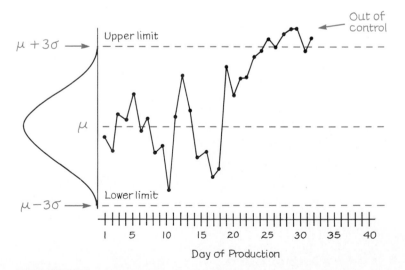

FIGURE 5-15

5-4 FINDING SCORES WHEN GIVEN PROBABILITIES

All the examples and problems in Sections 5-2 and 5-3 involved the determination of a probability or percentage based on the normal distribution data of Table A-2, a given mean, standard deviation, and relevant score(s). In this section we consider the same types of circumstances, but we will alter the known data so the computational procedure will change. These techniques will closely parallel some of the important procedures that will be introduced later in the book.

Let's begin with a practical problem. Based on real data from the National Health Survey, we know that for men aged 18 to 24, the mean weight is 165 lb and the standard deviation is 28 lb. Assuming that we want to identify the heaviest 33% so that we can test the effects of a special training program, what specific weight serves as the cutoff point that separates the top 33% from the lower 67%? Figure 5-16 shows the relevant normal distribution.

We can find the z score that corresponds to the x value we seek after first noting that the region containing 17% corresponds to a probability of 0.1700. Referring to Table A-2 we can see that a probability of 0.1700 corresponds to a z score of 0.44. This means that the desired value of x is 0.44 standard deviation away from the mean. Since the standard deviation is given as 28, we can conclude that 0.44 standard deviation is $0.44 \times 28 = 12.32$. The score x is above the mean and

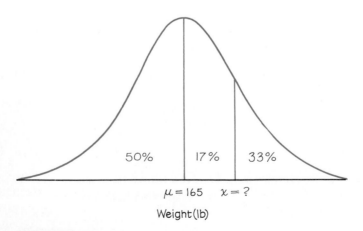

FIGURE 5-16

is 12.32 away. Since the mean is given as 165, the score x must be $165 + 12.32 = 177$ (rounded off). That is, a weight of 177 lb separates the top 33% from the lower 67%. We could also conclude that 177 is the 67th percentile, or P_{67}.

We could have achieved the same results by noting that

$$z = \frac{x - \mu}{\sigma} \quad \text{becomes} \quad 0.44 = \frac{x - 165}{28}$$

when we substitute the given values for the mean μ, the standard deviation σ, and the z score corresponding to a probability of 0.1700. We can solve this last equation for x by multiplying both sides by 28 and then adding 165 to both sides. We get $x = 177$.

EXAMPLE

A clothing manufacturer finds it unprofitable to make clothes for very tall or very short adult males. The executives decide to discontinue production of goods for the tallest 7.5% and the shortest 7.5% of the adult male population. Find the minimum and maximum heights they will continue to serve. The heights of adult males are normally distributed with a mean of 69.0 in. and a standard deviation of 2.8 in. (The figures are based on data from the National Health Survey, USDHEW publication 79-1659.)

Solution

See Figure 5-17. We note that the two outer regions total 15%, so the two equal inner regions must comprise the remaining 85%. This implies that each of the two inner regions must represent 42.5%, which is equivalent to a probability of 0.425. We can use Table A-2 to find the z score that yields a probability of 0.425. With $z = 1.44$, the corresponding probability of 0.4251 is close enough so we conclude that the upper and lower cutoff scores will correspond to $z = 1.44$ and $z = -1.44$. With $\mu = 69.0$ and $\sigma = 2.8$, we can use both values of z in $z = (x - \mu)/\sigma$ to get

$$1.44 = \frac{x - 69.0}{2.8} \quad \text{and} \quad -1.44 = \frac{x - 69.0}{2.8}$$

Solving both of these equations results in the values of 73.032 and 64.968. That is, the company will make clothing only for adult males between 65.0 in. and 73.0 in. tall.

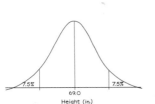

7.5%   7.5%

69.0
Height (in.)

FIGURE 5-17

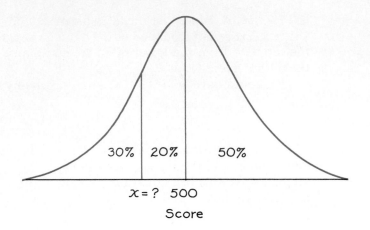

FIGURE 5-18

In the preceding example we can see that the thoughtless use of Equation 5-2 will produce the maximum height of 73.0 in., but determination of the minimum height requires an adjustment that comes only with an understanding of the whole situation. The moral is clear: Don't plug in numbers blindly. Instead, develop an understanding of the underlying meaning. Always draw a graph of the normal distribution with the relevant labels and apply common sense to guarantee that the results are reasonable. The graph and the common-sense check help reduce the incidence of errors.

The following example shows how blind application of Equation 5-2 leads to serious errors in the case of a negative z score.

EXAMPLE

As one of its admissions criteria; a college requires an entrance examination score that is among the top 70% of all scores. Assuming a normal distribution with $\mu = 500$ and $\sigma = 100$, find the minimum acceptable score.

Solution

From Figure 5-18 we see that the area bounded by the unknown relevant score x and the mean constitutes 20% of the total area. If that 20% region were to the right of the mean, we could use

continued ▶

> **Solution, continued**
>
> Table A-2 directly with no difficulty. Let's proceed by pretending that this is the case. Twenty percent, or 0.20, is approximated by a z score of 0.52 (see Table A-2). We can now reason as follows. The value of x differs from the mean of 500 by 0.52 standard deviation. With $\sigma = 100$, 0.52 standard deviation becomes $0.52 \cdot 100 = 52$, so x must be 52 *below* 500. That is, x must be 448. We can see from Figure 5-18 that x is less than 500, so the difference of 52 must be subtracted from 500 instead of added to 500. If this solution were attempted through a superficial application of Equation 5-2, we would get the wrong answer of 552. But 552 is unreasonable because the given data require a result less than 500. This problem could have been quickly solved by simply letting $z = -0.52$ as follows:
>
> $$z = \frac{x - \mu}{\sigma} \quad \text{becomes} \quad -0.52 = \frac{x - 500}{100}$$
>
> which implies that $x = 448$. However, it would be easy to make the mistake of forgetting the negative sign in -0.52. This illustrates the importance of drawing a picture when dealing with the normal distribution. In solving problems of this type, we should *always* draw a figure—such as Figure 5-18—so that we can see exactly what is happening.

When using Table A-2 we can usually avoid interpolation by simply selecting the closest value. However, there are two special cases that involve important values commonly used. The two cases are dealt with by using the interpolated values shown below. Except for these two special cases, we can select the closest value in the table. (If a desired value is midway between two table values, select the larger value.) Also, for z scores above 3.09, we can use 0.4999 as an approximation of the corresponding area.

z score	Area
1.645	0.4500
2.575	0.4950

5-4 EXERCISES A

In Exercises 5-93 through 5-104, assume that the readings on a scale (in meters) are normally distributed with a mean of 0 m and a standard deviation of 1 m.

5-93   Ninety-five percent of the errors are below what value?

5-94   Ninety-nine percent of the errors are below what value?

5-95   Ninety-five percent of the errors are above what value?

5-96   Ninety-nine percent of the errors are above what value?

5-97   If the top 5% and the bottom 5% of all errors are unacceptable, find the minimum and maximum acceptable errors.

5-98   If the top 0.5% and the bottom 0.5% of all errors are unacceptable, find the minimum and maximum acceptable errors.

5-99   If the top 10% and the bottom 5% of all errors are unacceptable, find the minimum and maximum acceptable errors.

5-100   If the top 15% and the bottom 20% of all errors are unacceptable, find the minimum and maximum acceptable errors.

5-101   Find the value that separates the top 40% of all errors from the bottom 60%.

5-102   Find the value that separates the top 82% of all errors from the bottom 18%.

5-103   Find the value of the third quartile (Q_3), which separates the top 25% of all errors from the bottom 75%.

5-104   Find the value of P_{18} (18th percentile), which separates the bottom 18% of all errors from the top 82%.

In Exercises 5-105 through 5-120, answer the given question. In each case, draw a graph.

5-105   Using the Burnout Measure, subjects have measures that are normally distributed with a mean of 2.97 and a standard deviation of 0.60 (based on "Moderating Effect of Social Support on the Stress-Burnout Relationship," by Etzion, *Journal of Applied Psychology*, Vol. 69, No. 4). Find the 95th percentile for these measures of burnout. That is, find the score separating the top 5% from the lower 95%.

5-106   The heights of six-year-old girls are normally distributed with a mean of 117.80 cm and a standard deviation of 5.52 cm (based on data from the National Health Survey, USDHEW publication 73-1605). Find the 90th percentile. That is, find the height that separates the tallest 10% from the shortest 90%.

5-107 A company manager does a study of the lengths of time clients are kept on hold when calling for information. Those times (in seconds) are found to be normally distributed with a mean of 72.0 s and a standard deviation of 15.0 s. Find D_7, the seventh decile. That is, find the time that separates the longest 30% from the shortest 70%.

5-108 For a certain population, scores on the Thematic Apperception Test are normally distributed with a mean of 22.83 and a standard deviation of 8.55 (based on "Relationships Between Achievement-Related Motives, Extrinsic Conditions, and Task Performance," by Schroth, *Journal of Social Psychology*, Vol. 127, No. 1). Find the third quartile Q_3. That is, find the score separating the upper 25% from the lower 75%.

5-109 The heights of adult males are normally distributed with a mean of 69.0 in. and a standard deviation of 2.8 in. (based on data from the National Health Survey, USDHEW publication 79-1659). If 95% of all males satisfy a minimum height requirement for police officers, what is that minimum height requirement?

5-110 Scores on a college entrance examination are normally distributed with a mean of 500 and a standard deviation of 100. If 70% of all examinees pass, find the passing grade.

5-111 In tests conducted on jet pilots, it has been found that their blackout thresholds are normally distributed with a mean of 4.7 G and a standard deviation of 0.8 G. If the Air Force were to establish the rule that a pilot's blackout threshold must be in the top 67%, what would be the lowest acceptable level?

5-112 A manufacturer of bulbs for movie projectors finds that the lives (in hours) of the bulbs are normally distributed with a mean of 61.0 h and a standard deviation of 6.3 h. The manufacturer will guarantee the bulbs so that only 3% will be replaced because of failure before the guaranteed number of hours. For how many hours should the bulbs be guaranteed?

5-113 A standard IQ test produces normally distributed results with a mean of 100 and a standard deviation of 15. A class of high school science students is grouped homogeneously by excluding students with IQ scores in either the top 5% or the bottom 5%. Find the lowest and highest possible IQ scores of students remaining in the class.

5-114 A machine fills sugar boxes in such a way that the weights (in grams) of the contents are normally distributed with a mean of 2260 g and a standard deviation of 20 g. Another machine checks the weights and rejects packages in the top 1% or bottom 1%. Find the minimum and maximum acceptable weights.

5-115   A study shows that Michigan teachers have measures of job dissatisfaction that are normally distributed with a mean of 3.80 and a standard deviation of 0.95. (See "Stress and Strain from Family Roles and Work-Role Expectations," by Cooke and Rousseau, *Journal of Applied Psychology*, Vol. 69, No. 2.) Find the value of the 10–90 percentile range. That is, find the difference between the 10th percentile and the 90th percentile.

5-116   For a certain population, scores on the Miller Analogies Test are normally distributed with a mean of 58.84 and a standard deviation of 15.94. (The data are based on "Equivalencing MAT and GRE Scores Using Simple Linear Transformation and Regression Methods," by Kagan and Stock, *Journal of Experimental Education*, Vol. 49, No. 1.) Find the interquartile range. That is, find the value of $Q_3 - Q_1$ where Q_3 is the third quartile and Q_1 is the first quartile.

5-117   A manufacturer has contracted to supply ball bearings. Product analysis reveals that the diameters are normally distributed with a mean of 25.1 mm and a standard deviation of 0.2 mm. The largest 7% of the diameters and the smallest 13% of the diameters are unacceptable. Find the limits for the diameters of the acceptable ball bearings.

5-118   A manufacturer of color television sets tests competing brands and finds that the amounts of energy they require are normally distributed with a mean of 320 kwh and a standard deviation of 7.5 kwh. If the lowest 30% and the highest 20% are not included in a second round of tests, what are the limits for the energy amounts of the remaining sets?

5-119   A particular x-ray machine gives radiation dosages (in milliroentgens) that are normally distributed with a mean of 4.13 and a standard deviation of 1.27. A dosimeter is set so that it displays yellow for radiation levels that are not in the top 10% or bottom 30%. Find the lowest and highest "yellow" radiation levels.

5-120   In a study of coliform contamination in streams, an environmentalist finds that one region has normally distributed coliform levels (number of cells per 100 ml) with a mean of 122 and a standard deviation of 14. The contamination levels are classified into three equal groups of low, medium, and high. Find the minimum and maximum levels for the "medium" category.

5-4 EXERCISES B

5-121   A city sponsored cross-country race has 4830 applicants, but only 200 are allowed to run in the final race. A qualifying run was held two weeks before the final, and the times are normally distributed with a mean of 36.2 min and a standard deviation of 3.8 min. If the 200 fastest times qualify, what is the cutoff time?

5-122 A teacher gives a test and gets normally distributed results with a mean of 50 and a standard deviation of 10. Grades are to be assigned according to the following scheme. Find the numerical limits for each letter grade.

  A: Top 10%

  B: Scores above the bottom 70% and below the top 10%

  C: Scores above the bottom 30% and below the top 30%

  D: Scores above the bottom 10% and below the top 70%

  F: Bottom 10%

5-123 Assume that the following scores are representative of a normally distributed list of test scores.
a. Find the mean of $\bar{x}$ of this sample.
b. Find the standard deviation s of this sample.
c. If a grade of A is given to the top 5%, find the minimum numerical score that corresponds to A in this sample.
d. Find the theoretical score that separates the top 5% by using the sample mean and standard deviation as estimates for the population mean and standard deviation.

59	69	64	60	77	72	74	64
76	61	77	47	69	82	76	69
60	72	66	92	80	74	78	54
82	59	64	77	69	66	56	53
72	79	58	59	59	50	72	76

5-124 Using recent data from the College Entrance Examination Board, the mean math SAT score is 475 and 17.0% of the scores are above 600. Find the standard deviation and then use that result to find the 99th percentile.

5-5 NORMAL AS APPROXIMATION TO BINOMIAL

The title of this section should really be "The Normal Distribution Used as an Approximation to the Binomial Distribution," but we wanted something a little snappier. This longer version does better

How Large Was Shakespeare's Vocabulary?

According to Bradley Effron and Ronald Thisted, Shakespeare's writings included 31,534 different words. (See "Estimating the Number of Unseen Species: How Many Words Did Shakespeare Know?" in *Biometrika*, vol. 63, no. 3.) They used probability theory to conclude that Shakespeare probably knew at least another 35,000 words that he did not use in his writings. The problem of estimating the size of a population is an important problem often considered in studies of ecology, but the result given here is another interesting application.

express the true intent of this section: We want to solve certain binomial probability problems by approximating the binomial probability distribution by the normal probability distribution. Consider the following situation.

Northwest Airlines recently reported a 72% on-time rate for flight arrivals. Find the probability that among 80 randomly selected flights, at least 52 arrive on time.

This probability problem is binomial with $n = 80$, $p = 0.72$, $q = 0.28$, and x assuming the values of 52, 53, 54, . . . , 80. Table A-1 stops at $n = 25$ so it can't be used here, because we have $n = 80$. In theory, we could apply the binomial probability formula 29 times beginning with

$$P(52) = \frac{80!}{(80 - 52)! \, 52!} \cdot 0.72^{52} \cdot 0.28^{80 - 52}$$

The resulting 29 probabilities can be added to produce the correct result. However, these computations would be *extremely* time consuming, tedious, and generally hazardous to mental health. Fortunately, we will use a much simpler method.

In some cases, which will be described soon, the normal distribution serves as a good approximation to the binomial distribution. Reexamine the graph of Figure 4-7 on page 240, where we illustrate the binomial distribution for $n = 50$ and $p = 0.5$. The graph of that figure strongly resembles a normal distribution. Similar observations have led mathematicians to recognize that the normal distribution may be used as an approximation to the binomial distribution. Although results will be approximate, they are usually good and we can circumvent the lengthy computation of the binomial probability function and use normal distribution computations instead. This explains why we bother with an approximation when results can be exact. The price of exactness (namely, very lengthy computations) is too great. We settle for the normal distribution approximation for reasons of time and effort. However, availability of suitable computer software will make the direct approach better in some cases.

Thus far we have justified the use of the normal distribution as an approximation to the binomial distribution simply because of a strong resemblance between the two graphs as in Figure 4-7. Actually, this is not a sound justification. There are other distributions (such as the t distribution to be examined later) with the same basic bell shape, yet they cannot be approximated by the normal distribution since unacceptable errors result. The justification that allows us to use the normal distribution as an approximation to the binomial distribution results from more advanced mathematics. Specifically, the central limit theorem (discussed in Section 5-6) tells us that

---

if $np \geq 5$ and $nq \geq 5$, then the binomial random variable is approximately normally distributed with the mean and standard deviation given as

$$\mu = np$$
$$\sigma = \sqrt{npq}$$

Results of higher mathematics are used to show that the sampling distribution of sample means from *any* population tends to be normally distributed as long as the sample size is large enough. This will be discussed in more detail later. Relative to binomial experiments, past experience has shown that the normal distribution is a reasonable approximation to the binomial distribution as long as $np \geq 5$ and $nq \geq 5$. In effect, we are stating that there is a formal theorem that justifies our approximation. Unfortunately (or fortunately, depending on your perspective), it is not practical to outline here the details of the proof for that theorem. For the present, you should accept the intuitive evidence of the strong resemblance between the binomial and normal distributions and take the more rigorous evidence on faith.

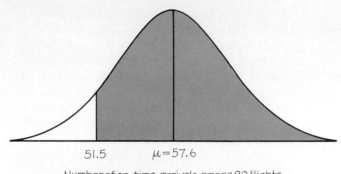

51.5 μ=57.6

Number of on-time arrivals among 80 flights

FIGURE 5-19

Figure 5-19 shows the normal distribution that approximates the binomial experiment involving the 80 flights. The mean of 57.6 was obtained by applying the formula $\mu = np$, which describes the mean for any binomial experiment. (From Section 4-5 we know that $\mu = np$ for binomial experiments.)

$$\mu = np = 80 \cdot 0.72 = 57.6$$

Similarly, the standard deviation can be calculated as

$$\sigma = \sqrt{npq} = \sqrt{80 \cdot 0.72 \cdot 0.28} = 4.0$$

We seek the probability of getting *at least* 52 on-time flights among 80, so we must include the probability of *exactly* 52. But the discrete value of 52 is approximated in the continuous normal distribution by the interval from 51.5 to 52.5. Such conversions from a discrete to a continuous distribution are called **continuity corrections**. See Figure 5-20 for an illustration of how the discrete value of 52 is corrected for continuity when represented in the continuous normal distribution. If we ignore or forget the continuity correction, the additional error will be very small as long as n is large. However, we should neither ignore nor forget the continuity correction. We should always use it when approximating the binomial distribution by the normal distribution.

Reverting to standard procedures associated with normal distributions and Table A-2, we recognize that the probability corresponding to the shaded region of Figure 5-19 cannot be obtained directly. We must find the probability representing the area immediately to the left of the line through the mean and add that value to 0.5.

$$z = \frac{x - \mu}{\sigma} = \frac{51.5 - 57.6}{4.0} = -1.53$$

Magazine Survey Results Reflect Readership

Magazines often boost sales through surveys of their readers, but such surveys are typically biased and reflect only the views of the respondents. A *Time* article on magazine surveys noted that when wives were asked if they have ever had an extramarital affair, the results were 21% yes for *Ladies Home Journal*, 34% yes for *Playboy*, and 54% yes for *Cosmopolitan*. One pollster suggested that a *Reader's Digest* survey "would probably find that *nobody* had any extramarital affairs." Readers should realize that the results of such surveys are typically very biased because the surveys do not follow good techniques for sampling. While the results may be interesting, they are generally unreliable.

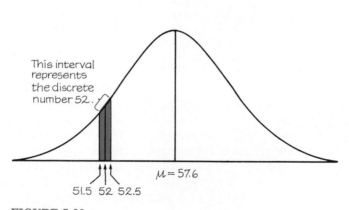

FIGURE 5-20

Table A-2 indicates that a z score of -1.53 yields a probability of 0.4370 so that the shaded region has a probability of $0.4370 + 0.5 = 0.9370$. Without the continuity correction, we would have used 52 in place of 51.5 and an answer of 0.9192 would have resulted, but the probability of 0.9370 is the better result.

Listed below is a computer printout from the statistics package **STATDISK**. One program in that package computes binomial probabilities, and the sample display shows one page of the output that results when we request the binomial probabilities corresponding to $n = 80$ and $p = 0.72$. The table shows that the cumulative probability corresponding to $x = 51$ is 0.06693, so that the probability of getting any value 52 and above is $1 - 0.06693 = 0.93307$, which is very close to our result of 0.9370. For this particular example, the STATDISK program is a good alternative to the normal approximation method, but every program will have some limitations that do not apply to the approximation method.

STATDISK DISPLAY

```
X      P(X)    Cum.Prob.        X      P(X)    Cum.Prob.
-------------------------        -------------------------
44   0.00048   0.00086         55   0.07810   0.29603
45   0.00098   0.00185         56   0.08965   0.38569
46   0.00193   0.00377         57   0.09707   0.48275
47   0.00358   0.00736         58   0.09898   0.58173
48   0.00633   0.01369         59   0.09491   0.67664
49   0.01064   0.02433         60   0.08542   0.76206
50   0.01696   0.04128         61   0.07201   0.83407
51   0.02565   0.06693         62   0.05675   0.89082
52   0.03678   0.10372         63   0.04169   0.93251
53   0.04997   0.15369         64   0.02848   0.96099
54   0.06425   0.21793         65   0.01803   0.97901
```

TABLE 5-2
Minimum sample required to approximate a binomial distribution by a normal distribution

p	n must be at least
0.001	5000
0.01	500
0.1	50
0.2	25
0.3	17
0.4	13
0.5	10
0.6	13
0.7	17
0.8	25
0.9	50
0.99	500
0.999	5000

In this example, the normal distribution does serve as a good approximation to the binomial experiment, but this is not always the case. For large values of n with p not too close to 0 or 1, the normal distribution approximates the binomial fairly accurately, but for *small values of n with p near 0 or 1, the binomial distribution is approximated very poorly by the normal distribution*. The approximation is suitable only when $np \geq 5$ and $nq \geq 5$; both conditions must be met. Table 5-2 lists some specific values of p along with the corresponding minimum values of n.

Suppose, for example, that a couple plans to have four children and they seek the probability of having three girls and one boy. With $n = 4$, $p = 0.5$, $q = 0.5$, and $x = 3$, we see that $np = 4 \cdot 0.5 = 2$ and $nq = 4 \cdot 0.5 = 2$, so the normal distribution should *not* be used. Even though the distribution is symmetrical and mound shaped, it does not fit a normal distribution well. Table 5-2 indicates that they should plan on at least 10 children if they really want to use the normal distribution as an approximation to the binomial. (Admittedly, this is not much of an inducement for a large family.) The binomial probability formula can be applied easily to the given situation.

In Figure 5-21 we summarize the procedure for using the normal distribution as an approximation to the binomial distribution. The following example illustrates this procedure.

FIGURE 5-21
*Solving binomial
probability problems*

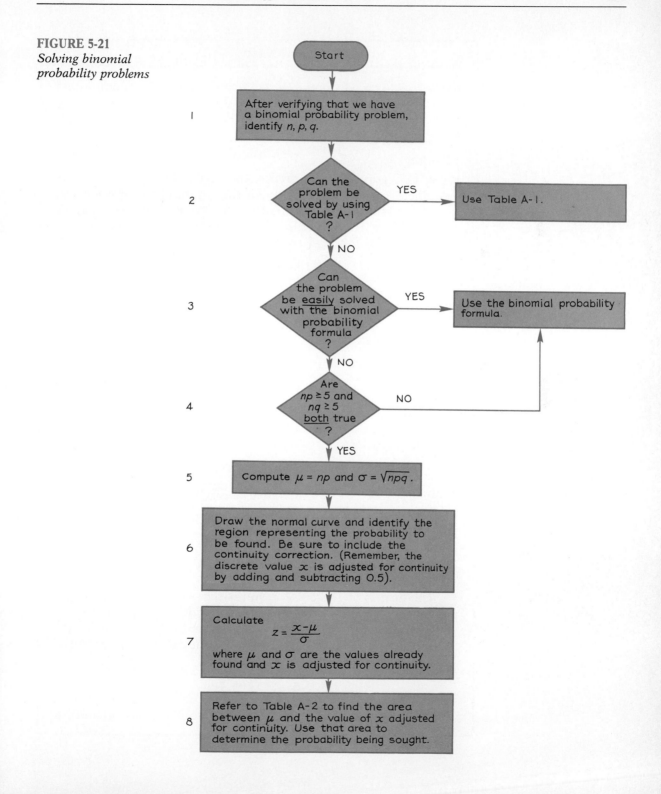

EXAMPLE

According to data from the Hertz Corporation, 80% of commuters use their own vehicle. Find the probability that of 100 randomly selected commuters, *exactly* 85 use their own vehicles.

Solution

Refer to Figure 5-21. In step 1 we verify that the conditions described do satisfy the criteria for the binomial distribution and $n = 100$, $p = 0.80$, $q = 0.20$, and $x = 85$. Proceeding to step 2, we see that Table A-1 cannot be used because n is too large. In step 3, the binomial probability formula applies, but

$$P(85) = \frac{100!}{(100 - 85)!\, 85!} \cdot 0.80^{85} \cdot 0.20^{100 - 85}$$

is too difficult to compute. Most calculators cannot evaluate anything above 70!, but the availability of calculators or a computer would simplify this approach. In step 4 we get

$$np = 100 \cdot 0.80 = 80 \geq 5$$

$$nq = 100 \cdot 0.20 = 20 \geq 5$$

and since np and nq are both at least 5, we conclude that the normal approximation to the binomial is satisfactory. We now go on to step 5, where we obtain the values of μ and σ as follows.

$$\mu = np = 100 \cdot 0.80 = 80.0$$

$$\sigma = \sqrt{npq} = \sqrt{100 \cdot 0.80 \cdot 0.20} = 4.0$$

Now we go to step 6, where we draw the normal curve shown in Figure 5-22. The shaded region of Figure 5-22 represents the probability we want. Use of the continuity correction results in the representation of 85 by the region extending from 84.5 to 85.5. We now proceed to step 7.

The format of Table A-2 requires that we first find the probability corresponding to the region bounded on the left by the vertical line through the mean of 80.0 and on the right by the vertical line through 85.5, so that one of the calculations required in step 7 is as follows.

$$z_2 = \frac{x - \mu}{\sigma} = \frac{85.5 - 80.0}{4.0} = 1.38$$

continued ▶

Solution, continued

We also need the probability corresponding to the region bounded by 80.0 and 84.5, so we calculate

$$z_1 = \frac{x - \mu}{\sigma} = \frac{84.5 - 80.0}{4.0} = 1.13$$

Finally, in step 8 we use Table A-2 to find that a probability of 0.4162 corresponds to $z_2 = 1.38$ and 0.3708 corresponds to $z_1 = 1.13$. Consequently, the entire shaded region of Figure 5-22 depicts a probability of $0.4162 - 0.3708 = 0.0454$. Using the software package STATDISK, we find that the probability is 0.04806; the discrepancy of 0.00266 is very small.

As one last example that can be checked with Table A-1, let's assume that an experiment consists of randomly selecting 16 births and recording the number of boys. With $n = 16$, $p = 0.5$, and $q = 0.5$, we could let x assume the values 0, 1, 2, . . . , 15, 16 and apply the binomial probability formula to obtain the results summarized in Table 5-3. The graph of Table 5-3 consists of the rectangles shown in Figure 5-23. (Figure 5-23 provides a visual comparison of a binomial distribution and the approximating normal distribution.)

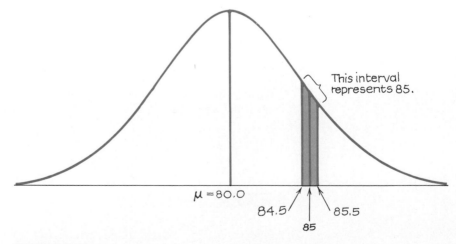

FIGURE 5-22

TABLE 5-3 The values of $P(x)$ are computed with the binomial probability formula or from Table A-1	
x	$P(x)$
0	0+
1	0+
2	0.002
3	0.009
4	0.028
5	0.067
6	0.122
7	0.175
8	0.196
9	0.175
10	0.122
11	0.067
12	0.028
13	0.009
14	0.002
15	0+
16	0+

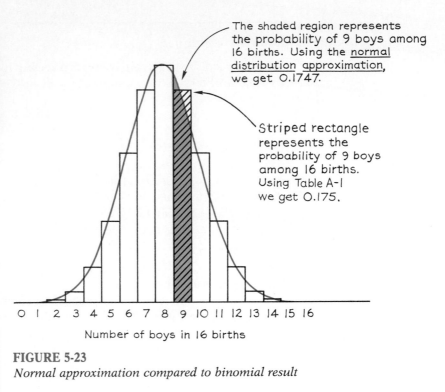

The shaded region represents the probability of 9 boys among 16 births. Using the normal distribution approximation, we get 0.1747.

Striped rectangle represents the probability of 9 boys among 16 births. Using Table A-1 we get 0.175.

Number of boys in 16 births

FIGURE 5-23
Normal approximation compared to binomial result

The approximating normal curve is characterized by $\mu = np = 16 \cdot 0.5 = 8$ and $\sigma = \sqrt{npq} = \sqrt{16 \cdot 0.5 \cdot 0.5} = \sqrt{4} = 2$. As a basis for comparison, we find the probability of getting exactly nine boys using the binomial distribution and the normal distribution. Using Table 5-3 we see that $P(9) = 0.175$. Using the normal distribution with the continuity correction, $P(9)$ is computed to be 0.1747 and the two results agree.

We now have three different methods for determining probabilities in binomial experiments. These three methods are summarized in Figure 5-21.

5-5 EXERCISES A

In Exercises 5-125 through 5-132, check that np ≥ 5 and nq ≥ 5 in order to determine whether the normal distribution is a suitable approximation. In each case, also find the values of μ and σ.

5-125 $n = 25, p = 0.250$ **5-126** $n = 50, p = 0.333$

5-127 $n = 90, p = 0.048$ **5-128** $n = 63, p = 0.068$

5-129 $n = 250, p = 0.010$ **5-130** $n = 657, p = 0.994$

5-131 $n = 84, p = 0.950$ **5-132** $n = 125, p = 0.961$

In Exercises 5-133 through 5-140, find the indicated binomial probabilities by using (a) Table A-1 in Appendix A and (b) the normal distribution as an approximation to the binomial probability distribution.

5-133 With $n = 12$ and $p = 0.50$, find $P(8)$.

5-134 With $n = 15$ and $p = 0.40$, find $P(7)$.

5-135 With $n = 20$ and $p = 0.70$, find $P(12)$.

5-136 With $n = 25$ and $p = 0.30$, find $P(2)$.

5-137 With $n = 12$ and $p = 0.50$, find P(at least 8).

5-138 With $n = 15$ and $p = 0.40$, find P(at least 7).

5-139 With $n = 20$ and $p = 0.70$, find P(at most 12).

5-140 With $n = 25$ and $p = 0.30$, find P(at most 2).

5-141 Find the probability of getting at least 60 girls in 100 births.

5-142 Find the probability of getting exactly 40 girls in 80 births.

5-143 Find the probability of passing a true-false test of 100 questions if 65% is passing and all responses are random guesses.

5-144 A multiple-choice test consists of 40 questions with possible answers of *a, b, c, d, e*. Find the probability of getting at most 30% correct if all answers are random guesses.

5-145 Northwest Airlines recently reported that its on-time arrival rate is 72%. Find the probability that among 80 randomly selected flights, fewer than 70 arrive on time.

5-146 Based on U.S. Bureau of Justice data, 16% of those arrested are women. If 250 arrested people are randomly selected, find the probability that the number of women is at least 35.

5-147 A survey conducted by the U.S. Department of Transportation showed that 25% of New York City drivers wear seat belts. If 400 New York City drivers are randomly selected, find the probability that at least 125 of them wear seat belts.

5-148 Based on U.S. Census Bureau data, 12% of the men in the United States have earned bachelor's degrees. If 140 U.S. men are randomly selected, find the probability that at least 20 of them have a bachelor's degree.

5-149 *Popular Science* magazine reported that in a study of chain saw accidents, it was found that 25% were due to kickback. Find the probability of getting exactly 21 kickback accidents in 84 randomly selected chain saw accidents.

5-150   *Glamour* magazine reported that your chances of being hit by lightning are 1 in 600,000. If 6 million people are randomly selected, find the probability that fewer than 6 are hit by lightning.

5-151   According to Census Bureau data, among men aged 18 to 24, 60% live at home with their parents. If 500 men aged 18 to 24 are randomly selected, find the probability that more than 325 of them live at home with their parents.

5-152   According to Helen Fisher of the American Museum of Natural History, among couples who divorce, 40% have no children. If 250 divorce cases are randomly selected, find the probability that more than 80 involve couples with no children.

5-153   Among women aged 18 to 24, 75% are more than 159 cm tall (based on data from the National Health Survey, USDHEW publication 79-1659). If 320 women aged 18 to 24 are randomly selected, find the probability that more than 250 of them are more than 159 cm tall.

5-154   An airline company experiences a 7% rate of no-shows on advance reservations. Find the probability that of 250 randomly selected advance reservations, there will be at least 10 no-shows.

5-155   A certain genetic characteristic appears in one-quarter of all offspring. Find the probability that of 40 randomly selected offspring, fewer than 5 exhibit the characteristic in question.

5-156   The IRS finds that of all taxpayers whose returns are audited, 70% end up paying additional taxes. Find the probability that of 500 randomly selected returns, at least 400 end up paying additional taxes.

5-157   Of those who commute to southern Manhattan, 6.5% use commuter railroads (based on data from the New York Metropolitan Transportation Council). If we randomly select 175 people who commute to southern Manhattan, find the probability that the number who use railroads is between 10 and 15 inclusive.

5-158   In Illinois, 17% of men surveyed were found to have at least two alcoholic drinks per day (based on data from the National Centers for Disease Control). Assuming that this rate is correct, find the probability that among 125 randomly selected Illinois men, the number who average at least two alcoholic drinks per day is between 20 and 25 inclusive.

5-159   Forty-five percent of us have group O blood, according to data provided by the Greater New York Blood Program. If 400 subjects are randomly selected, find the probability that the number with group O blood is between 200 and 205 inclusive.

5-160   Among workers aged 20 to 24, 26% work more than 40 hours per week (based on data from the U.S. Department of Labor). If we randomly select 350 workers aged 20 to 24, find the probability that the number who work more than 40 hours per week is between 80 and 90 inclusive.

5-5 EXERCISES B

5-161   In a binomial experiment with $n = 25$ and $p = 0.4$, find P(at least 18) using
a. The table of binomial probabilities (Table A-1)
b. The binomial probability formula
c. The normal distribution approximation

5-162   The cause of death is related to heart disease in 52% of the cases studied. Find the probability that in 500 randomly selected cases, the number of heart-disease–related deaths differs from the mean by more than 2 standard deviations.

5-163   An airline company works only with advance reservations and experiences a 7% rate of no-shows. How many reservations could be accepted for an airliner with a capacity of 250 if there is at least a 0.95 probability that all reservation holders who show will be accommodated?

5-164   A company manufactures integrated circuit chips with a 23% rate of defects. What is the minimum number of chips that must be manufactured if there must be at least a 90% chance that 5000 good chips can be supplied?

5-6 THE CENTRAL LIMIT THEOREM

Now that we know something about distributions, let's consider the following question: If samples are randomly selected from a population that is *uniformly* distributed, what will be the distribution of the resulting sample means? To be more specific, suppose that we randomly select 10 telephone numbers and record only the last digit of each number. Now we can find the mean of these 10 sample scores. Repeat that process to generate more sample means. What will be the distribution of these sample means? Many people *incorrectly* conclude that the distribution of these sample means will be uniform, since they come from a uniform distribution. Wrong. The **central limit theorem** will provide us with important knowledge of the correct distribution.

The central limit theorem is one of the most important topics in the study of statistics, but students traditionally have some difficulty

Juxtaposition

It often happens that a statistic is uninteresting and lifeless when presented alone. However, that same statistic may become fascinating and more understandable when accompanied by another statistic that could be used for a comparison or contrast. For example, *Harper's Magazine* listed these statistics:

"Amount the Reagan Administration budgeted for military bands in 1987: $154,200,000

Amount it budgeted for the National Endowment for the Arts: $144,900,000"

with it. Before considering the statement of the central limit theorem, we will first try to develop an intuitive understanding of one of its consequences.

The sampling distribution of sample means tends to be a normal distribution. This implies that if we collect samples all of the same size, compute their means, and then develop a histogram of those means, it will tend to assume the bell shape of a normal distribution. This is true regardless of the shape of the distribution of the original population. The central limit theorem qualifies the preceding remarks and includes additional aspects, but stop and try to understand the thrust of these remarks before continuing.

Let's begin with some concrete numbers. Table 5-4 contains a block of data consisting of 300 sample scores. These scores were generated through a computer simulation, but they could have been extracted from a telephone directory or a book of random numbers. (Now *there's* exciting reading, although the plot is a little thin.) In Figure 5-24 we illustrate the histogram of the 300 sample scores, and we can see that their distribution is essentially uniform. Now consider the 300 scores to be 30 samples, with 10 scores in each sample. The resulting 30 sample means are listed in Table 5-4 and illustrated in the histogram of Figure 5-25. Note that the shape of Figure 5-25 is roughly that of a normal distribution. It is important to observe that even though the original population has a uniform distribution, the sample means seem to have a normal distribution. It was observations exactly like this that led to the formulation of the central limit theorem. If our sample means were based on samples of size larger than 10, Figure 5-25 would more closely resemble a normal distribution. We are now ready to consider the central limit theorem.

Let's assume that the variable x represents scores that may or may not be normally distributed, and that the mean of the x values is μ while the standard deviation is σ. Suppose we collect a sample of size n and calculate the sample mean $\bar{x}$. What do we know about the collection of all sample means that we produce by repeating this experiment, collecting a sample of size n to get the sample mean? The central limit theorem tells us that as the sample size n increases, the sample means will tend to approach a normal distribution with mean μ and standard deviation $\sigma/\sqrt{n}$. See Figure 5-26. The distribution of sample means *tends* to be a normal distribution in the sense that as n becomes larger, the distribution of sample means gets closer to a normal distribution. This conclusion is not intuitively obvious, and it was arrived at through extensive research and analysis. The formal rigorous proof requires advanced mathematics and is beyond the scope of this text, so we can only illustrate the theorem and give examples of its use.

TABLE 5-4 Thirty collections of samples with ten random numbers between 1 and 9 in each sample. The right column consists of the corresponding sample means.

Sample	Data										Sample mean
1	2	7	5	5	2	1	7	7	9	4	4.9
2	5	8	1	1	5	7	1	4	1	4	3.7
3	7	6	9	8	5	1	6	4	7	9	6.2
4	7	3	1	7	3	6	7	9	4	3	5.0
5	9	7	7	6	1	6	8	3	4	7	5.8
6	5	3	3	4	2	5	9	9	1	9	5.0
7	5	5	3	9	5	3	1	9	1	5	4.6
8	4	3	9	5	5	9	1	7	7	8	5.8
9	2	1	7	8	6	7	7	9	8	3	5.8
10	3	4	5	6	8	4	8	3	4	5	5.0
11	5	3	2	2	6	8	1	5	5	9	4.6
12	7	5	9	6	8	2	2	7	2	1	4.9
13	3	1	4	1	7	9	3	2	3	8	4.1
14	6	2	7	4	4	5	2	6	8	6	5.0
15	9	6	2	9	4	2	6	3	5	5	5.1
16	9	2	2	3	6	2	6	6	8	3	4.7
17	5	4	2	1	9	4	2	9	4	2	4.2
18	8	1	2	1	4	3	2	8	5	4	3.8
19	5	8	9	6	2	7	9	3	8	5	6.2
20	5	6	8	7	5	9	6	4	8	7	6.5
21	7	9	9	8	3	5	5	1	4	6	5.7
22	8	4	7	8	7	8	7	7	1	8	6.5
23	5	5	1	7	5	7	7	2	9	8	5.6
24	9	5	2	5	9	2	5	3	5	8	5.3
25	4	5	8	4	2	9	2	6	6	1	4.7
26	1	7	7	3	4	7	7	2	8	7	5.3
27	8	1	1	7	6	2	2	1	4	9	4.1
28	9	4	3	7	3	7	8	4	3	2	5.0
29	1	2	9	3	8	2	4	6	2	8	4.5
30	2	9	3	3	1	2	6	7	8	7	4.8

↑ See Figure 5-24   ↑ See Figure 5-25

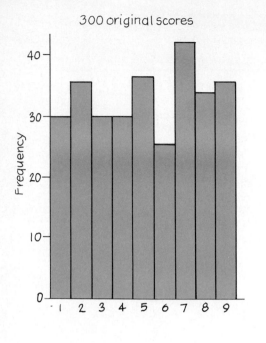

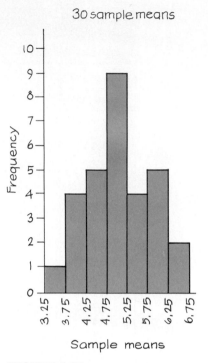

FIGURE 5-24
Histogram of the 300 original scores randomly selected (between 1 and 9)

FIGURE 5-25
Histogram of the 30 sample means. Each sample mean is based on 10 raw scores randomly selected between 1 and 9, inclusive.

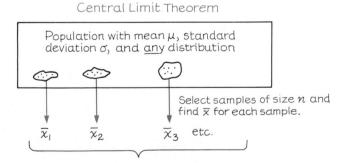

FIGURE 5-26

CENTRAL LIMIT THEOREM

Given:

1. The random variable x has some distribution with mean μ and standard deviation σ.
2. Samples of size n are randomly selected from this population.

Conclusions:

1. The distribution of all possible sample means $\bar{x}$ will approach a *normal* distribution.
2. The *mean* of the sample means will be μ.
3. The *standard deviation* of the sample means will be $\sigma/\sqrt{n}$.

In the first conclusion above, when we say that the sample means *approach* a normal distribution, we are employing these commonly used rules:

1. For samples of size n larger than 30, the sample means can be approximated reasonably well by a normal distribution. The approximation gets better as the sample size n becomes larger.
2. If the original population is itself normally distributed, then the sample means will be normally distributed, regardless of the sample size n.

NOTATION

If all possible random samples of size n are selected from a population with mean μ and standard deviation σ, the mean of the sample means is denoted by $\mu_{\bar{x}}$ so that

$$\mu_{\bar{x}} = \mu$$

Also, the standard deviation of the sample means is denoted by $\sigma_{\bar{x}}$ so that

$$\sigma_{\bar{x}} = \frac{\sigma}{\sqrt{n}}$$

$\sigma_{\bar{x}}$ is often called the **standard error of the mean**.

Comparison of Figures 5-24 and 5-25 should confirm that the original numbers have a nonnormal distribution, while the sample means approximate a normal distribution. The central limit theorem also indicates that the mean of *all* such sample means should be μ (the mean of the original population) and the standard deviation of all such sample means should be $\sigma/\sqrt{n}$ (where σ is the standard deviation of the original population and n is the sample size of 10). We can find μ and σ for the original population of numbers between 1 and 9 by noting that, if those numbers occur with equal frequency as they should, then the population mean μ is given by

$$\mu = \frac{1 + 2 + 3 + 4 + 5 + 6 + 7 + 8 + 9}{9} = 5.0$$

Similarly, we can find σ by again using 1, 2, 3, 4, 5, 6, 7, 8, 9 as an ideal or theoretical representation of the population. Following this course, σ is computed to be 2.58. The mean and standard deviation of the sample means can now be found as follows.

$$\mu_{\bar{x}} = \mu = 5.0$$

$$\sigma_{\bar{x}} = \frac{\sigma}{\sqrt{n}} = \frac{2.58}{\sqrt{10}} = \frac{2.58}{3.16} = 0.82$$

The preceding results represent *all* sample means of size $n = 10$. For the 30 sample means shown in Table 5-4, we have a mean of 5.08 and a standard deviation of 0.75. We can see that our real data conform quite well to the theoretically predicted values for $\mu_{\bar{x}}$ and $\sigma_{\bar{x}}$. Many important and practical problems can be solved with the central limit theorem.

EXAMPLE

The Goodenough-Harris Drawing Test is used to measure the intellectual maturity of young people. For twelve-year-old girls, the scores are normally distributed with a mean of 34.8 and a standard deviation of 7.02. (The figures are based on data from the National Health Survey.) If 36 twelve-year-old girls are randomly selected, find the probability that their mean score is between 34.8 and 37.0.

continued ▶

Example, continued

Solution

After noting that the sample size $n = 36$ is greater than 30, we apply the central limit theorem and conclude that the distribution of sample means is the normal distribution. We proceed to find $\mu_{\bar{x}}$ and $\sigma_{\bar{x}}$ as follows.

$$\mu_{\bar{x}} = \mu = 34.8$$

$$\sigma_{\bar{x}} = \frac{\sigma}{\sqrt{n}} = \frac{7.02}{\sqrt{36}} = 1.17$$

In Figure 5-27 we show the shaded area corresponding to the probability we seek. We find that area by first determining the value of the z score.

$$z = \frac{\bar{x} - \mu_{\bar{x}}}{\sigma_{\bar{x}}} = \frac{\bar{x} - \mu_{\bar{x}}}{\dfrac{\sigma}{\sqrt{n}}} = \frac{37.0 - 34.8}{1.17} = 1.88$$

From Table A-2 we find that $z = 1.88$ corresponds to an area of 0.4699, so that the probability we seek is 0.4699. That is, $P(34.8 < \bar{x} < 37.0) = P(0 < z < 1.88) = 0.4699$. Note that the key variable is denoted as $\bar{x}$ because we are dealing with a mean.

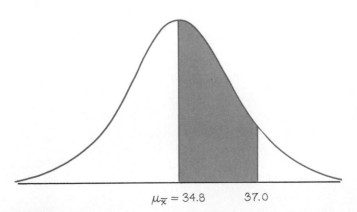

FIGURE 5-27

EXAMPLE

Given the same set of normally distributed scores with mean 34.8 and standard deviation 7.02, assume that 100 twelve-year-old girls are randomly selected. Find the probability that this sample group has a mean score greater than 36.9.

Solution

Since the sample size is greater than 30, we apply the central limit theorem and conclude that the distribution of sample means is the normal distribution, with $\mu_{\bar{x}}$ and $\sigma_{\bar{x}}$ determined as follows.

$$\mu_{\bar{x}} = \mu = 34.8$$

$$\sigma_{\bar{x}} = \frac{\sigma}{\sqrt{n}} = \frac{7.02}{\sqrt{100}} = 0.702$$

In Figure 5-28 we show the graph that describes the distribution of sample means, where the samples are all of size $n = 100$ and they are randomly selected from the given population. We find the area of the shaded region by first finding the area of the region to its immediate left.

$$z = \frac{\bar{x} - \mu_{\bar{x}}}{\sigma_{\bar{x}}} = \frac{\bar{x} - \mu_{\bar{x}}}{\dfrac{\sigma}{\sqrt{n}}} = \frac{36.9 - 34.8}{\dfrac{7.02}{\sqrt{100}}} = 2.99$$

From Table A-2 we see that a z score of 2.99 corresponds to an area of 0.4986, so the shaded region is $0.5 - 0.4986 = 0.0014$. The probability of getting a sample mean greater than 36.9 is $P(\bar{x} > 36.9) = P(z > 2.99) = 0.0014$.

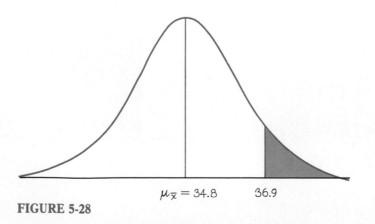

FIGURE 5-28

It is interesting to note that, as the sample size increases, the sample means tend to vary less, since $\sigma_{\bar{x}} = \sigma/\sqrt{n}$ gets smaller as n gets larger. For example, assume that IQ scores have a mean of 100 and a standard deviation of 15. Samples of 36 will produce means with $\sigma_{\bar{x}} = 15/\sqrt{36} = 2.5$, so that 99% of all such samples will have means between 93.6 and 106.4. If the sample size is increased to 100, $\sigma_{\bar{x}}$ becomes $15/\sqrt{100}$, or 1.5, so that 99% of the samples will have means between 96.1 and 103.9. (These particular computations and concepts are explained in Section 7-2.)

These results are supported by common sense: As the sample size increases, the corresponding sample mean will tend to be closer to the true population mean. The effect of an unusual or outstanding score tends to be dampened as it is averaged in as part of a sample.

Our use of $\sigma_{\bar{x}} = \sigma/\sqrt{n}$ assumes that the population is infinite. When we sample with replacement of selected data, for example, the population is effectively infinite. Yet realistic applications involve sampling without replacement, so successive samples depend on previous outcomes.

NOTATION

Just as n denotes the *sample* size, N denotes the size of a *population*. For finite populations of size N, we should incorporate the **finite population correction factor** $\sqrt{(N - n) \div (N - 1)}$ so that $\sigma_{\bar{x}}$ is found as follows.

$$\sigma_{\bar{x}} = \frac{\sigma}{\sqrt{n}} \sqrt{\frac{N - n}{N - 1}}$$

If the sample size n is small in comparison to the population size N, the finite population correction factor will be close to 1. Consequently its impact will be negligible and it can therefore be ignored. Statisticians have devised the following rule of thumb.

RULE

Use the finite population correction factor when computing $\sigma_{\bar{x}}$ if the population is finite and $n > 0.05N$. That is, use the correction factor only if the sample size is greater than 5% of the population size.

Consumer Price Index as a Measure of Inflation

The government uses the consumer price index (CPI) as a measure of inflation. Each month, the cost of certain goods and services is determined for a typical family. The CPI was constructed so that the cost of mortgages and home purchase prices accounted for about 25% of the total monthly cost. The effect was that radical changes in the cost of houses and mortgages caused radical changes in the monthly CPI, even though a small percentage of families actually buy or mortgage a house in any given month. The Bureau of Labor Statistics revised the formula for determining CPI so that home purchase costs were replaced by housing rental costs. In one particular month, that change lowered the CPI from 15.4% to 10%.

Since almost 50 million Americans receive wages or benefits affected by the CPI, its value has dramatic effects. For example, an increase in the CPI of only 1 percentage point will cost the government about $3 billion in increased spending.

EXAMPLE

For 1000 fuses, the breaking points (in amperes) have a mean of 7.5 A, a standard deviation of 1.0 A, and their distribution is approximately normal.

a. If one fuse is randomly selected, find the probability that its breaking point is between 7.5 A and 7.6 A.

b. If 36 different fuses are randomly selected without replacement, find the probability that their mean is between 7.5 A and 7.6 A.

c. If 150 different fuses are randomly selected without replacement, find the probability that their mean is between 7.5 A and 7.6 A.

Solution

a. *Approach: Use the same methods presented in Section 5-3.* See Figure 5-29(a). We seek the area of the shaded region that can be found from Table A-2 with the following z score.

$$z = \frac{x - \mu}{\sigma} = \frac{7.6 - 7.5}{1.0} = 0.10$$

From Table A-2 we find that an area of 0.0398 corresponds to $z = 0.10$, so that the probability is 0.0398; that is,

$$P(7.5 < x < 7.6) = P(0 < z < 0.10) = 0.0398.$$

continued ▶

Solution, continued

b. *Approach: Use the central limit theorem without the finite population correction factor (since the sample size of 36 does not exceed 5% of 1000).* We seek the shaded area shown in Figure 5-29(b), which can be found by first determining the z score as follows.

$$z = \frac{\bar{x} - \mu_{\bar{x}}}{\sigma_{\bar{x}}} = \frac{\bar{x} - \mu_{\bar{x}}}{\dfrac{\sigma}{\sqrt{n}}} = \frac{7.6 - 7.5}{\dfrac{1.0}{\sqrt{36}}} = 0.60$$

From Table A-2 we find that an area of 0.2257 corresponds to $z = 0.60$, so that the probability is $P(7.5 < \bar{x} < 7.6) = P(0 < z < 0.60) = 0.2257$. Note that in applying the central limit theorem, we let $\sigma_{\bar{x}} = \sigma/\sqrt{n}$.

c. *Approach: Use the central limit theorem with the finite population correction factor (since the sample size of 150 is greater than 5% of 1000).* With $\mu = 7.5$ A, $\sigma = 1.0$ A, $N = 1000$, and $n = 150$, we compute

$$\mu_{\bar{x}} = \mu = 7.5$$

$$\sigma_{\bar{x}} = \frac{\sigma}{\sqrt{n}} \sqrt{\frac{N-n}{N-1}} = \frac{1.0}{\sqrt{150}} \sqrt{\frac{1000 - 150}{1000 - 1}} = 0.075$$

The finite population correction factor was used because $n > 0.05N$, or $150 > 0.05(1000)$. This means that the sample size is large in comparison to the population size. Having determined the values of $\mu_{\bar{x}}$ and $\sigma_{\bar{x}}$, we can apply the central limit theorem by using the methods associated with the normal distribution. Figure 5-29(c) illustrates the normal distribution and the region representing the desired probability.

$$z = \frac{\bar{x} - \mu_{\bar{x}}}{\sigma_{\bar{x}}} = \frac{7.6 - 7.5}{0.075} = \frac{0.1}{0.075} = 1.33$$

From Table A-2 we see that $z = 1.33$ corresponds to a probability of 0.4082 so that $P(7.5 < \bar{x} < 7.6) = P(0 < z < 1.33) = 0.4082$. If we did not use the finite population correction factor, our answer would have been 0.3888 instead of the better result of 0.4082.

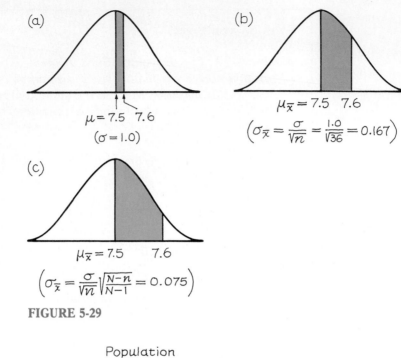

FIGURE 5-29

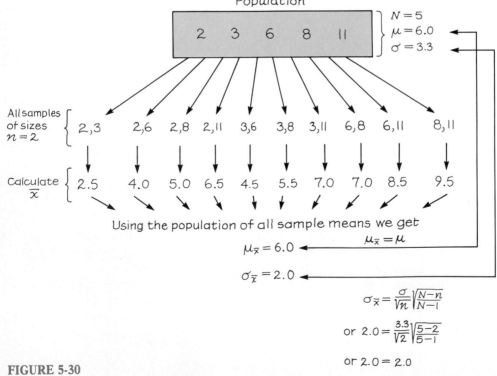

FIGURE 5-30

While a formal derivation for $\sigma_{\bar{x}}$ and the finite population correction factor is beyond the scope of this book, we can illustrate it with a particular set of data. Let's begin with this *population* of scores: 2, 3, 6, 8, 11. From that population we determine that $N = 5$, $\mu = 6.0$, $\sigma = 3.3$. From that population we select samples of size $n = 2$. There are only 10 such samples and they are shown in Figure 5-30. After finding the population of the 10 sample means, we can proceed to find their mean and standard deviation. We get the results shown in Figure 5-30, where we also verify that

$$\mu_{\bar{x}} = \mu \quad \text{and} \quad \sigma_{\bar{x}} = \frac{\sigma}{\sqrt{n}} \sqrt{\frac{N - n}{N - 1}}$$

In Figure 5-31 on the next page we outline the key points presented in this section.

5-6 EXERCISES A

5-165 A large normally distributed population has a mean of 50 and a standard deviation of 10.
   a. Find the probability that a randomly selected score is between 50 and 53.
   b. If a sample of size $n = 36$ is randomly selected, find the probability that the sample mean $\bar{x}$ will be between 50 and 53.

5-166 A large normally distributed population has a mean of 150 and a standard deviation of 20.
   a. Find the probability that a randomly selected score is between 150 and 155.
   b. If a sample of size $n = 100$ is randomly selected, find the probability that the sample mean $\bar{x}$ will be between 150 and 155.

5-167 A large normally distributed population has a mean of 4.50 and a standard deviation of 1.05.
   a. Find the probability that a randomly selected score is less than 5.00.
   b. If a sample of size 40 is randomly selected, find the probability that the sample mean is less than 5.00.

5-168 A large normally distributed population has a mean of 640 and a standard deviation of 53.
   a. Find the probability that a randomly selected score is greater than 630.
   b. If a sample of size 65 is randomly selected, find the probability that the sample mean is greater than 630.

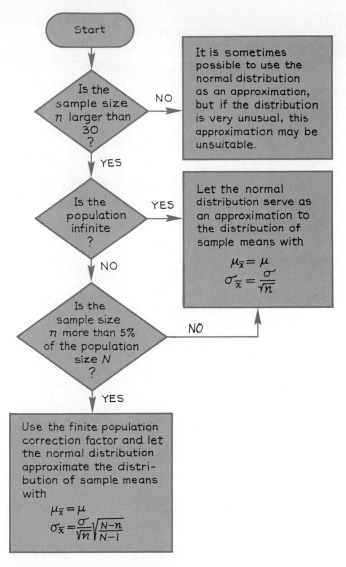

FIGURE 5-31
*Flow chart summarizing the decisions to be made
when considering a distribution of sample means*

5-169 Scores on an IQ test are normally distributed with a mean of 100 and a standard deviation of 15.
a. If someone is randomly selected, find the probability that his or her IQ score will be between 100 and 105.
b. If 36 people are randomly selected, find the probability that their mean will be between 100 and 105.

5-170 Scores on a test of extrasensory perception are normally distributed with a mean of 48.0 and a standard deviation of 8.0.
a. Find the probability of a randomly selected subject scoring between 48.0 and 50.0.
b. If 64 people are randomly selected, find the probability that their mean will be between 48.0 and 50.0.

5-171 A study of the time (in seconds) required for airplanes to land at a certain airport shows that the mean is 280 s with a standard deviation of 20 s. (The times are not normally distributed.) For a sample of 36 randomly selected incoming airplanes, find the probability that their mean landing time is greater than 275 s.

5-172 In a study of the time (in hours) college freshmen use to study each week, it is found that the mean is 7.06 h and the standard deviation is 5.32 h (based on data from *The American Freshman*). If 55 freshmen are randomly selected, find the probability that their mean weekly study time exceeds 7.00 h.

5-173 In a study of the time high school students spend working each week at a job, it is found that the mean is 10.7 h and the standard deviation is 11.2 h (based on data from the National Federation of State High School Associations). If 42 high school students are randomly selected, find the probability that their mean weekly work time is less than 12.0 h.

5-174 An aircraft strobe light is designed so that the times between flashes have a mean of 10.15 s and a standard deviation of 0.40 s. A sample of 50 times is randomly selected. Find the probability that the sample mean is greater than 10.00 s.

5-175 For 800 subjects given a test for reaction times the mean is 0.820 s with a standard deviation of 0.180 s. If a sample of 33 is selected from this population of 800 subjects, find the probability that the sample mean is less than 0.825 s.

5-176 A population consists of 900 beginning students at a certain college. For this group, the mean IQ score is 117.0 and the standard deviation is 8.0. If a sample of 40 of these students is randomly selected, find the probability that the mean of this sample group is below 120.0.

5-177   In the process of conducting the National Health Survey, 1524 women aged 18 to 24 were measured for height. These scores have a mean of 64.3 in. and a standard deviation of 2.5 in., and they approximate a normal distribution. If 49 of these women are randomly selected, find the probability that their mean height is less than 64.0 in.

5-178   In a study of 218,344 births in upstate New York, it was found that the gestation times had a mean of 39.4 weeks and a standard deviation of 2.43 weeks, and those times had a distribution that is approximately normal (based on data from the New York State Department of Health). If 1000 of these births are randomly selected, find the probability that the mean gestation time exceeds 39.6 weeks.

5-179   A total of 630 students were measured for "burnout." The resulting scores have a mean of 2.97 and a standard deviation of 0.60 (based on data from "Moderating Effects of Social Support on the Stress-Burnout Relationship," by Etzion, *Journal of Applied Psychology*, Vol. 69, No. 4). If 31 of these subjects are randomly selected, find the probability that their mean burnout score is between 3.00 and 3.10.

5-180   In a study of work patterns, data from 669 subjects were collected. For the time spent in child care, the mean is 23.08 h and the standard deviation is 15.58 h (based on data from "Nonstandard Work Schedules and Family Life," by Staines and Pleck, *Journal of Applied Psychology*, Vol. 69, No. 3). If 32 of these subjects are randomly selected, find the probability that their mean time in child care is between 20.00 h and 22.00 h.

5-181   A population has a standard deviation of 20. Samples of size n are taken randomly and the means of the samples are computed. What happens to the standard error of the mean if the sample size is increased from 100 to 400?

5-182   A population has a standard deviation of 20. Samples of size n are randomly selected and the means of the samples are computed. What happens to the standard error of the mean if the sample size is decreased from 64 to 16?

In Exercises 5-183 and 5-184, assume that samples of size n *are randomly selected from a finite population of size* N. *Also assume that* σ = 15. *Find the value of* $\sigma_{\bar{x}}$. *(Be sure to include the finite population correction factor whenever* n > 0.05N.)

5-183   a. $N = 5000, n = 200$
        b. $N = 12{,}000, n = 1000$
        c. $N = 4000, n = 500$
        d. $N = 8000, n = 3000$
        e. $N = 1500, n = 50$

5-184   a. $N = 750, n = 50$
        b. $N = 673, n = 32$
        c. $N = 866, n = 73$
        d. $N = 50{,}000, n = 10{,}000$
        e. $N = 8362, n = 935$

In Exercises 5-185 through 5-188, be sure to check for the use of the finite population correction factor and use it whenever necessary.

5-185   In a study of Reye's syndrome (by Holtzhauer and others, *American Journal of Diseases of Children*, Vol. 140), 160 children had a mean age of 8.5 years, standard deviation of 3.96 years, and their ages approximated a normal distribution. If 36 of those children are randomly selected, find the probability that their mean age is between 7.0 years and 10.0 years.

5-186   A study involves a population of 300 women who are 5 ft tall and are between 18 and 24 years of age. This population has a mean weight of 121.5 lb and a standard deviation of 6.5 lb. If 50 members of this population are randomly selected, find the probability that the mean weight of this sample group is greater than 120.0 lb.

5-187   A psychologist collects a population of 250 volunteers who will participate in a behavior modification experiment. A pretest of this population produces behavior indices with a mean of 436 and a standard deviation of 24. If 20% of the population is randomly selected for one phase of treatment, find the probability that the mean for this sample group is between the desired limits of 430 and 440.

5-188   In doing an economic impact study, a sociologist identifies a population of 1200 households with a mean annual income of $23,460 and a standard deviation of $3750. If 10% of these households are randomly selected for a more detailed survey, find the probability that the mean for this sample group will fall between the acceptable limits of $23,000 and $24,000.

5-6 EXERCISES B

5-189   A population consists of these scores:

$$2 \quad 3 \quad 6 \quad 8 \quad 11 \quad 18$$

a. Find μ and σ.
b. List all samples of size $n = 2$.
c. Find the population of all values of $\bar{x}$ by finding the mean of each sample from part (b).
d. Find the mean $\mu_{\bar{x}}$ and standard deviation $\sigma_{\bar{x}}$ for the probability of sample means found in part (c).
e. Verify that

$$\mu_{\bar{x}} = \mu \quad \text{and} \quad \sigma_{\bar{x}} = \frac{\sigma}{\sqrt{n}} \sqrt{\frac{N - n}{N - 1}}$$

x	f
0– 4	2
5– 9	0
10–14	5
15–19	8
20–24	12
25–29	17
30–34	20
35–39	14
40–44	6
45–49	3

5-190 Assume that a population is infinite. Find the probability that the mean of a sample of 100 differs from the population mean by more than $\sigma/4$.

5-191 The accompanying frequency table summarizes the number of defective units produced by a machine on 87 different days. Find μ and σ for this population. If 32 of the 87 days are randomly selected, use the central limit theorem to find the probability that the mean for the 32 days is greater than 30.0.

5-192 A sample of size 50 is randomly selected from a population of size N, with the result that the standard error of the mean is one-tenth the value of the population standard deviation. Find the size of the population.

VOCABULARY LIST

Define and give an example of each term.

normal distribution
standard normal
   distribution
z score

standard score
central limit theorem
standard error of the
   mean

finite population
   correction factor
continuity correction

REVIEW

The main concern of this chapter is the concept of a **normal distribution**, the most important of all continuous probability distributions. Many real and natural occurrences yield data that are normally distributed or can be approximated by a normal distribution. The normal distribution, which appears to be bell-shaped when graphed, can be described algebraically by an equation, but the complexity of that equation usually forces us to use a table of values instead.

Table A-2 represents the **standard normal distribution**, which has a mean of 0 and a standard deviation of 1. This table relates deviations away from the mean with areas under the curve. Since the total area under the curve is 1, those areas correspond to probability values.

In the early sections of this chapter, we worked with the standard procedures used in applying Table A-2 to a variety of different situations. We saw that Table A-2 can be applied indirectly to normal distributions that are nonstandard. (That is, μ and σ are not 0 and 1, respectively.) We were able to find the number of standard deviations that a score x is away from the mean μ by computing $z = (x - \mu)/\sigma$.

In Sections 5-3 and 5-4 we considered real and practical examples as we converted from a nonstandard to a standard normal distribution. In Section 5-5 we saw that we can sometimes approximate a binomial probability distribution by a normal distribution. If both $np \geq 5$ and $nq \geq 5$, the binomial random variable x is approximately normally distributed with the mean and standard deviation given as $\mu = np$ and $\sigma = \sqrt{npq}$. Since the binomial probability distribution deals with discrete data while the normal distribution deals with continuous data, we introduced the **continuity correction**, which should be used in normal approximations to binomial distributions if n is small. Finally, in Section 5-6, we considered the distribution of sample means that can come from normal or nonnormal populations. The **central limit theorem** asserts that the distribution of sample means $\bar{x}$ (based on random samples of size n) will, as n increases, approach a normal distribution with mean μ and standard deviation $\sigma/\sqrt{n}$. This means that if samples are of size n where $n > 30$, we can approximate the distribution of those sample means by a normal distribution. The **standard error of the mean** is $\sigma/\sqrt{n}$ as long as the population is infinite or the sample size is not more than 5% of the population. But if the sample n exceeds 5% of the population N, then the standard error of the mean must be adjusted by the **finite population correction factor** with $\sigma/\sqrt{n}$ multiplied by $\sqrt{(N - n)/(N - 1)}$. Figure 5-31 summarizes these concepts.

In Chapter 6 we apply many of the concepts introduced in this chapter as we study the extremely important process of testing hypotheses. Since basic concepts of this chapter serve as critical prerequisites for the following material, it would be wise to master these ideas and methods now.

IMPORTANT FORMULAS

Standard normal distribution has $\mu = 0$ and $\sigma = 1$. Standard score or z score:

$$z = \frac{x - \mu}{\sigma}$$

Prerequisites for approximating binomial by normal:

$$np \geq 5 \qquad nq \geq 5$$

Parameters used when approximating binomial by normal:

$$\mu = np \qquad \sigma = \sqrt{npq}$$

Parameters used when applying central limit theorem: $\mu_{\bar{x}} = \mu$

$$\sigma_{\bar{x}} = \frac{\sigma}{\sqrt{n}} \qquad \text{(standard error of the mean)}$$

$$\sigma_{\bar{x}} = \frac{\sigma}{\sqrt{n}} \sqrt{\frac{N - n}{N - 1}} \qquad \text{(used when } n > 0.05N\text{)}$$

REVIEW EXERCISES

5-193 A population of 700 scores has a mean of 5.40, a standard deviation of 1.20, and its distribution is approximately normal.
a. If a score is randomly selected, find the probability that it is greater than 5.00.
b. If 32 different scores are randomly selected without replacement, find the probability that their mean is greater than 5.00.
c. If 36 different scores are randomly selected without replacement, find the probability that their mean is greater than 5.00.

5-194 A landfill operation collects daily waste amounts, which are normally distributed with a mean of 26.4 tons and a standard deviation of 2.4 tons.
a. For a randomly selected day, find the probability that the amount of waste is between 26.4 tons and 30.0 tons.
b. For a randomly selected day, find the probability that the amount of waste is less than 27.0 tons.
c. For a randomly selected day, find the probability that the amount of waste is between 26.0 tons and 28.0 tons.
d. If the landfill operation is overburdened for 5% of the days it is open, find the minimum amount that causes an overburden.
e. For 40 randomly selected days, find the probability that the mean amount of waste is less than 26.0 tons.

5-195 Errors from meter readings are normally distributed with a mean of 0 V and a standard deviation of 1 V. (The errors can be positive or negative.) One reading is randomly selected. Find the probability that the error is
a. Between 0 V and 1.42 V
b. Greater than −1.05 V
c. Between 0.50 V and 1.50 V

5-196 The probability of a particular computer component being defective is 0.2. If 500 of these components are randomly selected and tested, find the probability of getting at least 120 defective components.

5-197 A sociologist finds that for a certain segment of the population, the numbers of years of formal education are normally distributed with a mean of 13.20 years and a standard deviation of 2.95 years.
a. For a person randomly selected from this group, find the probability that he or she has between 13.20 and 13.50 years of education.
b. For a person randomly selected from this group, find the probability that he or she has at least 12.00 years of education.
c. Find the first quartile, Q_1. That is, find the value separating the lowest 25% from the highest 75%.   *(continued)*

d. If an employer wants to establish a minimum education requirement, how many years of education would be required if only the top 5% of this group would qualify?

e. If 35 people are randomly selected from this group, find the probability that their mean years of education is at least 12.00 years.

5-198   Errors on a scale are normally distributed with a mean of 0 kg and a standard deviation of 1 kg. One item is randomly selected and weighed. (The errors can be positive or negative.)

a. Find the probability that the error is between 0 and 0.74 kg.

b. Find the probability that the error is greater than 1.76 kg.

c. Find the probability that the error is greater than −1.08 kg.

5-199   It has been found that 4% of the respondents fail to answer a particular question on a survey. Find the probability that among 150 respondents, at least five fail to answer the question.

5-200   Scores on a hearing test are normally distributed with a mean of 600 and a standard deviation of 100.

a. If one subject is randomly selected, find the probability that the score is between 600 and 735.

b. If one subject is randomly selected, find the probability that the score is more than 450.

c. If one subject is randomly selected, find the probability that the score is between 500 and 800.

d. If a job requires a score in the top 80%, find the lowest acceptable score.

e. If 50 subjects are randomly selected, find the probability that their mean score is between 600 and 635.

5-201   Scores on a standard IQ test are normally distributed with a mean of 100 and a standard deviation of 15.

a. Find the probability that a randomly selected subject will achieve a score between 90 and 120.

b. Find the probability that a randomly selected subject will achieve a score above 105.

c. If 30 subjects are randomly selected and tested, find the probability that their mean IQ score is above 105.

d. Find P_{95}, the IQ score separating the top 5% from the lower 95%.

e. Find P_{15}, the IQ score separating the bottom 15% from the top 85%.

5-202   Suppose that 800 scores have a mean of $\mu = 120.0$, a standard deviation of $\sigma = 4.0$, and a distribution that is approximately normal.

a. If a score is randomly selected, find the probability that it is above 125.0.

b. If 36 scores are randomly selected, find the probability that their mean is between 120.0 and 121.0.

c. If 49 scores are randomly selected, find the probability that their mean is between 120.0 and 121.0.

5-203   A study has shown that among people without any preschool education, 32% were employed at age 19. If these figures are correct, find the probability that for a group of 100 people without any preschool education, 40 or fewer are employed at age 19.

5-204   Among Americans aged 18 and older, 8% are divorced (based on U.S. Census Bureau data). If 225 Americans aged 18 or older are randomly selected, find the probability that at least 20 of them are divorced.

COMPUTER PROJECT

Listed below are two BASIC programs, which may require some minor modification in order to run on certain computers. The first program will produce 36 randomly generated numbers. The second program will produce a mean of 36 randomly generated numbers.

a. Enter and run the first program and manually construct a histogram of these 36 values.

b. Enter the second program and run it 36 times to get 36 sample means. Then manually construct a histogram of these 36 values. *Hint:* Instead of entering RUN 36 times, we can run the program 36 times by including these two lines:

5 FOR J = 1 to 36
65 NEXT J

c. Compare the two resulting histograms.

```
10 RANDOMIZE
20 FOR I = 1 to 36
30   PRINT INT(100*RND(X))
40 NEXT I
50 END

10 RANDOMIZE
20 LET T = 0
30 FOR I = 1 to 36
40   LET T = T + INT(100*RND(X))
50 NEXT I
60 PRINT T/36
70 END
```

CASE STUDY ACTIVITY

In Table 5-4, we listed 30 collections of samples with ten random numbers between 1 and 9 in each sample. Use a table of random digits, or a telephone book, or a computer to observe or generate 50 collections of samples with ten random numbers between 0 and 9. Construct a frequency table and histogram of the combined sample of 500 scores, and find the mean and standard deviation. Then find the mean of each sample and, using the 50 sample means, construct a frequency table and histogram and find the mean and standard deviation. Compare the results obtained from the combined sample of 500 scores to those obtained for the 50 sample means.

DATA PROJECT

Refer to the 92 pulse rates (number of beats in one minute) listed in the data set found in Appendix B.

a. Find the mean and standard deviation of those 92 pulse rates.

b. Identify any pulse rates that are more than 3 standard deviations away from the mean.

c. Is it likely that the "outliers" found in part (b) are correct, or does it seem more likely that they are mistakes?

d. For any normally distributed population, find the percentage of values that will be more than 3 standard deviations away from the mean.

e. Among 92 scores, how many would you expect to be more than 3 standard deviations away from the mean? (Use the result from part (d).)

f. After removing the "outliers" identified in part (b), calculate the mean and standard deviation for the remaining pulse rates. Did the outliers have much of an effect on these statistics?

g. Use the mean and standard deviation from part (f) and let x represent the pulse rate of someone randomly selected from the same population that the 92 sample scores came from. Find $P(x > 60)$ and $P(x < 75)$.

h. Use the mean and standard deviation from part (f). If 50 people are randomly selected from the same population that the 92 sample scores came from, find the probability that the mean for the group of 50 is greater than 72.0.

Chapter Six

CHAPTER CONTENTS

6-1 Overview
We define chapter **objectives**.

6-2 Testing a Claim About a Mean
We present the **general procedure** for testing hypotheses. The hypotheses considered in this section relate to claims made about a population mean.

6-3 *P*-Values
We present the ***P*-value approach** commonly used in computer software packages.

6-4 *t* Test
We present the *t* **distribution**, which is used in certain hypothesis tests instead of the normal distribution.

6-5 Tests of Proportions
We describe the method of testing a hypothesis made about a population **proportion** or **percentage**.

6-6 Tests of Variances
We describe the method of testing a hypothesis made about the **standard deviation** or **variance** of a population. In such cases, we use the **chi-square distribution**.

6 Testing Hypotheses

CHAPTER PROBLEM

A fundamental concept in this chapter is that of significance. In Section 6-2, we consider a manufacturer's claim that a new type of tire has a skid distance less than the old tire's mean of 152 ft. A sample of 36 tires yields a mean skid distance of 148 ft, but is that value *significantly* less than 152 ft? Do these sample results really support the manufacturer's claim of a shorter skid distance? Could these sample results be attributed to chance fluctuations? This chapter provides us with the ability to answer such questions so that decisions can be made about a variety of different claims.

6-1 OVERVIEW

Our main objective in this chapter is to develop an understanding of the concepts that underlie hypothesis testing. We also want you to develop the skills required to successfully execute the test of a hypothesis relating to a mean, proportion, or variance.

A hypothesis is a statement that something is true. The following statements are examples of hypotheses that can be tested by the procedures that we develop in this chapter:

- A consumer claims that a brewery gives less than 32 oz of beer per bottle.
- A manufacturer claims that a new type of snow tire has a shorter skid distance.
- A tobacco company claims that its cigarettes contain less than 40 mg of nicotine.
- A senator claims that 60% of her constituents favor a gun control bill.
- A sociologist claims that the unemployment rate is 9.2%.

Before beginning with Section 6-2, it would be very helpful to have a general sense of the thinking used in **hypothesis tests**, also called **tests of significance**. Try to follow the reasoning behind the following example.

Suppose you take a dime from your pocket and claim that it favors heads when it is flipped. That claim is a hypothesis, and we can test it by flipping the dime 100 times. We would expect to get around 50 heads with a fair coin. If heads occur 94 times out of 100 tosses, most people would agree that the coin favors heads. If heads occur 51 times out of 100 tosses, we should not conclude that the dime favors heads since we could easily get 51 heads with a fair and unbiased coin. Here is the key point: We should conclude that the dime favors heads only if we get *significantly* more heads than we would expect with an unbiased dime.

This intuitive discussion should reveal a fundamental concept that underlies the method of testing hypotheses. That method involves a variety of standard terms and conditions in the context of an organized procedure. We suggest that you begin the study of this chapter by first reading Section 6-2 casually to obtain a general idea of its concepts. Then read the material more carefully to gain familiarity with the terminology. Subsequent readings should incorporate the details and refinements into the basic procedure. You are not expected to master the principles of hypothesis testing in one reading. It may take several readings for the material to become understandable.

Drug Approval Requires Strict Procedure

The Pharmaceutical Manufacturing Association has reported that the development and approval of a new drug will cost around $87 million and will take around eight years. Extensive laboratory and animal testing is followed by FDA approval for human testing, which is done in three phases. Phase I human testing involves about 80 people, phase II involves about 250 people, and phase III involves between 1000 and 3000 volunteers. Overseeing such a complex, expensive, and time-consuming process would be enough to give anyone a headache, but the process does protect us from dangerous or worthless drugs.

6-2 TESTING A CLAIM ABOUT A MEAN

The coin example mentioned in the introduction is largely intuitive and devoid of some of the components that are required in a formal statistical test of a hypothesis. For example, we implied that the occurrence of 94 heads in 100 tosses of a fair coin is a very unusual event, but we made no attempt to specify the exact criteria used to identify unusual events. We will begin this section with another illustrative example, but we will include more of the necessary details. We will then proceed to define the terms, conditions, and procedures that constitute the formal method of a standard hypothesis test.

Consider the claim that a new and more expensive type of snow tire has a shorter skid distance. Because of the costs involved, the owner of a large fleet of cars will purchase these tires only if strongly convinced that the tires really do skid less. The only way to prove or disprove the claim is to test all such tires, but that is clearly impractical. Instead, sample data must be used to form a conclusion. Let's assume that we have obtained sample test results from an independent testing laboratory. We learn that 36 of these tires were tested and, under standard conditions, the mean skid distance for that sample group is found to be $\bar{x} = 148$ ft. The testing laboratory also informs us that skid distances of the traditional type of snow tire are normally distributed with a mean of 152 ft and a standard deviation of 12 ft. We might be tempted to conclude that the hypothesis of a shorter skid distance is correct simply because the sample mean of 148 ft is less than the population mean of 152 ft. But let's analyze this critically. We know that sample data fluctuate and display errors of various amounts. Tires produced by the same workers and the same machines do not necessarily provide identical performances. Recognizing this, we formulate a key question: Does the sample mean score of 148 represent a statistically *significant* decrease from the population mean of 152, or is the difference more likely due to chance variations in the skid distances? Let's summarize these key points.

- A traditional type of snow tire has a mean skid distance of 152 ft and a standard deviation of 12 ft.
- A *sample* of 36 new tires is tested and $\bar{x} = 148$ ft.
- Claim: The population of new tires has a mean μ that is less than 152 ft.

On the following pages we use this example to illustrate the complete method of testing hypotheses.

In Figure 6-1 we outline the general procedure for testing hypotheses, and we demonstrate how this procedure applies to a specific example. The following are some of the standard terms used in this procedure.

- **Null hypothesis** (denoted by H_0): The statement of a zero or null difference that is directly tested. This will correspond to the original claim if that claim includes the condition of no change or difference (such as $=, \leq, \geq$). Otherwise, the null hypothesis is the negation of the original claim. We test the null hypothesis directly in the sense that the final conclusion will be either rejection of H_0 or failure to reject H_0.

- **Alternative hypothesis** (denoted by H_1): The statement that must be true if the null hypothesis is false.

- **Type I error**: The mistake of rejecting the null hypothesis when it is true.

- **Type II error**: The mistake of failing to reject the null hypothesis when it is false.

- α **(alpha)**: Symbol used to represent the probability of a type I error.

- β **(beta)**: Symbol used to represent the probability of a type II error.

- **Test statistic**: A sample statistic or a value based on the sample data. It is used in making the decision about the rejection of the null hypothesis.

- **Critical region**: The set of all values of the test statistic that would cause us to reject the null hypothesis.
- **Critical value(s)**: The value(s) that separates the critical region from the values of the test statistic that would not lead to rejection of the null hypothesis. The critical value(s) depends on the nature of the null hypothesis, the relevant sampling distribution, and the level of significance α.
- **Significance level**: The probability of rejecting the null hypothesis when it is true. Typical values selected are 0.05 and 0.01. That is, the values of $\alpha = 0.05$ and $\alpha = 0.01$ are typically used. (We use the symbol α to represent the significance level.)
- **Elation**: The feeling experienced when the techniques of hypothesis testing are mastered.

In addition to knowing the general procedure outlined in Figure 6-1 and the preceding terms, you should be aware of some other details. We will first consider these topics and then present two additional examples.

Null and Alternative Hypotheses

From steps 1, 2, and 3 of Figure 6-1 we see how to determine the null and alternative hypotheses. Note that the original claim may be the null or alternative hypothesis, depending on how it is stated. If we are making our own claims, we should arrange the null and alternative hypotheses so that the most serious error is a type I error (rejecting a true null hypothesis). In this text we assume that we are testing a claim made by someone else. Ideally, all claims would be made so that they would all be null hypotheses. Unfortunately, our real world is not ideal. There is poverty, war, crime, and the presence of people who make claims that are actually alternative hypotheses. This text was written with the understanding that not all original claims are as they should be. As a result, some examples and exercises involve claims that are null hypotheses whereas others involve claims that are alternative hypotheses.

In conducting a formal statistical hypothesis test, we are *always* testing the *null hypothesis*, whether it corresponds to the original claim or not. Sometimes the null hypothesis corresponds to the original claim and sometimes it corresponds to the opposite of the original claim. Since we always test the null hypothesis, we will be testing the original claim in some cases and the opposite of the original claim in other cases. Carefully examine the examples in the box on page 336.

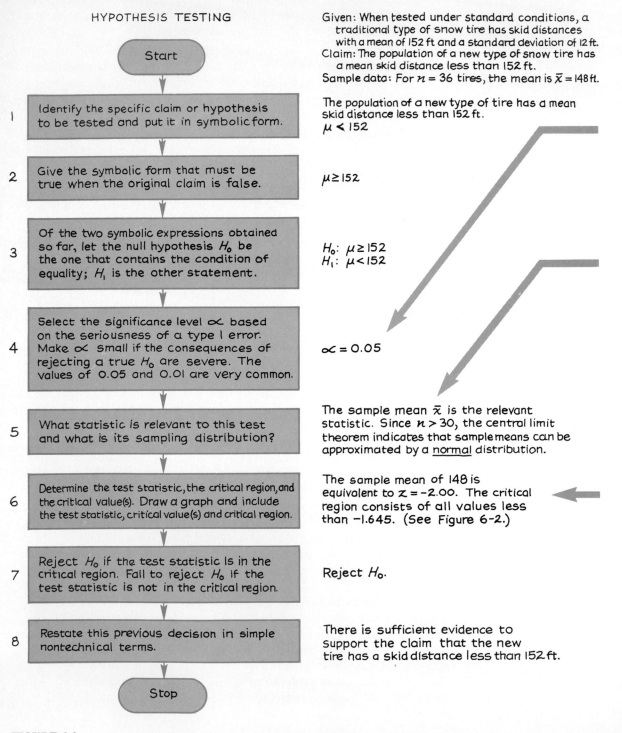

HYPOTHESIS TESTING

Start

1   Identify the specific claim or hypothesis to be tested and put it in symbolic form.

2   Give the symbolic form that must be true when the original claim is false.

3   Of the two symbolic expressions obtained so far, let the null hypothesis H_0 be the one that contains the condition of equality; H_1 is the other statement.

4   Select the significance level α based on the seriousness of a type I error. Make α small if the consequences of rejecting a true H_0 are severe. The values of 0.05 and 0.01 are very common.

5   What statistic is relevant to this test and what is its sampling distribution?

6   Determine the test statistic, the critical region, and the critical value(s). Draw a graph and include the test statistic, critical value(s) and critical region.

7   Reject H_0 if the test statistic is in the critical region. Fail to reject H_0 if the test statistic is not in the critical region.

8   Restate this previous decision in simple nontechnical terms.

Stop

Given: When tested under standard conditions, a traditional type of snow tire has skid distances with a mean of 152 ft and a standard deviation of 12 ft.
Claim: The population of a new type of snow tire has a mean skid distance less than 152 ft.
Sample data: For $n = 36$ tires, the mean is $\bar{x} = 148$ ft.

The population of a new type of tire has a mean skid distance less than 152 ft.
$\mu < 152$

$\mu \geq 152$

H_0: $\mu \geq 152$
H_1: $\mu < 152$

$\alpha = 0.05$

The sample mean $\bar{x}$ is the relevant statistic. Since $n > 30$, the central limit theorem indicates that sample means can be approximated by a <u>normal</u> distribution.

The sample mean of 148 is equivalent to $z = -2.00$. The critical region consists of all values less than -1.645. (See Figure 6-2.)

Reject H_0.

There is sufficient evidence to support the claim that the new tire has a skid distance less than 152 ft.

FIGURE 6-1

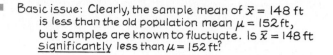

Basic issue: Clearly, the sample mean of $\bar{x} = 148$ ft is less than the old population mean $\mu = 152$ ft, but samples are known to fluctuate. Is $\bar{x} = 148$ ft <u>significantly</u> less than $\mu = 152$ ft?

Significance level: What exactly do we mean by "<u>significantly</u> less than 152 ft?" We will arbitrarily select 5% (or a probability of 0.05) as the level that separates a significant difference from a chance fluctuation. (The choice of 0.05 is very common.) We will conclude that $\bar{x} = 148$ ft is significantly less than 152 ft only if its probability is less than 0.05. That is, with $\mu = 152$ ft, $P(\bar{x} \leq 148$ ft) must be less than 0.05.

Distribution: From the central limit theorem (Section 5-6), we know that means tend to be normally distributed, and we get

$$\mu_{\bar{x}} = \mu = 152$$

$$\sigma_{\bar{x}} = \frac{\sigma}{\sqrt{n}} = \frac{12}{\sqrt{36}} = 2.00$$

Region of significant decreases (5% of total area)

Test statistic: Using the <u>sample data</u> we get

$$z = \frac{\bar{x} - \mu_{\bar{x}}}{\sigma_{\bar{x}}} = \frac{148 - 152}{12/\sqrt{36}} = -2.00$$

Critical value: With a significance level of 0.05, the left tail of Figure 6-2 is the region of significant differences and its boundary corresponds to $z = -1.645$, which is found from Table A-2.

Critical region

$z = -1.645$
Critical value

$\mu = 152$
or
$z = 0$

Sample: $\bar{x} = 148$
or
$z = -2.00$ Test statistic

Conclusion: Figure 6-2 shows that $\bar{x} = 148$ ft is significantly less than 152 ft. We support the claim that the new tires have a shorter skid distance. <u>There is sufficient sample evidence to support the claim that the skid distance is less than 152 ft.</u>

FIGURE 6-2

Original Claim				
	The mean grade is 75.	The mean grade is not 75.	The mean grade is at least 75.	The mean grade is above 75.
Step 1: Symbolic form of original claim.	$\mu = 75$	$\mu \neq 75$	$\mu \geq 75$	$\mu > 75$
Step 2: Symbolic form that is true when original claim is false.	$\mu \neq 75$	$\mu = 75$	$\mu < 75$	$\mu \leq 75$
Step 3: Null hypothesis H_0 (must contain equality).	$H_0: \mu = 75$	$H_0: \mu = 75$	$H_0: \mu \geq 75$	$H_0: \mu \leq 75$
Alternative hypothesis H_1 (cannot contain equality).	$H_1: \mu \neq 75$	$H_1: \mu \neq 75$	$H_1: \mu < 75$	$H_1: \mu > 75$

Type I and Type II Errors

From Table 6-1 we see that the conclusion in a hypothesis test may be correct or wrong. A type I error is the mistake of rejecting a true null hypothesis. A type II error is the mistake of failing to reject a false null hypothesis. The probability of a type I error is the significance level α and the probability of a type II error is denoted by β.

One step in our procedure for testing hypotheses involves the selection of $P(\text{type I error}) = \alpha$, but we don't select $P(\text{type II error}) = \beta$. We could select both α and β—the required sample size would then be determined—but the usual procedure used in research and industry is to determine in advance the values of α and n, so that the value of β is determined. The following practical considerations may be relevant to some experiments:

1. To decrease both α and β, increase the sample size n.
2. For any fixed sample size n, a decrease in α will cause an increase in β. Conversely, an increase in α will cause a decrease in β. To further consider β, see Exercise 6-32.

Why Professional Articles Are Rejected

In an editorial published in the *Journal of Applied Psychology*, John P. Campbell cited the major reasons for rejecting submitted articles. After citing meaningless topics, unclear writing, and inappropriate methodology, Campbell stated that "a much less frequent disqualifier was low statistical power." In some instances too small a sample was a reason for rejection." He also stated that a frequent reason for rejection was that "the procedure (not the statistical analysis) used in the study could not answer the question(s) that were asked. For example, there was every reason to believe that the measures used in a study had no reliability or validity or that alternative explanations for the results were much more likely." He went on to say that the acceptance or rejection of a submitted paper depended heavily on expertise, sound measurement, suitable methodology, and clear writing.

TABLE 6-1		
	The null hypothesis is true.	The null hypothesis is false.
We decide to reject the null hypothesis.	Type I error	Correct decision
We fail to reject the null hypothesis.	Correct decision	Type II error

Conclusions

We always test the null hypothesis and our initial conclusion will always be one of the following:

1. Fail to reject the null hypothesis, H_0.
2. Reject the null hypothesis, H_0.

Some texts say that we "accept the null hypothesis" instead of "fail to reject the null hypothesis." Whether we use the term *accept* or the term *fail to reject*, we should recognize that *we are not proving the null hypothesis*; we are merely saying that the sample evidence is not strong enough to warrant rejection of the null hypothesis. It's like a jury saying that there is not enough evidence to convict a suspect. The term *accept* is somewhat misleading since it seems incorrectly to imply that the null hypothesis has been proved. The phrase *fail to reject* says, more correctly, "let's withhold judgment because the available evidence isn't strong enough." In this text, we will use *fail to reject the null hypothesis* instead of *accept the null hypothesis*.

We either fail to reject the null hypothesis or we reject the null hypothesis. That type of conclusion is fine for those of us with the wisdom to take a statistics course, but it's usually necessary to use simple nontechnical terms in stating what the conclusion suggests. Figure 6-3 shows how to formulate the correct wording of the final

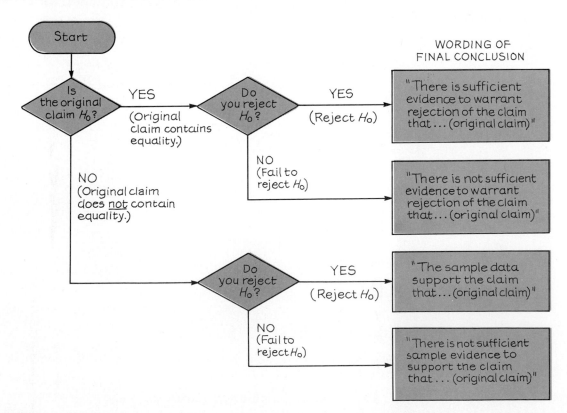

FIGURE 6-3

conclusion. Note that only one case leads to wording where the sample data actually *support* the conclusion. If you want to justify some claim, state it in such a way that it becomes the *alternative* hypothesis, and then hope that the null hypothesis gets rejected.

Left-Tailed, Right-Tailed, Two-Tailed

Sign used in H_1	Type of test
>	Right-tailed
<	Left-tailed
≠	Two-tailed

Our first example of hypothesis testing involved the claim that the new type of tire has a skid distance less than 152 ft. The test used is called a **left-tailed** test because the critical region of Figure 6-2 is in the extreme left region under the curve. We reject the null hypothesis H_0 if our test statistic is in the critical region, because that indicates a significant conflict between the null hypothesis and the sample data. Some tests will be **right-tailed**, with the critical region located in the extreme right region under the curve. Other tests may be **two-tailed**, because the critical region is comprised of two components located in the two extreme regions under the curve. Figure 6-4 illustrates these possibilities. *In the two-tailed case, α is divided equally* between the two components comprising the critical region. In each case, we convert the claim into symbolic form and then determine the symbolic alternative. The null hypothesis H_0 becomes the symbolic statement containing the condition of equality. The alternative hypothesis becomes the other symbolic statement. We reject H_0 if there is significant evidence supporting H_1. For this reason, critical regions correspond to the extremes indicated by H_1. In Figure 6-4, the inequality sign points to the critical region. The symbol ≠ is often expressed in programming languages as $<>$, and this reminds us that an alternative hypothesis such as $\mu \neq 100$ corresponds to a *two*-tailed test.

EXAMPLE

The engineering department of a car manufacturer claims that the fuel consumption rate of one model is equal to 35 mi/gal. The advertising department wants to test this claim to see if the announced figure should be higher or lower than 35 mi/gal. The quality control group suggests that $\sigma = 4$ mi/gal, and a sample of 50 cars yields $\bar{x} = 33.6$ mi/gal. Test the claim of the engineering department by using a 0.05 level of significance.

continued on page 341 ▶

Two-tailed test

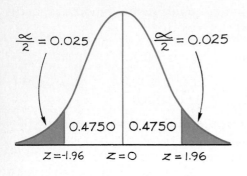

$\frac{\alpha}{2} = 0.025$ $\frac{\alpha}{2} = 0.025$

0.4750   0.4750

$z = -1.96$ $z = 0$ $z = 1.96$

Claim: Prisoners have a mean I Q
       equal to 100.

$H_0: \mu = 100$
$H_1: \mu \neq 100$ (Sign suggests $<>$ that
$\alpha = 0.05$ points to both tails)

Reject H_0 if the sample mean $\bar{x}$ is
significantly above 100 (right tail)
or below 100 (left tail).

Left-tailed test

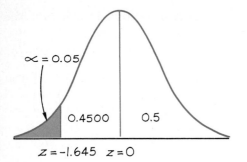

$\alpha = 0.05$

0.4500   0.5

$z = -1.645$ $z = 0$

Claim: Boxers have a mean I Q
       less than 100.

$H_0: \mu \geq 100$
$H_1: \mu < 100$ (Sign points to left)
$\alpha = .05$

Reject H_0 if the sample mean $\bar{x}$ is
significantly below 100 (left tail).

Right-tailed test

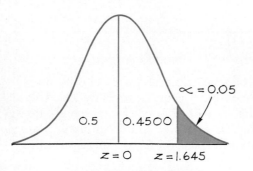

0.5   0.4500 $\alpha = 0.05$

$z = 0$ $z = 1.645$

Claim: Teachers have a mean I Q
       greater than 100.

$H_0: \mu \leq 100$
$H_1: \mu > 100$ (Sign points to right)
$\alpha = 0.05$

Reject H_0 if the sample mean $\bar{x}$ is
significantly greater than 100
(right tail).

FIGURE 6-4
Critical regions are shaded.

Example, continued

Solution

We outline our test of the manufacturer's claim by following the scheme in Figure 6-1. The result is shown in Figures 6-5 and 6-6. The test is two-tailed because a sample mean significantly greater than 35 mi/gal (right tail) or less than 35 mi/gal (left tail) is strong evidence against the null hypothesis that $\mu = 35$ mi/gal. Our sample mean of 33.6 mi/gal is found to be equivalent to $z = -2.47$ through the following computation:

$$z = \frac{\bar{x} - \mu_{\bar{x}}}{\sigma_{\bar{x}}} = \frac{33.6 - 35}{4/\sqrt{50}} = -2.47$$

The critical z values are found by distributing $\alpha = 0.05$ equally between the two tails to get 0.025 in each tail. We then refer to Table A-2 (since we are assuming a normal distribution) to find the z value corresponding to $0.5 - 0.025$ or 0.4750. After finding $z = 1.96$, we use the property of symmetry to conclude that the left critical value is -1.96.

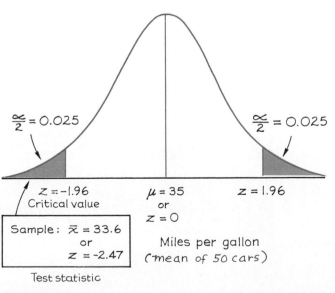

FIGURE 6-5

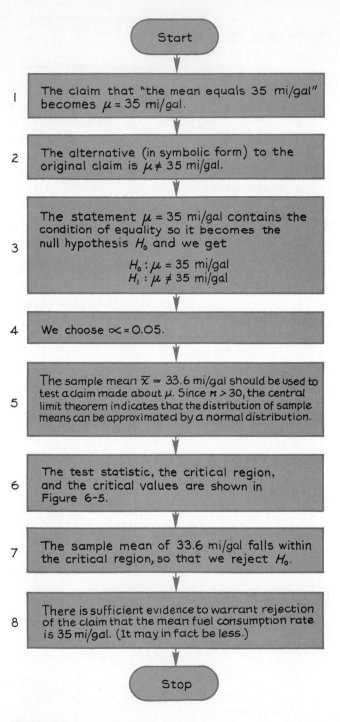

Start

1   The claim that "the mean equals 35 mi/gal" becomes $\mu = 35$ mi/gal.

2   The alternative (in symbolic form) to the original claim is $\mu \neq 35$ mi/gal.

3   The statement $\mu = 35$ mi/gal contains the condition of equality so it becomes the null hypothesis H_0 and we get

$$H_0 : \mu = 35 \text{ mi/gal}$$
$$H_1 : \mu \neq 35 \text{ mi/gal}$$

4   We choose $\alpha = 0.05$.

5   The sample mean $\overline{x} = 33.6$ mi/gal should be used to test a claim made about μ. Since $n > 30$, the central limit theorem indicates that the distribution of sample means can be approximated by a normal distribution.

6   The test statistic, the critical region, and the critical values are shown in Figure 6-5.

7   The sample mean of 33.6 mi/gal falls within the critical region, so that we reject H_0.

8   There is sufficient evidence to warrant rejection of the claim that the mean fuel consumption rate is 35 mi/gal. (It may in fact be less.)

Stop

FIGURE 6-6

EXAMPLE

A brewery distributes beer in bottles labeled 32 oz. The local Bureau of Weights and Measures randomly selects 50 of these bottles, measures their contents, and obtains a sample mean of 31.0 oz. Assuming that σ is known to be 0.75 oz, is it valid at the 0.01 significance level to conclude that the brewery is cheating the consumer?

Solution

The brewery is cheating the consumer if they give significantly less than 32 oz of beer. We outline the test of the claim that the mean is less than 32 oz by again following the model of Figure 6-1. The results are presented in Figures 6-7 and 6-8. The z value of -9.43 is computed as follows:

$$z = \frac{\bar{x} - \mu_{\bar{x}}}{\sigma_{\bar{x}}} = \frac{31 - 32}{0.75/\sqrt{50}} = -9.43$$

The critical z value is found in Table A-2 as the z value corresponding to an area of 0.4900.

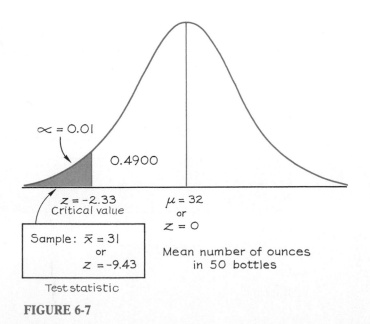

FIGURE 6-7

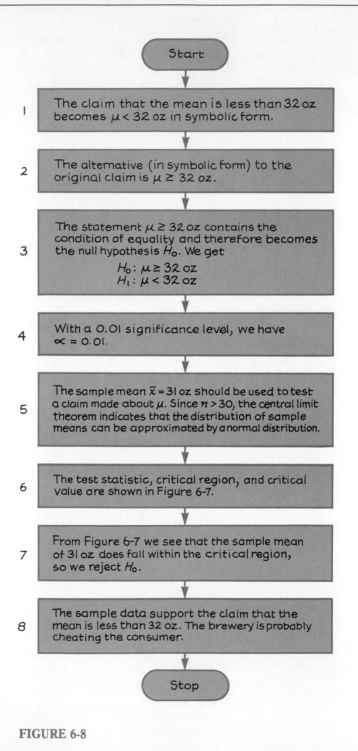

FIGURE 6-8

In presenting the results of a hypothesis test, it is not always necessary to show all of the steps included in Figure 6-8. However, the results should include the null hypothesis, the alternative hypothesis, the calculation of the test statistic, a graph such as Figure 6-7, and the initial conclusion (reject H_0 or fail to reject H_0) and the final conclusion stated in nontechnical terms. The graph should show the test statistic, critical value(s), critical region, and significance level.

Assumptions

For the examples and exercises of this section, we are working with these assumptions.

1. The claim is made about the mean of a single population.
2. a. The sample is large ($n > 30$), so that the central limit theorem applies and we can use the normal distribution.
      or
   b. If the sample is small ($n \leq 30$), then the population is normally distributed and the value of the population standard deviation σ is known.

A potentially unrealistic feature of some examples and exercises from this section is the assumption that σ is known. Realistic tests of hypotheses must often be made without knowledge of the population standard deviation. *If the sample is large* ($n > 30$), *we can compute the sample standard deviation and we may be able to use that value of* s *as an estimate of* σ. When σ is not known and the sample is small ($n \leq 30$), we may be able to use the t statistic discussed in Section 6-4.

It is easy to become entangled in a web of cookbook-type steps without ever understanding the underlying rationale for the procedure. The key to that understanding lies with recognition of this concept: **If an event can easily occur, we attribute it to chance, but if the event appears to be unusual, we attribute that significant departure to the presence of characteristics different from those assumed to be true.** If we keep this idea in mind as we examine various examples, the method of hypothesis testing will become meaningful instead of a rote mechanical process.

There. Wasn't that easy?

6-2 EXERCISES A

In Exercises 6-1 and 6-2, read the given claim and identify H_0 *and* H_1 *as in the following example: The mean IQ of doctors is greater than 110.*

$$H_0: \mu \leq 110$$
$$H_1: \mu > 110$$

6-1　a. The mean age of professors is more than 30 years.
　　 b. The mean IQ of criminals is above 100.
　　 c. The mean IQ of college students is at least 100.
　　 d. The mean annual household income is at least $12,300.
　　 e. The mean monthly maintenance cost of an aircraft is $3271.

6-2　a. The mean height of females is 1.6 m.
　　 b. The mean annual salary of air traffic controllers is more than $26,000.
　　 c. The mean annual salary of college presidents is under $50,000.
　　 d. The mean weight of girls at birth is at most 3.2 kg.
　　 e. The mean life of a car battery is not more than 46 months.

6-3　Identify the type I error and the type II error corresponding to each claim in Exercise 6-1.

6-4　Identify the type I error and the type II error for each claim in Exercise 6-2.

6-5　For each claim in Exercise 6-1, categorize the hypothesis test as a right-tailed test, a left-tailed test, or a two-tailed test.

6-6　For each claim in Exercise 6-2, categorize the hypothesis test as a right-tailed test, a left-tailed test, or a two-tailed test.

In Exercises 6-7 and 6-8, find the critical z value for the given conditions. In each case assume that the normal distribution applies, so that Table A-2 can be used. Also, draw a graph showing the critical value and critical region.

6-7　a. Right-tailed test; $\alpha = 0.05$　6-8　a. Left-tailed test; $\alpha = 0.02$
　　 b. Right-tailed test; $\alpha = 0.01$　　　 b. Two-tailed test; $\alpha = 0.10$
　　 c. Two-tailed test; $\alpha = 0.05$　　　　 c. Right-tailed test; $\alpha = 0.005$
　　 d. Two-tailed test; $\alpha = 0.01$　　　　 d. Right-tailed test; $\alpha = 0.025$
　　 e. Left-tailed test; $\alpha = 0.05$　　　　 e. Left-tailed test; $\alpha = 0.025$

In each of the following exercises, test the given hypotheses by following the procedure suggested by Figure 6-1. Draw the appropriate graph, as in Figure 6-2.

6-9　Test the claim that $\mu \leq 100$ given a sample of $n = 81$ for which $\bar{x} = 100.8$. Assume that $\sigma = 5$, and test at the $\alpha = 0.01$ significance level.

6-10   Test the claim that $\mu \leq 40$ given a sample of $n = 150$ for which $\bar{x} = 41.6$. Assume that $\sigma = 9$, and test at the $\alpha = 0.01$ significance level.

6-11   Test the claim that $\mu \geq 20$ given a sample of $n = 100$ for which $\bar{x} = 18.7$. Assume that $\sigma = 3$, and test at the $\alpha = 0.05$ significance level.

6-12   Test the claim that $\mu \geq 15.5$ given a sample of $n = 45$ for which $\bar{x} = 14.3$. Assume that $\sigma = 5.5$, and test at the $\alpha = 0.05$ significance level.

6-13   Test the claim that a population mean equals 500. You have a sample of 300 items for which the sample mean is 510. Assume that $\sigma = 100$, and test at the $\alpha = 0.10$ significance level.

6-14   Test the claim that a population mean equals 65. You have a sample of 50 items for which the sample mean is 66.1. Assume that $\sigma = 4$, and test at the $\alpha = 0.05$ significance level.

6-15   Test the claim that a population mean exceeds 40. You have a sample of 50 items for which the sample mean is 42. Assume that $\sigma = 8$, and test at the $\alpha = 0.05$ significance level.

6-16   Test the claim that a population mean is less than 75.0. You have a sample of 32 items for which the sample mean is 73.8. Assume that $\alpha = 4.2$ and test at the $\sigma = 0.10$ significance level.

6-17   In a study of distances traveled by buses before the first major engine failure, a sample of 191 buses results in a mean of 96,700 mi and a standard deviation of 37,500 mi (based on data in *Technometrics*, Vol. 22, No. 4). At the 0.05 level of significance, test the claim that mean distance traveled before a major engine failure is more than 90,000 mi. (Assume that the sample standard deviation can be used for σ.)

6-18   When 150 randomly selected boys aged 6 to 11 are given the reading portion of the Wide Range Achievement Test, their mean score is 52.4 and the standard deviation is 13.14 (based on data from the National Health Survey, USDHEW publication 72-1011). At the 0.05 level of significance, test the claim that this sample is from a population with a mean greater than 51.0. (Assume that the sample standard deviation can be used for σ.)

6-19   When 100 randomly selected car owners are polled, it is found that the mean length of time they plan to keep their car is 7.01 years, and the standard deviation is 3.74 years (based on data from a Roper poll). At the 0.01 significance level, test the claim that the mean for all car owners is less than 7.5 years. (Assume that the sample standard deviation can be used for σ.)

6-20   When SAT results are obtained for 302 high school students, the mean and standard deviation are computed to be 511.4 and 126.8, respectively (based on data from Roy C. Ketcham High School in upstate New York). At the 0.02 level of significance, test the claim that these students come from a population with a mean math SAT score that is at least 550. (Assume that the sample standard deviation can be used for σ.)

6-21   A sociologist finds that for a certain population, the mean number of years of education is 13.20, while the standard deviation is 2.95. In one region, a random sample of 60 people is drawn from this population, and the sample mean is 13.87 years. At the 0.05 level of significance, test the claim that the mean for this region is the same as the mean of the population.

6-22   A paint is applied to tin panels and baked for 1 h so that the mean index of hardness is 35.2. Suppose 38 test panels are painted and baked for 3 h, producing a sample mean index of hardness equal to 35.9. Assuming that $\sigma = 2.7$, test (at the $\alpha = 0.05$ significance level) the claim that longer baking does not affect hardness of the paint.

6-23   The mean time between failures (in hours) for a certain type of radio used in light aircraft is 420 h. Suppose 35 new radios have been modified for more reliability, and tests show that the mean time between failures for this sample is 385 h. Assume that σ is known to be 24 h and let $\alpha = 0.05$. Test the claim that the modifications improved reliability. (Note that improved reliability should correspond to a *longer* mean time between failures.)

6-24   A certain nighttime cold medicine bears a label indicating the presence of 600 mg of acetaminophen in each fluid ounce of the drug. The Food and Drug Administration randomly selects 65 1-oz samples and finds that the mean acetaminophen content is 589 mg, while the standard deviation is 21 mg. With $\alpha = 0.01$, test the claim that the population mean is equal to 600 mg. (Assume that the sample standard deviation can be used for σ.)

6-25   The mean and standard deviation for fuel consumption in miles per gallon for a given car are 22.6 and 3.4, respectively. A revolutionary new spark plug is used and a sample of 33 tests results in a mean of 20.1 mi/gal. At the $\alpha = 0.05$ significance level, test the claim that the new spark plug does not change fuel consumption as measured in miles per gallon.

6-26   Tests on automobile braking reaction times for normal young men have produced a mean and standard deviation of 0.610 s and 0.123 s, respectively. When 40 young male graduates of a driving school were randomly selected and tested for their braking reaction times, a mean of 0.587 s resulted. At the $\alpha = 0.10$ significance level, test the claim of the driving instructor that his graduates had faster reaction times.

6-27   In the article "Multiple Spans in Transcription Typing" (*Journal of Applied Psychology*, Vol. 72, No. 2), data was given for a sample of 45 typists. Their mean normal typing score is 182, while the standard deviation is 52. Test the claim that the sample is from a population with a mean of 180. (Assume that the sample standard deviation can be used for σ.)

6-28 When 2000 randomly selected Dutchess County residents are surveyed, it is found that their mean age is 31.7, while the standard deviation is 21.8 (based on data from the *Cornell Community and Resource Development Series*). Test the claim that the mean age of Dutchess County residents is greater than 30.0. (Assume that the sample standard deviation can be used for σ.)

6-2 EXERCISES B

6-29 At the $\alpha = 0.03$ significance level, test the claim that the following IQ scores come from a special group in which the mean is above 100. Use the sample standard deviation as an estimate for σ.

101	110	114	105	79	144	111	99	101
107	103	82	107	90	91	99	95	117
93	103	120	82	123	112	107	89	105
130	106	103	100	118	98	101		

6-30 A brewery claims that consumers are getting a mean volume equal to 32 oz of beer in their quart bottles. The Bureau of Weights and Measures randomly selects 36 bottles and obtains the following measures in ounces:

32.09	31.89	31.06	32.03	31.42	31.39	31.75
31.53	32.42	31.56	31.95	32.00	31.39	32.09
31.67	31.47	32.45	32.14	31.86	32.09	32.34
32.00	30.95	33.53	32.17	31.81	31.78	32.64
31.06	32.64	32.20	32.11	31.42	32.09	33.00
32.06						

Using the sample standard deviation as an estimate for σ, test the claim of the brewery at the 0.05 significance level.

x	f
0–4	2
5–9	0
10–14	5
15–19	8
20–24	12
25–29	17
30–34	20
35–39	14
40–44	6
45–49	3

6-31 The accompanying frequency table summarizes the numbers of defective units produced by a machine for a sample of randomly selected days. Find $\bar{x}$ and s for this sample and use s as an estimate of σ. Then test the claim that the mean number of defective parts per day is equal to 30.0. Use a 0.04 level of significance.

6-32 **The probability β of a type II error:** For a given hypothesis test, the probability α of a type I error is fixed, whereas the probability β of a type II error depends on the particular value of μ that is used as an alternative to the null hypothesis. For hypothesis tests of the type found in this section, we can find β as follows.

1. Find the value of $\bar{x}$ that corresponds to the critical value. In

$$z = \frac{\bar{x} - \mu_{\bar{x}}}{\sigma_{\bar{x}}}$$

substitute the critical score for z, enter the values for $\mu_{\bar{x}}$ and $\sigma_{\bar{x}}$, then solve for $\bar{x}$.
2. Given a particular value of μ that is an alternative to the null hypothesis H_0, draw the normal curve with this new value of μ at the center. Also plot the value of $\bar{x}$ found in step 1.
3. Refer to the graph from step 2 and find the area of the new critical region bounded by $\bar{x}$. This is the probability of rejecting the null hypothesis given that the new value of μ is correct.
4. The value of β is 1 minus the area from step 3. This is the probability of failing to reject the null hypothesis given that the new value of μ is correct.

The preceding steps allow you to find the probability of failing to reject H_0 when it is false. You are finding the area under the curve that *excludes* the critical region where you reject H_0; this area therefore corresponds to a failure to reject H_0 that is false, since we use a particular value of μ that goes against H_0. Refer to the snow tire example discussed in this section (see Figures 6-1 and 6-2) and find the value of β corresponding to

a. $\mu = 151$ b. $\mu = 150$ c. $\mu = 145$

6-3   P-VALUES

In Sections 6-1 and 6-2 we introduced the basic procedure used to test a hypothesis or claim made about a population mean. We saw that the conclusion involved a decision either to reject or to fail to reject the null hypothesis, and that decision was determined by a comparison of the test statistic and the critical value. The decision criterion is essentially this: If the test statistic falls in the critical region bounded by the critical value, we reject the null hypothesis. Otherwise, we fail to reject the null hypothesis.

In addition to many professional articles, many software packages refer to another decision criterion, which involves a **probability-value** or *P*-**value**. Listed below are excerpts from the displayed results of three typical software packages.

```
P-value = 0.0228    (from STATDISK)
```

```
THE TEST IS STATISTICALLY
SIGNIFICANT AT ALPHA = 0.0228
```    (from Minitab)

```
2-TAIL PROB. 0.0456    (from SPSS)
```

All of these displays express, in different formats, results using the *P*-value approach.

While the basic procedure for testing hypotheses remains the same, the decision criterion appears to be different from the one presented in Section 6-2. In this section we will examine this alternative approach.

Let's begin by recalling the essential components of the snow tire example from Section 6-2.

Claim: The new tires have a mean skid distance less than 152 ft.

Null hypothesis: H_0: $\mu \geq 152$

Alternative hypothesis: H_1: $\mu < 152$

Significance level: $\alpha = 0.05$

Critical value: $z = -1.645$ (see Figure 6-9)

Test statistic: $z = -2.00$ (see Figure 6-9)

Conclusion: Reject the null hypothesis; that is, there is sufficient evidence to support the claim that the new tire has a skid distance less than 152 ft.

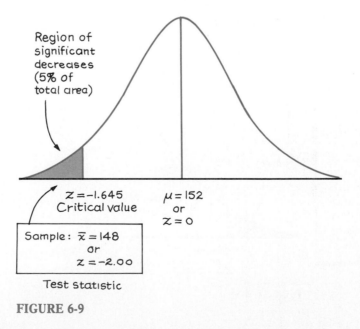

FIGURE 6-9

Beware of *P*-Value Misuse

John P. Campbell, editor of the *Journal of Applied Psychology*, wrote that "books have been written to dissuade people from the notion that smaller *P*-values mean more important results or that statistical significance has anything to do with substantive significance. It is almost impossible to drag authors away from their *P*-values, and the more zeros after the decimal point, the harder people cling to them." While it might be necessary to provide a statistical analysis of the results of a study, we should place strong emphasis on the significance of the results themselves.

With the *P*-value approach, we find the probability of observing a value of $\bar{x}$ that is at least as extreme as the $\bar{x}$ found from the sample data. Examine Figure 6-10 and observe that the area to the left of $\bar{x} = 148$ (or $z = -2.00$) can be found by using Table A-2. With $z = 2.00$ in Table A-2, we get the area of 0.4772 as shown in Figure 6-10. That area is subtracted from 0.5 to yield the left-tail area of 0.0228.

Now that we have a *P*-value of 0.0228, what do we do with it? Some statisticians prefer simply to report the *P*-value and leave the conclusion to the reader. Others prefer to use this decision criterion in deciding to reject or fail to reject the null hypothesis:

1. *Reject* the null hypothesis if the *P*-value is less than or equal to the significance level α.
2. *Fail to reject* the null hypothesis if the *P*-value is greater than the significance level α.

In this example, the *P*-value of 0.0228 is less than the significance level of 0.05, so we reject the null hypothesis as we did in Section 6-2. The procedure used in Section 6-2 is sometimes called the **classical** or **traditional approach**, while the procedure of this section is called the probability-value or *P*-value approach. The classical approach involves a comparison of the test statistic and critical value, while the *P*-value approach involves either a comparison of the *P*-value and significance level or a conclusion based on the *P*-value alone.

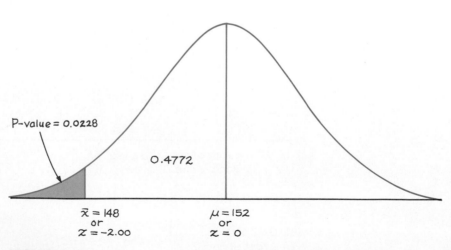

FIGURE 6-10

If the conclusion is based on the *P*-value alone, the following guide may be helpful.

P-value	Interpretation
Less than 0.01	Highly statistically significant Very strong evidence against the null hypothesis
0.01 to 0.05	Statistically significant Adequate evidence against the null hypothesis
Greater than 0.05	Insufficient evidence against the null hypothesis

In a right-tailed test, the *P*-value is obtained by finding the area to the right of the test statistic. However, we must be careful to note that in a two-tailed test, the *P*-value is *twice* the area of the extreme region bounded by the test statistic. This makes sense when we recognize that the *P*-value gives us the probability of getting a sample mean that is *at least as extreme* as the sample mean actually obtained, and the two-tailed case has critical or extreme regions in *both* tails. See Figure 6-11.

In Section 6-2 we included an example of a two-tailed hypothesis test, and that example used the classical approach to hypothesis testing. We have extracted the essential components of that example so that they may be compared to the *P*-value approach. (We use the decision criterion that involves a comparison of the significance level α and the *P*-value.) Note that the only real difference is the decision criterion, which leads to the same conclusion in both cases.

In Section 6-2, we stated that the significance level α should be selected *before* a hypothesis test is conducted. Many statisticians consider this to be a good practice since it helps to prevent us from using the data to support subjective conclusions or beliefs. They feel that this practice becomes especially important with the *P*-value approach because we may be tempted to adjust the significance level based on the resulting *P*-value. With a 0.05 level of significance and a *P*-value of 0.06, we should fail to reject the null hypothesis, but it is sometimes tempting to say that a probability of 0.06 is small enough to warrant rejection of the null hypothesis. Consequently, we should always select the significance level first. Other statisticians feel that prior selection of a significance level reduces the usefulness of *P*-values. They contend that no significance level should be specified, and the conclusion should be left to the reader. We shall use the decision criterion that involves a comparison of a significance level and the *P*-value.

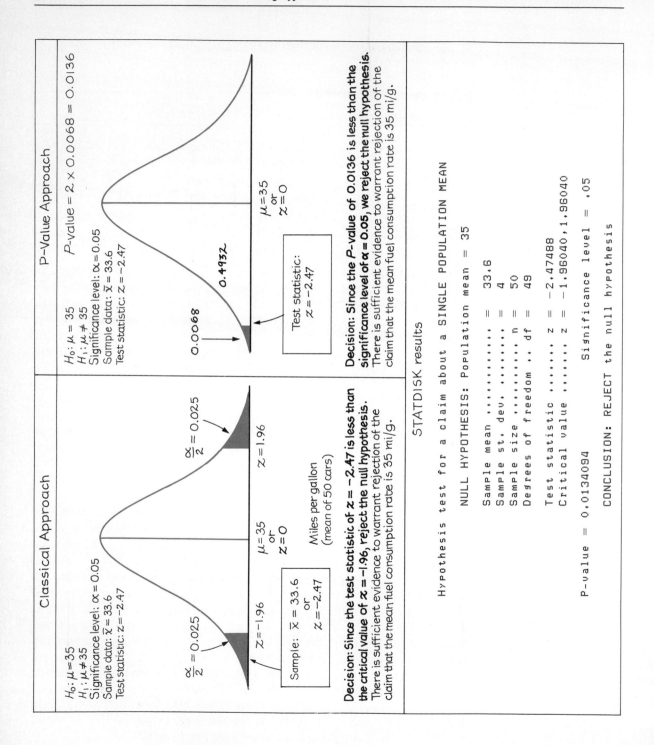

Classical Approach

$H_0: \mu = 35$
$H_1: \mu \neq 35$
Significance level: $\alpha = 0.05$
Sample data: $\bar{x} = 33.6$
Test statistic: $z = -2.47$

$\frac{\alpha}{2} = 0.025$

$\frac{\alpha}{2} = 0.025$

$z = -1.96$

$z = 1.96$

$\mu = 35$
or
$z = 0$

Miles per gallon
(mean of 50 cars)

Sample: $\bar{x} = 33.6$
or
$z = -2.47$

Decision: Since the test statistic of $z = -2.47$ is less than the critical value of $z = -1.96$, we reject the null hypothesis. There is sufficient evidence to warrant rejection of the claim that the mean fuel consumption rate is 35 mi/g.

P-Value Approach

$H_0: \mu = 35$
$H_1: \mu \neq 35$
Significance level: $\alpha = 0.05$
Sample data: $\bar{x} = 33.6$
Test statistic: $z = -2.47$

P-value $= 2 \times 0.0068 = 0.0136$

0.0068

0.4932

$\mu = 35$
or
$z = 0$

Test statistic:
$z = -2.47$

Decision: Since the P-value of 0.0136 is less than the significance level of $\alpha = 0.05$, we reject the null hypothesis. There is sufficient evidence to warrant rejection of the claim that the mean fuel consumption rate is 35 mi/g.

STATDISK results

Hypothesis test for a claim about a SINGLE POPULATION MEAN

NULL HYPOTHESIS: Population mean = 35

Sample mean .... = 33.6
Sample st. dev. .... = 4
Sample size .... n = 50
Degrees of freedom .. df = 49

Test statistic .... z = -2.47488
Critical value .... z = -1.96040,1.96040

P-value = 0.0134094   Significance level = .05

CONCLUSION: REJECT the null hypothesis

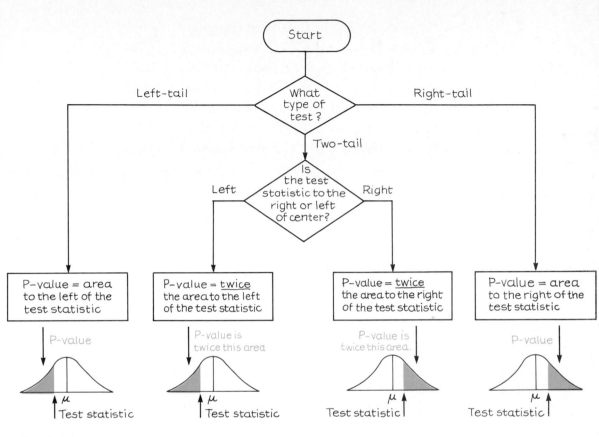

FIGURE 6-11

6-3 EXERCISES A

In Exercises 6-33 through 6-36, use the P-value *and significance level to choose between rejecting the null hypothesis or failing to reject the null hypothesis.*

6-33 *P*-value: 0.03; significance level: $\alpha = 0.05$

6-34 *P*-value: 0.04; significance level: $\alpha = 0.01$

6-35 *P*-value: 0.405; significance level: $\alpha = 0.10$

6-36 *P*-value: 0.09; significance level: $\alpha = 0.10$

In Exercises 6-37 through 6-44, first find the P-value. *Then either reject or fail to reject the null hypothesis by assuming a significance level of* $\alpha = 0.05$.

6-37 $H_0: \mu \geq 152$; $H_1: \mu < 152$; test statistic: $z = -1.40$

6-38 H_0: $\mu \geq 100$; H_1: $\mu < 100$; test statistic: $z = -0.46$

6-39 H_0: $\mu \leq 15.7$; H_1: $\mu > 15.7$; test statistic: $z = 1.94$

6-40 H_0: $\mu \leq 428$; H_1: $\mu > 428$; test statistic: $z = 1.66$

6-41 H_0: $\mu = 75.0$; H_1: $\mu \neq 75.0$; test statistic: $z = 1.66$

6-42 H_0: $\mu = 1365$; H_1: $\mu \neq 1365$; test statistic: $z = -1.30$

6-43 H_0: $\mu = 2.53$; H_1: $\mu \neq 2.53$; test statistic: $z = -1.94$

6-44 H_0: $\mu = 12.8$; H_1: $\mu \neq 12.8$; test statistic: $z = 3.00$

In Exercises 6-45 through 6-52, use the P-value approach to test the given hypotheses.

6-45 Test the claim that $\mu \geq 100$, given a sample of $n = 45$ for which $\bar{x} = 95$. Assume that $\sigma = 15$, and test at the $\alpha = 0.05$ significance level.

6-46 Test the claim that $\mu \leq 500$, given a sample of $n = 35$ for which $\bar{x} = 508$. Assume that $\sigma = 90$, and test at the $\alpha = 0.01$ significance level.

6-47 Test the claim that $\mu = 75.6$, given a sample of $n = 81$ for which $\bar{x} = 78.8$. Assume that $\sigma = 12.0$, and test at the $\alpha = 0.01$ significance level.

6-48 Test the claim that $\mu = 98.6$, given a sample of $n = 200$ for which $\bar{x} = 97.1$. Assume that $\sigma = 2.35$, and test at the $\alpha = 0.10$ significance level.

6-49 At the 0.05 level of significance, test the claim that the mean lengths of long-distance telephone calls are at least 10.00 min. Sample data consist of 60 randomly selected long distance calls, and the mean for this sample group is 8.94 min. Assume that the standard deviation for all such calls is 8.53 min.

6-50 The manager of a moving company claims that the average (mean) load weighs more than 10,000 lb. Test the claim at the 0.01 level of significance. A sample of 540 loads has a mean of 10,410 lb, and assume that $\sigma = 4227$ lb.

6-51 In a survey of 950 patrons attending a certain movie, the mean age is found to be 26.1 years. Test the claim that the mean age of people seeing this movie is 25 years. Use a 0.01 level of significance and assume that $\sigma = 13.08$ years.

6-52 At the 0.05 level of significance, test the claim that the mean asking price of a used car is under $4000. Use sample data consisting of 80 cars with a mean price of $3780. Assume that $\sigma = \$3922$.

6-3 EXERCISES B

6-53 What do you know about the null hypothesis if a *P*-value is found to be greater than 0.5?

6-54   For a random sample of 40 people, what is the mean IQ score if a *P*-value of 0.04 is obtained when testing the claim that $\mu \geq 100$? Assume that $\sigma = 15$.

6-55   For a random sample of 60 adults, what is the mean number of years of education if a *P*-value of 0.2005 is obtained when testing the claim that $\mu \leq 13.20$ years? Assume that $\sigma = 2.95$ years.

6-56   Find the smallest mean above $23,460 that leads to a rejection of the claim that the mean annual household income is $23,460. Assume a 0.02 significance level, and assume that $\sigma = \$3750$. The sample consists of 50 randomly selected households.

6-4   *t* TEST

In Sections 6-2 and 6-3 we introduced the general method for testing hypotheses, but all the examples and exercises involved situations in which the central limit theorem applies so that the normal distribution can be used. The population standard deviations were given and/or the samples were large, and each hypothesis tested related to a population mean. In those cases, we can apply the central limit theorem and use the normal distribution as an approximation to the distribution of sample means. A very unrealistic feature of those examples and exercises is the assumption that σ is known. If σ is unknown and the sample is large, we can treat s as if it were σ and proceed as in Section 6-2 or 6-3. This estimation of σ by s is reasonable because large random samples tend to be representative of the population. But small random samples may exhibit unusual behavior, and they cannot be so trusted. Here we consider tests of hypotheses about a population mean, where the samples are small and σ is unknown. We begin by referring to Figure 6-12, which outlines the theory we are describing.

Starting at the top of Figure 6-12, we see that our immediate concerns lie only with hypotheses made about one population mean. (In following sections we will consider hypotheses made about population parameters other than the mean.) Figure 6-12 summarizes the following observations:

1. In *any* population, the distribution of sample means can be approximated by the normal distribution as long as the random samples are large. This is justified by the central limit theorem.

2. In populations with distributions that are essentially normal, samples of *any* size will yield means having a distribution that is approximately normal. The value of μ would correspond to the

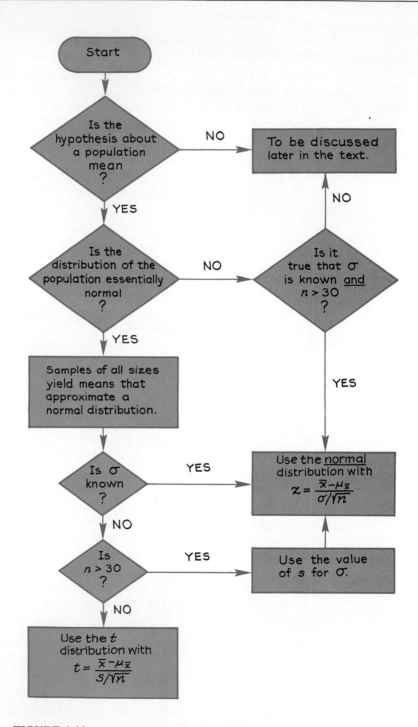

FIGURE 6-12

null hypothesis, and the value of σ must be known. If σ is unknown and the samples are large, we can use s as a substitute for σ, since large random samples tend to be representative of the populations from which they come.

3. In populations with distributions that are essentially normal, assume that we randomly select *small* samples and we do not know the value of σ. For this case we can use the *Student t distribution*, which is described in this section.

4. If our random samples are small, σ is unknown, and the population is grossly nonnormal, then we can use nonparametric methods, some of which are discussed in Chapter 11.

We refer to large and small samples, but the number that separates large and small samples is not derived with absolute exactness. It is somewhat arbitrary. However, there is widespread agreement that samples of size $n > 30$ are large enough so that the distribution of their means can be approximated by a normal distribution.

Around 1908, William S. Gosset (1876–1937) developed specific distributions for small samples in response to certain applied problems he encountered while working for a brewery. The Irish brewery where he worked did not allow the publication of research results, so Gosset published under the pseudonym "Student." The small sample results are extremely valuable because there are many real and practical cases for which large samples are impossible or impractical and for which σ is unknown. Gosset, in his original research that led to small sample techniques, found that temperature and ingredient changes associated with brewing often allowed experimentation that provided only small samples. Factors such as cost and time often severely limited the size of a sample, so the normal distribution could not be an appropriate approximation to the distribution of these small sample means. As a result of those earlier experiments and studies of small samples, we can now use the Student t distributions instead.

TEST STATISTIC

If a population is essentially normal, then the distribution of

$$t = \frac{\bar{x} - \mu}{s/\sqrt{n}}$$

is essentially a **Student t distribution** for all samples of size n. (The Student t distribution is often referred to as the ***t* distribution**.)

We will not discuss the complicated mathematical equations that correspond to this Student t distribution. Instead, we list critical values for this distribution in Table A-3. (Note that the Student t distribution of Table A-3 involves only the critical values of t corresponding to common choices of α.) The critical t value is obtained by locating the proper value for degrees of freedom in the left column and then proceeding across that corresponding row until reaching the number directly below the applicable value of α. Roughly stated, degrees of freedom correspond to the number of values that may vary after certain restrictions have been imposed on all values. For example, if 10 scores must total 50, we can freely assign values to the first 9 scores, but the tenth score would then be determined so that there would be 9 degrees of freedom. In tests on a mean, the number of degrees of freedom is simply the sample size minus 1.

---

degrees of freedom = $n - 1$

For example, suppose we are testing at the $\alpha = 0.05$ significance level the hypothesis that $\mu = 100$ for some population, and we have only a sample of 20 for which $\bar{x} = 102$ and $s = 5$. (We assume that the sample is random and that the population is essentially normal.) The critical t value is found from Table A-3 by noting the following:

1. The test involves two tails since H_0 is $\mu = 100$ and H_1 becomes $\mu \neq 100$.
2. $\alpha = 0.05$ for the two-tailed case.
3. The sample size is $n = 20$, so there are $20 - 1$, or 19, degrees of freedom.

Locating 19 at the left column and 0.05 (two tails) at the top row, we determine that the critical t value is 2.093. Actually, $t = \pm 2.093$ since the test is two-tailed. The test statistic is obtained by computing

$$t = \frac{\bar{x} - \mu}{s/\sqrt{n}} = \frac{102 - 100}{5/\sqrt{20}} = 1.789$$

We would therefore fail to reject H_0, since the test statistic would not be in the critical region. (We will soon present another example in much more detail and with the critical region, critical values, and test statistic depicted in an appropriate graph.)

To use the Student t distribution, we require that the parent population be essentially normal. The population may not be exactly normal, but if it has only one mode and is basically symmetric, we will

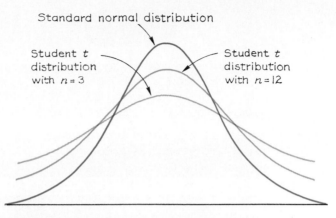

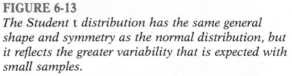

FIGURE 6-13
The Student t *distribution has the same general shape and symmetry as the normal distribution, but it reflects the greater variability that is expected with small samples.*

generally get good results if we use the Student *t* distribution. If there is strong evidence that the population has a very nonnormal distribution, then nonparametric methods (see Chapter 11) may apply.

Important Properties of the Student *t* Distribution

1. The Student *t* distribution is different for different sample sizes. See Figure 6-13 for the cases $n = 3$ and $n = 12$.

2. The Student *t* distribution has the same general bell shape of the normal distribution, but it reflects the greater variability that is expected with small samples.

3. The Student *t* distribution has a mean of $t = 0$ (just as the standard normal distribution has a mean of $z = 0$).

4. The standard deviation of the Student *t* distribution is greater than 1 (unlike the standard normal distribution, which has $\sigma = 1$).

5. As the sample size *n* gets larger, the Student *t* distribution gets closer to the normal distribution. For values of $n > 30$, the differences are so small that we can use the critical *z* values instead of developing a much larger table of critical *t* values. (The values in the bottom row of Table A-3 are equal to the corresponding critical *z* values from the normal distribution.)

Let's use a few examples to illustrate the use of the Student *t* distribution. Remember that the Student *t* distribution applies when

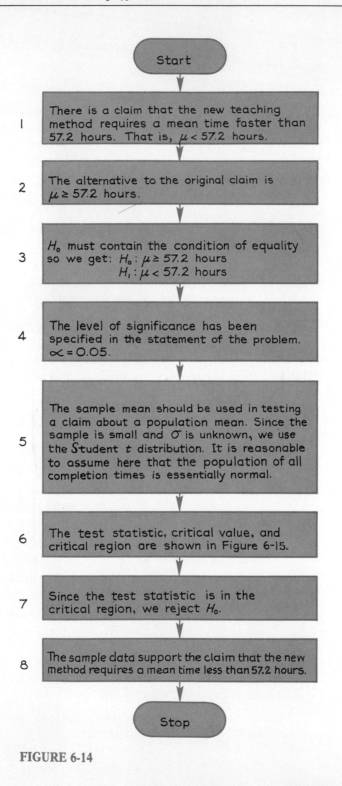

FIGURE 6-14

we test a claim about a population mean and all the following conditions are met:

CONDITIONS FOR STUDENT *t* DISTRIBUTION

1. The sample is small ($n \leq 30$); and
2. σ is unknown; and
3. the parent population is essentially normal.

EXAMPLE

A pilot training program usually takes an average of 57.2 h, but new teaching methods were used on the last class of 25 students. Computations reveal that for this experimental class, the completion times had a mean of 54.8 h and a standard deviation of 4.3 h. At the $\alpha = 0.05$ significance level, test the claim that the new teaching techniques reduce the instruction time.

Solution

Let μ represent the mean completion time for the new teaching method. The claim that it reduces instruction time is equivalent to the claim that $\mu < 57.2$ h. We use the format of Figure 6-1 to outline the test of this claim, and the results are shown in Figures 6-14 and 6-15. We compute the test statistic as follows:

$$t = \frac{\bar{x} - \mu}{s/\sqrt{n}} = \frac{54.8 - 57.2}{4.3/\sqrt{25}} = -2.791$$

We find the critical t value from Table A-3, where we locate $25 - 1$, or 24, degrees of freedom at the left column and $\alpha = 0.05$ (one tail) across the top. The critical t value of 1.711 is obtained; but since small values of $\bar{x}$ will cause the rejection of H_0, we recognize that $t = -1.711$ is the actual t value that is the boundary for the critical region.

It is easy to lose sight of the underlying rationale as we go through this hypothesis-testing procedure, so let's review the essence of the test. We set out to determine whether the sample mean of 54.8 h is *significantly* below the value of 57.2 h. Knowing the distribution of sample means (of which 54.8 is one) and choos-
continued ▶

Solution, continued

ing a level of significance (5% or $\alpha = 0.05$), we are able to determine the cutoff for what is a significant difference and what is not. Any sample mean equivalent to a t score below -1.711 represents a significant difference. The mean of 54.8 h is significantly below 57.2 h, so it appears that the new teaching method does reduce the instruction time.

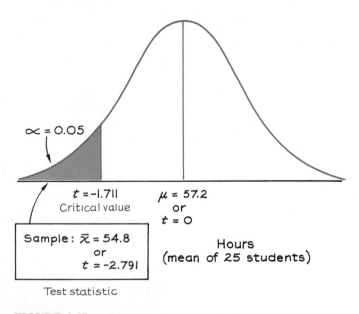

$\alpha = 0.05$

$t = -1.711$
Critical value

$\mu = 57.2$
or
$t = 0$

Sample: $\bar{x} = 54.8$
or
$t = -2.791$

Hours
(mean of 25 students)

Test statistic

FIGURE 6-15

EXAMPLE

A tobacco company claims that its best-selling cigarettes contain at most 40 mg of nicotine. Test this claim at the 1% significance level by using the results of 15 randomly selected cigarettes for which $\bar{x} = 42.6$ mg and $s = 3.7$ mg.

continued ▶

Example, continued

Solution

In this example we will list only certain key elements of the solution. The null and alternative hypotheses are as follows.

$$H_0: \mu \leq 40 \text{ mg} \qquad H_1: \mu > 40 \text{ mg}$$

The significance level of 1% corresponds to $\alpha = 0.01$. The sample mean should be used in testing a claim about the population mean, and we assume that such sample means have a Student *t* distribution because of the following:

1. It is reasonable to expect that the nicotine contents of all cigarettes of a certain brand are essentially a normal distribution.

2. σ is unknown.

3. The sample size ($n = 15$) is small. It is less than or equal to 30.

The test statistic is

$$t = \frac{\bar{x} - \mu}{s/\sqrt{n}} = \frac{42.6 - 40}{3.7/\sqrt{15}} = 2.722$$

In this right-tailed test, the critical value of 2.625 is found from Table A-3 by noting that $\alpha = 0.01$ (one tail) and degrees of freedom $= 15 - 1$, or 14. Since the test statistic of $t = 2.722$ does fall in the critical region, we reject H_0 and conclude that there is sufficient evidence to warrant rejection of the tobacco company's claim. These cigarettes appear to contain significantly more than 40 mg of nicotine.

P-Values

In the examples presented in this section, we used the classical approach to hypothesis testing. However, much of the literature and many computer packages will use the *P*-value approach. Shown here, for example, is the STATDISK display for the last example. The *P*-value of 8.27152E-03 is actually 0.00827152.

STATDISK DISPLAY

```
Hypothesis test for a claim about a SINGLE POPULATION MEAN

      NULL HYPOTHESIS: Population mean <= 40

      Sample mean ............... = 42.6
      Sample st. dev. .......... = 3.7
      Sample size ............ n = 15
      Degrees of freedom ... df = 14

      Test statistic ......... t = 2.72155
      Critical value ........ t = 2.62610

P-value = 8.27152E-03            Significance level = .01

      CONCLUSION: REJECT the null hypothesis
```

Since the t distribution table (Table A-3) includes only selected values of α, we cannot usually find the specific P-value from Table A-3. Instead, we can use that table to identify limits that contain the P-value. In the last example we found the test statistic to be $t = 2.722$, and we know that the test is one-tailed with 14 degrees of freedom. By examining the row of Table A-3 corresponding to 14 degrees of freedom, we see that the test statistic of 2.722 falls between the table values of 2.977 and 2.625, which, in a one-tailed test, correspond to $\alpha = 0.005$ and $\alpha = 0.01$. While we cannot determine the exact P-value, we do know that it must fall between 0.005 and 0.01, so that

$$0.005 < P\text{-value} < 0.01$$

With a significance level of 0.01 and a P-value less than 0.01, we would reject the null hypothesis as we did in the classical approach.

In the next section we test hypotheses relating to claims made about population proportions or percentages.

Product Testing Is Big Business

The United States Testing Company in Hoboken, New Jersey, is the world's largest independent product-testing laboratory. Senior Vice President Noel Schwartz says that they are often hired to verify advertising claims. "We have to produce the data to show that nine out of ten customers really did prefer X brand to Y brand. The television networks and the Federal Trade Commission demand it." Schwartz goes on to say that the most difficult part of his job is "telling a client when his product stinks. But if we didn't do that, we'd have no credibility." He says that there have been a few cases when a client wanted to have results fabricated, but most clients want honest results. Some clients sought other laboratories when they received unfavorable reports.

United States Testing evaluates laundry detergents, cosmetics, insulation materials, zippers, pantyhose, football helmets, toothpaste, fertilizer, and a variety of other products.

6-4 EXERCISES A

In Exercises 6-57 and 6-58, find the critical t *value suggested by the given data.*

6-57
 a. $H_0: \mu = 12$
    $n = 27$
    $\alpha = 0.05$
 b. $H_0: \mu \leq 50$
    $n = 17$
    $\alpha = 0.10$
 c. $H_0: \mu \geq 1.36$
    $n = 6$
    $\alpha = 0.01$
 d. $H_0: \mu = 1.36$
    $n = 6$
    $\alpha = 0.01$
 e. $H_0: \mu \geq 10.75$
    $n = 29$
    $\alpha = 0.01$

6-58
 a. $H_0: \mu \leq 100$
    $n = 27$
    $\alpha = 0.10$
 b. $H_1: \mu \neq 500$
    $n = 16$
    $\alpha = 0.05$
 c. $H_1: \mu < 67.5$
    $n = 12$
    $\alpha = 0.05$
 d. $H_1: \mu > 98.4$
    $n = 7$
    $\alpha = 0.05$
 e. $H_1: \mu \neq 75$
    $n = 24$
    $\alpha = 0.05$

In Exercises 6-59 through 6-64, assume that the population is normally distributed.

6-59   Test the claim that $\mu \leq 10$, given a sample of 9 for which $\bar{x} = 11$ and $s = 2$. Use a significance level of $\alpha = 0.05$.

6-60   Test the claim that $\mu \leq 32$, given a sample of 27 for which $\bar{x} = 33.5$ and $s = 3$. Use a significance level of $\alpha = 0.10$.

6-61   Test the claim that $\mu \geq 98.6$, given a sample of 18 for which $\bar{x} = 98.2$ and $s = 0.8$. Use a significance level of $\alpha = 0.025$.

6-62   Test the claim that $\mu \geq 100$, given a sample of 22 for which $\bar{x} = 95$ and $s = 18$. Use a 5% level of significance.

6-63   Test the claim that $\mu = 75$, given a sample of 15 for which $\bar{x} = 77.6$ and $s = 5$. Use a significance level of $\alpha = 0.05$.

6-64   Test the claim that $\mu = 500$, given a sample of 20 for which $\bar{x} = 541$ and $s = 115$. Use a significance level of $\alpha = 0.10$.

In Exercises 6-65 through 6-84, test the given hypothesis by following the procedure suggested by Figure 6-1. Draw the appropriate graph. In each case, assume that the population is approximately normal.

6-65   The skid properties of a snow tire have been tested and the mean skid distance of 154 ft has been established for standardized conditions. A new, more expensive tire is developed, but tests on a sample of 20 new tires yield a mean skid distance of 141 ft with a standard deviation of 12 ft. Because of the cost involved, the new tires will be purchased only if they skid less at the $\alpha = 0.005$ significance level. Based on the sample, will the new tires be purchased?

6-66   Teaching method A produces a mean score of 77 on a standard test. Teaching method B has been introduced, and the first sample of 19 students achieved a mean score of 79 with a standard deviation of 8. At the 5% level of significance, test the claim that method B is better.

6-67   An aircraft manufacturer randomly selects 12 planes of the same model and tests them to determine the distance (in meters) they require for takeoff. The sample mean and standard deviation are computed to be 524 m and 23 m, respectively. At the 5% level of significance, test the claim that the mean for all such planes is more than 500 m.

6-68   A pill is supposed to contain 20.0 mg of phenobarbitol. A random sample of 30 pills yields a mean and standard deviation of 20.5 mg and 1.5 mg, respectively. Are these sample pills acceptable at the $\alpha = 0.02$ significance level?

6-69   The Federal Aviation Administration randomly selects five light aircraft of the same type and tests the left wings for their loading capacities. The sample mean and the standard deviation are 16,735 lb and

978 lb, respectively. At the 5% level of significance, test the claim that the mean loading capacity for all such aircraft wings is equal to 17,850 lb.

6-70   A standard test for braking reaction times (in seconds) has produced an average of 0.75 s for young females. A driving instructor claims that his class of young females exhibits an overall reaction time that is below the average. Test his claim if it is known that 13 of his students (randomly selected) produced a mean of 0.71 s and a standard deviation of 0.06 s. Test at the 1% level of significance.

6-71   In a study of consumer credit, 25 randomly selected credit card holders were surveyed, and the mean amount they charged in the past 12 months was found to be $1756, while the standard deviation was $843. Use a 0.025 level of significance to test the claim that the mean amount charged by all credit card holders was greater than $1500.

6-72   A sociologist designs a test to measure prejudicial attitudes and claims that the mean population score is 60. The test is then administered to 28 randomly selected subjects and the results produce a mean and standard deviation of 69 and 12, respectively. At the 5% level of significance, test the sociologist's claim.

6-73   A biofeedback experiment involves measurement (in microvolts) of muscle tension by using an instrument attached to a person's forehead. For a sample group of 16 randomly selected subjects, the mean and standard deviation are found to be 6.3 mV and 1.8 mV, respectively. At the 5% level of significance, test the claim that the mean of all subjects equals 5.4 mV.

6-74   A long-range missile misses its target by an average of 0.88 mi. A new steering device is supposed to increase accuracy, and a random sample of eight missiles is equipped with this new mechanism and tested. These eight missiles miss by distances with a mean of 0.76 mi and a standard deviation of 0.04 mi. At $\alpha = 0.01$, does the new steering mechanism lower the miss distance?

6-75   At the 0.10 level of significance, test the claim that a brewery fills bottles with amounts having a mean greater than 32 oz. A sample of 27 bottles produces a mean of 32.2 oz and a standard deviation of 0.4 oz.

6-76   A high school principal is concerned with the amount of time her students devote to working at an after-school job. She randomly selects 25 students, obtains their working hours, and computes $\bar{x} = 12.3$ and $s = 11.2$. Both values are in hours for one week. She claims that this is significantly more than the mean of 10.7 h obtained from a study conducted by the National Federation of State High School Associations. Test her claim by using a 0.05 level of significance.

6-77   A high school senior is concerned about attending college because she knows that many college students require more than four years to earn a bachelor's degree. At the 0.10 level of significance, test the claim of a guidance counselor who states that the mean time is greater than five years. Sample data consist of 28 randomly selected college graduates who had a mean of 5.15 years and a standard deviation of 1.68 years (figures based on data from the National Center for Education Statistics).

6-78   A study was conducted to see the effects of mental training in organizations. (See "Routinization of Mental Training in Organizations: Effects on Performance and Well-Being," by Larsson, *Journal of Applied Psychology*, Vol. 72, No. 1.) For an experimental group of 20 subjects, a performance exam resulted in scores with a mean of 79.12 and a standard deviation of 17.49. At the 0.10 level of significance, test the claim that the experimental group comes from a population with a mean less than 85.70.

6-79   In a study of the long-term effects of promoting or holding back elementary school students, the following reading test results were obtained for a sample of 15 third grade students: $\bar{x} = 31.0$, $s = 10.5$. (The data are based on "A Longitudinal Study of the Effects of Retention/Promotion on Academic Achievement," by Peterson and others, *American Educational Research Journal*, Vol. 24, No. 1.) Does this third grade sample mean differ significantly from a first grade population mean of 41.9? Assume a 0.01 level of significance.

6-80   In a study of factors affecting hypnotism, visual analogue scale (VAS) sensory ratings were obtained for 16 subjects. For these sample ratings, the mean is 8.33 while the standard deviation is 1.96 (based on data from "An Analysis of Factors that Contribute to the Efficacy of Hypnotic Analgesia," by Price and Barber, *Journal of Abnormal Psychology*, Vol. 96, No. 1). At the 0.01 level of significance, test the claim that this sample comes from a population with a mean rating less than 10.00.

6-81   The residual lung volume is the amount of air left in the lung after breathing out as much air as possible. In a study of a method for estimating residual lung volumes, 20 subjects are tested for oxygen dilation and the results (in liters) yield a mean of 1.363 L and a standard deviation of 0.293 L. (The figures are based on "A Simplified Method for Determination of Residual Lung Volumes," by Wilmore, *Journal of Applied Physiology*, Vol. 27, No. 1.) At the 0.05 level of significance, test the claim that these sample values are drawn from a population with a mean residual lung volume equal to 1.000 L.

6-82   One effect of strenuous exercise was investigated in a study. For six subjects, the peak oxygen consumption levels were obtained. The mean and standard deviation are 3.98 and 0.49, respectively; both values are in liters for one minute (based on data from "Supramaximal Exercise After Training-Induced Hypervolemia," by Green and others, *Journal of Applied Physiology*, Vol. 62, No. 5). At the 0.05 level of significance, test the claim that this sample comes from a population with a mean that is less than 4.00 L for one minute.

6-83 A standard final examination in an elementary statistics course produces a mean score of 75. At the 5% level of significance, test the claim that the following sample scores reflect an above-average class:

79	79	78	74	82	89	74	75	78	73
74	84	82	66	84	82	82	71	72	83

6-84 A sample of beer cans labeled 16 oz is randomly selected and the actual contents accurately measured. The results (in ounces) are as follows. Is the consumer being cheated?

15.8	16.2	16.3	15.9	15.5
15.9	16.0	15.6	15.8	

6-4 EXERCISES B

6-85 For certain conditions, a hypothesis test requires the Student *t* distribution, as described in this section. Assume that the standard normal distribution is incorrectly used instead. Using the standard normal distribution, are you more likely to reject the null hypothesis, less likely, or does it make no difference? Explain.

6-86 What do you know about the *P*-value in each of the following cases?
a. $H_0: \mu \leq 5.00$; $n = 10$; test statistic: $t = 2.205$
b. $H_0: \mu = 5.00$; $n = 20$; test statistic: $t = 2.678$
c. $H_0: \mu \geq 5.00$; $n = 16$; test statistic: $t = -1.234$

6-87 Some computer programs approximate critical *t* values by

$$t = \sqrt{DF \cdot (e^{A^2/DF} - 1)}$$

where $DF = n - 1$

$e = 2.718$

$$A = z\left(\frac{8\,DF + 3}{8\,DF + 1}\right)$$

and *z* is the critical *z* score

Use this approximation to find the critical *t* score corresponding to $n = 10$ and a significance level of 0.05 in a right-tailed case. Compare the result to the critical *t* found in Table A-3.

6-88   Refer to the pilot training example on pp. 363. If the actual value of μ is 53.4 h, find β, the probability of a type II error. (See Exercise 6-32.)

6-5 TESTS OF PROPORTIONS

In Section 1-4 we presented the four different levels of measurement (nominal, ordinal, interval, ratio). We saw that data at the nominal level of measurement lacked any real numerical significance and was essentially qualitative in nature. One way to make a quantitative analysis using such qualitative data is to represent that data in the form of a proportion or percentage. For example, we cannot average a sample of 390 Democrats and 330 Republicans, but we can analyze the fact that 54.2% of this sample of 720 voters is comprised of Democrats. In this section we present the method used for testing hypotheses made about population proportions or percentages, a method that can be used with sample data at the nominal level of measurement. This method is very useful in a variety of applications, including surveys, polls, and quality control considerations involving the proportions of defective parts.

Throughout this section we assume that individual trials are independent so that the relevant probability remains constant for each trial. Recall that with a fixed number of independent trials having constant probabilities, as long as each trial has two outcomes, we have a binomial experiment (see Section 4-4). When we deal with proportions or percentages, we can usually make the assumptions that enable us to use a binomial distribution.

In Section 5-5 we saw that under suitable circumstances (namely, $np \geq 5$ and $nq \geq 5$) the binomial distribution can be approximated by a normal distribution with the mean and standard deviation given by $\mu = np$ and $\sigma = \sqrt{n \cdot p \cdot q}$. If $np \geq 5$ and $nq \geq 5$ are not both true, we may be able to use Table A-1 or the binomial probability formula described in Section 4-4, but this section deals only with situations in which the normal distribution is a suitable approximation for the distribution of sample proportions. Replacing μ and σ by their binomial counterparts, we get

$$z = \frac{x - \mu}{\sigma} = \frac{x - np}{\sqrt{n \cdot p \cdot q}}$$

where x is the number of successes in n trials.

Misleading Statistics

Money reported that among airlines, Air North had the second highest complaint rate in one particular year. But reporter George Bernstein investigated and found that Air North's high complaint rate of 38.6 per 100,000 passengers really represented only 22 complaints. (Air North flew 57,000 passengers that year.) "That many complaints could have come from a single delayed flight," wrote Bernstein. He went on to state that "if nothing else, this proves how dangerous statistics can be, and how easy it is for them to be blown out of proportion if you don't know exactly what's behind them." The following year, Air North had only three complaints.

If we divide the numerator and denominator by n and denote the sample proportion x/n by $\hat{p}$ (called "p hat"), we get

$$z = \frac{\dfrac{x}{n} - \dfrac{np}{n}}{\dfrac{\sqrt{npq}}{n}} = \frac{\hat{p} - p}{\sqrt{\dfrac{npq}{n^2}}} = \frac{\hat{p} - p}{\sqrt{\dfrac{pq}{n}}}$$

To test a hypothesis made about a population proportion or percentage, we will therefore follow the standard procedure described in Section 6-2, but the value of the test statistic will be found by computing z, as follows.

TEST STATISTIC

$$z = \frac{\hat{p} - p}{\sqrt{\dfrac{pq}{n}}}$$

where n = number of trials

p = population proportion (given in the null hypothesis)

$q = 1 - p$

$\hat{p} = x/n$ (sample proportion)

Let's assume that your county Republican party leader has claimed that the Democratic presidential candidate will receive no more than 48% of all votes cast in the county. Let us also assume that this claim is to be tested on the basis of the survey of 720 randomly selected registered voters of which 390 have indicated a preference for the Democratic candidate. The sample proportion of 390/720 (or 54.2%) is more than the claimed value of 48%, but we must now determine whether it is *significantly* more. We will assume a significance level of 5%. The sample data can now be summarized:

$$\left.\begin{array}{l} n = 720 \\ x = 390 \end{array}\right\} \hat{p} = \frac{x}{n} = \frac{390}{720} = 0.542$$

The population parameters p and q are suggested by the claim that we are testing:

$$p = 0.48$$

$$q = 1 - 0.48 = 0.52$$

The significance level of 5% indicates that $\alpha = 0.05$. Following the pattern developed in Section 6-2 (see Figure 6-1), we get the following null and alternative hypotheses:

$$H_0: p \le 0.48$$

$$H_1: p > 0.48$$

The statistic relevant to this test is found to be

$$z = \frac{\hat{p} - p}{\sqrt{\dfrac{pq}{n}}}$$

$$= \frac{0.542 - 0.48}{\sqrt{\dfrac{(0.48)(0.52)}{720}}} = 3.33$$

(In Section 4-4 we included a correction for continuity, but we ignore it here since its effect is negligible.) This test statistic, the critical region, and critical value are shown in Figure 6-16. The test statistic does fall in the critical region, and we therefore reject the null hypothesis H_0. This indicates that the sample result of a 54.2% Democratic preference among the 720 voters does represent a proportion of voters *significantly* greater than the 48% value claimed by the county Republican leader. If the 48% value is correct, there is less than a 5% chance

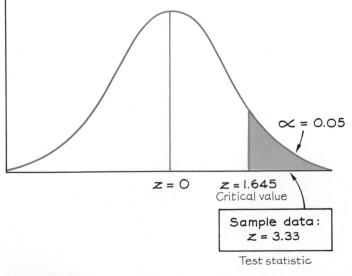

FIGURE 6-16

Have Atomic Tests Caused Cancer?

Conquerors was a 1954 movie made in Utah. A few years ago a team of investigative reporters attempted to locate members of the cast and crew. Of the 79 people they found (some living and some dead), 27 had developed cancer, including John Wayne, Susan Hayward, and Dick Powell. Two key questions arise:

1. Is a cancer rate of 27 out of 79 significant, or could it be a coincidence?

2. Was the cancer caused by fallout from the nuclear tests previously conducted in the area?

We can use statistics to answer the first question, but not the second.

of getting the sample results used in this example. Rather than concluding that an unusual sample has been obtained, we conclude that the true population proportion is more likely to be greater than 0.48 or 48%. This conclusion may in reality be wrong, even though it follows from accepted standard statistical techniques. But that is the nature of statistics. We are led to likely, but not certain, conclusions.

If the original problem is stated with a percentage, convert the percentage to the equivalent decimal form. For example, the claim that at least 45% of all married people are women would suggest the null hypothesis H_0: $p \geq 0.45$. In our computations, we should assume that $p = 0.45$ and $q = 0.55$. In addition to proportions and percentages, the methods described here also apply to tests of hypotheses made about probabilities. Whether we have a proportion, a percentage, or probability, the value of p must be between 0 and 1, and the sum of p and q must be exactly 1. (Yogi Berra revealed that he lacked formal training in statistics when he said that "baseball is 90% mental; the other half is physical.")

EXAMPLE

A television executive claims that "less than half of all adults are annoyed by the violence on television." (That is, violence in television shows, not atop television sets.) Test this claim using the sample data from a Roper poll in which 1998 surveyed adults resulted in 48% who indicated their annoyance with television violence. Use a 0.05 significance level.

Solution

We summarize the key components of the hypothesis test.

$$H_0: p \geq 0.5$$

$$H_1: p < 0.5 \text{ (from the claim that "less than half are annoyed")}$$

$$\text{Test statistic:} \quad z = \frac{\hat{p} - p}{\sqrt{\dfrac{pq}{n}}} = \frac{0.48 - 0.5}{\sqrt{\dfrac{(0.5)(0.5)}{1998}}} = -1.79$$

The test statistic, critical value, and critical region are shown in Figure 6-17. Since the test statistic is in the critical region, we reject the null hypothesis. There is sufficient sample evidence to support the claim that less than half of all adults are annoyed by the violence on television.

The Year Was (Safe) (Unsafe)

Impressions can be manipulated by the statistics that are presented. For a recent year, there were 31 air traffic accidents among planes operated by scheduled airlines in the United States. That was the highest number of accidents since 1974. This all makes it sound like a bad year. But the Air Transport Association called this the seventh safest year in airline aviation history because the fatality rate was only 0.43 per 100,000 flights.

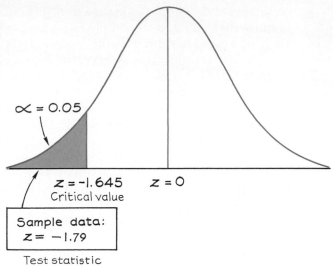

FIGURE 6-17

P-Values

The examples of this section followed the traditional approach to hypothesis testing, but it would be easy to use the P-value approach since the test statistic is a z score. The P-value is obtained by using the same procedure described in Section 6-3. In a right-tailed test, the P-value is the area to the right of the test statistic. In a left-tailed test, the P-value is the area to the left of the test statistic. In a two-tailed test, the P-value is twice the area of the extreme region bounded by the test statistic. (See Figure 6-11.) We reject the null hypothesis if the P-value is less than the significance level.

The last example was left-tailed, so the P-value is the area to the left of the test statistic $z = -1.79$. Table A-2 indicates that the area between $z = 0$ and $z = -1.79$ is 0.4633, so the P-value is $0.5 - 0.4633 = 0.0367$. Since the P-value of 0.0367 is less than the significance level of 0.05, we reject the null hypothesis and again conclude that there is sufficient sample evidence to support the claim that less than half of all adults are annoyed with television violence. Again, the P-value approach is another way of arriving at the same conclusion.

6-5 EXERCISES A

In Exercises 6-89 through 6-108, test the given hypotheses. Include the steps listed in Figure 6-1, and draw the appropriate graph.

6-89 At the 0.05 significance level, test the claim that the proportion of defects p for a certain product equals 0.3. Sample data consist of $n = 100$ randomly selected products, of which 45 are defective.

6-90 At the 0.05 significance level, test the claim that the proportion of females p at a given college equals 0.6. Sample data consist of $n = 80$ randomly selected students, of which 54 are females.

6-91 At the 0.01 level of significance, test the claim that the proportion of voters p who favor nuclear disarmament is less than 0.7. Sample data consist of 50 randomly selected voters, of whom 31 favor nuclear disarmament.

6-92 Test the claim that the proportion of adults who have credit cards is greater than 0.75. Sample data consist of 400 randomly selected adults, of whom 325 have credit cards. Use a 0.01 level of significance.

6-93 Test the claim that more than 1/4 of all white collar criminals have attended college. Sample data (from U.S. Bureau of Justice statistics) consist of 1400 randomly selected white collar criminals, with 33% of them having attended college. Use a 0.02 level of significance.

6-94 Recently, the College Entrance Examination Board claimed that the percentage of math SAT scores above 600 is equal to 17.0%. Test that claim using the sample results consisting of 512 randomly selected math SAT scores. Among those 512 scores, 19.1% are above 600. Use a 0.05 level of significance.

6-95 According to a Harris Poll, 71% of Americans believe that the overall cost of lawsuits is too high. If a random sample of 500 people results in 74% who hold that belief, test the claim that the actual percentage is 71%. Use a 0.10 significance level.

6-96 In a study of 500 aircraft accidents involving spatial disorientation of the pilot, it was found that 91% of those accidents resulted in fatalities (based on data from the Department of Transportation). At the 0.05 level of significance, test the claim that three-fourths of all such accidents will result in fatalities.

6-97 Test the claim that less than 10% of U.S. senior medical students prefer pediatrics. Sample data consist of 1068 randomly selected medical school seniors, with 64 of them choosing pediatrics (based on data reported by the Association of American Medical Colleges). Use a 0.01 significance level.

6-98   The U.S. Bureau of the Census recently claimed that 46.5% of registered voters are males (based on 54,060,000 registered males and 62,118,000 registered females). If a random sample of 1225 registered voters includes 539 males, test that claim. Use a 0.05 level of significance.

6-99   Recently, TWA reported an on-time arrival rate of 78.4%. Assume that a later random sample of 750 flights results in 630 that are on time. If TWA were to claim that its on-time arrival rate is now higher than 78.4%, would that claim be supported at the 0.01 level of significance?

6-100 A study by the Environmental Protection Agency led to the claim that 4.4% of catalytic converters are removed from cars originally installed with them. Test this claim if a study of 200 cars built with catalytic converters reveals that 8.0% of them were removed. Use a 0.01 level of significance.

6-101 An airline reservations system suffers from a 7% rate of no-shows. A new procedure is instituted whereby reservations are confirmed on the day preceding the actual flight, and a study is then made of 5218 randomly selected reservations made under the new system. If 333 no-shows are recorded, test the claim that the no-show rate is lower with the new system. Use a 5% level of significance.

6-102 In a genetics experiment, the Mendelian law is followed as expected if one-eighth of the offspring exhibit a certain recessive trait. Analysis of 500 randomly selected offspring indicates that 83 exhibited the necessary recessive trait. Is the Mendelian law being followed as expected? Use a 2% level of significance.

6-103 At the 0.01 level of significance, test the claim that more than 1/3 of all adults smoke. A random sample of 324 adults shows that 120 smoke.

6-104 In a Gallup poll of 1553 randomly selected adult Americans, 92% recognize the name Christopher Columbus. At the 0.01 significance level, test the claim that 95% of all adult Americans recognize the name Columbus.

6-105 A manufacturer considers her production process to be out of control when defects exceed 3%. A random sample of 500 items includes exactly 22 defects, but the manager claims that this represents a chance fluctuation and that production is not really out of control. Test the manager's claim at the 5% level of significance.

6-106 Test the claim that among those with preschool education training, 2/3 went on to graduate from high school. Sample data consist of 295 randomly selected subjects with preschool training, and 71% of them graduated from high school. Use a 0.05 level of significance.

6-107 Test the claim that fewer than 1/2 of San Francisco residential telephones have unlisted numbers. A random sample of 400 such phones results in an unlisted rate of 39%. Use a 0.01 level of significance.

6-108   The Kennedy-Nixon presidential race was extremely close. Kennedy
         won by 34,227,000 votes to Nixon's 34,108,000 votes. At the 0.01 level
         of significance, test the claim that the true population proportion for
         Kennedy exceeded 0.5. Assume that the voters represent a random
         sampling of those eligible.

6-5 EXERCISES B

6-109   A supplier of chemical waste containers finds that 3% of a sample of
         500 units are defective. Being somewhat devious, he wants to make a
         claim that the defective rate is no more than some specified percentage,
         and he doesn't want that claim rejected at the 0.05 level of significance
         if the sample data are used. What is the *lowest* defective rate he can
         claim under these conditions?

6-110   A reporter claims that 10% of the residents of her city feel that the
         mayor is doing a good job. Test her claim if it is known that, in a
         random sample of 15 residents, there are none who feel that the mayor
         is doing a good job. Use a 5% level of significance. Since $np = 1.5$ and
         is not at least 5, the normal distribution is not a suitable approximation
         to the distribution of sample proportions.

6-111   In a study of 500 aircraft accidents involving spatial disorientation of
         the pilot, it was found that 91% of those accidents resulted in fatalities.
         Someone with a vested interest wants to claim that the percentage is
         at least some particular value, and they don't want that claim rejected
         at the 0.01 level of significance if the sample data are used. What is
         the *highest* percentage that can be claimed under these conditions?

6-112   Refer to the example on p. 375 that relates to television violence. If
         the true value of p is 0.45, find β, the probability of a type II error.
         (See Exercise 6-32.) *Hint:* In step 3 use the values of $p = 0.45$ and
         $\sqrt{pq/n} = \sqrt{(0.45)(0.55)/1998}$.

6-6   TESTS OF VARIANCES

The preceding sections of this chapter deal with tests of hypotheses made about means and proportions. In testing these hypotheses, we use the normal and Student t distributions in ways that are very similar. They have the same basic bell shapes, they are both symmetric about zero, and the test statistics involve comparable computations. Here, we encounter a very different distribution as we test hypotheses made about a population variance or standard deviation. Recall that the standard deviation is simply the square root of the variance, so if we know the value of one, we also know the value of the other.

NOTATION	
σ	Population standard deviation
σ^2	Population variance
s	Sample standard deviation
s^2	Sample variance

For this reason, the comments we make about variances will also apply to standard deviations.

Many real and practical situations demand decisions or inferences about variances. In manufacturing, quality control engineers want to ensure that a product is, on the average, acceptable. But the engineers also want to produce items of *consistent* quality so there are only a few defective products. This consistency is measured by the variance.

As a specific example, let us consider aircraft altimeters. Due to mass-production techniques and a variety of other factors, these altimeters do not all give exact readings. (Some errors are built in.) It would be easy to change the overall average reading by simply shifting the scale on the altimeters, but it would be very difficult to change the variance of the readings. Even if the overall average were perfect, a very large variance would indicate that some altimeters give excessively high or low readings and seriously jeopardize safety. In this case, quality control engineers want to keep the variance below some tolerable level. When the variance exceeds that level, production is considered to be out of control, and corrective action must be taken.

In this section, **we assume that the population in question has normally distributed values**. We made this same assumption earlier in this chapter, but it is a more critical assumption here. In using the

Student t distribution of Section 6-4, for example, we require that the population of values be approximately normal, and we can accept deviations away from normality that are not too severe. However, when we deal with variances by using the distribution to be introduced shortly, departures away from normality will lead to gross errors. Consequently the assumption of a normally distributed population must be followed much more strictly.

In a normally distributed population with variance σ^2, we randomly select independent samples of size n and compute the variance s^2 for each sample. The random variable $(n - 1)s^2/\sigma^2$ has a distribution called the **chi-square distribution**. We denote chi-square by χ^2 and we pronounce it "kigh square." The specific mathematical equations used to define this distribution are not given here since they are beyond the scope of this text. Instead, you can refer to Table A-4 for the critical values of the chi-square distribution. We should also note that the general form of the chi-square distribution is $(df)s^2/\sigma^2$, where df represents degrees of freedom. In this section we have $n - 1$ degrees of freedom.

The test statistic used in tests of hypotheses about variances is χ^2 (chi-square):

TEST STATISTIC

$$\chi^2 = \frac{(n - 1)s^2}{\sigma^2}$$

where n = sample size

s^2 = sample variance

σ^2 = population variance (given in the null hypothesis)

The chi-square distribution resembles the Student t distribution in that there is actually a different distribution for each sample size n. In Figure 6-18 we see that the chi-square distribution does not have the same symmetric bell shape of the normal and Student t distributions. The chi-square distribution has a longer right tail. Unlike the normal and Student t distributions, the chi-square distribution does not include negative numbers. In Figure 6-19 we see that the chi-square distribution is different for each sample size n.

In using Table A-4 to determine critical values of the chi-square distribution, we must first determine the degrees of freedom, which are $n - 1$ as in the Student t distribution.

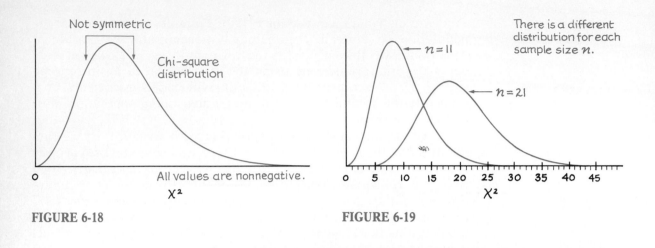

FIGURE 6-18   **FIGURE 6-19**

---

degrees of freedom $= n - 1$

We can use this expression for Section 6-4 and this section, but in later chapters we will encounter situations in which the degrees of freedom are not $n - 1$. For that reason, we should not universally equate degrees of freedom with $n - 1$.

   Once we have determined degrees of freedom, the significance level α, and the type of test (left-tailed, right-tailed, or two-tailed), we can use Table A-4 to find the critical chi-square values. An important feature of this table is that each critical value separates an area to the *right* that corresponds to the value given in the top row.

EXAMPLE

Find the critical values of χ^2 that determine critical regions containing areas of 0.025 in each tail. Assume that the relevant sample size is 10 so that the degrees of freedom are $10 - 1$, or 9.

Solution

See Figure 6-20 and refer to Table A-4. The critical value to the right (19.023) is obtained in a straightforward manner by locating 9 in the degrees-of-freedom column at the left and 0.025 across the top. The left critical value of 2.700 once again corresponds to 9 in the degrees-of-freedom column, but we must locate 0.975 across the top since the values in the top row are always *areas to the right* of the critical value.

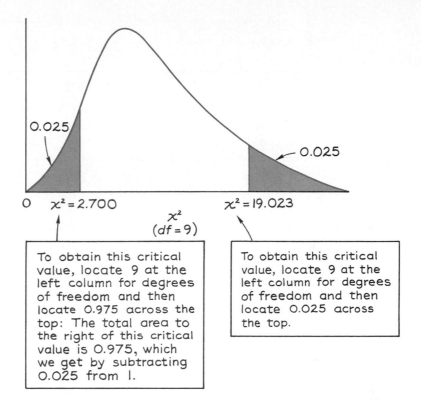

FIGURE 6-20
Finding critical values of the chi-square distribution using Table A-4.

In a right-tailed test, the value of α will correspond exactly to the areas given in the top row of Table A-4. In a left-tailed test, the value of $1 - \alpha$ will correspond exactly to the areas given in the top row of Table A-4. In a two-tailed test, the values of $\alpha/2$ and $1 - \alpha/2$ will correspond exactly to the areas given in the top row of Table A-4.

Having discussed the underlying theory and the mechanics of using Table A-4, we now give an example of a hypothesis test.

EXAMPLE

Test the claim that scores on a standard IQ test have a variance equal to 225 if a sample of 41 randomly selected subjects achieve scores with a variance of 258. Use a significance level of $\alpha = 0.05$.

continued ▶

Example, continued

Solution

We again use the basic method for testing hypotheses outlined in Figure 6-1. The claim that the variance equals 225 is $\sigma^2 = 225$ in symbolic form. The alternative is $\sigma^2 \neq 225$ and we get the following null and alternative hypotheses.

$$H_0: \sigma^2 = 225$$

$$H_1: \sigma^2 \neq 225$$

We can see that this is a two-tailed test since a sample variance significantly above or below 225 will be a basis for rejecting H_0. The significance level of $\alpha = 0.05$ has already been stipulated. The sample variance should obviously be used in testing a claim about the population variance, and the χ^2 distribution is therefore appropriate since IQ scores tend to be normally distributed. We compute the test statistic as follows.

$$\chi^2 = \frac{(n-1)s^2}{\sigma^2} = \frac{(41-1) \cdot 258}{225} = 45.867$$

We find the right critical χ^2 value of 59.342 in Table A-4 by locating 40 degrees of freedom and an area of 0.025. We find the left critical χ^2 value of 24.433 in Table A-4 by locating 40 degrees of freedom and an area of 0.975. (Since $\alpha = 0.05$, there is an area of 0.025 in each tail, but the left critical value is found by locating the area to its right. See Figure 6-21.) The test statistic based on the sample data is not in the critical region, so we fail to reject H_0. There is not sufficient evidence to warrant rejection of the claim that the variance equals 225.

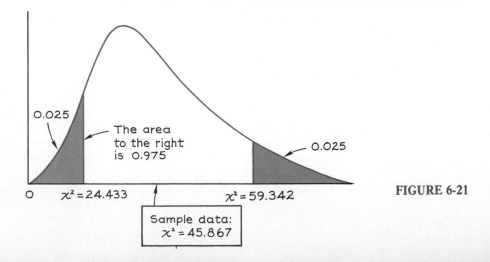

FIGURE 6-21

One example of variance occurs in the waiting lines of banks. In the past, customers traditionally entered a bank and selected one of several lines formed at different windows. A different system growing in popularity involves one main waiting line, which feeds the various windows as vacancies occur. The mean waiting time isn't reduced, but the variability among waiting times is decreased and the irritation of being caught in a slow line is also diminished.

EXAMPLE

With individual lines at its various windows, a bank finds that the standard deviation for normally distributed waiting times on Friday afternoons is 6.2 min. The bank experiments with a single main waiting line and finds that for a random sample of 25 customers, the waiting times have a standard deviation of 3.8 min. At the $\alpha = 0.05$ significance level, test the claim that a single line causes lower variation among the waiting times.

Solution

We wish to test $\sigma < 6.2$ based on a sample of $n = 25$ for which $s = 3.8$. We begin by identifying the null and alternative hypotheses. (We will use σ^2 instead of σ since the χ^2 distribution involves variances directly.)

$$H_0: \sigma^2 \geq (6.2)^2$$

$$H_1: \sigma^2 < (6.2)^2$$

The significance level of $\alpha = 0.05$ has already been selected, so we proceed to compute the value of χ^2 based on the given data:

$$\chi^2 = \frac{(n-1)s^2}{\sigma^2} = \frac{(25-1)(3.8)^2}{(6.2)^2} = 9.016$$

This test is left-tailed since H_0 will be rejected only for small values of χ^2; with $\alpha = 0.05$ and $n = 25$, we go to Table A-4 and align 24 degrees of freedom with an area of 0.95 to obtain the critical χ^2 value of 13.848 (see Figure 6-22). Since the test statistic falls within the critical region, we reject H_0 and conclude that the 3.8-min standard deviation is significantly less than the 6.2-min standard deviation that corresponds to multiple waiting lines. The sample data support the claim of lower variation. That is, the single main line does appear to lower the variation among waiting times.

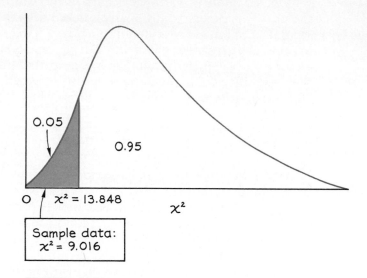

FIGURE 6-22

In this section we have illustrated an important use of the chi-square distribution. Other valuable uses, such as situations in which we wish to analyze the significance of differences between expected frequencies and the frequencies that actually occur, are considered in Chapter 10.

6-6 EXERCISES A

In Exercises 6-113 and 6-114, use Table A-4 to find the critical values of χ^2 based on the given data.

6-113
  a. $\alpha = 0.05$
    $n = 20$
    $H_0: \sigma^2 = 256$
  b. $\alpha = 0.05$
    $n = 20$
    $H_0: \sigma^2 \geq 256$
  c. $\alpha = 0.01$
    $n = 23$
    $H_0: \sigma^2 = 10$
  d. $\alpha = 0.01$
    $n = 23$
    $H_0: \sigma^2 \leq 10$
  e. $\alpha = 0.005$
    $n = 15$
    $H_1: \sigma^2 < 23.4$

6-114
  a. $\alpha = 0.10$
    $n = 6$
    $H_1: \sigma^2 < 100$
  b. $\alpha = 0.05$
    $n = 40$
    $H_1: \sigma^2 > 500$
  c. $\alpha = 0.025$
    $n = 81$
    $H_0: \sigma^2 \geq 144$
  d. $\alpha = 0.01$
    $n = 50$
    $H_0: \sigma^2 = 225$
  e. $\alpha = 0.05$
    $n = 75$
    $H_1: \sigma^2 \neq 31.5$

In Exercises 6-115 through 6-136, test the given hypotheses. Follow the pattern outlined in Figure 6-1 and draw the appropriate graph. In all cases, assume that the population is normally distributed.

6-115   At the $\alpha = 0.05$ significance level, test the claim that $\sigma^2 > 100$ if a random sample of 27 yields $s^2 = 194$.

6-116   At the $\alpha = 0.05$ significance level, test the claim that $\sigma^2 > 225$ if a random sample of 30 yields $s^2 = 380$.

6-117   At the $\alpha = 0.05$ significance level, test the claim that $\sigma^2 \geq 416$ if a random sample of 17 yields a variance of 247.

6-118   At the $\alpha = 0.10$ significance level, test the claim that $\sigma^2 \geq 90$ if a random sample of 21 yields a variance of 53.

6-119   At the $\alpha = 0.01$ significance level, test the claim that $\sigma^2 = 2.38$ if a random sample of 12 yields a variance of 5.00.

6-120   At the $\alpha = 0.05$ significance level, test the claim that $\sigma^2 = 100$ if a random sample of 27 yields a variance of 57.

6-121   At the $\alpha = 0.01$ significance level, test the claim that a population has a standard deviation of 10.0. A random sample of 18 items yields a standard deviation of 14.5.

6-122   At the $\alpha = 0.05$ significance level, test the claim that a population has a standard deviation of 52.0. A random sample of 18 items yields a standard deviation of 71.2.

6-123   At the $\alpha = 0.05$ significance level, test the claim that a population has a variance less than or equal to 9.00. A random sample of 81 items yields a variance of 12.25.

6-124   At the $\alpha = 0.025$ level of significance, test the claim that a population has a standard deviation less than 98.6. A random sample of 51 items yields a standard deviation of 79.000.

6-125   If the standard deviation for the weekly downtimes of a computer is low, then availability of the computer is predictable and planning is facilitated. If 12 weekly downtimes for a computer are randomly selected and the standard deviation is computed to be 2.85 h, at the 0.025 significance level, test the claim that $\sigma > 2.00$ h.

6-126   Based on recent data from the College Entrance Examination Board, math SAT scores have a standard deviation of 130. For a random sample of 100 students in Dutchess County, New York, the math SAT scores have a standard deviation of 127. This county is a residential suburban community with a large percentage of college-educated professionals, so that we might expect a more homogeneous group with a lower standard deviation. At the 0.05 significance level, test the claim that this community has math SAT scores with a standard deviation less than the national value of 130.

6-127   A machine pours medicine into a bottle in such a way that the standard deviation of the weights is 0.15 oz. A new machine is tested on 71 bottles, and the standard deviation for this group is 0.12 oz. At the 0.05 significance level, test the claim that the new machine produces less variance.

6-128   Test the claim that scores on a standard IQ test have a standard deviation equal to 15 if a sample of 24 randomly selected subjects yields a standard deviation of 10. Use a significance level of $\alpha = 0.01$.

6-129   In comparing systolic blood pressure levels of men and women, a medical researcher obtains readings for a random sample of 50 women. The sample mean and standard deviation are found to be 130.7 and 23.4, respectively. If systolic blood pressure levels for men are known to have a mean and standard deviation of 133.4 and 19.7 respectively, test the claim that women have a larger standard deviation. Use a 0.05 level of significance. (All readings are in millimeters of mercury, and the data are based on the National Health Survey, USDHHS publication 81-1671.)

6-130   A software firm finds that the times required to run one particular computer program have a standard deviation of 52 h. A sample of 30 new computer programs produces a standard deviation of 60 h. At the 0.05 significance level, test the claim that the standard deviation for the new computer programs is equal to 52 h.

6-131   When 22 bolts are tested for hardness, their indices have a standard deviation of 65.0. Test the claim that the standard deviation of the hardness indices for all such bolts is greater than 50.0. Test at the 0.025 level of significance.

6-132   In a study of birth weights (in grams), a random sample of 30 baby girls produces a mean of 3264 g and a standard deviation of 485 g. Assume that the mean and standard deviation for birth weights of boys are known to be 3393 g and 470 g, respectively. (The figures are based on data from the New York State Department of Health, Monograph No. 11.) At the 0.05 level of significance, test the claim that boys and girls have birth weights with the same standard deviation.

6-133   In a study of test scores, a sample of 68 women received Graduate Record Examination Verbal (GREV) scores with a standard deviation of 114.16. Assume that males are known to have scores with a standard deviation of 97.23. (See "Equivalencing MAT GRE Scores Using Simple Linear Transformation and Regression Methods," by Kagan and Stock, *Journal of Experimental Education*, Vol. 49, No. 1.) At the 0.10 significance level, test the claim that the standard deviation of women is equal to the 97.23 standard deviation for men.

6-134   In a study of the wide ranges in the academic success of college freshmen, an obvious factor is the amount of time spent studying. At the

0.05 significance level, test the claim that the standard deviation is more than 4.00 h. Sample data consist of 70 randomly selected freshmen who have a standard deviation of 5.33 h (based on data reported by *USA Today*).

6-135   The caffeine contents (in mg) for a dozen randomly selected cans of a soft drink are given below. At the 0.025 level of significance, test the claim that the standard deviation for all such cans is less than 2.0 mg.

34.2	33.7	31.9	34.3	31.6	32.7
33.1	35.2	31.6	32.9	33.0	32.4

6-136   Based on data from the National Health Survey (USDHEW publication 79-1659), men aged 25 to 34 have heights with a standard deviation of 2.9 in. Test the claim that men aged 45 to 54 have heights with a standard deviation less than 2.9 in. Use a 0.05 significance level. The heights of 25 randomly selected men in the 45 to 54 age bracket are listed below.

66.80	71.22	65.80	66.24	69.62	70.49	70.00
71.46	65.72	68.10	72.14	71.58	66.85	69.88
68.69	72.77	67.34	68.40	68.96	68.70	72.69
68.67	67.79	63.97	67.19			

6-6 EXERCISES B

6-137   For large numbers of degrees of freedom, we can approximate values of χ^2 as follows.

$$\chi^2 = \frac{1}{2}[z + \sqrt{2k - 1}]^2$$

where k = number of degrees of freedom and z = the critical score, found from Table A-2.

    For example, if we want to approximate the two critical values of χ^2 in a two-tailed hypothesis test with $\alpha = 0.05$ and a sample size of 150, we let $k = 149$ with $z = -1.96$, followed by $k = 149$ and $z = 1.96$.

a. Use this approximation to estimate the critical values of χ^2 in a two-tailed hypothesis test when $n = 101$ and $\alpha = 0.05$. Compare the results to those found in Table A-4.

b. Use this approximation to estimate the critical values of χ^2 in a two-tailed hypothesis test when $n = 150$ and $\alpha = 0.05$.

6-138   Do Exercise 6-137 using the approximation

$$\chi^2 = k\left[1 - \frac{2}{9k} + z\sqrt{\frac{2}{9k}}\right]^3$$

where k and z are as described in that exercise.

kwh	f
100–119	5
120–139	12
140–159	20
160–179	15
180–199	8

6-139   A consumer's testing company tests 60 different randomly selected coffee makers and records the estimated annual energy consumption as measured in kilowatt hours. The results are summarized in the accompanying frequency table. At the 0.025 level of significance, test the claim that the standard deviation for all such coffee makers is greater than 20 kwh.

6-140   Refer to the last example presented in this section. Assuming that σ is actually 4.0, find β (the probability of a type II error). See Exercise 6-32 and modify the procedure so that it applies to a hypothesis test involving σ instead of μ.

VOCABULARY LIST

Define and give an example of each term.

hypothesis
hypothesis test
test of significance
null hypothesis
alternative
   hypothesis
type I error
type II error

test statistic
critical region
critical value
significance level
left-tailed test
right-tailed test
two-tailed test

P-value
Student t
   distribution
t distribution
degrees of freedom
chi-square
   distribution

REVIEW

In this chapter, we studied standard methods used in statistical tests of hypotheses made about the values of population means, proportions, percentages, variances, and standard deviations. (A hypothesis is simply a statement that something is true.) When sample data conflict with the given hypothesis, we decide whether the differences are due to chance fluctuations or whether the differences are so significant that they are not likely to occur by chance. We are able to select exact **levels of significance**; 0.05 and 0.01 are common values. Sample results are said to reflect significant differences when their occurrences have probabilities less than the chosen level of significance.

Section 6-2 presented in detail the procedure for testing hypotheses. The essential steps are summarized in Figure 6-1. We defined **null hypothesis**, **alternative hypothesis**, **type I error**, **type II error**, **test statistic**, **critical region**, **critical value**, and **significance level**. All these standard terms are commonly used in discussing tests of hypotheses. We also identified the three basic types of tests: **right-tailed**, **left-tailed**, and **two-tailed** (see Figure 6-4).

We introduced the method of testing hypotheses in Section 6-2 by using examples in which only the normal distribution applies, and we introduced other distributions in subsequent sections. The brief table that follows summarizes the hypothesis tests covered in this chapter.

IMPORTANT FORMULAS				
Parameter to which hypothesis refers	Applicable distribution	Assumption	Test statistic	Table of critical values
μ (population mean)	Normal	σ is known and population is normally distributed	$z = \dfrac{\bar{x} - \mu}{\sigma/\sqrt{n}}$	Table A-2
	Normal	$n > 30$ (If σ is not known, use s for σ.)	$z = \dfrac{\bar{x} - \mu}{\sigma/\sqrt{n}}$	Table A-2
	Student t	σ is unknown and $n \leq 30$ and population is normally distributed	$t = \dfrac{\bar{x} - \mu}{s/\sqrt{n}}$	Table A-3
p (population proportion)	Normal	$np \geq 5$ and $nq \geq 5$	$z = \dfrac{\hat{p} - p}{\sqrt{pq/n}}$ where $\hat{p} = \dfrac{x}{n}$	Table A-2
σ^2 (population variance) σ (population standard deviation)	Chi-square	Population is normally distributed	$\chi^2 = \dfrac{(n-1)s^2}{\sigma^2}$	Table A-4

Section 6-2 outlines the **classical approach** to testing hypotheses, while the ***P*-value** approach was presented in Section 6-3. In the classical approach we make a decision about the null hypothesis by comparing the test statistic and critical value. With the P-value approach we base that decision on a comparison of the significance level and the P-value, which represents the probability of getting a sample that is at least as extreme as the one obtained.

In addition to presenting the method for testing hypotheses, we introduced the Student t and chi-square distributions. Chapter 7 will refer to these distributions again as we investigate estimations of parameters and ways to determine how large certain samples should be.

REVIEW EXERCISES

In Exercises 6-141 and 6-142, find the appropriate critical values.

6-141   a. $\alpha = 0.05$
          $n = 160$
          $H_0: p \geq 0.5$
   b. $\alpha = 0.01$
          $n = 35$
          $H_0: \mu \leq 16.5$
   c. $\alpha = 0.01$
          $n = 12$
          $H_0: \mu = 38.4$
   d. $\alpha = 0.01$
          $n = 25$
          $H_0: \sigma^2 \geq 225$
   e. $\alpha = 0.05$
          $n = 30$
          $H_0: \sigma^2 = 84.3$

6-142   a. $\alpha = 0.10$
          $n = 15$
          $H_0: \mu = 1.23$
   b. $\alpha = 0.10$
          $n = 15$
          $H_0: \sigma^2 = 123$
   c. $\alpha = 0.06$
          $n = 100$
          $H_0: \mu = 72.3$
   d. $\alpha = 0.05$
          $n = 10$
          $H_0: \sigma = 15$
   e. $\alpha = 0.01$
          $n = 30$
          $H_1: \sigma < 5.8$

In Exercises 6-143 and 6-144, respond to each of the following:

a. Give the null hypothesis in symbolic form.
b. Is this test left-tailed, right-tailed, or two-tailed?
c. In simple terms devoid of symbolism and technical language, describe the type I error.
d. In simple terms devoid of symbolism and technical language, describe the type II error.
e. What is the probability of making a type I error?

6-143   At the 0.01 level of significance, the claim is that the mean treatment time for a dentist is at least 20.0 min.

6-144   At the 5% level of significance, the claim is that the mean reading in a biofeedback experiment is 6.2 mV.

6-145   At the 0.05 level of significance, test the claim that 20% of adult Americans are unable to read. A random sample of 600 adult Americans included exactly 132 who could not read.

6-146   Test the claim that the mean time required for new employees to learn a certain task is 18.4 min. A random sample of 18 new employees produces a mean and standard deviation of 19.6 min and 2.7 min, respectively. Use a 0.10 significance level.

6-147   At the 0.05 level of significance, test the claim that $\sigma = 4.0$ g. A random sample includes 20 observations, for which $s = 2.8$ g.

6-148   At the 0.01 level of significance, test the claim that $\mu \leq 25.5$ ft. Sample data consist of 25 observations, for which $\bar{x} = 27.3$ ft and $s = 3.7$ ft.

6-149 At the 0.05 level of significance, test the claim that $\mu = 10.0$ s. Sample data consist of 100 scores, for which $\bar{x} = 8.2$ s. Assume that $\sigma = 6.0$ s.

6-150 In a study of blood pressure levels, a sample of 75 randomly selected women aged 25 to 34 results in a mean systolic value of 116.7 and a standard deviation of 12.5, with both values measured in millimeters of mercury. (The figures are based on data from the National Health Survey, USDHHS publication 81-1671.) At the 0.05 level of significance, test the claim that this sample comes from a population with a mean equal to 120.0 mm of mercury.

6-151 Of 200 females randomly selected and interviewed in the Midwest, 27% believed (incorrectly) that birth control pills prevent venereal disease. Use these sample results to test (at the 5% level of significance) the claim that more than one-fifth of all Midwestern women believe that birth control pills prevent venereal disease.

6-152 One type of pump is designed to remove 600 gallons of water per hour. It has been determined that the actual amounts removed have a standard deviation of 10.5 gal. A new regulator valve is installed and a sample of 20 modified pumps results in a standard deviation of 16.4 gal. At the 0.05 level of significance, test the claim that the modified pump has the same standard deviation.

6-153 Test the claim that a certain x-ray machine gives radiation dosages with a mean below 5.00 milliroentgens. Sample data consist of 36 observations with a mean of 4.13 milliroentgens and a standard deviation of 1.91 milliroentgens. Use a 0.01 level of significance.

6-154 Test the claim that 25% of those earning a bachelor's degree in business are women. A random sample of 200 business graduates earning the bachelor's degree includes 80 women and 120 men.

6-155 One large high school has found that students taking a standard college aptitude test earn scores with a variance of 6410. A counselor claims that the current group of test subjects includes a group with more varied aptitudes. Test the claim that the variance is larger than 6410 if a random sample of 60 students produces a variance of 8464. Use a 0.10 level of significance.

6-156 Use a 0.025 level of significance to test the claim that vehicle speeds at a certain location have a mean above 55.0 mi/h. A random sample of 50 vehicles produces a mean of 61.3 mi/h and a standard deviation of 3.3 mi/h.

6-157 Of 80 workers randomly selected and interviewed, 55 were opposed to an increase in social security taxes. At the 5% level of significance, test the claim that the majority (more than 50%) of such workers are opposed to the increase in taxes.

6-158 Test the claim that the mean female reaction time to a highway signal is less than 0.700 s. When 18 females are randomly selected and tested, their mean is 0.668 s. Assume that $\sigma = 0.100$ s, and use a 5% level of significance.

6-159 The following sample scores have been randomly selected from a normally distributed population. At the 0.01 significance level, test the claim that the population has a mean of 100.

101	106	98	92	97	80	89	88
110	112	100	100	103	97	97	

6-160 Using the same sample data from Exercise 6-159, test the claim that the standard deviation equals 10.0. Use a 0.05 level of significance.

COMPUTER PROJECT

1. Most statistics software packages include a procedure for testing a claim made about a single mean. Such a procedure is often referred to as a "t test". Use the t test option in a software package such as STATDISK or Minitab to solve Exercise 6-83.

2. Construct a program that takes as input the following items:
   a. A set of sample data.
   b. The value of the population mean μ.

(Assume that the sample data are randomly selected from a population with an essentially normal distribution.) Output should consist of the sample mean $\bar{x}$, the sample standard deviation s, the value of $s/\sqrt{n}$, and the value of test statistic z (if the sample is large, with $n > 30$) or t (if the sample is small, with $n \leq 30$). The test statistic should be identified as being a z score or a t score.

CASE STUDY ACTIVITY

Conduct a survey by asking the question "Do you favor or oppose the death penalty for people convicted of murder?" Survey at least 50 people and test the claim that the proportion in favor is less than or equal to 0.5. Identify the population from which you are sampling, and identify any factors that might suggest that your sample is not representative of the population. (If you record the response along with the sex of the respondent, you may be able to use the same data in Chapter 10.)

DATA PROJECT

Refer to the data sets in Appendix B.

1. Test the claim that the percentage of left-handed people is less than 15%.

2. For the same time period during which the 150 homes were sold, *U.S. Housing Markets* reported that the average price of a home in the United States was $121,000. Test the claim that the sample of the 150 homes comes from a population with a mean selling price that is greater than the national average of $121,000.

3. Test the claim that for women who are at least 66 in. tall, the mean weight is greater than 120 lb.

Chapter Seven

CHAPTER CONTENTS

7-1 Overview
We identify **objectives**.

7-2 Estimates and Sample Sizes of Means
We estimate the value of a population **mean** by the **point estimate** and **confidence interval** and present the method of determining the sample size.

7-3 Estimates and Sample Sizes of Proportions
We estimate the value of a population **proportion** by the **point estimate** and **confidence interval** and present the method of determining the sample size.

7-4 Estimates and Sample Sizes of Variances
We estimate the value of a population **variance** by the **point estimate** and **confidence interval** and present the method of determining sample size.

7

Estimates and Sample Sizes

CHAPTER PROBLEM

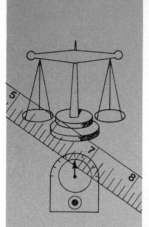

A *Newsweek* article described the use of "people meters" as a way of determining how many people are watching the different television programs. This article noted that some preliminary tests revealed a "fatigue factor" whereby the people using the meters get tired of pushing the buttons and they tend to quit. The article also noted that the technology of this television rating device tends to favor younger urban residents who are generally more adept at using more sophisticated technological devices. The article ended by stating that "statisticians have long argued that the household samples used by the rating services are simply too small to accurately determine what America is watching. In that light, it may be illuminating to note that the 4000 homes reached by the people meters constitute exactly 0.0045% of the wired nation."

Television ratings are important since they are used to determine which shows are canceled and what rates to charge for advertising. In this chapter we will analyze the implied claim that a sample size of 4000 homes is too small.

7-1 OVERVIEW

Chapter 6 introduced one aspect of inferential statistics, hypothesis testing. We used sample data to make **decisions** about claims or hypotheses. In this chapter we use sample data to make **estimates** of the values of population parameters. Section 7-2 begins by using the statistic $\bar{x}$ in estimating the value of the parameter μ. In subsequent sections we use sample proportions and sample variances to estimate the values of population proportions and population variances. We also identify some ways of determining how large samples should be.

We apply these methods of estimating and determining sample size to population means, proportions, percentages, variances, and standard deviations. The same distributions whose parameters you studied in Chapter 6 are included in the methods we develop in this chapter. Like hypothesis testing, the fundamental concepts of this chapter are very important and basic to the subject of inferential statistics. Practical applications will become apparent through the examples and exercises.

7-2 ESTIMATES AND SAMPLE SIZES OF MEANS

The following low temperatures (in degrees Fahrenheit) were recorded in Dutchess County for the month of July.

65	61	61	64	59	60	58	64	69	70	69
69	68	63	58	54	50	57	67	68	69	69
66	74	68	66	59	56	53	54	58		

Let's assume that these low temperatures are a representative sample for all July days in Dutchess County. Using only these sample scores, we want to estimate the mean low temperature of *all* July days in Dutchess County.

We could use statistics such as the sample median, midrange, or mode as estimates of μ, but the sample mean $\bar{x}$ usually provides the best estimate of μ. This is not simply an intuitive conclusion. It is based on careful study and analysis of the distributions of various estimators. For many populations, the distribution of sample means $\bar{x}$ has a smaller variance than the distribution of the other possible estimators, so $\bar{x}$ tends to be more consistent. For all populations we say that $\bar{x}$ is an **unbiased** estimator. This means that the distribution of $\bar{x}$ values tends to center about the value of μ. For these reasons, we will use $\bar{x}$ as the best estimate of μ. Because $\bar{x}$ is a single number that corresponds to a point on the number scale, we call it a point estimate.

DEFINITION

The sample mean $\bar{x}$ is the best **point estimate of the population mean μ.**

Computing the sample mean of the preceding 31 temperatures, we get $\bar{x} = 62.8$, which becomes our point estimate of the mean temperature for all July days. Even though 62.8 is our *best* estimate of μ, we really have no indication of just how good that estimate is. Sometimes even the best is very poor. Suppose that we have only the first two temperatures of 65 and 61. Their mean of 63.0 is the best point estimate of μ, but we cannot expect this best estimate to be very good since it is based on such a small sample. Mathematicians have developed an estimator that does reveal how good it it is. We will first present this estimator, illustrate its use, and then explain the underlying rationale.

NOTATION

$z_{\alpha/2}$ is the positive standard z value that separates an area of $\alpha/2$ in the right tail of the standard normal distribution (see Figure 7-1).

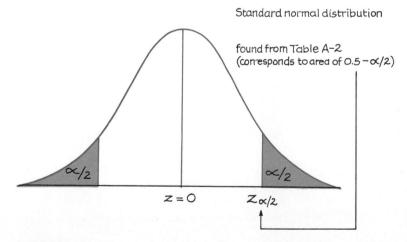

FIGURE 7-1

EXAMPLE

If $\alpha = 0.05$, then $z_{\alpha/2} = 1.96$; 1.96 is the standard z value that separates a right-tail region with an area of 0.05/2 or 0.025. We find $z_{\alpha/2}$ by noting that the region to its left must be $0.5 - 0.025$, or 0.475. That is, the area bounded by the centerline and $z_{\alpha/2}$ is 0.475. In Table A-2, an area of 0.4750 corresponds exactly to a z score of 1.96.

DEFINITION

The **maximum error of the estimate of** μ is given by

$$E = z_{\alpha/2} \cdot \frac{\sigma}{\sqrt{n}}$$

and there is a probability of $1 - \alpha$ that $\bar{x}$ differs from μ by less than E.

The maximum error of estimate E is sometimes called the **bound on the error estimate**. For a given α, there is a probability of $1 - \alpha$ that the point estimate $\bar{x}$ will miss μ by no more than the value of E. If $\alpha = 0.05$, for example, there is a 0.95 probability that $\bar{x}$ is in error (from μ) by at most E. In this sense, E is the maximum error of the point estimate $\bar{x}$.

DEFINITION

The **confidence interval** (or **interval estimate**) for the population mean is given by $\bar{x} - E < \mu < \bar{x} + E$.

We will use the preceding form of the confidence interval, but other equivalent forms are $\mu = \bar{x} \pm E$ and $(\bar{x} - E, \bar{x} + E)$.

The **degree of confidence** is the probability $1 - \alpha$ that the parameter μ is contained in the confidence interval. (The probability is often expressed as the equivalent percentage value.) The degree of confidence is also referred to as the **level of confidence** or the **confidence coefficient**.

Common choices for the degree of confidence are 95%, 99%, and 90%. The choice of 95% is most common, since it seems to represent a good balance between precision (as reflected in the width of the confidence interval) and reliability (as expressed by the degree of confidence).

In reality, the calculation of a value for E, as given above, leads to a major obstacle since it requires prior knowledge of the value of the population standard deviation σ. It's rare that we know σ for a population but we don't know μ. We overcome the obstacle of the unknown σ by using the same approach developed in Chapter 6:

In a normally distributed population with unknown σ, we can replace σ by s if $n > 30$. If $n \leq 30$ we can replace $z_{\alpha/2}$ by $t_{\alpha/2}$.

Later in this section we will consider small sample cases involving the t distribution, but the following example involves a sample large enough so that the sample standard deviation s can be used for the population standard deviation σ.

EXAMPLE

Referring to the temperature data given at the beginning of this section, we see that $n = 31$, $\bar{x} = 62.8$, and $s = 6.06$. With a 0.95 degree of confidence, use this data to find each of the following.
  a. The maximum error of estimate E
  b. The confidence interval for μ

Solution

a. The 0.95 degree of confidence implies that $\alpha = 0.05$, so that

$$E = z_{\alpha/2} \frac{\sigma}{\sqrt{n}} = 1.96 \frac{6.06}{\sqrt{31}} = 2.1$$

continued ▶

Solution, continued

(Note that since σ is unknown and $n > 30$, we used $s = 6.06$ for the value of σ. Also, recall from the previous example that when $\alpha = 0.05$, $z_{\alpha/2} = 1.96$.)

b. With $\bar{x} = 62.8$ and $E = 2.1$, we get

$$\bar{x} - E < \mu < \bar{x} + E$$

$$62.8 - 2.1 < \mu < 62.8 + 2.1$$

$$60.7 < \mu < 64.9$$

We interpret this result as follows: If we were to select different samples from the given population and use the above method for finding the corresponding intervals, in the long run 95% of those intervals would actually contain the value of μ.

Instead of expressing the result as $60.7 < \mu < 64.9$, two other equivalent forms are $\mu = 62.8 \pm 2.1$ and $(60.7, 64.9)$. We will not use these two forms in this book.

Assume that in the preceding example the temperatures actually come from a population with a mean of 63.0. Then the confidence interval obtained from the given sample data does contain the population mean since 63.0 is between 60.7 and 64.9. This is illustrated in Figure 7-2.

Note the wording of the interpretation of the confidence interval given in part (b) of the preceding example. We must be careful to interpret correctly the meaning of a confidence interval. Let's assume for the present discussion that we are using a 0.95 degree of confidence. Given the appropriate data, we can calculate the values of $\bar{x} - E$ and $\bar{x} + E$, which we will call the **confidence interval limits**. In general, these limits will have a 95% chance of enclosing μ. That is, based on sample data to be collected, there will be a 0.95 probability that the confidence interval will contain μ. If we recognize that different samples produce different confidence intervals, in the long run we will be correct 95% of the time when we say that μ is between the confidence interval limits. But once we use actual sample data to find specific limits, those limits either enclose μ or they do not, and we cannot determine if they do or don't without knowing the whole population. It is incorrect to state that μ has a 95% chance of falling within specific limits since μ is a constant, not a random variable, and it will either fall within the interval or it won't—and there's no probability involved in that. Although it's wrong to say that μ has a 95% chance of falling

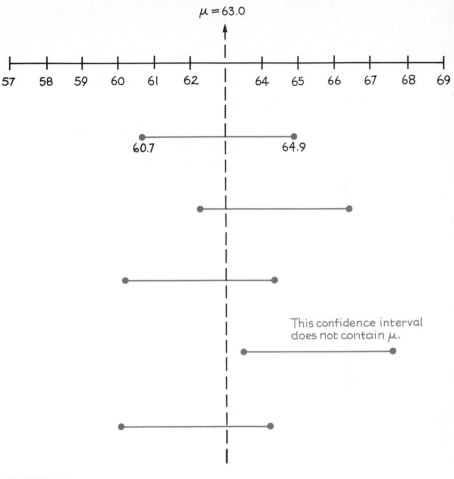

FIGURE 7-2

between the confidence interval limits, it is correct to say that these methods will result in confidence limits that, in the long run, will contain μ in 95% of the random samples collected.

So far we have defined the confidence interval for the mean and illustrated its use. We now explain why the confidence interval has the form given in the definition.

The basic underlying idea relates to the central limit theorem, which indicates that the distribution of sample means is approximately normal as long as the samples are large ($n > 30$). The central limit theorem was also used to determine that sample means have a mean of μ while the standard deviation of means from samples of size n is $\sigma/\sqrt{n}$. That is,

$$\mu_{\bar{x}} = \mu$$

$$\sigma_{\bar{x}} = \frac{\sigma}{\sqrt{n}}$$

Recall that $\sigma_{\bar{x}}$ is called the standard error of the mean. It is the standard deviation of means computed from samples of size n. Since a z score is the number of standard deviations a value is away from the mean, we conclude that $z_{\alpha/2}\, \sigma/\sqrt{n}$ represents a number of standard deviations away from μ. **There is a probability of $1 - \alpha$ that a sample mean will differ from μ by less than $z_{\alpha/2}\, \sigma/\sqrt{n}$.** See Figure 7-3 and note that the unshaded inner regions total $1 - \alpha$ and correspond to sample means that differ from μ by less than $z_{\alpha/2}\, \sigma/\sqrt{n}$. In other words, a sample mean error of $\mu - \bar{x}$ will be between $-z_{\alpha/2}\, \sigma/\sqrt{n}$ and $z_{\alpha/2}\, \sigma/\sqrt{n}$. This can be expressed as one inequality:

$$-z_{\alpha/2}\frac{\sigma}{\sqrt{n}} < \mu - \bar{x} < z_{\alpha/2}\frac{\sigma}{\sqrt{n}}$$

or

$$-E < \mu - \bar{x} < E$$

Adding $\bar{x}$ to each component of the inequality, we get the confidence interval $\bar{x} - E < \mu < \bar{x} + E$.

FIGURE 7-3
(a) There is a $1 - \alpha$ probability that a sample mean will be in error by less than E or $z_{\alpha/2}\, \sigma/\sqrt{n}$.
(b) There is a probability of α that a sample mean will be in error by more than E (in one of the shaded tails).

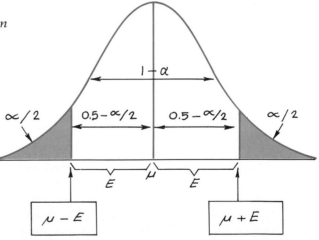

Applying these general concepts to the specific temperature data given at the beginning of this section, we can see that there is a 95% chance that a sample mean will be in error by less than

$$z_{\alpha/2}\frac{\sigma}{\sqrt{n}} = 1.96 \cdot \frac{6.06}{\sqrt{31}} = 2.1$$

Excerpts from a Department of Transportation Circular

The following excerpts from a Department of Transportation circular concern some of the accuracy requirements for navigation equipment used in aircraft. Note the use of the confidence interval.

"The total of the error contributions of the airborne equipment, when combined with the appropriate flight technical errors listed, should not exceed the following with a 95% confidence (2-sigma) over a period of time equal to the update cycle."

"The system of airways and routes in the United States has widths of route protection used on a VOR system use accuracy of ±4.5 degrees on a 95% probability basis."

Of the confidence intervals we construct by this method, in the long run an average of 95% of them will involve sample means that are in error by less than 2.1. If $\bar{x} = 62.8$ is really in error by less than 2.1, then μ would be between 60.7 and 64.9. Another way of stating that is $60.7 < \mu < 64.9$, which corresponds to our confidence interval. It should be clear that this interval estimate of μ gives us much more insight than the point estimate of 62.8 alone. The interval estimate gives us some sense of how much the sample mean might be in error.

Determining Sample Size

If we begin with the expression for E and solve for n, we get

FORMULA 7-1
$$n = \left[\frac{z_{\alpha/2}\ \sigma}{E}\right]^2$$

which may be used in determining the sample size necessary to produce results accurate to a desired degree of confidence. This equation should be used when we know the value of σ and we want to determine the sample size necessary to establish, with a probability of $1 - \alpha$, the value of μ to within $\pm E$. The existence of such an equation is somewhat remarkable, since it implies that the sample size does not depend on the size of the population.

EXAMPLE

We wish to be 99% sure that a random sample of IQ scores yields a mean that is within 2.0 of the true mean. How large should the sample be? Assume that σ is 15.

Solution

We seek n given that $\alpha = 0.01$ (from 99% confidence) and the maximum allowable error is 2.0. Applying the equation for sample size n, we get

$$n = \left[\frac{z_{\alpha/2}\sigma}{E}\right]^2 = \left[\frac{2.575 \times 15}{2.0}\right]^2$$

$$= (19.3125)^2 = 373 \quad \text{(rounded up)}$$

Therefore we should obtain at least 373 randomly selected IQ scores if we require 99% confidence that our sample mean is within 2.0 units of the true mean.

If we can settle for less accurate results and accept a maximum error of 4.0 instead of 2.0, we can see that the required sample size is reduced from 373 to 94. Direct application of the equation for *n* produces a value of 93.24, which is *rounded up* to 94. (We always round up in sample size computations so that the required number is at least adequate instead of being slightly inadequate.)

Doubling the maximum error from 2.0 to 4.0 caused the required sample size to decrease to one-fourth of its original value. Conversely, if we want to halve the maximum error, we must quadruple the sample size. In the equation, *n* is inversely proportional to the square of *E* and directly proportional to the square of $z_{\alpha/2}$. All of this implies that we can obtain more accurate results with greater confidence, but the sample size will be substantially increased. Since larger samples generally require more time and money, there may be a need for a tradeoff between the sample size and the confidence level.

Small Sample Cases

Unfortunately, the application of the equation for sample size (Formula 7-1) requires prior knowledge of σ. Realistically, σ is usually unknown unless previous research results are available. When we do sample from a population in which σ is unknown, a preliminary study must be conducted so that σ can be reasonably estimated. Only then can we determine the sample size required to meet our error tolerance and confidence demands. However, if we intend to construct a confidence interval, do not know σ, and do not plan a preliminary study, we can use the Student *t* distribution as long as the population is essentially normal. In a normally distributed population with unknown σ, we can replace σ by *s* if $n > 30$. If $n \leq 30$ we can replace $z_{\alpha/2}$ by $t_{\alpha/2}$ in the computation of *E* to get the maximum error *E* in the following expression.

MAXIMUM ERROR

$$E = t_{\alpha/2} \frac{s}{\sqrt{n}}$$

when *all* the following conditions are met:

1. $n \leq 30$; and
2. σ is unknown; and
3. the population is normally distributed.

EXAMPLE

In a study of lung capacities, 10 subjects were measured for their residual lung volumes and the following results (in liters) were obtained. (See "Validation of Esophageal Balloon Technique at Different Lung Volumes and Postures," by Baydur, Cha, Sassoon, *Journal of Applied Physiology*, Vol. 62, No. 1.)

$$
\begin{array}{ll}
1.36 & 3.16 \\
2.02 & 0.77 \\
1.59 & 1.31 \\
1.17 & 1.87 \\
1.10 & 1.86
\end{array}
\quad
\begin{array}{l}
n = 10 \\
\bar{x} = 1.621 \\
s = 0.668
\end{array}
$$

Recognizing that σ is unknown, construct the 95% confidence interval for the mean residual lung volume of all such subjects.

Solution

Since this sample involves scores that are physical measurements from "10 normal subjects," it is reasonable to assume that the population of all such scores is normally distributed. With σ unknown and a sample of $n = 10$ scores, we know from Section 6-4 that the distribution of such sample means is a Student t distribution. We therefore calculate E using the Student t distribution and the sample standard deviation s.

$$
E = t_{\alpha/2} \frac{s}{\sqrt{n}} = 2.262 \cdot \frac{0.668}{\sqrt{10}} = 0.478
$$

and

$$
\bar{x} - E < \mu < \bar{x} + E
$$

now becomes

$$
1.621 - 0.478 < \mu < 1.621 + 0.478
$$

or

$$
1.143 < \mu < 2.099
$$

For the given sample data, the point estimate of μ is $\bar{x} = 1.621$ and the interval estimate is as given above. In the long run, 95% of such samples will lead to confidence limits that actually do contain the true value of the population mean μ.

Shown below is the STATDISK computer display of the results from the data in the preceding example. The user enters the level of confidence, the sample size, sample mean, and sample standard deviation. (STATDISK allows entry of a population standard deviation if it is known, but this is not shown on the display below.)

STATDISK DISPLAY

```
        Confidence level  . . . . = .95
        Sample size . . . . . . . = 10
        Sample mean . . . . . . . = 1.621
        Sample standard deviation = .668

        The Confidence Interval is:
        ------------------------------

        1.1428 < Population mean < 2.0992
```

We also show the Minitab display that results from the data in the preceding example.

MINITAB DISPLAY

```
MTB > SET C1
DATA> 1.36 3.16 2.02 0.77 1.59 1.31 1.17 1.87 1.10 1.86
DATA> ENDOFDATA
MTB > TINTERVAL with 95 percent for data in C1

          N      MEAN     STDEV   SE MEAN    95.0 PERCENT C.I.

C1        10     1.621    0.668    0.211   (   1.143,    2.099)
```

7-2 EXERCISES A

7-1 a. If $\alpha = 0.05$, find $z_{\alpha/2}$.
   b. If $\alpha = 0.02$, find $z_{\alpha/2}$.
   c. Find $z_{\alpha/2}$ for the value of α corresponding to a confidence level of 96%.
   d. If $\alpha = 0.05$, find $t_{\alpha/2}$ for a sample of 20 scores.
   e. If $\alpha = 0.01$, find $t_{\alpha/2}$ for a sample of 15 scores.

7-2 a. If $\alpha = 0.10$, find $z_{\alpha/2}$.
   b. Find $z_{\alpha/2}$ for the value of α corresponding to a confidence level of 95%.
   c. Find $z_{\alpha/2}$ for the value of α corresponding to a confidence level of 80%.
   d. If $\alpha = 0.10$, find $t_{\alpha/2}$ for a sample of 10 scores.
   e. If $\alpha = 0.02$, find $t_{\alpha/2}$ for a sample of 25 scores.

In Exercises 7-3 through 7-8, use the given data to find the maximum error of estimate E. *Be sure to use the correct expression for* E.

7-3 $\alpha = 0.05$, $\sigma = 15$, $n = 100$ **7-4** $\alpha = 0.05$, $\sigma = 15$, $n = 64$
7-5 $\alpha = 0.01$, $\sigma = 40$, $n = 25$ **7-6** $\alpha = 0.05$, $s = 30$, $n = 25$
7-7 $\alpha = 0.01$, $s = 15$, $n = 100$ **7-8** $\alpha = 0.01$, $s = 30$, $n = 64$

7-9 Find the 95% confidence interval for μ if $\sigma = 5$, $\bar{x} = 70.4$, and $n = 36$.

7-10 Find the 95% confidence interval for μ if $\sigma = 7.3$, $\bar{x} = 84.2$, and $n = 40$.

7-11 Find the 99% confidence interval for μ if $\sigma = 2$, $\bar{x} = 98.6$, and $n = 100$.

7-12 Find the 90% confidence interval for μ if $\sigma = 5.5$, $\bar{x} = 123.6$, and $n = 75$.

7-13 In a study of physical attractiveness and mental disorders, 231 subjects were rated for attractiveness and the resulting sample mean and standard deviation are 3.94 and 0.75, respectively. (See "Physical Attractiveness and Self-Perception of Mental Disorder," by Burns and Farina, *Journal of Abnormal Psychology*, Vol. 96, No. 2.) Use this sample data to construct the 95% confidence interval for the population mean.

7-14 The U.S. Department of Health, Education, and Welfare collected sample data for 772 males between the ages of 18 and 24. That sample group has a mean height of 69.7 in. with a standard deviation of 2.8 in. (See USDHEW publication 79-1659.) Use this sample data to find the 99% confidence interval for the mean height of all males between the ages of 18 and 24.

7-15 The National Center for Education Statistics surveyed 4400 college graduates about the lengths of time required to earn their bachelor's degrees. The mean is 5.15 years and the standard deviation is 1.68 years. Based on this sample data, construct the 99% confidence interval for the mean time required by all college graduates.

7-16   The U.S. Department of Health, Education, and Welfare collected sample data for 1525 women aged 18 to 24. That sample group has a mean serum cholesterol level (measured in mg/100 ml) of 191.7 with a standard deviation of 41.0. (See USDHEW publication 78-1652.) Use this sample data to find the 97% confidence interval for the mean serum cholesterol level of all women in the 18 to 24 age bracket.

7-17   In a study of relationships between work schedules and family life, 29 subjects work at night. Their times (in hours) spent in child care are measured and the mean and standard deviation are 26.84 and 17.66, respectively. (See "Nonstandard Work Schedules and Family Life," by Staines and Pleck, *Journal of Applied Psychology*, Vol. 69, No. 3.) Use this sample data to construct the 95% confidence interval for the mean time in child care for all night workers.

7-18   The birth process of a newly discovered mammal is being studied, and the lengths of 18 observed pregnancies have been recorded. The mean and standard deviation for these 18 times is 97.3 days and 2.2 days, respectively. Find the 95% confidence interval for the mean time of pregnancy.

7-19   A country mints a coin that is supposed to weigh 1 oz. A random sample of 25 such coins produces a mean of 0.996 oz with a standard deviation of 0.004 oz. Construct the 98% confidence interval for the mean weight of all such coins.

7-20   A botanist measures the heights of 16 seedlings and obtains a mean and standard deviation of 72.5 cm and 4.5 cm, respectively. Find the 97% confidence interval for the mean height of seedlings in the population from which the sample was selected.

7-21   Find the sample size necessary to estimate a population mean to within three units if $\sigma = 16$ and we want 95% confidence in our results.

7-22   Find the sample size necessary to estimate a population mean. Assume that $\sigma = 20$, the maximum allowable error is 1.5, and we want 95% confidence in our results.

7-23   On a standard IQ test, σ is 15. How many random IQ scores must be obtained if we want to find the true population mean (with an allowable error of 0.5) and we want 99% confidence in the results?

7-24   Do Exercise 7-23 assuming that the maximum error is 0.25 unit instead of 0.5 unit.

7-25   In a study of factors affecting soldiers' decisions to reenlist, 320 subjects were measured for an index of satisfaction and the sample mean and standard deviation are 28.8 and 7.3, respectively. (See "Affective and Cognitive Factors in Soldiers' Reenlistment Decisions," by Motowidlo and Lawton, *Journal of Applied Psychology*, Vol. 69, No. 1.) Use the given sample data to construct the 98% confidence interval for the population mean.

7-26   A magazine reporter is conducting independent tests to determine the distance a certain car will travel while consuming only 1 gallon of gas. A sample of five cars is tested and a mean of 28.2 mi is obtained. Assuming that $\sigma = 2.7$ mi, find the 98% confidence interval for the mean distance traveled by all such cars using 1 gallon of gas.

7-27   A sample of seven monkeys is studied. Among the results we find that the times required for the monkeys to learn a task had a mean of 14.7 min and a standard deviation of 2.5 min. If previous testing has shown that $\sigma = 2.5$ min, find the 95% confidence interval for the mean time required to learn the task.

7-28   A sample of 25 employees exposed to loud noises is randomly selected for a hearing test and the resulting mean score is found to be 68.5. Assuming that σ is known to be 5.5, construct the 95% confidence interval for the mean of all such employees.

7-29   Do Exercise 7-28 assuming that σ is unknown and $s = 5.5$ is computed from the sample.

7-30   In sampling the caloric content of 24 bottles of "light" beer, the mean and standard deviation are found to be 107.3 calories and 3.9 calories, respectively. Find the 90% confidence interval for the mean caloric content of all such bottles.

7-31   A sample of computer connect times (in hours) is obtained for 18 randomly selected students. For this sample, the mean is 16.2 while the standard deviation is 3.4. Construct the 99% confidence interval for the mean connect time of all such students.

7-32   Ten randomly selected cars are tested for fuel consumption; the mean and standard deviation are found to be 25.1 mi/gal and 1.9 mi/gal respectively. Construct the 95% confidence interval for the mean of all such cars.

7-33   Assume that $\sigma = 2.8$ in. for the heights of adult males (based on data from the National Health Survey, publication 79-1659). Find the sample size necessary to estimate the mean height of all adult males to within 0.5 in. if we want 99% confidence in our results.

7-34   We want to estimate the mean weight of one type of coin minted by a certain country. How many coins must we sample if we want to be 99% confident that the sample mean is within 0.001 oz of the true mean? Assume that a pilot study has shown that $\sigma = 0.004$ oz can be used.

7-35   We want to determine the mean weight of all boxes of cereal labeled 400 grams. We need to be 98% confident that our sample mean is within 3 g of the population mean, and a pilot study suggests that $\sigma = 10$ g. How large must our sample be?

7-36 A psychologist has developed a new test of spatial perception, and she wants to estimate the mean score achieved by adult male pilots. How many people must she test if she wants the sample mean to be in error by no more than 2.0 points, with 95% confidence? A pilot study suggests that $\sigma = 21.2$.

7-37 In a study of lung capacities, 10 subjects were measured for their functional residual capacities and the following results (in liters) were obtained. (See "Validation of Esophageal Balloon Technique at Different Lung Volumes and Postures," by Baydur, Cha, and Sassoon, *Journal of Applied Physiology*, Vol. 62, No. 1.) Construct the 95% confidence interval for the mean functional residual capacity.

2.96	4.65	3.27	2.50	2.59
5.97	1.74	3.51	4.37	4.02

7-38 In a study of the use of hypnosis to relieve pain, sensory ratings were measured for 16 subjects with the results given below. (See "An Analysis of Factors That Contribute to the Efficacy of Hypnotic Analgesia," by Price and Barber, *Journal of Abnormal Psychology*, Vol. 96, No. 1.) Use this sample data to construct the 95% confidence interval for the mean sensory rating for the population from which the sample was drawn.

8.8	6.6	8.4	6.5	8.4	7.0	9.0	10.3
8.7	11.3	8.1	5.2	6.3	11.6	6.2	10.9

7-39 The Internal Revenue Service conducts a study of estates valued at more than $300,000 and determines the value of bonds for a randomly selected sample with the results (in dollars) given below. Find an interval estimate of the mean value of bonds for all such estates. Assume that the sample standard deviation can be used as an estimate of the population standard deviation.

45,300	36,200	72,500	50,500	15,300	58,500
26,200	97,100	74,200	83,700	72,000	10,000
63,000	15,000	49,200	37,500	81,000	24,000
145,000	27,900	53,100	27,500	94,000	23,800
74,600	36,800	65,900	29,400	86,300	25,600
53,200	47,200	61,800	33,200	18,200	75,000

7-40 In one region of a city, a random survey of households includes a question about the number of people in the household. The results are given in the accompanying frequency table. Construct the 90% confidence interval for the mean size of all such households. Assume that the sample standard deviation can be used as an estimate of the population standard deviation.

Household size	f
1	15
2	20
3	37
4	23
5	14
6	4
7	2

7-2 EXERCISES B

7-41 The confidence interval $75.63 < \mu < 77.97$ is found by using sample data for which $n = 35$, $\bar{x} = 76.8$, and $s = 4.2$. Find the degree of confidence.

7-42 Given a collection of sample data with $n = 400$, $\bar{x} = 50.0$, and $s = 10.0$, find confidence intervals for these degrees of confidence: 99%, 95%, 90%. In general, how is the confidence interval affected by the degree of confidence?

7-43 The development of Formula 7-1 assumes that the population is infinite, or we are sampling with replacement, or the population is very large. If we have a relatively small population and we sample without replacement, we should modify E to include the finite population correction factor as follows:

$$E = z_{\alpha/2} \frac{\sigma}{\sqrt{n}} \sqrt{\frac{N - n}{N - 1}} \qquad \text{where } N \text{ is the population size}$$

a. Show that the preceding expression can be solved for n to yield

$$n = \frac{N\sigma^2[z_{\alpha/2}]^2}{(N - 1)E^2 + \sigma^2[z_{\alpha/2}]^2}$$

b. Do Exercise 7-34 assuming that the coins are selected without replacement from a population of $N = 100$ coins.

7-44 The standard error of the mean is $\sigma/\sqrt{n}$ provided that the population size is infinite. If the population size is finite and is denoted by N, then the correction factor

$$\sqrt{\frac{N - n}{N - 1}}$$

should be used whenever $n > 0.05N$. This correction factor multiplies the standard error of the mean, as shown in Exercise 7-43. Find the 95% confidence interval for the mean of 100 IQ scores if a sample of 30 scores produces a mean and standard deviation of 132 and 10, respectively.

7-45 Solve for n and show all work:

$$E = \frac{z_{\alpha/2}\sigma}{\sqrt{n}}$$

7-46   Why *don't* we take

$$E = \frac{t_{\alpha/2}s}{\sqrt{n}}$$

and solve for n to get an equation that can be used to determine the sample size required for cases in which the Student t distribution applies?

7-47   Using the data from Exercise 7-39 as a pilot study, how many random estates must be surveyed in order to estimate the mean bond value? Assume that we want to be 95% confident that the sample mean is in error by at most $1000.

7-48   Using the data from Exercise 7-40 as a pilot study, how many random households must be surveyed if we want to estimate the mean household size? Assume that we want 99% confidence that the sample mean is in error by at most 0.1.

7-3 ESTIMATES AND SAMPLE SIZES OF PROPORTIONS

In this section we consider the same concepts of estimating and sample size determination that were in Section 7-2, but we apply those concepts to proportions instead of means. We already studied proportions in Section 6-5, where we presented a method of conducting tests of hypotheses about population proportions. As in Section 6-5, we again assume that the conditions for the binomial distribution are essentially satisfied. We consider binomial experiments for which $np \geq 5$ and $nq \geq 5$. These assumptions enable us to use the normal distribution as an approximation to the binomial distribution.

Although we make repeated references to proportions, you should remember that the theory and procedures also apply to probabilities and percents. Proportions and probabilities are both expressed in decimal or fraction form. If we intend to deal with percents we can easily convert them to proportions by deleting the percent sign and dividing by 100. The symbol p may therefore represent a proportion, a probability, or the decimal equivalent of a percent. We continue to use p as the population proportion in the same way that we use μ to represent the population mean. In Section 6-5, we represented a sample proportion by $\hat{p} = x/n$ where x was the number of successes in n trials.

Sample Size Too Small

The Children's Defense Fund was organized to promote the welfare of children. The group published *Children out of School in America*, which reported that in one area, 37.5% of the 16- and 17-year-old Puerto Rican children were out of school. This statistic received much media attention, but it was based on a sample of only 16 Puerto Rican children. The report also noted that among secondary school students suspended in another region, 33% were suspended twice and 67% were suspended at least three times, but these figures were based on a total sample size of only three. (See "Firsthand Report: How Flawed Statistics Can Make an Ugly Picture Look Even Worse," *American School Board Journal*, Vol. 162.)

NOTATION

$$\hat{p} = \frac{x}{n}$$

In this way, p represents the population proportion while $\hat{p}$ (called "p hat") represents the sample proportion. In previous chapters we stipulated that $q = 1 - p$, so it is natural to stipulate that $\hat{q} = 1 - \hat{p}$.

The term $\hat{p}$ denotes the sample proportion that is analogous to the relative frequency definition of a probability. As an example, suppose that a pollster is hired to determine the proportion of adult Americans who favor socialized medicine. Let's assume that 2000 adult Americans are surveyed with 1347 favorable reactions. The pollster seeks the value of p, the true proportion of all adult Americans favoring socialized medicine. Sample results indicate that $x = 1347$ and $n = 2000$, so that

$$\hat{p} = \frac{x}{n} = \frac{1347}{2000} = 0.6735$$

Just as $\bar{x}$ was selected as the point estimate of μ, we now select $\hat{p}$ as the best point estimate of p.

DEFINITION

The sample proportion $\hat{p}$ is the best **point estimate of the population proportion p**.

Of the various estimators that could be used for p, $\hat{p}$ is deemed best because it is unbiased and the most consistent. It is unbiased in the sense that the distribution of sample proportions tends to center about the value of p. It is most consistent in the sense that the variance of sample proportions tends to be smaller than the variance of any other unbiased estimators.

We assume in this section that the binomial conditions are essentially satisfied and that the normal distribution can be used as an approximation to the distribution of sample proportions. This allows us to draw from results established in Section 5-5 and to conclude that the mean number of successes μ and the standard deviation of the number of successes σ are given by

$$\mu = np$$
$$\sigma = \sqrt{npq}$$

where p is the probability of a success. Both of these parameters pertain to n trials, and we now convert them to a "per trial" basis simply by dividing by n:

$$\text{mean of sample proportions} = \frac{np}{n} = p$$

$$\text{standard deviation of sample proportions} = \frac{\sqrt{npq}}{n} = \sqrt{\frac{npq}{n^2}} = \sqrt{\frac{pq}{n}}$$

The first result may seem trivial since we have already stipulated that the true population proportion is p. The second result is nontrivial and very useful. In the last section, we saw that the sample mean $\bar{x}$ has a probability of $1 - \alpha$ of being within $z_{\alpha/2}\sigma/\sqrt{n}$ of μ. Similar reasoning leads us to conclude that $\hat{p}$ has a probability of $1 - \alpha$ of being within $z_{\alpha/2}\sqrt{pq/n}$ of p. But if we already know the value of p or q, we have no need for estimates or sample size determinations. Consequently, we must replace p and q by their point estimates of $\hat{p}$ and $\hat{q}$ so that an error factor can be computed in real situations. This leads to the following results.

DEFINITION

The **maximum error of the estimate of p** is given by

$$E = z_{\alpha/2}\sqrt{\frac{\hat{p}\hat{q}}{n}}$$

and the probability that $\hat{p}$ differs from p by less than E is $1 - \alpha$.

How One Telephone Survey Was Conducted

A *New York Times*/CBS News survey was based on 1417 interviews of adult men and women in the United States. It reported a 95% certainty that the sample results differ by no more than three percentage points from the percentage that would have been obtained if *every* adult American had been interviewed. A computer was used to select telephone exchanges in proportion to the population distribution. After selecting an exchange number, the computer generated random numbers to develop a complete phone number so that both listed and unlisted numbers could be included.

DEFINITION

The **confidence interval** (or interval estimate) for the population proportion p is given by

$$\hat{p} - E < p < \hat{p} + E$$

The following example illustrates the construction of a confidence interval for a proportion.

EXAMPLE

In a Roper Organization poll of 2000 adults, 1280 have money in regular savings accounts. Find the 95% confidence interval for the true proportion of adults who have money in regular savings accounts.

Solution

The sample results are $x = 1280$ and $n = 2000$, so that $\hat{p} = 1280/2000 = 0.640$ and $\hat{q} = 1 - 0.640 = 0.360$. A confidence level of 95% requires that $\alpha = 0.05$, so that $z_{\alpha/2} = 1.96$. We first calculate the maximum error of estimate E.

$$E = z_{\alpha/2} \sqrt{\frac{\hat{p}\hat{q}}{n}} = 1.96 \sqrt{\frac{(0.640)(0.360)}{2000}}$$

$$= 0.021$$

We can now find the confidence interval since we know that $\hat{p} = 0.640$ and $E = 0.021$.

$$\hat{p} - E < p < \hat{p} + E$$

$$0.640 - 0.021 < p < 0.640 + 0.021$$

$$0.619 < p < 0.661$$

If we wanted the 95% confidence interval for the true population *percent*, we could express the result as $61.9\% < p < 66.1\%$.

Sample Size

Having discussed point estimates and confidence intervals for p, we now consider the problem of determining how large a sample should be when we want to find the approximate value of a population proportion. In the previous section we started with the expression for the error E and solved for n. Following that reasonable precedent, we begin with

$$E = z_{\alpha/2} \sqrt{\frac{\hat{p}\hat{q}}{n}}$$

and we solve for n to get

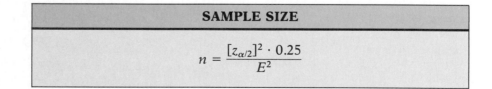

SAMPLE SIZE

$$n = \frac{[z_{\alpha/2}]^2 \hat{p}\hat{q}}{E^2}$$

$\hat{p}$	$\hat{q}$	$\hat{p}\hat{q}$
0.1	0.9	0.09
0.2	0.8	0.16
0.3	0.7	0.21
0.4	0.6	0.24
0.5	0.5	0.25
0.6	0.4	0.24
0.7	0.3	0.21
0.8	0.2	0.16
0.9	0.1	0.09

But if we are going to determine the necessary sample size, we can assume that the sampling has not yet taken place, so $\hat{p}$ and $\hat{q}$ are not known. Mathematicians have cleverly circumvented this problem by showing that, in the absence of $\hat{p}$ and $\hat{q}$, we can assign the value of 0.5 to each of those statistics and the resulting sample size will be at least sufficient. The underlying reason for the assignment of 0.5 is found in the conclusion that the product $\hat{p} \cdot \hat{q}$ achieves a maximum possible value of 0.25 when $\hat{p} = 0.5$ and $\hat{q} = 0.5$. See the accompanying table, which lists some values of $\hat{p}$ and $\hat{q}$. In practice, this means that no knowledge of $\hat{p}$ or $\hat{q}$ requires that the preceding expression for n evolves into

SAMPLE SIZE

$$n = \frac{[z_{\alpha/2}]^2 \cdot 0.25}{E^2}$$

where the occurrence of 0.25 reflects the substitution of 0.5 for each of $\hat{p}$ and $\hat{q}$. If we have evidence supporting specific known values of $\hat{p}$ or $\hat{q}$, we can substitute those values and thereby reduce the sample size accordingly. For example, if $\hat{p} = 0.6$, then $\hat{q} = 0.4$ and $\hat{p}\hat{q} = 0.24$, which is less than 0.25, so that the resulting value of n will be smaller. Such evidence about $\hat{p}$ or $\hat{q}$ could come from a pilot study, previous experience, or other such sources.

The Wisdom of Hindsight

The Roper Organization erred by 10 percentage points when it predicted the outcome of the Mondale-Reagan presidential race. Roper noted that a sample error was caused when eight-year-old census data were not updated. This resulted in a sample with a disproportionate number of Democrats. Also, respondents were first asked to identify the candidate they favored and then asked if they planned to vote. This order of questions results in an inflated figure for the second question: Respondents who identify a favorite candidate are more inclined to say that they will vote, even if they are not really inclined to do so. The Roper Organization is usually quite accurate in its results, and steps have been take to prevent such errors in the future.

EXAMPLE

We want to estimate, with a maximum error of 0.03, the true proportion of all TV households tuned in to a particular show, and we want 95% confidence in our results. We have no prior information suggesting a possible value of p. How many TV households must we survey?

Solution

With a confidence level of 95%, we have $\alpha = 0.05$, so that $z_{\alpha/2} = 1.96$. We are given $E = 0.03$, but in the absence of $\hat{p}$ or $\hat{q}$ we use the last expression for n. We get

$$n = \frac{[z_{\alpha/2}]^2 \cdot 0.25}{E^2} = \frac{[1.96]^2 \cdot 0.25}{0.03^2} = 1067.11 = 1068 \text{ (rounded up)}$$

To be 95% confident that we come within 0.03 of the true proportion of TV households who watch the show, we should poll 1068 randomly selected TV households.

At the beginning of this chapter we presented a brief excerpt from an article implying that a sample of 4000 may be too small, since it is only 0.0045% of all television households. We now know that this position is essentially incorrect, because the desired sample size doesn't depend on the population size; it depends on the levels of accuracy and confidence we desire. It is the absolute size of the sample that is important, not the size of the sample relative to the population. It is the sample size number that determines its credibility, not the percent of the population.

The previous example shows that with a sample of 1068 randomly selected TV households, we have 95% confidence that the sample proportion is within 0.03 of the true proportion. Based on Nielsen data, there are about 90 million TV households, so that a sample of size 1068 is a sample that is about 0.001% of the population. Although that percent is small, we are still about 95% confident that we are within 0.03 of the true value, so that the results will be quite good. Clearly, a sample of size 2000 or 4000 would be even better. In this sense, that article is misleading. Although Nielsen ratings might be criticized for being based on a sample that is not truly representative of the population, they should not be criticized because the sample is a small percent of the population.

Large Sample Size Isn't Good Enough

Biased sample data should not be used for inferences, no matter how large the sample is. For example, in *Women and Love: A Cultural Revolution in Progress*, Shere Hite bases her conclusions on 4500 replies that she received after mailing 100,000 questionnaires to various women's groups. A *random* sample of 4500 subjects would usually provide good results, but Hite's sample is biased. It is criticized for overrepresenting women who join groups and women who feel strongly about the issues addressed. Because Hite's sample is biased, her inferences are not valid, even though the sample size of 4500 might seem to be sufficiently large.

EXAMPLE

We want to estimate the proportion of those who believe that the extended use of marijuana weakens self-discipline. We want an error of no more than 0.02 and a confidence level of 96%. A previous survey indicates that p should be close to 0.85. How large should our sample be?

Solution

With a 96% confidence level, we have $\alpha = 0.04$ and $z_{\alpha/2} = 2.05$. We are given $E = 0.02$, and since $\hat{p}$ is around 0.85, we conclude that $\hat{q} = 1 - \hat{p} = 1 - 0.85 = 0.15$. We can now use our original expression for n to get

$$n = \frac{[z_{\alpha/2}]^2 \hat{p}\hat{q}}{E^2} = \frac{[2.05]^2(0.85)(0.15)}{(0.02)^2}$$

$$= 1339.55 = 1340 \qquad \text{(rounded up)}$$

Rounding *up*, we find that the sample size should be 1340. If we had no prior knowledge of the value of p, we would have used 0.25 for $\hat{p}\hat{q}$ and our required sample size would have been 2627, almost twice as large!

Newspaper, magazine, television, and radio reports often feature results of polls. Reporters frequently provide percentages without any indication of the sample size or degree of confidence. For example, one national newspaper reported that "the American Lung Association says that 64 percent of smokers agree they shouldn't light up near nonsmokers." Without knowing the sample size and degree of confidence, we have no real sense for how good that statistic is. In contrast, the *New York Times* published an article giving results of a poll conducted on the popularity of the president. The *Times* included a five-paragraph insert explaining that the results were "based on telephone interviews conducted Nov. 20 through Nov. 24 with 1553 adults around the U.S., excluding Alaska and Hawaii." The *Times* also explained how the telephone numbers were selected to be representative of the population. They explained how results were weighted to be representative of region, race, sex, age, and education. They explained that "in theory, in 19 cases out of 20 the results based on such samples will differ by no more than 3 percentage points in either direction from what would have been obtained by interviewing all adult Americans." In general, the five-paragraph insert provided information that allows informed readers to recognize the quality of the poll.

Polling is an important and common practice in the United States. Polls can affect the television shows we watch, the leaders we elect, the legislation that governs us, and the products we consume. Understanding the concepts of this section should remove much of the mystery and misunderstanding often created by polls.

7-3 EXERCISES A

In Exercises 7-49 through 7-52, a trial is repeated n *times with* x *successes. In each case find (a)* $\hat{p}$; *(b)* $\hat{q}$; *(c) the best point estimate for the value of* p; *(d) the maximum error of estimate* E *(assuming that* $\alpha = 0.05$*).*

7-49 $n = 500, x = 100$ **7-50** $n = 2000, x = 300$
7-51 $n = 1068, x = 325$ **7-52** $n = 1776, x = 50$

In Exercises 7-53 through 7-56, use the given data to find the appropriate confidence interval for the population proportion p.

7-53 $n = 400, x = 100$, 95% confidence

7-54 $n = 900, x = 400$, 95% confidence

7-55 $n = 512, x = 309$, 98% confidence

7-56 $n = 12,485, x = 3456$, 99% confidence

7-57   You want to determine the percentage of individual tax returns that include capital gains deductions. How many such returns must be randomly selected and checked? We want to be 90% confident that our sample percent is in error by no more than four percentage points.

7-58   You want to estimate the percentage of employees who have been with their current employer for one year or less. Data from the U.S. Department of Labor suggest that this percentage is around 29%, but you must conduct a survey to be 99% confident that your randomly selected sample of employees leads to a sample percentage that is off by no more than two percentage points. How many employees must you survey?

7-59   The *New York Times* reported that of 9261 people interviewed in six polls, 2307 said that they were not registered to vote. Find the 98% confidence interval for the percentage of people who are not registered to vote.

7-60   You want to estimate the proportion of home accident deaths that are caused by falls. How many home accident deaths must you survey in order to be 95% confident that your sample proportion is within 0.04 of the true population proportion?

7-61   You randomly select 650 home accident deaths and you find that 180 of them are caused by falls (based on data from the National Safety Council). Construct the 95% confidence interval for the true population proportion of all home accident deaths caused by falls.

7-62   A survey of 1002 voters by Martilla and Kiley, Inc., showed that 13% of them felt that it is very likely that we will get into a nuclear war within the next 10 years. Construct the 95% confidence interval for the true percentage of all voters who feel that way.

7-63   In an Airport Transit Association poll of 4664 adults, 72% indicated that they have flown in an airplane. Find the 99% confidence interval for the percentage of all adults who have flown in an airplane.

7-64   In a survey of 1500 people, 63% indicated that they listened to their favorite radio station because of the music (based on data from a Strategic Radio Research poll). Construct the 98% interval estimate of the true population percentage of all people who listen to their favorite radio station because of the music.

7-65   You plan to conduct a poll to estimate the percentage of consumers who are satisfied with long-distance phone service. You want to be 90% confident that your sample percentage is within 2.5 percentage points of the true population value, and a Roper poll suggests that this percentage should be about 85%. How large must your sample be?

7-66   A television manufacturer wants to estimate the percentage of TV households that have only black and white televisions. Based on past data collected by the A. C. Nielsen Company, there is reason to believe that the percentage is around 10%. How many TV households must be polled if we want 90% confidence that we are within two percentage points of the true population percentage?

7-67   A pollster is hired to determine the percentage of voters favoring the Republican presidential nominee. If we require 99% confidence that the estimated value is within two percentage points of the true value, how large should the random sample be?

7-68   A *New York Times* article about poll results states that "in theory, in 19 cases out of 20, the results from such a poll should differ by no more than one percentage point in either direction from what would have been obtained by interviewing all voters in the United States." Find the sample size suggested by this statement.

7-69   In a November 27 sobriety checkpoint run by the Dutchess County, New York, Sheriff's Department, 676 drivers were screened and 6 were arrested for DWI (driving while intoxicated). Find the 95% confidence interval for the true population proportion of drivers who were DWI.

7-70   A Roper poll of 1998 adults resulted in 53% saying that security was an important aspect of money. Based on this sample data, construct the 95% confidence interval for the true percentage of the entire adult population.

7-71   Researchers at the University of California at Berkeley interviewed 765 children aged 11 and 12. Of those interviewed, 81% indicated that they would like to spend more time in activities with their parents. Find the 96% confidence interval for the true percentage of all children aged 11 and 12 who feel that way.

7-72   Nordhaus Research, Inc., conducted interviews with 2000 people and found that 68% of them favor laws requiring seat belt use. Construct the 97% interval estimate for the true population percentage.

7-73   A multiple-choice test question is considered easy if at least 80% of the responses are correct. A sample of 6503 responses to one question indicates that 5463 of those responses were correct. Construct the 99% confidence interval for the true proportion of correct responses. Is it likely that this question is really easy?

7-74   Of 600 people who completed the first item on a questionnaire, 24% responded "always," 60% responded "sometimes," and the others responded "never." Construct the 99% confidence interval for the true proportion of people who respond "never."

7-75   Among New Yorkers over 18 years of age, 51.1% voted in a recent presidential election (based on data from the Committee for the Study of the American Electorate). If that figure is an estimate based on a sample of 5000, construct the 90% confidence interval for the true percentage.

7-76   Of 281 aviation accidents, 95 resulted in fatalities (based on data from the U.S. Department of Transportation). Using this sample data, find the 98% confidence interval for the true proportion of all aviation accidents that result in fatalities.

7-3 EXERCISES B

7-77   A rectangle has sides of length p and q, and the perimeter is two units. What values of p and q will cause the rectangle to have the largest possible area? To what concept of Section 7-3 does this problem relate?

7-78   In this section we developed two formulas used for determining sample size, and in both cases we assume that the population is infinite, or we are sampling with replacement, or the population is very large. If we have a relatively small population and we sample without replacement, we should modify E to include the finite population correction factor as follows:

$$E = z_{\alpha/2} \sqrt{\frac{\hat{p}\hat{q}}{n}} \sqrt{\frac{N - n}{N - 1}}$$

where N is the size of the population.

a. Show that the above expression can be solved for n to yield

$$n = \frac{N\hat{p}\hat{q}[z_{\alpha/2}]^2}{\hat{p}\hat{q}[z_{\alpha/2}]^2 + (N - 1)E^2}$$

b. Do Exercise 7-57 assuming that there is a finite population of size $N = 500$ tax returns and sampling is done without replacement.

7-79   A newspaper article indicates that an estimate of the unemployment rate involves a sample of 47,000 people. If the reported unemployment rate must have an error no larger than 0.2 percentage point and the rate is known to be aobut 8%, find the corresponding confidence level.

7-80   a. If IQ scores of adults are normally distributed with a mean of 100 and a standard deviation of 15, use the methods of Chapter 5 to find the percentage of IQ scores above 130.

     b. Now assume that you plan to test a sample of adults with the intention of estimating the percentage of IQ scores above 130. How many adults must you test if you want to be 98% confident that your error is no more than 2.5 percentage points? (Use the result from part (a).)

7-4 ESTIMATES AND SAMPLE SIZES OF VARIANCES (Optional)

In Section 6-6 we considered tests of hypotheses made about population variances or standard deviations. We noted that many real and practical situations, such as quality control in a manufacturing process, require inferences about variances or standard deviations. In addition to making products with good average quality, the manufacturer must make products of *consistent* quality that do not run the gamut from extremely poor to extremely good. This consistency can be measured by variance and standard deviation, so these statistics become important in maintaining the quality of products. There are many other situations in which variance and standard deviation are critically important.

As in Section 6-6 we assume here that the population in question has normally distributed values. This assumption is again a strict requirement since the chi-square distribution is so sensitive to departures from normality that gross errors can easily arise. We describe this sensitivity by saying that inferences about σ^2 (or σ) made on the basis of the chi-square distribution are not *robust* against departures from normality. In contrast, inferences made about μ based on the Student t distribution are reasonably robust since departures from normality that are not too extreme will not lead to gross errors.

In this section we extend the concepts of the previous sections to variances. These concepts are point estimates, confidence intervals, and sample size determination. (Because of the nature of the chi-square distribution, the techniques of this section will not closely parallel those of the preceding two sections.)

Since sample variances tend to center on the value of the population variance, we say that s^2 is an unbiased estimator of σ^2. Also, the variance of s^2 values tends to be smaller than the variance of the other unbiased estimators. For these reasons we decree that, among the various possible statistics we could use to estimate σ^2, the best is s^2.

DEFINITION

The sample variance s^2 is the best **point estimate of the population variance σ^2**.

Since s^2 is the best point estimate of σ^2, it would be natural to expect that s is the best point estimate of σ, but this is not the case. For reasons we will not pursue, s is a biased estimator of σ; if the sample size is large, however, the bias is small so that we can use s as a reasonably good estimate of σ.

Although s^2 is the best point estimate of σ^2, no indication exists of how good this best estimate is. To compensate for that deficiency, we develop a more informative interval estimate (or confidence interval).

Recall that in Section 6-6 we tested hypotheses about σ^2 by using the test statistic $(n - 1)s^2/\sigma^2$, which has a chi-square distribution. In Figure 7-4 we illustrate the chi-square distribution of $(n - 1)s^2/\sigma^2$ for samples of size $n = 20$. We also show the two critical values of χ^2

Commercials, Commercials, Commercials

Who regulates or tests advertising claims? Television networks have their own clearance departments, which require either verification or changes in commercials. The National Advertising Division, a branch of the Council of Better Business Bureaus, also investigates advertising claims. The Federal Trade Commission and local district attorneys also get into the act. As a result of these efforts, Firestone had to drop a claim that its tires resulted in 25% faster stops and Warner Lambert had to spend $10 million informing customers that Listerine doesn't prevent or cure colds. However, many deceptive ads are voluntarily discontinued as soon as any inquiry is started. Others escape scrutiny simply because the regulatory mechanisms cannot keep up with the thousands of claims that come and go.

corresponding to a 95% level of significance. In a normal or t distribution, the left and right critical values are the same numbers with opposite signs, but from Figure 7-4 we can see that this is not the case with the chi-square distribution. Because we will use those values in developing confidence intervals for standard deviations and variances, we introduce the following notation.

NOTATION

With a total area of α divided equally between the two tails of a chi-square distribution, χ_L^2 denotes the left-tailed critical value and χ_R^2 denotes the right-tailed critical value. (See Figure 7-5.)

A sample of 20 scores corresponds to 19 degrees of freedom, and if we refer to Table A-4 we find that the critical values shown in Figure 7-4 separate areas to the right of 0.025 and 0.975.

Figure 7-4 shows that, for a sample of 20 scores taken from a normally distributed population, the statistic $(n - 1)s^2/\sigma^2$ has a 0.95 probability of falling between 8.907 and 32.852. In general, there is a probability of $1 - \alpha$ that the statistic $(n - 1)s^2/\sigma^2$ will fall between χ_L^2 and χ_R^2. In other words (and symbols), there is a $1 - \alpha$ probability that both of the following are true:

$$\frac{(n - 1)s^2}{\sigma^2} < \chi_R^2 \quad \text{and} \quad \frac{(n - 1)s^2}{\sigma^2} > \chi_L^2$$

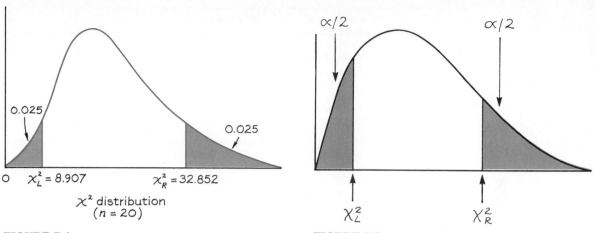

FIGURE 7-4   FIGURE 7-5

If we multiply both of the preceding inequalities by σ^2 and divide each inequality by the appropriate critical value of χ^2, we see that the two inequalities can be expressed in the equivalent forms

$$\frac{(n-1)s^2}{\chi_R^2} < \sigma^2 \quad \text{and} \quad \frac{(n-1)s^2}{\chi_L^2} > \sigma^2$$

These last two inequalities can be combined into one inequality,

$$\frac{(n-1)s^2}{\chi_R^2} < \sigma^2 < \frac{(n-1)s^2}{\chi_L^2}$$

There is a probability of $1 - \alpha$ that the population variance σ^2 is contained in the above interval.

These results provide the foundation for the following definition, which applies to normally distributed populations.

DEFINITION

The **confidence interval** (or **interval estimate**) for the population variance σ^2 is given by

$$\frac{(n-1)s^2}{\chi_R^2} < \sigma^2 < \frac{(n-1)s^2}{\chi_L^2}$$

The confidence interval (or interval estimate) for σ is found by taking the square root of each component of the preceding inequality:

$$\sqrt{\frac{(n-1)s^2}{\chi_R^2}} < \sigma < \sqrt{\frac{(n-1)s^2}{\chi_L^2}}$$

EXAMPLE

A bank experiments with a single waiting line that feeds all windows as openings occur. An employee observes 30 randomly selected customers, and their waiting times produce a standard deviation of 3.8 min. Construct the 95% confidence interval for the true value of σ^2.

Solution

With a sample size $n = 30$, we get $n - 1 = 29$ degrees of freedom. Since we seek 95% confidence, we divide the 5% chance of error between the two tails so that each tail contains a proportion of 0.025. In the 29th row of Table A-4, we find that 0.025 in the left and right tails indicates that $\chi_L^2 = 16.047$ and $\chi_R^2 = 45.722$. With these values of χ_L^2 and χ_R^2, with $n = 30$, and with $s = 3.8$, we use the preceding definition to obtain

$$\frac{(30 - 1)(3.8)^2}{45.722} < \sigma^2 < \frac{(30 - 1)(3.8)^2}{16.047}$$

which becomes $9.159 < \sigma^2 < 26.096$.

If the previous problem had required the 95% confidence interval for σ instead of σ^2, we could have taken the square root of each component to get $\sqrt{9.159} < \sigma < \sqrt{26.096}$, which can then be expressed as $3.026 < \sigma < 5.108$.

The problem of determining the sample size necessary to estimate σ^2 to within given tolerances and confidence levels becomes much more complex than it was in similar problems that dealt with means and proportions. Instead of developing very complicated procedures, we supply Table 7-1, which lists approximate sample sizes.

EXAMPLE

You wish to estimate σ^2 to within 10% and you need 99% confidence in your results. How large should your sample be? Assume that the population is normally distributed.

Solution

From Table 7-1, 99% confidence and an error of 10% for σ^2 correspond to a sample of size 1400. You should randomly select 1400 values from the population.

TABLE 7-1	
To be 95% confident that s^2 is within	of the value of σ^2, the sample size n should be at least
1%	77,210
5%	3,150
10%	806
20%	210
30%	97
40%	57
50%	38
To be 99% confident that s^2 is within	of the value of σ^2, the sample size n should be at least
1%	133,362
5%	5,454
10%	1,400
20%	368
30%	172
40%	101
50%	67
To be 95% confident that s is within	of the value of σ, the sample size n should be at least
1%	19,205
5%	767
10%	192
20%	47
30%	21
40%	12
50%	8
To be 99% confident that s is within	of the value of σ, the sample size n should be at least
1%	33,196
5%	1,335
10%	336
20%	85
30%	38
40%	22
50%	14

7-4 EXERCISES A

7-81 a. If a sample is described by the statistics $n = 100$, $\bar{x} = 146$, and $s^2 = 12$, find the best point estimate of σ^2.

b. Use Table 7-1 to find the approximate minimum sample size necessary to estimate σ^2 with a 30% maximum error and 95% confidence.

c. Find the χ_L^2 and χ_R^2 values for a sample of 25 scores and a confidence level of 99%.

7-82 a. If a sample is described by the statistics $n = 1087$, $\bar{x} = 77.3$, and $s = 4.0$, find the best point estimate of σ^2.

b. Use Table 7-1 to find the approximate minimum sample size needed to estimate σ with a 10% maximum error and 95% confidence.

c. Find the χ_L^2 and χ_R^2 values for a sample of 15 scores and a confidence level of 95%.

7-83 a. Find the χ_L^2 and χ_R^2 values for a sample of 11 scores and a confidence level of 95%.

b. Use Table 7-1 to find the approximate minimum sample size needed to estimate σ with a 5% maximum error and 99% confidence.

c. Find the best point estimate of σ^2 based on a sample for which $n = 17$, $\bar{x} = 69.2$, and $s = 1.2$.

7-84 a. Find the χ_L^2 and χ_R^2 values for a sample of 27 scores and a confidence level of 90%.

b. Use Table 7-1 to find the approximate minimum sample size needed to estimate σ with a 10% maximum error and 99% confidence.

c. Find the best point estimate of σ^2 based on a sample for which $n = 6$, $\bar{x} = 428.2$, and $s = 1.9$.

7-85 The statistics $n = 30$, $\bar{x} = 16.4$, and $s = 2.5$ are obtained from a random sample drawn from a normally distributed population. Construct the 95% confidence interval about σ^2.

7-86 The statistics $n = 16$, $\bar{x} = 12.37$, and $s = 1.05$ are obtained from a random sample drawn from a normally distributed population. Construct the 99% confidence interval about σ.

7-87 Construct a 95% confidence interval about σ^2 if a random sample of 21 scores is selected from a normally distributed population and the sample variance is 100.

7-88 Construct a 95% confidence interval about σ^2 if a random sample of 10 scores is selected from a normally distributed population and the sample variance is 225.

7-89 In a study of the relationships between work schedules and family life, 29 subjects work at night. Their times (in hours) spent in child care are measured and the mean and standard deviation are 26.84 h and 17.66 h, respectively. (See "Nonstandard Work Schedules and Family Life," by Staines and Pleck, *Journal of Applied Psychology*, Vol. 69, No. 3.) Use this sample data to construct the 95% confidence interval for the standard deviation of the times in child care for all night workers.

7-90 Car owners were randomly selected and asked about the length of time they plan to keep their cars. The sample mean and standard deviation are 7.01 years and 3.74 years, respectively (based on data from a Roper poll). Assume that the sample size is 100 and construct the 95% confidence interval for the population standard deviation.

7-91 When working high school students are randomly selected and surveyed about the time they work at after-school jobs, the mean and standard deviation are found to be 17.6 h and 9.3 h, respectively (based on data from the National Federation of State High School Associations). Assume that this data is from a sample of 50 subjects and construct the 99% confidence interval about the standard deviation for all working high school students.

7-92 A study involves 40 randomly selected cases of drivers arrested for intoxication, and their blood alcohol concentrations produce a standard deviation of 0.0306. Construct a 99% confidence interval for the standard deviation of all such arrested drivers.

7-93 A researcher finds that 40 randomly selected business students have computer connect times (in hours) that yield a variance of 77.6. Find the 99% confidence interval for the true standard deviation of all such business students at the same college.

7-94 An environmentalist randomly selects the pollution indices for 30 different days in the past five years and computes the variance as 91.3. Find the 98% confidence interval for the standard deviation of indices for all days within the past five years.

7-95 In a study of the ages of Dutchess County (New York) residents, 100 residents are randomly selected and surveyed, with the result that $\bar{x} = 31.7$ years and $s = 21.8$ years (based on data from the *Cornell Community and Resource Development Series*, Bulletin 6-13). Construct the 90% confidence interval for the standard deviation of the ages of all Dutchess County residents.

7-96 When 40 seventeen-year-old women are randomly selected and tested with the Goodenough-Harris Drawing Test, their mean and standard deviation are found to be 37.9 and 7.3, respectively (based on data from the National Health Survey, USDHEW publication 74-1620). Construct the 95% confidence interval for the standard deviation of the test scores for all seventeen-year-old women.

7-97 The following reaction times (in seconds) are randomly obtained from a normally distributed population: 0.60, 0.61, 0.63, 0.72, 0.91, 0.72. Find the 99% confidence interval for σ.

7-98   The following weights (in grams) are randomly obtained from a normally distributed population: 201, 203, 212, 222, 213, 215, 217, 230, 205, 208, 217, 225. Find the 95% confidence interval for σ.

7-99   In a study of lung capacities, 10 subjects were measured for their residual lung volumes and the following results (in liters) were obtained. (See "Validation of Esophageal Balloon Technique at Different Lung Volumes and Postures," by Baydur, Cha, and Sassoon, *Journal of Applied Physiology*, Vol. 62, No. 1.) Construct the 95% confidence interval about σ^2.

1.36	2.02	1.59	1.17	1.10
3.16	0.77	1.31	1.87	1.86

7-100   In a study of the use of hypnotism to relieve pain, sensory ratings were measured for 16 subjects with the results given below (based on data from "An Analysis of Factors that Contribute to the Efficacy of Hypnotic Analgesia," by Price and Barber, *Journal of Abnormal Psychology*, Vol. 96, No. 1). Use this set of sample data to construct the 95% confidence interval for the standard deviation of the sensory ratings for the population from which the sample was drawn.

8.8	6.6	8.4	6.5	8.4	7.0	9.0	10.3
8.7	11.3	8.1	5.2	6.3	11.6	6.2	10.9

7-4 EXERCISES B

7-101   A random sample is drawn from a normally distributed population and it is found that $n = 20$, $\bar{x} = 45.2$, and $s = 3.8$. Based on this sample, the confidence interval

$$2.8 < \sigma < 6.0$$

is constructed. Find the degree of confidence.

7-102   A random sample of 12 scores is drawn from a normally distributed population and the 95% confidence interval is found to be

$$19.1 < \sigma < 45.8$$

Find the standard deviation of the sample.

7-103   In constructing confidence intervals for σ or σ^2, we use Table A-4 to find χ_L^2 and χ_R^2, but that table applies only to cases where $n \leq 101$ so that the number of degrees of freedom is 100 or less. For large numbers of degrees of freedom, we can approximate χ_L^2 and χ_R^2 by

$$\chi^2 = \frac{1}{2}[\pm \ z_{\alpha/2} + \sqrt{2k - 1} \ ]^2$$

where k = number of degrees of freedom and $z_{\alpha/2}$ is as described in the preceding sections. (See also Exercise 6-137.) Construct the 95% confidence interval about σ by using the following sample data: For 772 males between the ages of 18 years and 24 years, their measured heights have a mean of 69.7 in. and a standard deviation of 2.8 in. (based on data from the National Health Survey, USDHEW publication 79-1659).

7-104   In a study of the factors that affect success in a calculus course, 235 students were given an algebra pretest. The mean and standard deviation for this sample group are 18.4 and 5.1, respectively. (See "Factors Affecting Achievement in the First Course in Calculus," by Edge and Friedberg, *Journal of Experimental Education*, Vol. 52, No. 3.) Construct the 99% confidence interval for σ. (See Exercise 7-103.)

VOCABULARY LIST

Define and give an example of each term.

point estimate   confidence interval   degree of confidence
maximum error of   interval estimate   confidence interval
   estimate   limits

REVIEW

In this chapter we continued our study of inferential statistics by introducing the concepts of **point estimate**, **confidence interval** (or **interval estimate**), and ways of determining the **sample size** necessary to estimate parameters to within given error factors. In Chapter 6 we used sample data to make **decisions** about hypotheses, but the central concern of this chapter is the **estimate** of parameter values. The parameters are population means, proportions, and variances.

The table on the next page summarizes some of the key results of this chapter.

IMPORTANT FORMULAS

Parameter	Point estimate	Confidence interval	Sample size
μ	$\bar{x}$	$\bar{x} - E < \mu < \bar{x} + E$ where $E = z_{\alpha/2}\dfrac{\sigma}{\sqrt{n}}$ (if σ is known or if $n > 30$, in which case we use s for σ) or $E = t_{\alpha/2}\dfrac{s}{\sqrt{n}}$ (if σ is unknown and $n \leq 30$)	$n = \left[\dfrac{z_{\alpha/2}\sigma}{E}\right]^2$
p	$\hat{p} = \dfrac{x}{n}$	$\hat{p} - E < p < \hat{p} + E$ where $E = z_{\alpha/2}\sqrt{\dfrac{\hat{p}\hat{q}}{n}}$	$n = \dfrac{[z_{\alpha/2}]^2\hat{p}\hat{q}}{E^2}$ or $n = \dfrac{[z_{\alpha/2}]^2 \cdot 0.25}{E^2}$
σ^2	s^2	$\dfrac{(n-1)s^2}{\chi_R^2} < \sigma^2 < \dfrac{(n-1)s^2}{\chi_L^2}$	See Table 7-1

REVIEW EXERCISES

7-105   Given $n = 60$, $\bar{x} = 83.2$ kg, $s = 4.1$ kg. Assume that the given statistics represent sample data randomly selected from a normally distributed population.
a. What is the best point estimate of μ?
b. Construct the 95% confidence interval about μ.

7-106   Use the sample data given in Exercise 7-105.
a. What is the best point estimate of σ^2?
b. Construct the 95% confidence interval about σ.

7-107   A newspaper editor plans to supervise a survey of regional opinions. How many people must be surveyed if, for the first question, she wants to be 94% confident that the percent of respondents saying "no" is in error by no more than five percentage points?

7-108   Of 1533 adults surveyed in a Gallup poll, 47% correctly identified Freud. Construct a 99% confidence interval for the true proportion of adults who can correctly identify Freud.

7-109   A medical researcher wishes to estimate the serum cholesterol level (in mg/100 ml) of all women aged 18 to 24. There is strong evidence suggesting that $\sigma = 41.0$ mg/100 ml (based on data from a survey of 1524 women aged 18 to 24, as part of the National Health Survey, USDHEW publication 78-1652). If the researcher wants to be 95% confident in obtaining a sample mean that is off by no more than four units, how large must the sample be?

7-110   A sociologist develops a test designed to measure a person's attitudes about disabled people, and 16 randomly selected subjects are given the test. Their mean is 71.2, while their standard deviation is 10.5. Construct the 99% confidence interval for the mean score of all subjects.

7-111   Using the sample data in Exercise 7-110, construct the 99% confidence interval for the standard deviation of the scores of all subjects.

7-112   A sociologist wants to determine the mean value of cars owned by retired people. If the sociologist wants to be 96% confident that the mean of the sample group is off by no more than $250, how many retired people must be sampled? A pilot study suggests that the standard deviation is $3050.

7-113   In a Roper survey of 308 adults, 17% listed jewelry among the romantic gifts men want to receive. Construct the 95% confidence interval for the proportion of all adults who include jewelry among their responses.

7-114   A psychologist wants to determine the proportion of students in a large school district who have divorced parents. How many students must be surveyed if the psychologist wants 96% confidence that the sample proportion is in error by no more than 0.06?

7-115   a. Evaluate $z_{\alpha/2}$ for $\alpha = 0.10$.
   b. Evaluate χ^2_L and χ^2_R for $\alpha = 0.05$ and a sample of 10 scores.
   c. Evaluate $t_{\alpha/2}$ for $\alpha = 0.05$ and a sample of 10.
   d. What is the largest possible value of $p \cdot q$?

7-116   Given $n = 16$, $\bar{x} = 83.2$ kg, $s = 4.1$ kg. Assume that the given statistics represent sample data randomly selected from a normally distributed population.
   a. What is the best point estimate of μ?
   b. Construct the 95% confidence interval about μ.

7-117   Use the sample data given in Exercise 7-116.
   a. What is the best point estimate of σ^2?
   b. Construct the 95% confidence interval about σ.

7-118   Based on recent data from the U.S. Bureau of the Census, the proportion of Americans below the poverty level is 0.140. A researcher wants to verify that figure by conducting an independent survey. Assuming that 0.140 is approximately correct, how many randomly selected Americans must be surveyed? The researcher wants to be 96% confident that the sample proportion is within 0.015 of the true population proportion.

7-119   In a Roper survey of 1998 adults, 24% included loud commercials among the annoying aspects of television. Construct the 99% confidence interval for the proportion of all adults who are annoyed by loud commercials.

7-120   A botanist wants to determine the mean diameter of pine trees in a forest. She conducts a preliminary study to establish that the standard deviation of all such trees is about 6.35 cm. How many randomly selected trees must be measured if she wants 98% confidence that the sample mean is in error by no more than 0.5 cm?

7-121   The president of the student body at a very large university wants to determine the percent of students who are registered voters. How many students must be surveyed if we want 90% confidence that the sample is in error by no more than five percentage points?

7-122   a. Evaluate $z_{\alpha/2}$ for a confidence level of 96%.
   b. Evaluate $t_{\alpha/2}$ for a confidence level of 99% and a sample size of 16.
   c. Evaluate χ^2_L and χ^2_R for a confidence level of 99% and a sample size of 20.
   d. For the same set of data, confidence intervals are constructed for the 95% and 99% confidence levels. Which interval has limits that are farther apart?

7-123 A psychologist is collecting data on the time it takes to learn a certain task. For 50 randomly selected adult subjects, the sample mean and standard deviation are computed to be 16.40 min and 4.00 min, respectively. Construct the 98% interval estimate for the mean time required by all adults.

7-124 Using the sample data from Exercise 7-123, construct the 99% confidence interval for σ.

7-125 An advertising firm wants to estimate the mean time spent by preschool children watching television on Saturday morning. A pilot study suggests that $\sigma = 0.8$ h. How many subjects must be surveyed for 98% confidence that the sample mean is off by no more than 0.02 h?

7-126 In a Gallup poll of 1004 adults, 93% indicated that restaurants and bars should refuse service to patrons who have had too much to drink. Construct the 98% confidence interval for the proportion of all adults who feel the same way.

7-127 Listed here are the measured total lung capacities (in liters) for 10 subjects (based on data from "Validation of Esophageal Balloon Techniques at Different Lung Volumes and Postures," by Baydur, Cha, and Sassoon, *Journal of Applied Physiology*, Vol. 62, No. 1). Construct the 95% confidence interval for the mean total lung capacity of the population from which this sample was selected.

| 6.04 | 8.36 | 6.96 | 4.51 | 4.87 |
| 10.59 | 4.92 | 7.52 | 7.77 | 7.63 |

7-128 Use the sample data in Exercise 7-127 to determine the 95% confidence interval for the standard deviation of the population from which the sample was drawn.

COMPUTER PROJECT

1. We know from this chapter that the formula

$$n = \frac{[z_{\alpha/2}]^2 \cdot 0.25}{E^2} = \frac{[1.96]^2 \cdot 0.25}{E^2} = \frac{0.9604}{E^2}$$

can be used to determine the sample size necessary to estimate a population proportion to within a maximum error E, and that result will correspond to a 95% confidence level. Develop a computer program that takes the value of E as input. Output should consist of the sample size corresponding to a 95% confidence level.

Run the program to determine the sample sizes corresponding to a variety of different values of E ranging from 0.001 to 0.150.

2. Computer software packages designed for statistics (such as STAT-DISK or Minitab) commonly provide programs for generating confidence intervals. Use such a program to find the confidence intervals referred to in Exercises 7-13, 7-37, 7-59, and 7-91.

CASE STUDY ACTIVITY

Determine the sample size needed to estimate the mean weight of a passenger car in your region. The calculation of that number requires the value of the population standard deviation, which is unknown. Conduct a preliminary pilot study by finding the weights of at least 30 different passenger cars. Those weights can usually be found on vehicle registration documents. List the year and make of each vehicle and identify any factors suggesting that your sample is not representative of the population of all passenger cars in your region. Calculate the standard deviation of the sample weights and use the result as an estimate of the population standard deviation so that the required sample size can be determined. Assume that you want to be 98% confident that the sample mean is within 25 lb of the true population mean.

DATA PROJECT

Refer to the data sets in Appendix B.

1. Construct 95% and 99% confidence intervals for the mean home selling price.

2. Construct 95% and 99% confidence intervals for the proportion of people who are left-handed.

3. Construct 95% and 99% confidence intervals for the living areas of homes recently sold in Dutchess County.

4. You want to estimate the mean selling price of all homes in Dutchess County. Use the sample of 150 selling prices in order to estimate the value of the standard deviation. Then use that result to find the required sample size. Assume that you want to be 95% confident that you are within $5000 of the population mean.

Chapter Eight

CHAPTER CONTENTS

8-1 Overview
We identify chapter **objectives**. This chapter deals with methods of testing hypotheses made about two population parameters.

8-2 Tests Comparing Two Variances
We present the method for testing hypotheses made about **two population variances** or standard deviations.

8-3 Tests Comparing Two Means
We present methods for testing hypotheses made about **two dependent means** or **two independent means** with equal or unequal variances.

8-4 Tests Comparing Two Proportions
We use the pooled estimate of p_1 and p_2 to test hypotheses made about the **population proportions** p_1 and p_2.

8

Tests Comparing Two Parameters

CHAPTER PROBLEM

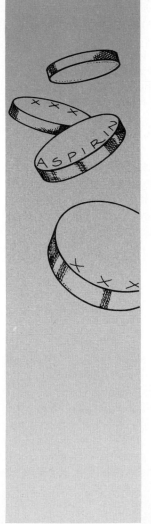

In a recent study of 22,000 male physicians, half were given regular doses of aspirin while the other half were given placebos with no effects. The study ran for six years at a cost of $4.4 million. Among those who took the aspirin doses, 104 suffered heart attacks. Among those who took the placebos, 189 suffered heart attacks. (The figures are based on data from *Time* and the *New England Journal of Medicine*, Vol. 318, No. 4.) Several criticisms and disagreements and many words of caution followed the report of these results, but we will focus on the central issue of statistics: Do these results show a significant decrease in heart attacks among the sample group who took aspirin? The issue is clearly quite important since it can affect many lives, not to mention the sales of aspirin.

In this chapter we will test the claim that the proportion of heart attacks among the aspirin group is significantly lower than the proportion for the placebo group.

8-1   OVERVIEW

In Chapter 6 we introduced the method of hypothesis testing, but that chapter involved only tests of claims made about a single population parameter. In this chapter we extend the methods of hypothesis testing to cases involving the comparison of two variances, two means, or two proportions. In reality, there are many cases where the main objective is a comparison of *two* groups of data instead of a comparison of *one* group to some known value. For example, a manufacturer may want to know which of two different production techniques yields better results. A college dean may be interested in the differences between the grades given at two different schools. A psychologist may need to know whether two different IQ tests produce similar or different results. A doctor may be interested in the comparative effectiveness of two different cold medicines. A sociologist may want to compare last year's college entrance examination scores to those of 20 years ago. A psychologist may wish to compare variability in reaction times between a group of nondrinkers and a group of people who have each consumed a double martini. In all of these cases, we want to compare two population parameters.

Chapters 6 and 7 started with means and then moved on to proportions, followed by variances. Here we depart from this order because we will sometimes use results of tests on two variances as a *prerequisite* for a test involving two means. Consequently, we begin this chapter with a discussion of the method for testing hypotheses involving a comparison of two variances.

8-2   TESTS COMPARING TWO VARIANCES

In many cases where we want to compare parameters from each of two separate populations, the **variances** are the most relevant parameters. (Recall that the variance is the square of the standard deviation.) For example, a manufacturer of automobile batteries may want to compare two different production methods of batteries that last, on the average, approximately four years. Consider the case in Table 8-1.

We assume throughout this section that we have two independent populations that are approximately normally distributed. A sample of size n_1 is drawn from the first population, while a sample of size n_2 is drawn from the second. Since we want to compare variances, we compute the respective sample variances s_1^2 and s_2^2 and use those statistics in testing for equality of σ_1^2 and σ_2^2.

TABLE 8-1 Life (in years) of car batteries	
Production method A	Production method B
2.0	3.7
2.1	3.9
2.5	3.9
3.0	3.9
3.3	4.0
4.2	4.0
4.2	4.0
4.3	4.1
6.8	4.2
7.6	4.3
$\bar{x}_1 = 4.00$	$\bar{x}_2 = 4.00$
$s_1^2 = 3.59$	$s_2^2 = 0.03$

From Table 8-1 we see that both methods seem to produce batteries that last the same length of time, but the batteries produced by method A exhibit much less consistency. The two production methods appear to differ radically in the variability of the battery lives, and this is reflected in the large difference between the two sample variances. Thus it is through a comparison of the two *variances* that we find the crucial difference between the two production methods. Since not all comparisons of variances involve such obvious differences, we need more standardized and objective procedures. Even for the data in Table 8-1, the difference between the variances of 3.59 and 0.03 must be weighed against the sample sizes to determine whether this "obvious" difference is statistically significant.

We stipulate that

s_1^2 represents the larger of the two sample variances

and s_2^2 represents the smaller of the two sample variances. We can do this because identification of the samples through subscript notation is arbitrary. That is, we can identify either data set as group 1 while the other becomes group 2. It will simplify computations if we identify the data set with the larger sample variance as group 1.

Extensive analyses have shown that **for two normally distributed populations with equal variances (that is, $\sigma_1^2 = \sigma_2^2$), the sampling distribution of**

$$F = \frac{s_1^2}{s_2^2}$$

is the *F* distribution shown in Figure 8-1 and described in Table A-5.

If the two populations really do have equal variances, then $F = s_1^2/s_2^2$ tends to be close to 1 since s_1^2 and s_2^2 tend to be close in value. But if the two populations have radically different variances, s_1^2 and s_2^2 tend to be very different numbers. Denoting the larger of the sample variances by s_1^2, we see that the ratio s_1^2/s_2^2 will be a large number whenever s_1^2 and s_2^2 are far apart in value. Consequently, a value of *F* near 1 will be evidence in favor of the conclusion that $\sigma_1^2 = \sigma_2^2$. A large value of *F* will be evidence against the conclusion of equality of the population variances. The critical *F* values are summarized in Table A-5.

When we use Table A-5 we obtain critical *F* values that are determined by the following three values:

1. The significance level α
2. The **degrees of freedom for the numerator**, $(n_1 - 1)$
3. The **degrees of freedom for the denominator**, $(n_2 - 1)$

When using Table A-5, be sure that n_1 corresponds to the sample having variance s_1^2, while n_2 is the size of the sample with variance s_2^2. Identify a level of significance and determine whether the test is one-tailed or two-tailed. For a one-tailed test use the significance level found in Table A-5. (Since we stipulate that the larger sample variance is s_1^2, all one-tailed tests will be right-tailed.) For a two-tailed test, first divide the area of the critical region (equal to the significance level) equally between the two tails and refer to that part of Table A-5 that represents *one-half* of the significance level. In that part of Table A-5,

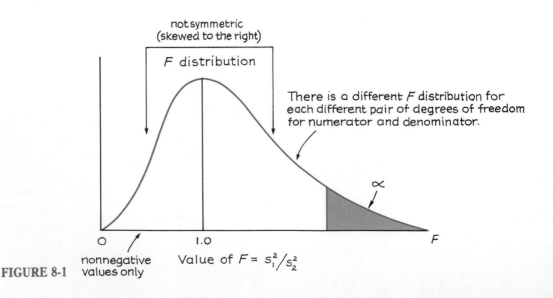

FIGURE 8-1

The Power of Your Vote

In the electoral college system, the power of a voter in a large state exceeds that of a voter in a small state. When voting power is measured as the ability to affect the outcome of an election, we see that a New Yorker has 3.312 times the voting power of a resident of the District of Columbia. This result is included in John Banzhof's article, "One Man, 3.312 Votes."

As an example, the outcome of the 1916 Presidential election could have been changed by shifting only 1,983 votes in California. If the same number of votes were changed in a much smaller state, the resulting change in electoral votes would not have been sufficient to alter the outcome.

intersect the column representing the degrees of freedom for s_1^2 with the row representing the degrees of freedom for s_2^2. (Unlike the normal and Student t distributions, the F distribution is not symmetric and does not have 0 at its center. Consequently, left-tail critical values *cannot* be found by using the negative of the right-tail critical values. Instead, left-tail critical values can be found by using the reciprocal of the right-tail value with the numbers of degrees of freedom reversed. See Exercise 8-25.)

The following example illustrates the method of testing hypotheses about two population variances.

EXAMPLE

A department store manager experiments with two methods for checking out customers, and the following sample data are obtained. At the 0.02 significance level, test the claim that $\sigma_1^2 = \sigma_2^2$. Assume that the sample data came from normally distributed populations.

Sample A	Sample B
$n_1 = 16$	$n_2 = 10$
$s_1^2 = 225$	$s_2^2 = 100$

Solution

The solution is summarized in Figures 8-2 and 8-3. With the claim of $\sigma_1^2 = \sigma_2^2$ and the alternative of $\sigma_1^2 \neq \sigma_2^2$, we have a two-tailed test, so the significance level of $\alpha = 0.02$ leads to an area of 0.01 in the left-tailed critical region and an area of 0.01 in the right-tailed critical region. But as long as we stipulate that the larger of the two sample variances be placed in the numerator, we need to find only the right-tailed critical value. With an area of 0.01 in the right tail, with $n_1 = 16$ and $n_2 = 10$, we locate the critical F value of 4.9621, which corresponds to 15 degrees of freedom in the numerator and 9 degrees of freedom in the denominator. The test statistic is computed as

$$F = \frac{s_1^2}{s_2^2} = \frac{225}{100} = 2.2500$$

From Figure 8-3 we see that the test statistic does not fall within the critical region, so we fail to reject the null hypothesis. There is not sufficient sample evidence to warrant rejection of the claim that the two variances are equal. Refer to Figures 8-2 and 8-3 for the complete hypothesis test.

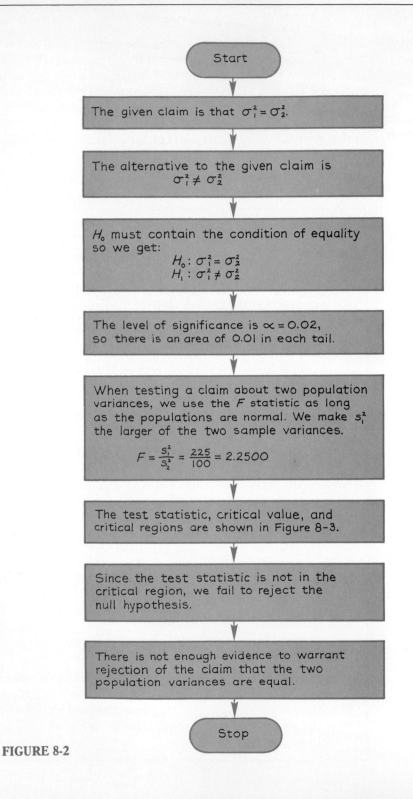

FIGURE 8-2

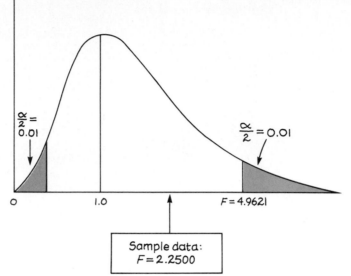

$\frac{\alpha}{2} = 0.01$

$\frac{\alpha}{2} = 0.01$

O   1.0   F = 4.9621

Sample data:
F = 2.2500

FIGURE 8-3

The preceding example involved data chosen so that the calculations would be simple and obvious. The next example involves real data. Because the variances of the sample scores would be very large, we divide all of the original scores by 1000. This causes both variances to be divided by 1,000,000, but the value of the test statistic $F = s_1^2/s_2^2$ is not affected. Since both numerator and denominator are divided by the same number, the result is the same.

EXAMPLE

A prospective home buyer is investigating home selling price differences between zone 1 (southern Dutchess County) and zone 7 (northern Dutchess County). The given random sample data are in thousands of dollars. (This will not affect the value of the test statistic F, but it allows us to work with more manageable numbers.) At the 0.05 significance level, test the claim that both zones have the same variance. Assume that the sample data come from normally distributed populations. (In many cases, home selling prices might not be normally distributed. But histograms show that the assumption of normal distributions is reasonable here.)

continued ▶

Example, continued

Zone 7 (north)		Zone 1 (south)	
270.000		115.000	
107.000		136.900	
148.000		121.000	
125.000		164.000	
127.500	$n = 11$	175.000	
125.500	$\bar{x} = 142.32$	128.500	$n = 14$
126.000	$s^2 = 2122$	147.500	$\bar{x} = 138.24$
109.000		147.000	$s^2 = 455$
113.500		105.000	
147.000		163.750	
167.000		115.000	
		149.165	
		120.500	
		147.000	

Solution

We begin by noting that the larger sample variance (2122) is found in zone 7, so we denote that set of data as group 1. The other zone becomes group 2. We now proceed to test the claim of $\sigma_1^2 = \sigma_2^2$ given these sample results:

Group 1	Group 2	
$n_1 = 11$	$n_2 = 14$	(The larger variance is s_1^2.)
$s_1^2 = 2122$	$s_2^2 = 455$	

This is a two-tailed test, so the significance level of 0.05 leads to an area of 0.025 in the left tail and 0.025 in the right tail. Following the stipulation that the larger of the two variances is placed in the numerator, we need to find only the right-tailed critical value. With an area of 0.025 in the right tail, with $n_1 = 11$ and $n_2 = 14$, we locate the critical F value of 3.2497, which corresponds to 10 degrees of freedom in the numerator and 13 degrees of freedom in the denominator. The test statistic is computed as

$$F = \frac{s_1^2}{s_2^2} = \frac{2122}{455} = 4.6637$$

Since the test statistic of $F = 4.6637$ exceeds the critical value of $F = 3.2497$, we reject the null hypothesis. See Figure 8-4. There is sufficient sample evidence to reject the claim that both zones have the same variance. It appears that the selling prices vary more in zone 7.

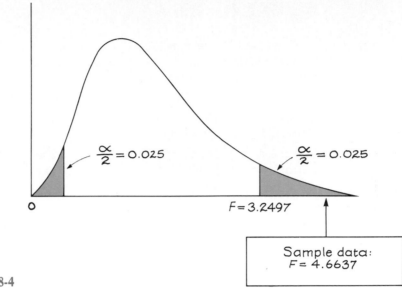

FIGURE 8-4

So far in this section we have discussed only tests comparing two population variances, but the same methods and theory can be used to compare two population standard deviations. Any claim about two population standard deviations can be easily restated in terms of the corresponding population variances. For example, suppose we want to test the claim that $\sigma_1 = \sigma_2$, and we are given the following sample data taken from normal populations.

Sample 1	Sample 2
$n_1 = 16$	$n_2 = 10$
$s_1 = 15$	$s_2 = 10$

Restating the claim as $\sigma_1^2 = \sigma_2^2$, we get $s_1^2 = 15^2 = 225$ and $s_2^2 = 10^2 = 100$, and we can proceed with the F test in the usual way. Rejection of $\sigma_1^2 = \sigma_2^2$ is equivalent to rejection of $\sigma_1 = \sigma_2$. Failure to reject $\sigma_1^2 = \sigma_2^2$ implies failure to reject $\sigma_1 = \sigma_2$.

Note that in all tests of hypotheses made about population variances and standard deviations, the values of the means are irrelevant. In the next section we consider tests comparing two population means, and we will see that those tests sometimes depend on the results of a hypothesis test about population variances. Some of the hypothesis tests of the following section will require the hypothesis tests described in this section.

8-2 EXERCISES A

In Exercises 8-1 through 8-8, test the claim that the two samples come from populations having equal variances. Use a significance level of $\alpha = 0.05$ and assume that all populations are normally distributed. Follow the pattern suggested by Figure 6-1 and draw the appropriate graphs.

8-1   Sample A: $n = 10$, $s^2 = 50$
      Sample B: $n = 10$, $s^2 = 25$

8-2   Sample A: $n = 10$, $s^2 = 50$
      Sample B: $n = 15$, $s^2 = 25$

8-3   Sample A: $n = 20$, $s^2 = 45.2$
      Sample B: $n = 5$, $s^2 = 5.8$

8-4   Sample A: $n = 8$, $s^2 = 8.47$
      Sample B: $n = 12$, $s^2 = 1.33$

8-5   Sample A: $n = 20$, $\bar{x} = 136$, $s^2 = 110$
      Sample B: $n = 25$, $\bar{x} = 142$, $s^2 = 265$

8-6   Sample A: $n = 41$, $\bar{x} = 49.3$, $s^2 = 15.1$
      Sample B: $n = 16$, $\bar{x} = 57.4$, $s^2 = 33.2$

8-7   Sample A: $n = 5$, $\bar{x} = 372$, $s = 14.3$
      Sample B: $n = 15$, $\bar{x} = 298$, $s = 1.1$

8-8   Sample A: $n = 25$, $\bar{x} = 583$, $s = 3.9$
      Sample B: $n = 10$, $\bar{x} = 648$, $s = 6.3$

In Exercises 8-9 through 8-12, test the claim that the variance of population A exceeds that of population B. Use a 5% level of significance and assume that all populations are normally distributed. Follow the pattern outlined in Figure 6-1 and draw the appropriate graphs.

8-9   Sample A: $n = 10$, $\bar{x} = 200$, $s^2 = 48$
      Sample B: $n = 10$, $\bar{x} = 180$, $s^2 = 12$

8-10   Sample A: $n = 50$, $\bar{x} = 75.3$, $s^2 = 18.2$
      Sample B: $n = 20$, $\bar{x} = 75.9$, $s^2 = 8.7$

8-11   Sample A: $n = 16$, $\bar{x} = 124$, $s^2 = 225$
      Sample B: $n = 200$, $\bar{x} = 128$, $s^2 = 160$

8-12   Sample A: $n = 35$, $\bar{x} = 238$, $s^2 = 42.3$
      Sample B: $n = 25$, $\bar{x} = 254$, $s^2 = 16.2$

8-13   On a test of manual strength, a sample of 25 randomly selected men earn scores that have a variance of 130. When 27 randomly selected women take the same test, the variance of their scores is 75. At the 0.02 significance level, test the claim that the population variances are not equal.

8-14   When 15 randomly selected adult males are given a test on reaction times, their scores produce a variance of 1.04. When 17 other randomly selected adult males are given a double martini before taking the same test, their scores produce a variance of 3.26. At the 0.05 significance level, test the claim that the population of all drinkers will have a variance larger than the population of all nondrinkers.

8-15 A scientist wants to compare two delicate weighing instruments by repeated weighings of the same object. The first scale is used 30 times and the weights have a standard deviation of 72 mg. The second scale is used 41 times and the weights have a standard deviation of 98 mg. At the 0.05 significance level, test the claim that the second scale produces greater variance.

8-16 An experiment is devised to study the variability of grading procedures among college professors. Two different professors are asked to grade the same set of 25 exam solutions, and their grades have variances of 103.4 and 39.7, respectively. At the 0.05 significance level, test the claim that the first professor's grading exhibits greater variance.

8-17 In a study of the effect of job previews on work expectation, subjects from different groups were tested. For 60 subjects given specific job previews, the mean promotion expectation score is 19.14 and the standard deviation is 6.56. For a "no booklet" sample group of 40 subjects, the mean is 20.81 and the standard deviation is 4.90. (The data are based on "Effects of Realistic Job Previews on Hiring Bank Tellers," by Dean and Wanous, *Journal of Applied Psychology*, Vol. 69, No. 1.) At the 0.10 level of significance, test the claim that the two sample groups come from populations with the same standard deviation.

8-18 In a study of stress and burnout, men were measured on separate scales. The 273 men achieved burnout scores with a mean of 2.83 and a standard deviation of 0.58. Their work stress scores had a mean of 3.98 and a standard deviation of 0.90. (See "Moderating Effect of Social Support on the Stress-Burnout Relationship," by Etzion, *Journal of Applied Psychology*, Vol. 69, No. 4.) At the 0.10 significance level, test the claim that scores from the two scales come from populations with different standard deviations.

8-19 A study was conducted to investigate relationships between different types of standard test scores. On the Graduate Record Examination Verbal test, 68 females had a mean of 538.82 and a standard deviation of 114.16, while 86 males had a mean of 525.23 and a standard deviation of 97.23. (See "Equivalencing MAT and GRE Scores Using Simple Linear Transformation and Regression Methods," by Kagan and Stock, *Journal of Experimental Education*, Vol. 49, No. 1.) At the 0.02 significance level, test the claim that the two groups come from populations with the same standard deviation.

8-20 As part of the National Health Survey, data were collected on the weights of men. For 804 men aged 25 to 34, the mean is 176 lb and the standard deviation is 35.0 lb. For 1657 men aged 65 to 74, the mean and standard deviation are 164 lb and 27.0 lb, respectively. (The data are based on the National Health Survey, USDHEW publication 79-1659.) At the 0.01 significance level, test the claim that the older men come from a population with a standard deviation less than that for men in the 25 to 34 age bracket.

8-21 The effectiveness of a mental training program was tested in a military training program. In an antiaircraft artillery examination, scores for an experimental group and a control group were recorded. Use the given data to test the claim that both groups come from populations with the same variance. Use a 0.05 significance level. (The data are based on "Routinization of Mental Training in Organizations: Effects on Performance and Well-Being," by Larsson, *Journal of Applied Psychology*, Vol. 72, No. 1.)

Experimental				Control			
60.83	117.80	44.71	75.38	122.80	70.02	119.89	138.27
73.46	34.26	82.25	59.77	118.43	54.22	118.58	74.61
69.95	21.37	59.78	92.72	121.70	70.70	99.08	120.76
72.14	57.29	64.05	44.09	104.06	94.23	111.26	121.67
80.03	76.59	74.27	66.87				

8-22 In a study involving motivation and test scores, data were obtained for females and males. Use the data given below to test the claim that the two groups come from populations with the same standard deviation. Use a 0.02 significance level. (See "Relationships Between Achievement-Related Motives, Extrinsic Conditions, and Task Performance," by Schroth, *Journal of Social Psychology*, Vol. 127, No. 1.)

Female				Male			
31.13	18.71	14.34	23.90	12.27	39.53	32.56	23.93
13.96	13.88	29.85	20.15	19.54	25.73	32.20	19.84
6.66	19.20	15.89		20.20	23.01	25.63	17.98
				22.99	22.12	12.63	18.06

8-23 Sample data were collected in a study of calcium supplements and the effects on blood pressure. A placebo group and a calcium group began the study with measures of blood pressures. At the 0.05 significance level, test the claim that the two sample groups come from populations with the same standard deviation. (See "Blood Pressure and Metabolic Effects of Calcium Supplementation in Normotensive White and Black Men," by Lyle and others, *Journal of the American Medical Association*, Vol. 257, No. 13.)

Placebo				Calcium			
124.6	104.8	96.5	116.3	129.1	123.4	102.7	118.1
106.1	128.8	107.2	123.1	114.7	120.9	104.4	116.3
118.1	108.5	120.4	122.5	109.6	127.7	108.0	124.3
113.6				106.6	121.4	113.2	

8-24 The arrangement of test items was studied for its effect on anxiety. At the 0.05 significance level, test the claim that the two given samples come from populations with the same variance. (See "Item Arrangement, Cognitive Entry Characteristics, Sex and Test Anxiety as Predictors of Achievement Examination Performance," by Klimko, *Journal of Experimental Education*, Vol. 52, No. 4.)

Easy to difficult					Difficult to easy			
24.64	39.29	16.32	32.83		33.62	34.02	26.63	30.26
28.02	33.31	20.60	21.13		35.91	26.68	29.49	35.32
26.69	28.90	26.43	24.23		27.24	32.34	29.34	33.53
7.10	32.86	21.06	28.89		27.62	42.91	30.20	32.54
28.71	31.73	30.02	21.96					
25.49	38.81	27.85	30.29					
30.72								

8-2 EXERCISES B

8-25 For hypothesis tests in this section that were two-tailed, we found only the upper critical value. Let's denote that value by F_R, where the subscript suggests the right side. The lower critical value F_L (for the left side) can be found by first interchanging the degrees of freedom and then taking the reciprocal of the resulting F value found in Table A-5. Find the critical values F_L and F_R for two-tailed hypothesis tests in which

a. $n_1 = 10$, $n_2 = 10$, $\alpha = 0.05$
b. $n_1 = 10$, $n_2 = 7$, $\alpha = 0.05$
c. $n_1 = 7$, $n_2 = 10$, $\alpha = 0.05$
d. $n_1 = 25$, $n_2 = 10$, $\alpha = 0.02$
e. $n_1 = 10$, $n_2 = 25$, $\alpha = 0.02$

8-26 In addition to testing claims involving σ_1^2 and σ_2^2, we can also construct interval estimates of the ratio σ_1^2/σ_2^2. Use

$$\frac{s_1^2}{s_2^2} \cdot F_L < \frac{\sigma_1^2}{\sigma_2^2} < \frac{s_1^2}{s_2^2} \cdot F_R$$

where F_L and F_R are as described in Exercise 8-25. Construct the 95% interval estimate for the ratio of the experimental group variance to the control group variance for the data in Exercise 8-21.

8-27 Sample data consist of temperatures recorded for two 'ifferent groups of items that were produced by two different production techniques. A quality control specialist plans to analyze the results. She begins by testing for equality of the two population standard deviations.

(*continued*)

a. If she adds the same constant to every temperature from both groups, is the value of the test statistic F affected? Explain.
b. If she uses the same constant to multiply every score from both groups, is the value of the test statistic F affected? Explain.
c. If she converts all temperatures from the Fahrenheit scale to the Celsius scale, is the value of the test statistic F affected? Explain.

8-28 a. Two samples of equal size produce variances of 37 and 57. At the 0.05 significance level, we test the claim that the variance of the second population exceeds that of the first, and that claim is upheld by the data. What is the approximate minimum size of each sample?

b. A sample of 21 scores produces a variance of 67.2 and another sample of 25 produces a variance that causes rejection of the claim that the two populations have equal variances. If this test is conducted at the 0.02 level of significance, find the maximum variance of the second sample if you know that it is smaller than that of the first sample.

8-3 TESTS COMPARING TWO MEANS

In this section we consider tests of hypotheses made about two population means. Many real and practical situations use such tests successfully. For example, an educator may want to compare mean test scores produced by two teaching methods. A manager may want to test for a difference in the mean weight of cereal loaded into boxes by two machines. A car manufacturer may want to test for a difference in the mean longevity of batteries produced by two suppliers. A psychologist may want to test for a difference in mean reaction times between men and women. A farmer may want to test for a difference in mean crop production for two irrigation methods. A medical researcher may want to test for a difference between a new drug and one currently in use.

The way in which we compare means using sample data taken from two populations is affected by the presence or absence of a relationship between those samples.

DEFINITION

Two samples are **dependent** if the values in one are related to the values in the other in some way. Two samples are **independent** if the values in one are *not* related to the values in the other.

Consider the sample data given below. We would expect the sample of pretraining weights and the sample of posttraining weights to be two *dependent* samples, since each pair is matched according to the person involved.

Pretraining weights (kg)	99 62 74 59 70
Posttraining weights (kg)	94 62 66 58 70

(The table is based on data from the *Journal of Applied Psychology*, Vol. 62, No. 1.) However, for the data given below, the two samples are *independent* since the sample of females is completely independent of the sample of males. The data are not matched as they are in the table above.

Weights of females (lb)	115 107 110 128 130
Weights of males (lb)	128 150 160 140 163 155 175

(The table is based on data from Appendix B.)

When dealing with two dependent samples, it is very wasteful to reduce the sample data to $\bar{x}_1, s_1, n_1, \bar{x}_2, s_2,$ and n_2 since the relationship between matched pairs of values would be completely lost. Instead, we compute the *differences* (d) between the pairs of data as follows:

x	22 48 27 29 32
y	23 51 25 29 32
$d = x - y$	−1 −3 2 0 0

NOTATION

Let $\bar{d}$ denote the mean value of d or $x - y$ for the paired sample data.

Let s_d denote the standard deviation of the d values for the paired sample data.

Let n denote the number of *pairs* of data.

EXAMPLE

For the d values of $-1, -3, 2, 0, 0$ taken from the preceding table, we get

$$\bar{d} = \frac{\Sigma d}{n} = \frac{(-1) + (-3) + 2 + 0 + 0}{5} = \frac{-2}{5} = -0.4$$

For the d values of $-1, -3, 2, 0, 0$, we get $s_d = 1.8$ as follows.

$$s_d = \sqrt{\frac{\Sigma(d - \bar{d})^2}{n - 1}} = \sqrt{\frac{13.2}{4}} = 1.8$$

For the data of the last table, $n = 5$.

In repeated random sampling from two normal and dependent populations in which the mean of the paired differences is μ_d, the following test statistic possesses a Student t distribution with $n - 1$ degrees of freedom:

TEST STATISTIC

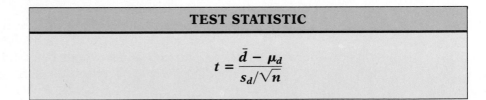

$$t = \frac{\bar{d} - \mu_d}{s_d/\sqrt{n}}$$

Note that the involved populations must be normally distributed. If the populations depart radically from normal distributions, we should not use the methods of this section. Instead, we may be able to apply the sign test (Section 11-2) or the Wilcoxon signed-ranks test (Section 11-3).

If we claim that there is no difference between the two population means, then we are claiming that $\mu_d = 0$. This makes sense if we recognize that $\bar{d}$ should be around zero if there is no difference between the two population means.

In the following example we illustrate a complete hypothesis test for a situation involving dependent samples.

EXAMPLE

A study was conducted to investigate the effectiveness of hypnotism in reducing pain. Results for randomly selected subjects are given below. At the 0.05 significance level, test the claim that the sensory measurements are lower after hypnotism. (The values are before and after hypnosis. The measurements are in centimeters on the mean visual analog scale, and the data are based on "An Analysis of Factors That Contribute to the Efficacy of Hypnotic Analgesia," by Price and Barber, *Journal of Abnormal Psychology*, Vol. 96, No. 1.)

Subject	A	B	C	D	E	F	G	H
Before	6.6	6.5	9.0	10.3	11.3	8.1	6.3	11.6
After	6.8	2.4	7.4	8.5	8.1	6.1	3.4	2.0
Difference	−0.2	4.1	1.6	1.8	3.2	2.0	2.9	9.6

Solution

Since each pair of scores is matched for one particular person, we can conclude that the values are dependent. Each difference is the "before" score minus the "after" score. If the hypnotism is effective, we would expect the after scores to be lower so that $\bar{d}$ would be positive and *significantly* greater than 0. That is, the claim of lower after scores is equivalent to $\mu_d > 0$. We therefore have the null hypothesis H_0: $\mu_d \leq 0$ and the alternative hypothesis H_1: $\mu_d > 0$. See Figures 8-5 and 8-6 for a summary of the key components of this hypothesis test.

The mean and standard deviation of the differences are found as follows.

$$\bar{d} = \frac{\Sigma d}{n} = \frac{-0.2 + 4.1 + \cdots + 9.6}{8}$$

$$= \frac{25}{8} = 3.125$$

continued ▶

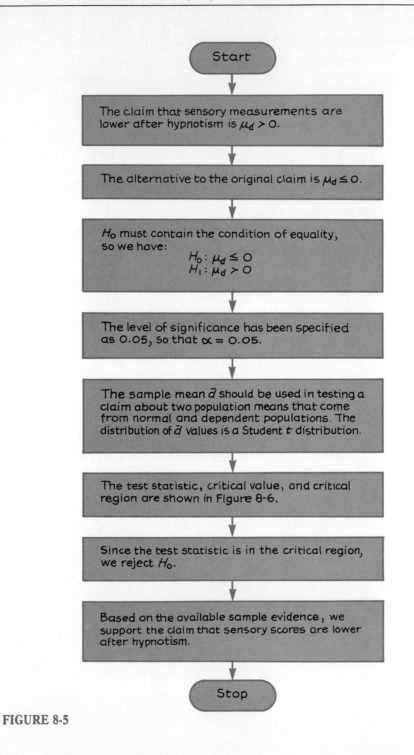

FIGURE 8-5

Solution, continued

$$s_d = \sqrt{\frac{\Sigma(d - \bar{d})^2}{n - 1}}$$

$$= \sqrt{\frac{(-0.2 - 3.125)^2 + (4.1 - 3.125)^2 + \cdots + (9.6 - 3.125)^2}{7}}$$

$$= \sqrt{\frac{59.335}{7}} = 2.911$$

With $\bar{d} = 3.125$, $s_d = 2.911$, and $n = 8$, we calculate the value of the test statistic as

$$t = \frac{\bar{d} - \mu_d}{s_d/\sqrt{n}} = \frac{3.125 - 0}{2.911/\sqrt{8}} = 3.036$$

The critical value of $t = 1.895$ can be found in Table A-3; this is a right-tailed test, $\alpha = 0.05$, and there are $8 - 1 = 7$ degrees of freedom.

Since the test statistic of $t = 3.036$ falls within the critical region (see Figure 8-6), we reject the null hypothesis. There is sufficient evidence to support the claim that the after scores are significantly lower than the before scores. It appears that hypnosis does have a significant effect on pain, as measured by the sensory scores.

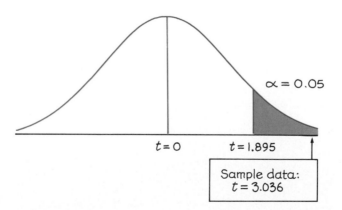

FIGURE 8-6

This test could also be conducted by using the STATDISK software package or any other software package designed to treat the case of two dependent samples. After the user selects the appropriate case from the menu of options, STATDISK prompts him or her to enter the necessary data, and the results are shown on the next page.

STATDISK DISPLAY

```
Hypothesis test for a claim about two DEPENDENT populations

         NULL HYPOTHESIS: First mean <= Second mean

             Mean of differences ... = 3.1250
             St. dev. of differences = 2.9114
             Degrees of freedom .... = 7

             Test statistic ...... t = 3.0359
             Critical value ...... t = 1.8955

    P-value =  9.4779E-03              Significance level = .05

             CONCLUSION: REJECT the null hypothesis
```

This test could also be run as a t test for a single population mean. Enter the data as differences and use zero as the value of the claimed population mean. Shown below is the Minitab display that begins with the original before-after data. The differences are calculated and assigned to the variable C3. The test is then run with the option "ALTERNATIVE=+1"—the Minitab code for a hypothesis test that is right-tailed.

MINITAB DISPLAY

```
MTB > SET C1
DATA> 6.6 6.5 9.0 10.3 11.3 8.1 6.3 11.6
DATA> ENDOFDATA
MTB > SET C2
DATA> 6.8 2.4 7.4 8.5 8.1 6.1 3.4 2.0
DATA> ENDOFDATA
MTB > LET C3=C1-C2
MTB > TTEST of mu=0 for data in C3;
SUBC> ALTERNATIVE=+1 .

TEST OF mu=0.000 VS mu G.T. 0.000

        N     MEAN    STDEV   SE MEAN     T    P VALUE
C3      8    3.125    2.911    1.029    3.04    0.0095
```

Note that the *P*-value of 0.0095 is less than the significance level of 0.05. This indicates that such sample results are not likely to occur by chance, assuming that the null hypothesis is true.

As we consider other tests of hypotheses made about two population means, we begin to encounter a maze, which can easily lead to confusion. Questions must be answered regarding the independence of samples, knowledge of σ_1 and σ_2, and sample size before the correct procedure can be selected. Most of the confusion can be avoided by referring to Figure 8-7, which summarizes the procedures discussed in this section. We illustrate the use of Figure 8-7 through specific examples and then present the underlying theory that led to its development. Since we have already presented an example involving dependent populations, our next examples will involve independent populations.

EXAMPLE

Two machines fill packages, and samples are selected from each machine. Denoting the results from machine A as group 1 and denoting the results from machine B as group 2, we have

Machine A	Machine B
$n_1 = 50$	$n_2 = 100$
$\bar{x}_1 = 4.53$ kg	$\bar{x}_2 = 4.01$ kg

If the standard deviations of the contents filled by machine A and machine B are 0.80 kg and 0.60 kg, respectively, test the claim that the mean contents produced by machine A equal the mean for machine B. Assume a 0.05 significance level.

Solution

The two means are independent and σ_1 and σ_2 are known. Referring to Figure 8-7, we see that we should use a normal distribution test with

$$z = \frac{(\bar{x}_1 - \bar{x}_2) - (\mu_1 - \mu_2)}{\sqrt{\dfrac{\sigma_1^2}{n_1} + \dfrac{\sigma_2^2}{n_2}}} = \frac{(4.53 - 4.01) - 0}{\sqrt{\dfrac{0.80^2}{50} + \dfrac{0.60^2}{100}}} = 4.06$$

With the null and alternative hypotheses described as

continued ▶

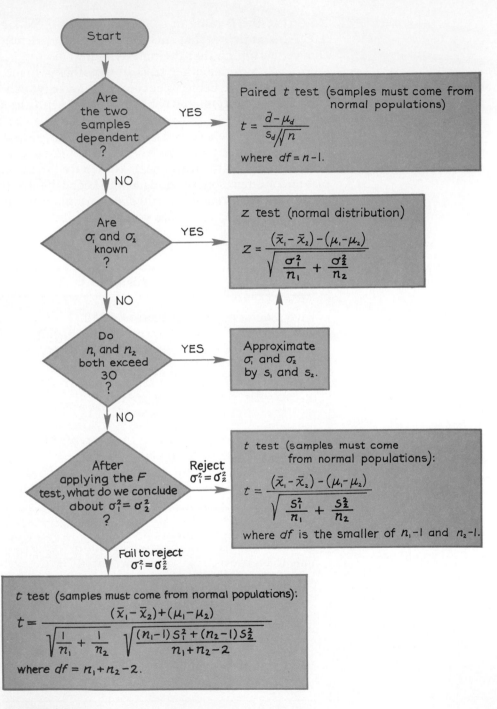

FIGURE 8-7
Testing hypotheses made about the means of two populations.

Solution, continued

$$H_0: \mu_1 = \mu_2 \quad (\text{or } \mu_1 - \mu_2 = 0)$$
$$H_1: \mu_1 \neq \mu_2 \quad (\text{or } \mu_1 - \mu_2 \neq 0)$$

and $\alpha = 0.05$, we conclude that the test involves two tails. From Table A-2 we extract the critical z values of 1.96 and -1.96. The test statistic of 4.06 is well into the critical region, and we therefore reject H_0 and conclude that the population means corresponding to the two machines are not equal. It appears that machine A fills with amounts that are significantly greater than those of machine B.

In the preceding example, we arbitrarily identified the machine A results as group 1, while those of machine B are identified as group 2. For this example, the two large sample sizes allow us to avoid the F test of $\sigma_1^2 = \sigma_2^2$. Although not always necessary, it is often a good strategy to **identify the data set with the larger sample variance as group 1**, and identify the data set with the smaller sample variance as group 2. This is the same procedure followed in Section 8-2. If it becomes necessary to use the F test of $\sigma_1^2 = \sigma_2^2$, this identification will be helpful.

The preceding example is somewhat contrived because we seldom know the values of σ_1 and σ_2. It is rare to sample from two populations with unknown means but known standard deviations. To cover more realistic cases involving independent samples with unknown standard deviations or variances, we next examine the sizes of the two samples, as suggested by Figure 8-7. In this next example and the remaining examples of this section, it is a requirement that the samples come from normally distributed populations. If this condition is not satisfied, we may be able to use the Wilcoxon rank-sum test described in Section 11-4.

If both samples are large (greater than 30), we can estimate σ_1 and σ_2 by s_1 and s_2. We can then proceed as in the last example. However, if either sample is small, we must apply the F test to determine whether the two sample variances are significantly different. The next example illustrates these points since the populations are independent, both population standard deviations are unknown, and both samples are small (less than or equal to 30). We see that the F test suggests that the two population variances are not equal and, from Figure 8-7, we see that the circumstances of this next example cause us to turn right at the last diamond.

EXAMPLE

Random samples of home selling prices are obtained from two zones in Dutchess County. Results are summarized below.

Zone 7 (north)	Zone 1 (south)
$n = 11$	$n = 14$
$\bar{x} = \$142{,}318$	$\bar{x} = \$138{,}237$
$s = \$46{,}068$	$s = \$21{,}336$

At the 0.05 significance level, test the claim that the two zones have the same mean selling price.

Solution

Since the zone 7 results have a larger variance, we identify zone 7 as group 1 and we identify zone 1 as group 2, so that we have

Group 1	Group 2
$n_1 = 11$	$n_2 = 14$
$\bar{x}_1 = \$142{,}318$	$\bar{x}_2 = \$138{,}237$
$s_1 = \$46{,}068$	$s_2 = \$21{,}336$

Since the samples are independent, neither standard deviation is known, and both samples are small, Figure 8-7 indicates that we should begin by applying the F test discussed in Section 8-2. See Section 8-2, where we presented a complete hypothesis test of the claim that $\sigma_1^2 = \sigma_2^2$. We want to decide whether $\sigma_1^2 = \sigma_2^2$, so we formulate the following null and alternative hypotheses:

$$H_0: \sigma_1^2 = \sigma_2^2$$

$$H_1: \sigma_1^2 \neq \sigma_2^2$$

With $\alpha = 0.05$, we do a complete hypothesis test to decide whether or not $\sigma_1^2 = \sigma_2^2$. We then do another complete hypothesis test of the claim that $\mu_1 = \mu_2$ using the appropriate Student t distribution. For the preliminary test we get

$$F = \frac{s_1^2}{s_2^2} = \frac{(46{,}068)^2}{(21{,}336)^2} = 4.6620$$

(In Section 8-2 we obtained a test statistic of 4.6637 because we used rounded values for the sample variances.) The critical F value obtained from Table A-5 is 3.2497. (The test involves two

continued ▶

Solution, continued

tails with $\alpha = 0.05$, and the degrees of freedom for the numerator and the denominator are 10 and 13, respectively.) These results cause us to reject the null hypothesis of equal variances and we reject $\sigma_1^2 = \sigma_2^2$. In Figure 8-7 we test the claim that $\mu_1 = \mu_2$ by using the Student t distribution and the test statistic given in the box to the *right* of the fourth diamond. With

$$H_0: \mu_1 = \mu_2 \qquad (\text{or } \mu_1 - \mu_2 = 0)$$

$$H_1: \mu_1 \neq \mu_2 \qquad (\text{or } \mu_1 - \mu_2 \neq 0)$$

$$\alpha = 0.05$$

we compute the test statistic based on the sample data.

$$t = \frac{(\bar{x}_1 - \bar{x}_2) - (\mu_1 - \mu_2)}{\sqrt{\dfrac{s_1^2}{n_1} + \dfrac{s_2^2}{n_2}}} = \frac{(142,318 - 138,237) - 0}{\sqrt{\dfrac{46068^2}{11} + \dfrac{21336^2}{14}}}$$

$$= 0.272$$

This is a two-tailed test with $\alpha = 0.05$ and 10 degrees of freedom, so the critical t values obtained from Table A-3 are $t = 2.228$ and $t = -2.228$. The computed t value of 0.272 does not fall within the critical region and we fail to reject the null hypothesis of equal means. Based on the available sample data, we cannot reject the claim that the two zones have the same mean selling price.

The next example illustrates a hypothesis test comparing two means for a situation featuring the following characteristics:

- The two samples come from independent and normal populations.
- σ_1 and σ_2 are unknown.
- Both sample sizes are small (≤ 30).
- The sample variances suggest, through the F test, that $\sigma_1^2 = \sigma_2^2$.

The conditions inherent in this next example cause us to follow the path leading to the bottom of the flowchart in Figure 8-7.

EXAMPLE

Samples of girls aged 6 and 7 are given the Wide Range Achievement Test with the results summarized below (based on data from the National Health Survey, USDHEW publication 72-1011). At the 0.05 level of significance, test the claim that the mean score for girls seven years old is greater than the mean for six-year-old girls. (We identify the age 7 group as group 1 because the variance is larger.)

Age 7	Age 6
$n_1 = 16$	$n_2 = 12$
$\bar{x}_1 = 44.0$	$\bar{x}_2 = 27.5$
$s_1 = 13.2$	$s_2 = 10.2$

Solution

Referring to Figure 8-7, we begin by questioning the independence of the two populations and conclude that they are independent because separate groups of subjects are used. We continue with the flow chart by noting that σ_1 and σ_2 are not known and that neither n_1 nor n_2 exceeds 30. At this stage, the flow chart brings us to the last diamond, which requires application of the F test. With H_0: $\sigma_1^2 = \sigma_2^2$, H_1: $\sigma_1^2 \neq \sigma_2^2$, and $\alpha = 0.05$, we compute

$$F = \frac{s_1^2}{s_2^2} = \frac{13.2^2}{10.2^2} = 1.6747$$

With $\alpha = 0.05$ in a two-tailed F test and with 15 and 11 as the degrees of freedom for the numerator and denominator, respectively, we use Table A-5 to obtain the critical F value of 3.3299. Since the computed test statistic of $F = 1.6747$ is not within the critical region, we fail to reject the null hypothesis of equal variances. Leaving the last diamond in Figure 8-7, we proceed downward to the bottom and apply the required t test as follows.

$$H_0: \mu_1 \leq \mu_2 \quad \text{(or } \mu_1 - \mu_2 \leq 0)$$

$$H_1: \mu_1 > \mu_2 \quad \text{(or } \mu_1 - \mu_2 > 0)$$

$$\alpha = 0.05$$

continued ▶

Solution, continued

$$t = \frac{(\bar{x}_1 - \bar{x}_2) - (\mu_1 - \mu_2)}{\sqrt{\dfrac{1}{n_1} + \dfrac{1}{n_2}} \sqrt{\dfrac{(n_1 - 1)s_1^2 + (n_2 - 1)s_2^2}{n_1 + n_2 - 2}}}$$

$$= \frac{(44.0 - 27.5) - 0}{\sqrt{\dfrac{1}{16} + \dfrac{1}{12}} \sqrt{\dfrac{(16 - 1)(13.2)^2 + (12 - 1)(10.2)^2}{16 + 12 - 2}}}$$

$$= 3.594$$

Noting that $\mu_1 > \mu_2$ is equivalent to $\mu_1 - \mu_2 > 0$, we can better see that the test is right-tailed. With $\alpha = 0.05$ in this right-tailed t test, and with $n_1 + n_2 - 2 = 16 + 12 - 2 = 26$ degrees of freedom, we obtain a critical t value of 1.706. The computed test statistic of $t = 3.594$ falls within the critical region, so we reject the null hypothesis. The given sample data support the claim that the mean for seven-year-old girls is greater than that for six-year-old girls.

The calculations in this last example might seem somewhat complex, but we should recognize that calculators and computers can be used to ease that burden. STATDISK, for example, can be used to conduct this test. The user selects the option indicating a hypothesis test involving the means of two independent samples, and the user is then prompted for the necessary data. The resulting display is shown on page 468. Note that the results of the prerequisite F test are included in the display. Also note the inclusion of the P-values in both tests. Using the given sample statistics, it takes about 60 seconds to run this test on STATDISK.

One of the most difficult aspects of tests comparing two means is the determination of the correct test to be used. Careful and consistent use of Figure 8-7 should help us avoid the common error of using the wrong procedures for a situation involving a hypothesis test. There is a danger of being overwhelmed by the overall complexity of the work when dealing with the five different cases considered here. However, we can use Figure 8-7 to decompose a complex problem into simpler components that can be treated individually.

It is not practical to outline in full detail the derivations leading to the general test statistics given in Figure 8-7, but we can give some reasons for their existence. We have already discussed the case for

STATDISK DISPLAY

```
Hypothesis test for a claim about two INDEPENDENT populations

          NULL HYPOTHESIS: Mean 1 <= Mean 2

    Test statistic F = 1.67474    P-value = 0.39218

    F Test CONCLUSION: FAIL TO REJECT equality of variances
-------------------------------------------------------------
  Test statistic... t = 3.59386    Degrees of freedom = 26

              Critical value... t = 1.70600

    P-value = 6.67751E-04    Significance level = .05

        t Test CONCLUSION: REJECT the null hypothesis
```

dependent populations having normal distributions. Actual experiments and mathematical derivations show that, in repeated random samplings from two normal and dependent populations, the values of $\bar{d}$ possess a Student t distribution with mean μ_d and standard deviation $\sigma_d/\sqrt{n}$.

Two cases of Figure 8-7 lead to the normal distribution with the test statistic given by

$$z = \frac{(\bar{x}_1 - \bar{x}_2) - (\mu_1 - \mu_2)}{\sqrt{\sigma_1^2/n_1 + \sigma_2^2/n_2}}$$

This expression is essentially an application of the central limit theorem, which tells us that sample means $\bar{x}$ are normally distributed with mean μ and standard deviation $\sigma/\sqrt{n}$.

In Section 5-6 we saw that, when samples are size 31 or larger, the normal distribution serves as a reasonable approximation to the distribution of sample means. By similar reasoning, the values of $\bar{x}_1 - \bar{x}_2$ also tend to approach a normal distribution with mean $\mu_1 - \mu_2$. When both samples are large, we conclude that the values of $\bar{x}_1 - \bar{x}_2$ will have a standard deviation of

$$\sqrt{\frac{\sigma_1^2}{n_1} + \frac{\sigma_2^2}{n_2}}$$

by using a property of variances: **the variance of the differences between two independent random variables equals the variance of the**

TABLE 8-2		
x	y	$x - y$
1	7	−6
3	0	3
1	0	1
3	4	−1
2	6	−4
8	3	5
3	9	−6
3	7	−4
9	0	9
1	1	0
4	2	2
9	4	5
4	9	−5
9	6	3
5	2	3
6	7	−1
5	6	−1
5	8	−3
3	3	0
3	6	−3
1	3	−2
0	2	−2
4	4	0
3	1	2
3	9	−6
6	4	2
9	5	4
5	1	4
2	2	0
9	6	3
2	2	0
6	1	5
1	9	−8
7	0	7
9	1	8
6	5	1

first random variable plus the variance of the second random variable. That is, the variance of values $\bar{x}_1 - \bar{x}_2$ will tend to equal $\sigma_{\bar{x}_1}^2 + \sigma_{\bar{x}_2}^2$ provided that $\bar{x}_1$ and $\bar{x}_2$ are independent. This is a difficult concept, and we therefore illustrate it by the specific data given in Table 8-2. The x and y scores were independently and randomly selected as the last digits of numbers in a telephone book. The variance of the x values is 7.68, the variance of the y values is 8.48, and the variances of the $x - y$ values is 17.22.

$$\left.\begin{array}{rl} s_x^2 &= 7.68 \\ s_y^2 &= 8.48 \\ s_{x-y}^2 &= 17.22 \end{array}\right] \quad s_{x-y}^2 \approx s_x^2 + s_y^2$$

We can see that $s_x^2 + s_y^2$ is roughly equal to s_{x-y}^2. By comparing the values of x, y, and $x - y$, we see that the variance is largest for the $x - y$ values. The x values range from 0 to 9, the y values range from 0 to 9, but the $x - y$ values exhibit greater variation by ranging from −8 to 9. If our x, y, and $x - y$ sample sizes were much larger than the sample sizes of 36 for Table 8-2, we would approach these theoretical values: $\sigma_x^2 = 8.25$, $\sigma_y^2 = 8.25$, and $\sigma_{x-y}^2 = 16.50$.

These theoretical values illustrate that $\sigma_{x-y}^2 = \sigma_x^2 + \sigma_y^2$ when x and y are independent random variables. When we deal with means of large random sample sizes, the standard deviation of those sample means is $\sigma/\sqrt{n}$, so the variance is σ^2/n.

We can now combine our additive property of variances with the central limit theorem's expression for variance of sample means to obtain the following result:

$$\sigma_{\bar{x}_1 - \bar{x}_2}^2 = \sigma_{\bar{x}_1}^2 + \sigma_{\bar{x}_2}^2 = \frac{\sigma_1^2}{n_1} + \frac{\sigma_2^2}{n_2}$$

In the preceding expression, we assume that we have population 1 with variance σ_1^2 and population 2 with variance σ_2^2. Samples of size n_1 are randomly drawn from population 1 and the mean $\bar{x}$ is computed. The same is done for population 2. Here, $\sigma_{\bar{x}_1 - \bar{x}_2}^2$ denotes the variance of $\bar{x}_1 - \bar{x}_2$ values. This result shows that the standard deviation of $\bar{x}_1 - \bar{x}_2$ values is

$$\sqrt{\frac{\sigma_1^2}{n_1} + \frac{\sigma_2^2}{n_2}}$$

Since z is a standard score that corresponds in general to

$$z = \frac{\text{(sample statistic)} - \text{(population mean)}}{\text{(population standard deviation)}}$$

we get

Survey Solicits Contributions

The American Institute for Cancer Research distributed a survey on diet and breast cancer. The stated purpose of the survey was to develop a "statistical profile of the eating habits of American women—and to help find the link between diet and breast cancer." However, the survey ended with an option for enclosing a gift ranging from $5 to $500. Two fundamental questions arise. What was the real purpose of the survey? Does the contribution request affect the validity of the results they receive?

$$z = \frac{(\bar{x}_1 - \bar{x}_2) - (\mu_1 - \mu_2)}{\sqrt{\sigma_1^2/n_1 + \sigma_2^2/n_2}}$$

by noting that the sample values of $\bar{x}_1 - \bar{x}_2$ will have a mean of $\mu_1 - \mu_2$ and the standard deviation just given.

The test statistic located at the bottom of Figure 8-7 is appropriate for tests of hypotheses about two means from independent and normal populations when the samples are small and the population variances appear to be equal. The numerator of $(\bar{x}_1 - \bar{x}_2) - (\mu_1 - \mu_2)$ again describes the difference between the sample statistic $(\bar{x}_1 - \bar{x}_2)$ and the mean of all such values $(\mu_1 - \mu_2)$. The denominator represents the standard deviation of $\bar{x}_1 - \bar{x}_2$ values, which come from repeated sampling when the stated assumptions are satisfied. Since these assumptions include equality of population variances, the denominator should be an estimate of

$$\sqrt{\frac{\sigma^2}{n_1} + \frac{\sigma^2}{n_2}} = \sigma\sqrt{\frac{1}{n_2} + \frac{1}{n_2}}$$

Both n_1 and n_2 will be known, so we need to estimate only σ, and we pool (combine) the sample variances to get the best possible estimate. The pooled estimate of σ is

$$\sqrt{\frac{(n_1 - 1)s_1^2 + (n_2 - 1)s_2^2}{n_1 + n_2 - 2}}$$

where the expression under the square root sign is just a weighted average of s_1^2 and s_2^2. (Weights of $n_1 - 1$ and $n_2 - 1$ are used.) See Exercise 8-64.

We use the test statistic

$$t = \frac{(\bar{x}_1 - \bar{x}_2) - (\mu_1 - \mu_2)}{\sqrt{\frac{s_1^2}{n_1} + \frac{s_2^2}{n_2}}}$$

in tests of hypotheses about two means coming from independent and normal populations when the samples are small and the two population variances appear to be different. The form of that test statistic seems to follow from similar reasoning used in the other situations, but that is not the case. In fact, no exact test has been found for testing the equality of two means when all the following conditions are met:

1. The two population variances are unknown.
2. The two population variances are unequal.
3. At least one of the samples is small ($n \leq 30$).
4. The two populations are independent and normal.

The approach given in Figure 8-7 for this case is only an approximate test, which is widely used. Also, in Figure 8-7 we indicate that the number of degrees of freedom for this case is found by selecting the smaller of $n_1 - 1$ and $n_2 - 1$. This is a more conservative and simplified alternative to computing the number of degrees of freedom, as follows:

$$df = \frac{(A + B)^2}{\dfrac{A^2}{n_1 - 1} + \dfrac{B^2}{n_2 - 1}} \qquad \text{where } A = \frac{s_1^2}{n_1} \text{ and } B = \frac{s_2^2}{n_2}$$

The two bottom cases of Figure 8-7 begin with a preliminary F test. If we apply the F test with a certain level of significance and then do a t test at that same level of significance, the overall result will not be at that same level of significance. Also, in addition to being sensitive to differences in population variances, the F statistic is also sensitive to departures from normal distributions so that it is possible to reject a null hypothesis for the wrong reason. Because of these factors, some statisticians do not recommend a preliminary F test, while others consider this approach to be better. In any event, there is no universal agreement.

8-3 EXERCISES A

In Exercises 8-29 through 8-32, use a 0.05 significance level to test the claim that $\mu_1 = \mu_2$. In each case, the two samples are independent, and they are randomly selected from populations with normal distributions.

8-29

Control Group	Experimental Group
$n_1 = 40$	$n_2 = 40$
$\bar{x}_1 = 79.6$	$\bar{x}_2 = 84.2$
$s_1 = 12.4$	$s_2 = 12.2$

8-30

Brand X	Brand Y
$n_1 = 16$	$n_2 = 14$
$\bar{x}_1 = 64.3$	$\bar{x}_2 = 65.1$
$s_1 = 2.50$	$s_2 = 2.50$

8-31

Treated	Untreated
$n_1 = 16$	$n_2 = 16$
$\bar{x}_1 = 98.6$	$\bar{x}_2 = 97.8$
$s_1 = 8.60$	$s_2 = 4.20$

8-32

Production Method A	Production Method B
$n_1 = 20$	$n_2 = 25$
$\bar{x}_1 = 127.4$	$\bar{x}_2 = 108.3$
$s_1 = 15.6$	$s_2 = 14.3$

8-33 In a study of techniques used to measure lung volumes, physiological data were collected for 10 subjects. The values given in the table are in liters, representing the measured forced vital capacities of the 10 subjects in a sitting position and in a supine (lying) position. (Table is based on data from "Validation of Esophageal Balloon Technique at Different Lung Volumes and Postures," by Baydur, Cha, and Sassoon, *Journal of Applied Physiology*, Vol. 62, No. 1.) At the 0.05 significance level, test the claim that both positions have the same mean.

Sitting	4.66	5.70	5.37	3.34	3.77	7.43	4.15	6.21	5.90	5.77
Supine	4.63	6.34	5.72	3.23	3.60	6.96	3.66	5.81	5.61	5.33

8-34 In a study of techniques used to measure lung volumes, physiological data were collected for 10 subjects. The values given in the table are in liters, representing the measured functional residual capacities of the 10 subjects in a sitting position and in a supine (lying) position. (Table is based on data from "Validation of Esophageal Balloon Technique at Different Lung Volumes and Postures," by Baydur, Cha, and Sassoon, *Journal of Applied Physiology*, Vol. 62, No. 1.) At the 0.05 significance level, test the claim that both positions have the same mean.

Sitting	2.96	4.65	3.27	2.50	2.59	5.97	1.74	3.51	4.37	4.02
Supine	1.97	3.05	2.29	1.68	1.58	4.43	1.53	2.81	2.70	2.70

8-35 Samples of two competing cold medicines are tested for the amount of acetaminophen, and the results follow. At the 0.05 significance level, test the claim that the mean amount of acetaminophen is the same in each brand.

Brand X	Brand Y
$n = 25$	$n = 35$
$\bar{x} = 503$ mg	$\bar{x} = 520$ mg
$s = 14$ mg	$s = 18$ mg

8-36 Researchers study commercial air-filtering systems for noise pollution with sample results as follows. At the 5% level of significance, test the claim that there is no difference in the mean noise levels.

Unit A	Unit B
$n = 8$	$n = 6$
$\bar{x} = 87.5$	$\bar{x} = 91.3$
$s = 0.8$	$s = 1.1$

8-37   In a study of the effect of job previews on work expectation, subjects from different groups were tested. For 60 subjects given specific job previews, the mean promotion expectation score is 19.14 and the standard deviation is 6.56. For a "no booklet" sample group of 40 subjects, the mean is 20.81 and the standard deviation is 4.90. (See "Effects of Realistic Job Previews on Hiring Bank Tellers," by Dean and Wanous, *Journal of Applied Psychology*, Vol. 69, No. 1.) At the 0.10 level of significance, test the claim that the two sample groups come from populations with the same mean.

8-38   At the 0.05 significance level, test the claim that two ambulance services have the same mean response time. A sample of 50 responses from the first firm produces a mean of 12.2 min and a standard deviation of 1.5 min. A sample of 50 responses from the second firm produces a mean of 14.0 min with a standard deviation of 2.1 min.

8-39   A small commuter airline owns two jets and has subcontracted its maintenance operations to two separate firms. Samples of the monthly downtimes of these subcontractors follow. At the 0.05 significance level, test the claim that the mean monthly downtimes (in hours) of both firms are equal.

Firm A	Firm B
$n = 12$	$n = 14$
$\bar{x} = 14.4$ h	$\bar{x} = 16.3$ h
$s = 3.1$ h	$s = 1.6$

8-40   Samples of two different car models are tested for fuel economy by determining the miles traveled using 1 gallon of gas. The results follow. At the 0.05 significance level, test the claim that the means of the numbers of miles traveled by the two models are equal.

Car A	Car B
$n_1 = 10$	$n_2 = 12$
$\bar{x}_1 = 27.2$ mi	$\bar{x}_2 = 31.6$ mi
$s_1 = 4.1$ mi	$s_2 = 2.1$ mi

8-41   A study was conducted to investigate the effectiveness of hypnotism in reducing pain. Results for randomly selected subjects are given below. At the 0.05 significance level, test the claim that the affective responses to pain are the same before and after hypnosis. (The data are based on "An Analysis of Factors That Contribute to the Efficacy of Hypnotic Analgesia," by Price and Barber, *Journal of Abnormal Psychology*, Vol. 96, No. 1.)

Before	-5.5	-5.0	-6.6	-9.7	-4.0	-7.0	-7.0	-8.4
After	-1.4	-0.5	0.7	1.0	2.0	0.0	-0.6	-1.8

8-42   A number of pregnant women agree to participate in an experiment to see if vitamin pills will affect the weights of their newborn children. Some women are given the vitamins, while others are given a placebo of no medicinal value. To eliminate variations due to sex, only the male births are included in the sample data given below. At the 0.05 level of significance, test the claim that there is no difference between the mean weights of the two groups.

Vitamin	Placebo
$n_1 = 40$	$n_2 = 35$
$\bar{x}_1 = 3.39$ kg	$\bar{x}_2 = 3.18$ kg
$s_1 = 0.44$ kg	$s_2 = 0.53$ kg

8-43   Cigarettes randomly selected from two brands are analyzed to determine their nicotine content. The results follow. At the $\alpha = 0.05$ significance level, test the claim that both brands have equal nicotine content.

Brand X	Brand Y
$n_1 = 20$	$n_2 = 40$
$\bar{x}_1 = 35.1$ mg	$\bar{x}_2 = 36.5$ mg
$s_1 = 4.2$ mg	$s_2 = 3.3$ mg

8-44   A large firm collects sample data on the lengths of telephone calls (in minutes) made by employees in two different divisions, and the results are given below. At the 0.02 level of significance, test the claim that there is no difference between the mean times of all long distance calls made in the two divisions.

Sales division	Customer service division
$n_1 = 40$	$n_2 = 20$
$\bar{x}_1 = 10.26$	$\bar{x}_2 = 6.93$
$s_1 = 8.65$	$s_2 = 4.93$

8-45   A study was conducted to investigate relationships between different types of standard test scores. On the Graduate Record Examination Verbal test, 68 females had a mean of 538.82 and a standard deviation of 114.16, while 86 males had a mean of 525.23 and a standard deviation of 97.23 (based on data from "Equivalencing MAT and GRE Scores Using Simple Linear Transformation and Regression Methods," by Kagan and Stock, *Journal of Experimental Education*, Vol. 49, No. 1). At the 0.02 significance level, test the claim that the two groups come from populations with the same mean.

8-46   The effectiveness of a mental training program was tested in a military training program. In an antiaircraft artillery examination, scores for an experimental group and a control group were recorded. Use the given data to test the claim that both groups come from populations with the same mean. Use a 0.05 significance level. (See "Routinization of Mental Training in Organizations: Effects on Performance and Well-Being," by Larsson, *Journal of Applied Psychology*, Vol. 72, No. 1.)

Experimental				Control			
60.83	117.80	44.71	75.38	122.80	70.02	119.89	138.27
73.46	34.26	82.25	59.77	118.43	54.22	118.58	74.61
69.95	21.37	59.78	92.72	121.70	70.70	99.08	120.76
72.14	57.29	64.05	44.09	104.06	94.23	111.26	121.67
80.03	76.59	74.27	66.87				

8-47 Two procedures are used for controlling the aircraft traffic at an airport. Sample results based on each of the two procedures follow. At the 0.05 significance level, test the claim that the use of system 2 results in a mean number of operations per hour exceeding the mean for system 1.

System 1	System 2
$n_1 = 24$ hours	$n_2 = 24$ hours
$\bar{x}_1 = 63.0$ operations per hour	$\bar{x}_2 = 60.1$ operations per hour
$s_1 = 5.2$ operations per hour	$s_2 = 3.2$ operations per hour

8-48 A dose of the drug captopril, designed to lower systolic blood pressure, is administered to 10 randomly selected volunteers. The results follow. At the $\alpha = 0.05$ significance level, test the claim that systolic blood pressure is not affected by the pill.

Before pill	120	136	160	98	115	110	180	190	138	128
After pill	118	122	143	105	98	98	180	175	105	112

8-49 Sample data were collected in a study of calcium supplements and the effects on blood pressure. A placebo group and a calcium group began the study with measures of blood pressures. At the 0.05 significance level, test the claim that the two sample groups come from populations with the same mean. The data are based on "Blood Pressure and Metabolic Effects of Calcium Supplementation in Normotensive White and Black Men," by Lyle and others, *Journal of the American Medical Association*, Vol. 257, No. 13.)

Placebo				Calcium			
124.6	104.8	96.5	116.3	129.1	123.4	102.7	118.1
106.1	128.8	107.2	123.1	114.7	120.9	104.4	116.3
118.1	108.5	120.4	122.5	109.6	127.7	108.0	124.3
113.6				106.6	121.4	113.2	

8-50   An investor is considering two possible locations for a new restaurant and commissions a study of the pedestrian traffic at both sites. At each location, the pedestrians are observed in 1-hour units and, for each hour, an index of desirable characteristics is compiled. The sample results are given below. At the 0.05 level of significance, test the claim that both sites have the same mean.

East	West
$n = 35$	$n = 50$
$\bar{x} = 421$	$\bar{x} = 347$
$s = 122$	$s = 85$

8-51   A course is designed to increase readers' speed and comprehension. To evaluate the effectiveness of this course, a test is given both before and after the course, and sample results follow. At the 0.05 significance level, test the claim that the scores are higher after the course.

Before	100	110	135	167	200	118	127	95	112	116
After	136	160	120	169	200	140	163	101	138	129

8-52   In writing an antiunion article, a management consultant presents statistics that purportedly show nonunion masons are more productive than their union counterparts. At the 0.05 significance level, test the consultant's claim that the mean number of bricks set in 1 hour by nonunion workers exceeds the corresponding mean for union masons. The following data are based on random samples consisting of bricks set in 1 hour.

Nonunion	Union
$n = 15$	$n = 10$
$\bar{x} = 24.3$	$\bar{x} = 23.3$
$s = 3.6$	$s = 1.8$

8-53   A test of driving ability is given to a random sample of 10 student drivers before and after they completed a formal driver education course. The results follow. At the $\alpha = 0.05$ significance level, test the claim that the mean score is not affected by the course.

Before course	100	121	93	146	101	109	149	130	127	120
After course	136	129	125	150	110	138	136	130	125	129

8-54 Two separate counties use different procedures for selecting jurors. We want to test the claim that the mean waiting time of prospective jurors is the same for both counties. The 40 randomly selected subjects from one county produce a mean of 183.0 min and a standard deviation of 21.0 min. The 50 randomly selected subjects from the other county produce a mean of 253.1 min and a standard deviation of 29.2 min. Test the claim at the 0.05 significance level.

8-55 In a study involving motivation and test scores, data were obtained for females and males. Use the data given below to test the claim that the two groups come from populations with the same mean. Use a 0.02 significance level. (See "Relationships Between Achievement-Related Motives, Extrinsic Conditions, and Task Performance," by Schroth, *Journal of Social Psychology*, Vol. 127, No. 1.)

Female				Male			
31.13	18.71	14.34	23.90	12.27	39.53	32.56	23.93
13.96	13.88	29.85	20.15	19.54	25.73	32.20	19.84
6.66	19.20	15.89		20.20	23.01	25.63	17.98
				22.99	22.12	12.63	18.06

8-56 An elementary school principal is confronted by an irate group of parents, who charge that the sixth grade students in their school are not reading as well as the sixth graders in a nearby school. From each of the two schools, the principal randomly selects 15 scores on a standard reading test and computes the sample data for those two groups of sixth grade students given below. At the 0.05 level of significance, test the principal's claim that both groups have the same mean.

A	B
$\bar{x} = 96.8$	$\bar{x} = 103.5$
$s = 14.2$	$s = 16.3$

8-57 As part of the National Health Survey, data were collected on the weights of men. For 804 men aged 25 to 34, the mean is 176 lb and the standard deviation is 35.0 lb. For 1657 men aged 65 to 74, the mean and standard deviation are 164 lb and 27.0 lb, respectively. (See the National Health Survey, USDHEW publication 79-1659.) At the 0.01 significance level, test the claim that the older men come from a population with a mean less than that for men in the 25 to 34 age bracket.

8-58 A study was conducted to investigate some effects of physical training. Sample data are listed at left. (See "Effect of Endurance Training on Possible Determinants of VO_2 During Heavy Exercise," by Casaburi and others, *Journal of Applied Physiology*, Vol. 62, No. 1.) At the 0.05 level of significance, test the claim that mean pretraining weight equals mean posttraining weight. All weights are given in kilograms.

Pre-training	Post-training
99	94
57	57
62	62
69	69
74	66
77	76
59	58
92	88
70	70
85	84

8-59   Two different firms design their own IQ tests, and a psychologist administers both versions to randomly selected subjects. The results are given below. At the 0.02 level of significance, test the claim that both versions produce the same mean score.

Subject	A	B	C	D	E	F	G	H	I	J
Test I	98	94	111	102	108	105	92	88	100	99
Test II	105	103	113	98	112	109	97	95	107	103

8-60   The arrangement of test items was studied for its effect on anxiety. At the 0.05 significance level, test the claim that the two given samples come from populations with the same mean. (See "Item Arrangement, Cognitive Entry Characteristics, Sex and Test Anxiety as Predictors of Achievement Examination Performance," by Klimko, *Journal of Experimental Education*, Vol. 52, No. 4.)

Easy to difficult				Difficult to easy			
24.64	39.29	16.32	32.83	33.62	34.02	26.63	30.26
28.02	33.31	20.60	21.13	35.91	26.68	29.49	35.32
26.69	28.90	26.43	24.23	27.24	32.34	29.34	33.53
7.10	32.86	21.06	28.89	27.62	42.91	30.20	32.54
28.71	31.73	30.02	21.96				
25.49	38.81	27.85	30.29				
30.72							

8-3 EXERCISES B

8-61   A confidence interval for the difference between (independent) population means μ_1 and μ_2 can be constructed for the case involving known values of σ_1 and σ_2 as follows.

$$(\bar{x}_1 - \bar{x}_2) - E < (\mu_1 - \mu_2) < (\bar{x}_1 - \bar{x}_2) + E$$

where

$$E = z_{\alpha/2} \sqrt{\frac{\sigma_1^2}{n_1} + \frac{\sigma_2^2}{n_2}}$$

Construct the 95% confidence interval to estimate the difference between the mean weight of men aged 25 to 34 and the mean weight of men aged 65 to 74. Use the sample data given in Exercise 8-57 and assume that the sample standard deviations can be used as estimates of the corresponding population standard deviations.

8-62 In Exercise 8-61, we presented the confidence interval for the difference between two independent population means. That confidence interval applies to the case where both σ_1 and σ_2 are known.

a. Construct the general form of the confidence interval for the difference between two independent population means, where the following conditions are satisfied:
   1. σ_1 and σ_2 are unknown.
   2. $n_1 \le 30$ and $n_2 \le 30$.
   3. The values of s_1 and s_2 suggest that $\sigma_1^2 = \sigma_2^2$.
b. Use the results of part (a) to construct the 95% confidence interval for $\mu_1 - \mu_2$, where the sample data is found in Exercise 8-55.

8-63 A confidence interval for the mean μ_d of the differences between matched pairs of dependent data can be constructed as follows.

$$\bar{d} - t_{\alpha/2}\frac{s_d}{\sqrt{n}} < \mu_d < \bar{d} + t_{\alpha/2}\frac{s_d}{\sqrt{n}}$$

Construct the 95% confidence interval for the paired data given in Exercise 8-59.

8-64 If

$$s_1^2 = \frac{\Sigma(x_1 - \bar{x}_1)^2}{n_1 - 1}$$

and

$$s_2^2 = \frac{\Sigma(x_2 - \bar{x}_2)^2}{n_2 - 1}$$

show that

$$s^2 = \frac{\Sigma(x_1 - \bar{x}_1)^2 + \Sigma(x_2 - \bar{x}_2)^2}{n_1 + n_2 - 2}$$

is equivalent to

$$s = \sqrt{\frac{(n_1 - 1)s_1^2 + (n_2 - 1)s_2^2}{n_1 + n_2 - 2}}$$

This result is used in developing the pooled estimate of σ for the test statistic found at the bottom of Figure 8-7.

8-4 TESTS COMPARING TWO PROPORTIONS

In this section we consider tests of hypotheses made about two population proportions. The concepts and procedures we develop can be used to answer questions such as the following:

- Is there a difference between the proportion of homicides among 17-year-olds and the proportion of homicides in the general population?
- Is there a difference between the proportion of men and women in management positions?
- Is there a difference between the percentage of students who passed mathematics courses and the percentage of students who passed history courses?
- Is there a difference between the proportion of IBM computers sold in California and the proportion sold in Texas?

Throughout this section we assume that our sample data come from two independent populations.

NOTATION

For population 1 we let:

p_1 denote the population proportion
n_1 denote the size of the sample
x_1 denote the number of successes

$$\hat{p}_1 = \frac{x_1}{n_1}$$

The corresponding meanings are attached to p_2, n_2, x_2, and $\hat{p}_2$ which come from population 2.

Just as $\hat{p}_1$ represents the sample estimate of p_1, the sample estimate of p_2 is represented by $\hat{p}_2$. We know from Section 5-5 that if $n_1 p_1 \geq 5$ and $n_1 q_1 \geq 5$, then the proportions $\hat{p}_1$ have a distribution that is approximately normal. The differences $\hat{p}_1 - \hat{p}_2$ also have a distribution that is approximately normal. Since the means of $\hat{p}_1$ and $\hat{p}_2$ are p_1 and p_2, respectively, it follows that the mean of the differences $\hat{p}_1 - \hat{p}_2$ will be $p_1 - p_2$. In Section 7-3, we established the fact that

Survey Medium Can Affect Results

In a survey of Catholics in Boston, the subjects were asked if contraceptives should be made available to unmarried women. In personal interviews, 44% of the respondents said yes. But for a similar group contacted by mail or telephone, 75% of the respondents answered yes to the same question.

sample proportions $\hat{p}$ have a standard deviation of $\sqrt{pq/n}$, and this implies that the variance of the sample proportion $\hat{p}_1$ is $p_1 q_1/n_1$. By similar reasoning, the variance of the $\hat{p}_2$ sample values is $p_2 q_2/n_2$. In Section 8-3, we established the fact that the variance of the differences between two independent random variables is the sum of their individual variances, and we use this property to get

$$\sigma^2_{(\hat{p}_1 - \hat{p}_2)} = \sigma^2_{\hat{p}_1} + \sigma^2_{\hat{p}_2} = \frac{p_1 q_1}{n_1} + \frac{p_2 q_2}{n_2}$$

However, we are often unable to use this expression, since we usually do not know the values of p_1, p_2, q_1, and q_2. But if we assume that $p_1 = p_2$, we can estimate their common value by pooling the sample data as follows.

DEFINITION

The **pooled estimate of p_1 and p_2** is denoted by $\bar{p}$ and is given by

$$\bar{p} = \frac{x_1 + x_2}{n_1 + n_2}$$

$$\bar{q} = 1 - \bar{p}$$

Using this pooled estimate in place of p_1 and p_2, we find that the standard deviation of the differences between the sample proportions becomes

$$\sigma_{(\hat{p}_1 - \hat{p}_2)} = \sqrt{\frac{\bar{p}\bar{q}}{n_1} + \frac{\bar{p}\bar{q}}{n_2}} = \sqrt{\bar{p}\bar{q}\left(\frac{1}{n_1} + \frac{1}{n_2}\right)}$$

We now know that the sample proportion differences $\hat{p}_1 - \hat{p}_2$ have a distribution that is approximately normal with mean $p_1 - p_2$ and standard deviation as given above. We now summarize these results: As long as $n_1 p_1$, $n_1 q_1$, $n_2 p_2$, and $n_2 q_2$ each has a value of at least 5, the test statistic given on page 483 has a sampling distribution that is approximately the standard normal distribution. This test statistic applies only to cases where the null hypothesis is $p_1 = p_2$, or $p_1 \geq p_2$, or $p_1 \leq p_2$. For testing claims that the difference $p_1 - p_2$ is equal to a nonzero constant, a different test statistic is used. (See Exercise 8-91.) We consider only hypotheses that lead to a null hypothesis of $p_1 = p_2$, or $p_1 \geq p_2$, or $p_1 \leq p_2$, so our tests will be conducted under the assumption that $p_1 = p_2$ or $p_1 - p_2 = 0$. We can then use the given test statistic with $p_1 - p_2$ replaced by 0.

Polio Experiment

In 1954 a vast medical experiment was conducted to test the effectiveness of the Salk vaccine as a protection against the devastating effects of polio. Previously developed polio vaccines had been used, but it was discovered that some of those earlier vaccinations actually *caused* paralytic polio. Researchers justifiably developed a cautious and conservative approach to approving new vaccines for general use, and they decided to conduct a large-scale experiment of the Salk vaccine with volunteers.

Because of a variety of physiological factors, the vaccine could not be 100% effective, so its effectiveness had to be proved by a lowered incidence of polio among inoculated children. It was hoped that this experiment would result in a lower incidence of polio among vaccinated children that would be so significant as to be overwhelmingly convincing.

The number of children involved in this experiment was necessarily large because a small sample would not provide the conclusive evidence required by these cautious researchers. Approximately 200,000 children were injected with an ineffective salt solution, while 200,000 other children were injected with the Salk vaccine. Assignments of the real vaccine and the useless salt solution were made on a random basis. The children being injected did not know whether they were given the real vaccine or the salt solution. Even the doctors giving the injections and evaluating subsequent results did not know which injections contained the real Salk vaccine. Only 33 of the 200,000 vaccinated children later developed paralytic polio, while 115 of the 200,000 injected with the salt solution later developed paralytic polio. Statistical analysis of these and other results led to the conclusion that the Salk vaccine was indeed effective against paralytic polio.

The polio experiment is one example of a *double blind* experiment, which is characterized by the fact that neither the subjects nor those who evaluate results are aware of the true nature of the treatment. In a *single blind* experiment, only the subjects are not aware of the treatment they receive. Medical experiments are generally double blind because doctors are often more inclined to recognize "cures" when they know that the patient has been given the real medicine and not the placebo.

Better Results with Smaller Class Size

In an experiment at the State University of New York at Stony Brook, it was found that students did significantly better in classes limited to 35 students than in large classes with 150 to 200 students. For a calculus course, failure rates were 19% for the small classes compared to 50% for the large classes. The percentages of A's were 24% for the small classes and 3% for the large classes. The smaller classes allow for more direct interaction between students and teachers.

TEST STATISTIC

(For $H_0: p_1 = p_2$, $H_0: p_1 \geq p_2$, or $H_0: p_1 \leq p_2$)

$$z = \frac{(\hat{p}_1 - \hat{p}_2) - (p_1 - p_2)}{\sqrt{\bar{p}\bar{q}\left(\frac{1}{n_1} + \frac{1}{n_2}\right)}}$$

where

$$\hat{p}_1 = \frac{x_1}{n_1} \qquad \hat{p}_2 = \frac{x_2}{n_2}$$

$$\bar{p} = \frac{x_1 + x_2}{n_1 + n_2}$$

$$\bar{q} = 1 - \bar{p}$$

EXAMPLE

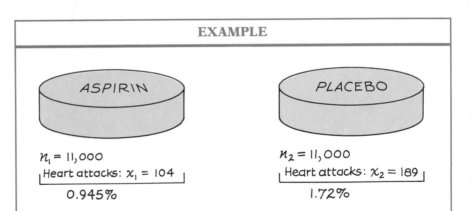

In a study of 22,000 male physicians, half were given regular doses of aspirin while the other half were given placebos with no effects. (The study was reported in *Time* and the *New England Journal of Medicine*, Vol. 318, No. 4.) Among those who took the aspirin doses, 104 suffered heart attacks. Among those who took the placebos, 189 suffered heart attacks. At the 0.01 significance level, test the claim that the aspirin group has a significantly lower heart attack rate.

The fundamental question is really this: Under the given circumstances, is the 0.945% rate significantly lower than the 1.72% rate? The claim of a lower heart attack rate for the aspirin group can be represented by $p_1 < p_2$. We therefore have

continued ▶

Example, continued

$$H_0: p_1 \geq p_2$$
$$H_1: p_1 < p_2$$

With $\alpha = 0.01$, we use the given sample data to get

$$\hat{p}_1 = \frac{x_1}{n_1} = \frac{104}{11,000} = 0.00945$$

$$\hat{p}_2 = \frac{x_2}{n_2} = \frac{189}{11,000} = 0.0172$$

$$\bar{p} = \frac{x_1 + x_2}{n_1 + n_2} = \frac{104 + 189}{11,000 + 11,000} = \frac{293}{22,000} = 0.0133$$

$$\bar{q} = 1 - \bar{p} = 1 - 0.0133 = 0.9867$$

$$z = \frac{(\hat{p}_1 - \hat{p}_2) - 0}{\sqrt{\bar{p}\bar{q}\left(\frac{1}{n_1} + \frac{1}{n_2}\right)}} = \frac{(0.00945 - 0.0172) - 0}{\sqrt{(0.0133)(0.9867)}\sqrt{\frac{1}{11,000} + \frac{1}{11,000}}}$$

$$= -5.02$$

In Figure 8-8 we show the test statistic of $z = -5.02$ along with the critical value of $z = -2.33$. The test statistic is within the critical region so we reject the null hypothesis. The sample data support the claim that the heart attack rate is lower for the aspirin group.

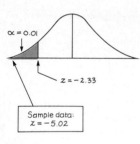

$\alpha = 0.01$

$z = -2.33$

Sample data:
$z = -5.02$

FIGURE 8-8

Does aspirin really prevent heart attacks? The preceding results show that the rate of 104 heart attacks among 11,000 subjects is significantly less than the rate of 189 heart attacks among 11,000 subjects. The media reports of these results included many words of caution and criticism. At the very least, these results suggest that medical researchers should continue with further research. While we cannot use statistics to *prove* that aspirin *causes* fewer heart attacks, we do have strong evidence that helps to guide future medical research.

The symbols x_1, x_2, n_1, n_2, $\hat{p}_1$, $\hat{p}_2$, $\bar{p}$, and $\bar{q}$ should have become more meaningful through this example. In particular, you should recognize that, under the assumption of equal proportions, the best estimate of the common proportion is obtained by pooling both samples into one larger sample. Then

$$\bar{p} = \frac{x_1 + x_2}{n_1 + n_2}$$

More Police, Fewer Crimes?

Does an increase in the number of police officers result in lower crime rates? The question was studied in a New York City experiment that involved a 40% increase in police officers in one precinct while adjacent precincts maintained a constant level of officers. Statistical analysis of the crime records showed that crimes visible from the street (such as auto thefts) did decrease, but crimes not visible from the street (such as burglaries) were not significantly affected.

becomes a more obvious estimate of the common population proportion.

So far we have discussed only proportions in this section, but probabilities are already in decimal or fractional form, so they can directly replace proportions in the preceding discussion. Percentages can also be dealt with by using the corresponding decimal equivalents, as illustrated in the following example.

EXAMPLE

In a study of people who stop to help drivers with disabled cars, researchers hypothesized that more people would stop if they first saw someone else getting help. In one experiment, 2000 drivers first saw a woman being helped with a flat tire and 2.90% of them stopped to help a second woman with a flat tire. Among 2000 drivers who did not see the first helper, only 1.75% stopped to help a woman with a flat tire. (See "Help On the Highway," by McCarthy, *Psychology Today*, July 1987.) At the 0.05 significance level, test the claim that the two proportions are equal.

Solution

The claim of equal proportions leads to $H_0: p_1 = p_2$ and $H_1: p_1 \neq p_2$. We can summarize the given data as follows.

Saw earlier helper	Saw no earlier helper
$n_1 = 2000$	$n_2 = 2000$
$x_1 = 2.90\%$ of $2000 = 58$	$x_2 = 1.75\%$ of $2000 = 35$

With $\hat{p}_1 = x_1/n_1 = 0.0290$ and $\hat{p}_2 = x_2/n_2 = 0.0175$, we can now find $\bar{p}$ and $\bar{q}$.

$$\bar{p} = \frac{x_1 + x_2}{n_1 + n_2} = \frac{58 + 35}{2000 + 2000} = \frac{93}{4000} = 0.02325$$

$$\bar{q} = 1 - \bar{p} = 0.97675$$

We continue by computing the value of the test statistic.

$$z = \frac{(\hat{p}_1 - \hat{p}_2) - 0}{\sqrt{\bar{p}\bar{q}\left(\dfrac{1}{n_1} + \dfrac{1}{n_2}\right)}} = \frac{(0.0290 - 0.0175) - 0}{\sqrt{(0.02325)(0.97675)\left(\dfrac{1}{2000} + \dfrac{1}{2000}\right)}}$$

$$= 2.41$$

continued ▶

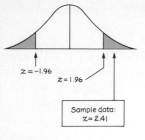

$z = -1.96$

$z = 1.96$

Sample data:
$z = 2.41$

FIGURE 8-9

Solution, continued

With $\alpha = 0.05$ in this two-tailed test, we use Table A-2 to obtain the critical z values of -1.96 and 1.96. See Figure 8-9. Since the test statistic falls within the critical region, we reject the null hypothesis. We conclude that there is sufficient sample evidence to warrant rejection of the claim that the two proportions are equal.

The P-value can be found by using Table A-2. The test statistic of $z = 2.41$ corresponds to an area of 0.4920. Since this test is two-tailed, the P-value is $2 \times (0.5000 - 0.4920) = 0.016$. Since the P-value of 0.016 is less than the significance level of 0.05, we should reject the null hypothesis of equal proportions.

Shown below is the STATDISK display for the data in the preceding example.

STATDISK DISPLAY

```
Hypothesis test for a claim about TWO POPULATION PROPORTIONS

            NULL HYPOTHESIS: p1 = p2

   First sample: x1 = 58    n1 = 2000    x1/n1 = 0.0290

  Second sample: x2 = 35    n2 = 2000    x2/n2 = 0.0175

           Test statistic ... z =   2.4132

           Critical value ... z = -1.9604 , 1.9604

   P-value = 0.0159      Significance level = .05

           CONCLUSION: REJECT the null hypothesis
```

8-4 EXERCISES A

In Exercises 8-65 through 8-68, use the given sample data to determine the values of $\hat{p}_1$, $\hat{p}_2$, $\bar{p}$, $\bar{q}$, and the z test statistic.

8-65 Sample A: Of 200 voters polled, 67 are Democrats.
Sample B: Of 400 voters polled, 148 are Democrats.

8-66 Sample A: In a sample of 250 adults, 38% smoked during the last week.
Sample B: In a sample of 300 adults, 46% smoked during the last week.

8-67 When 500 randomly selected subjects are divided into two equal groups, 20% of the first group passes a physical fitness test, while 24% of the second group passes the same test.

8-68 In a random sample of 300 men and 400 women, exactly 53% of each group believe in life on other planets.

8-69 Let samples from two populations be such that $x_1 = 45$, $n_1 = 100$, $x_2 = 115$, and $n_2 = 200$.
a. Compute the z test statistic based on the given data.
b. If the significance level is 0.05 and the test is two-tailed, find the critical z values.
c. Test the claim that the two populations have equal proportions using the significance level of $\alpha = 0.05$.
d. Find the P-value.

8-70 Samples taken from two populations yield the data $x_1 = 30$, $n_1 = 250$, $x_2 = 44$, $n_2 = 800$. The hypothesis of equal proportions is to be tested at the $\alpha = 0.02$ significance level.
a. Compute the z test statistic based on the data.
b. Find the critical z values.
c. What conclusion do you reach?

8-71 A manufacturer uses two different production methods. With the first method, 15 defects are present in a random sample of 300 items. With the second method, 20 defects are present in a random sample of 200 items.
a. Compute the value of the z test statistic based on the sample data.
b. If a manager claims that the first method has a lower rate of defects, find the critical value corresponding to a 0.01 significance level.
c. What conclusion do you reach?

8-72 In one county, a random sample of 500 surveyed seniors shows that 350 plan to attend college. In a second county, a random sample of 1000 surveyed seniors shows that 600 plan to attend college.
a. Compute the value of the z test statistic based on the sample data.
b. An educator claims that the first county has a larger percentage of seniors going to college. Find the critical value corresponding to a 0.05 significance level.
c. What conclusion do you reach?

8-73   Test the claim that the sample proportion of 0.26 does not differ significantly from the sample proportion of 0.20 if both sample sizes are 50.

8-74   Test the claim that the sample proportion 0.25 is significantly greater than the sample proportion of 0.20 if both sample sizes are 100. Use a 5% level of significance.

8-75   The New York State Department of Motor Vehicles provided the following motor vehicle conviction data for a recent year.

	Albany County	Monroe County
Total convictions	24,384	60,961
Speeding convictions	10,292	26,074

At the 0.10 significance level, test the claim that the proportion of speeding convictions is the same for both counties. Assume that the given data represent random samples drawn from a larger population.

8-76   In a *New York Times*/CBS News survey, 35% of 552 Democrats felt that the government should regulate airline prices, compared to 41% of the 417 Republicans surveyed. At the 0.05 significance level, test the claim that there is no difference between the proportions of Democrats and Republicans who feel that way.

8-77   When 200 female physicians are randomly selected, it is found that 2.5% of them are surgeons. When 250 male physicians are randomly selected, it is found that 19.2% of them are surgeons. (The data are based on information provided by the American Medical Association.) At the 0.05 significance level, test the claim that the percentage of male surgeons is greater than the percentage of female surgeons.

8-78   For 2750 randomly selected arrests of criminals under 21 years of age, 4.25% involve violent crimes. For 2200 randomly selected arrests of criminals 21 years of age or older, 4.55% involve violent crimes (based on data from Uniform Crime Reports). At the 0.05 significance level, test the claim that both age groups have the same rate of violent crimes.

8-79   In the past year, two professors each taught four sections of a statistics course with 25 students in each section. If 12 students of the first professor failed and 18 students of the second professor failed, test the claim that the rate of failure is the same for the two professors. Assume a significance level of $\alpha = 0.05$.

8-80   In a survey, 500 men and 500 women were randomly selected. Among the men, 52 were ticketed for speeding within the last year, while 27 of the women were ticketed for speeding (based on data from R. H. Bruskin Associates). At the 0.01 significance level, test the claim that women have a lower proportion of speeding tickets.

8-81   In a recent survey of 500 males aged 14 to 24, 3.6% of them were living alone. In a 1960 survey of 750 males aged 14 to 24, 1.6% of them were living alone. (These statistics are based on data from the U.S. Bureau of the Census.) At the 0.01 significance level, test the claim that the more recent rate is greater than the rate in 1960.

8-82   A study was conducted to investigate the use of seat belts in taxicabs. Among 72 taxis observed in Pittsburgh, 36 had seat belts at least partially visible. Among 129 taxis observed in Chicago, 77 had seat belts at least partially visible. (See "The Phantom Taxi Seat Belt," by Welkon and Reisinger, *American Journal of Public Health*, Vol. 67, No. 11.) At the 0.05 level of significance, test the claim that Pittsburgh and Chicago have the same proportion of taxis with seat belts at least partially visible.

8-83   The American College Testing Program provides data showing that 30% of four-year public college freshmen drop out, while the dropout rate at four-year private colleges is 26%. Assume that these results are based on observations of 1000 four-year public college freshmen and 500 four-year private college freshmen. At the 0.05 significance level, test the claim that four-year public and private colleges have the same freshman dropout rate.

8-84   In initial tests of the Salk vaccine, 33 of 200,000 vaccinated children later developed polio. Of 200,000 children vaccinated with a placebo, 115 later developed polio. At the 1% level of significance, test the claim that the Salk vaccine is effective.

8-85   The New York State Department of Motor Vehicles provided the following motor vehicle conviction data for a recent year.

	Albany County	Queens County
Total convictions	24,384	166,197
DWI convictions	558	1,214

At the 0.01 level of significance, test the claim that the proportion of DWI (driving while intoxicated) convictions is lower in Queens County. Assume that the given data represent random samples drawn from a larger population.

8-86   About 10 years ago, a survey of 2000 adults showed that 65% were concerned about nuclear power plants. A recent survey of 2000 adults showed that 82% expressed that same concern (based on data from a Roper poll). At the 0.01 significance level, test the claim that the proportion of concerned adults was lower 10 years ago.

8-87   An advertiser studies the proportion of radio listeners who prefer country music. In region A, 38% of the 250 listeners surveyed indicated a preference for country music. In region B, country music was preferred by 14% of the 400 listeners surveyed. At the 0.02 level of significance, test the claim that region A has a greater proportion of listeners who prefer country music.

8-88   When 300 women aged 18 to 29 are randomly selected, it is found that 31 gave birth during the last year. When 400 women aged 30 to 44 are randomly selected, it is found that 15 gave birth during the last year (based on data from the U.S. Census Bureau). At the 0.01 significance level, test the claim that the birth rate is higher for women in the 18 to 29 age group.

8-4 EXERCISES B

8-89   Refer to the sample data summarized in the accompanying table, which is based on a recent *New York Times*/CBS News voter exit poll. Use a 0.05 level of significance to test the claim that there is a difference between the proportion of males who voted for the Democrat and the proportion of females who voted for the Democrat.

	Democrat	Republican
Male	128	122
Female	324	276

8-90   A confidence interval for the difference between population proportions p_1 and p_2 can be constructed by evaluating

$$(\hat{p}_1 - \hat{p}_2) - E < (p_1 - p_2) < (\hat{p}_1 - \hat{p}_2) + E$$

where

$$E = z_{\alpha/2} \sqrt{\frac{\hat{p}_1 \hat{q}_1}{n_1} + \frac{\hat{p}_2 \hat{q}_2}{n_2}}$$

Construct the 95% confidence interval for the difference between the two proportions of surgeons. (See Exercise 8-77.)

8-91   To test the null hypothesis that the difference between two population proportions is equal to a nonzero constant c, use

$$z = \frac{(\hat{p}_1 - \hat{p}_2) - c}{\sqrt{\dfrac{\hat{p}_1(1 - \hat{p}_1)}{n_1} + \dfrac{\hat{p}_2(1 - \hat{p}_2)}{n_2}}}$$

As long as n_1 and n_2 are both large, the sampling distribution of the above test statistic z will be approximately the standard normal distribution. Suppose a winery is conducting market research in New York and California. In a sample of 500 New Yorkers, 120 like the wine, while a sample of 500 Californians shows that 210 like the wine. Use a 0.05 level of significance to test the claim that the percentage of Californians who like the wine is 25% more than the percentage of New Yorkers who like it.

8-92   Sample data are randomly drawn from three independent populations. The sample sizes and the numbers of successes follow.

Population 1	Population 2	Population 3
$n = 100$	$n = 100$	$n = 100$
$x = \;\;40$	$x = \;\;30$	$x = \;\;20$

a. At the 0.05 significance level, test the claim that $p_1 = p_2$.
b. At the 0.05 significance level, test the claim that $p_2 = p_3$.
c. At the 0.05 significance level, test the claim that $p_1 = p_3$.
d. In general, if hypothesis tests lead to the decisions that $p_1 = p_2$ and $p_2 = p_3$, does it follow that the decision $p_1 = p_3$ will be reached under the same conditions?

VOCABULARY LIST

Define and give an example of each term.

F distribution	denominator	independent
numerator degrees	degrees of	samples
of freedom	freedom	pooled estimate of
	dependent samples	p_1 and p_2

REVIEW

In this chapter we extended to two populations the method of testing hypotheses introduced in Chapter 6. We began by developing a test for comparing two population variances (or standard deviations) that come from two independent populations having normal distributions.

Exit Polls on Their Way Out?

The major television networks have been using exit polls to make early projections of winners in races for senate and gubernatorial seats as well as the presidency. In an exit poll, voters are randomly selected in various precincts as they exit the polling area after voting. These voters are surveyed on their election choices, age, sex, and race. Among other statistical flaws, such polls have no way to verify if the data supplied by the voters is correct. They can easily produce substantial errors. This lack of reliability, combined with pressure to discontinue the practice of making early projections, may cause the exit of exit polls.

We began with a test for comparing two variances or standard deviations since such a test is sometimes used as part of an overall test for comparing two population means. In Section 8-2 we saw that the sampling distribution of $F = s_1^2/s_2^2$ is the F distribution for which Table A-5 was computed.

In Section 8-3 we considered various situations that can occur when we want to use a hypothesis test for comparing two means. We should begin such a test by determining whether or not the two populations are **dependent** in the sense that they are related in some way. When comparing population means that come from two dependent and normal populations, we compute the differences between corresponding pairs of values. Those differences have a mean and standard deviation denoted by $\bar{d}$ and s_d, respectively. In repeated random samplings, the values of $\bar{d}$ possess a Student t distribution with mean μ_d and standard deviation $\sigma_d/\sqrt{n}$.

When using hypothesis tests to compare two population means from independent populations, we encounter four situations that can be summarized best by Figure 8-7. These cases incorporate standard deviations reflecting the property that, if one random variable x has variance σ_x^2 and another independent random variable y has variance σ_y^2, the random variable $x - y$ will have variance $\sigma_x^2 + \sigma_y^2$.

In Section 8-4 we considered hypothesis tests that can be used to compare proportions, probabilities, or percentages that come from two independent populations. We saw that the sample proportions have differences

$$\frac{x_1}{n_1} - \frac{x_2}{n_2} \quad \text{or} \quad \hat{p}_1 - \hat{p}_2$$

that tend to have a distribution that is approximately normal with mean $p_1 - p_2$ and a standard deviation estimated by

$$\sqrt{\bar{p}\bar{q}\left(\frac{1}{n_1} + \frac{1}{n_2}\right)}$$

when $p_1 = p_2$.

Also, $\bar{q} = 1 - \bar{p}$ and $\bar{p}$ is the pooled proportion $(x_1 + x_2)/(n_1 + n_2)$.

Figure 8-10 provides a reference chart for locating the appropriate test. One of the most difficult aspects of hypothesis testing involves the identification of the most appropriate distribution and the selection of the proper test statistic, and Figure 8-10 should help in that determination.

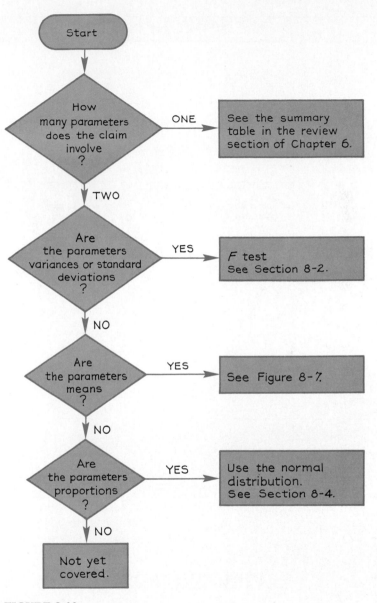

FIGURE 8-10

		IMPORTANT FORMULAS	
Parameters to which hypothesis refers	Applicable distribution	Test statistic	Table of critical values
σ_1, σ_2 (two standard deviations) or σ_1^2, σ_2^2 (two variances)	F	$F = \dfrac{s_1^2}{s_2^2}$ where $s_1^2 \geq s_2^2$	Table A-5
μ_1, μ_2 (two means): dependent samples	Student t	$t = \dfrac{\bar{d} - \mu_d}{s_d/\sqrt{n}}$	Table A-3
independent samples (use Figure 8-7 to determine the correct case)	Normal or Student t	$z = \dfrac{(\bar{x}_1 - \bar{x}_2) - (\mu_1 - \mu_2)}{\sqrt{\dfrac{\sigma_1^2}{n_1} + \dfrac{\sigma_2^2}{n_2}}}$	Table A-2
		$t = \dfrac{(\bar{x}_1 - \bar{x}_2) - (\mu_1 - \mu_2)}{\sqrt{\dfrac{s_1^2}{n_1} + \dfrac{s_2^2}{n_2}}}$	Table A-3
		$t = \dfrac{(\bar{x}_1 - \bar{x}_2) - (\mu_1 - \mu_2)}{\sqrt{\dfrac{1}{n_1} + \dfrac{1}{n_2}} \sqrt{\dfrac{(n_1 - 1)s_1^2 + (n_2 - 1)s_2^2}{n_1 + n_2 - 2}}}$	Table A-3
p_1, p_2 (two proportions)	Normal	$z = \dfrac{(\hat{p}_1 - \hat{p}_2) - (p_1 - p_2)}{\sqrt{\bar{p}\bar{q}\left(\dfrac{1}{n_1} + \dfrac{1}{n_2}\right)}}$	Table A-2

REVIEW EXERCISES

8-93   A manufacturer of shock absorbers operates one plant in the West and one in the East. A random selection of 400 shock absorbers is tested in each plant and 6 defective units are found in the Western plant while 11 defective units are found in the Eastern plant. At the 0.02 level of significance, test the claim that both plants have the same rate of defects.

8-94   In order to test the effectiveness of a lesson, a teacher gives randomly selected students a pretest and a follow-up test. The results are given below. At the 0.025 level of significance, test the claim that the lesson was effective.

Student	A	B	C	D	E	F	G	H
Before	6	8	5	4	3	5	4	7
After	9	10	8	7	6	8	7	10

8-95   Samples of similar wires are randomly selected from two different manufacturers and tested for their breaking strengths. The results are as follows:

Company A   Company B

$n = 10$ $n = 12$
$\bar{x} = 82.0$ kg $\bar{x} = 77.6$ kg
$s = 6.0$ kg $s = 8.1$ kg

At the 0.05 significance level, test the claim that the standard deviations of the breaking strengths of both brands are equal.

8-96   Use the sample data from Exercise 8-95 to test the claim that the two companies produce wire having the same mean breaking strength. Use a 0.05 level of significance.

8-97   A hearing sensitivity test is given to two groups of employees. Group A consists of clerical personnel working in a quiet environment, while group B consists of assembly workers constantly exposed to loud noises. The sample results follow. At the 0.05 significance level, test the claim that there is no difference between the two population means.

Group A   Group B

$n = 50$ $n = 40$
$\bar{x} = 76$ $\bar{x} = 64$
$s = 12$ $s = 16$

8-98   A test question is considered good if it discriminates between good and poor students. The first question on a test is answered correctly by 62 of 80 good students, while 23 of 50 poor students give correct answers. At the 5% level of significance, test the claim that this question is answered correctly by a greater proportion of good students.

8-99   Automobiles are selected at random and tested for fuel economy with each of two different carburetors. The following results show the distance traveled on 1 gallon of gas. At the 5% level of significance, test the claim that both carburetors produce the same mean mileage.

	Distance with carburetors A and B								
Car	1	2	3	4	5	6	7	8	9
A	16.1	21.3	19.2	14.8	29.3	20.2	18.6	19.7	16.4
B	18.2	23.4	19.7	14.7	28.7	23.4	19.0	21.2	18.2

8-100   Two different firms manufacture garage door springs that are designed to produce a tension of 68 kg. Random samples are selected from each of these two suppliers and tension test results are as follows:

Firm A	Firm B
$n = 20$	$n = 32$
$\bar{x} = 66.0$ kg	$\bar{x} = 68.3$ kg
$s = 2.1$ kg	$s = 0.4$ kg

At the 5% level of significance, test the claim that both firms produce the same standard deviation.

8-101   Use the sample data from Exercise 8-100 to test the claim that both firms' springs produce the same mean tension. Use a 0.05 level of significance.

8-102   Do Exercise 8-100 after changing the sample size for firm A to $n = 40$.

8-103   At the 10% level of significance, test the claim that the sample proportion of 0.4 differs significantly from the sample proportion of 0.6. Assume that both sample sizes are 75.

8-104   In a study of the effects of alcohol on driving, randomly selected subjects are given a test of coordination before and after having consumed several drinks. The results are given below. At the 0.05 level of significance, test the claim that alcohol has no effect.

Subject	A	B	C	D	E	F	G	H
Before	16	13	12	15	14	14	17	16
After	9	6	5	9	7	8	11	6

8-105   Two different production methods are used to make batteries for hearing aids. Batteries produced by both methods are randomly selected and tested for longevity (in hours). Eighteen batteries produced by the first method have a standard deviation of 42 h, while 12 batteries produced by the second method have a standard deviation of 78 h. At the 5% level of significance, test the claim that the two production methods yield batteries whose lives have equal standard deviations.

8-106   A bank has branches in two different cities, and it uses a standard credit-rating system for all loan applicants. Randomly selected applicants are chosen from each branch and the results are summarized below. At the 0.05 level of significance, test the claim that both populations have the same mean.

City A	City B
$n = 40$	$n = 60$
$\bar{x} = 43.7$	$\bar{x} = 48.2$
$s = 16.2$	$s = 16.5$

8-107   Do Exercise 8-106 after changing the sample size of City B to $n = 20$.

8-108   A poll reveals that 47.0% of 1500 randomly selected voters in Ohio favor a certain candidate, as do 48.0% of 500 randomly selected voters in Maryland. At the 1% level of significance, test the claim that the candidate is favored by the same percentage in both states.

8-109   To test the effectiveness of a physical training program, researchers asked randomly selected participants to run as far as possible in 5 minutes. This test was conducted before and after the training program and the results follow. The numbers represent distances in meters. At the 5% level of significance, test the claim that the training program was effective.

Before course	510	620	705	590	800	1450	790	830	1220	680
After course	1130	680	810	780	1275	1410	970	1050	1380	1050

8-110   Stores and theaters are randomly selected and their inside temperatures are measured in degrees Celsius. At the 5% level of significance, test the claim that theaters are warmer than stores.

Stores	Theaters
$n = 40$	$n = 32$
$\bar{x} = 18.3$	$\bar{x} = 22.2$
$s = 0.8$	$s = 0.9$

8-111   The manager of a movie theater conducts a study of the ages of those who view two different movies and sample results are given below. At the 0.025 level of significance, test the claim that there is no difference between the two population means.

Movie X	Movie Y
$n = 45$	$n = 65$
$\bar{x} = 22.6$ years	$\bar{x} = 31.0$ years
$s = 5.8$ years	$s = 4.7$ years

8-112   Two types of string are tested for strength, and the sample data follow. Test the claim that there is no difference between the two types of string.

	Breaking load of strings A and B in kilograms
A	23 25 25 28 19 31 35 30 26
B	18 17 16 24 20 21 25 15 15 16 18 21

COMPUTER PROJECT

1. Select any one of the tests listed below and develop a computer program that takes the two sets of data as input and gives as output the value of the appropriate test statistic.

   • Test comparing two variances
   • Test comparing two proportions
   • Test comparing two means for two sets of dependent data
   • Test comparing two means when either σ_1 and σ_2 are both known or n_1 and n_2 are both larger than 30

2. Use existing software that can be run with two sets of data and use it to solve Exercises 8-33, 8-49, and 8-77. (STATDISK can be used for these exercises.)

CASE STUDY ACTIVITY

Select two sample groups as follows. Let one group consist of 15 adults who do not attend college, and let the other group consist of 20 college students. Ask everyone in both groups to estimate the present age of the president. At the 0.05 level of significance, test the claim that both populations have the same mean.

DATA PROJECT

Refer to the data sets in Appendix B.

1. Test the claim that men have a mean pulse rate equal to that for women.

2. Find the number of times the digit 0 occurs among the 276 digits used in the numbers selected as random. Also find the number of times the digit 0 occurs among the 276 digits reported for social security numbers. Test the claim that these two samples come from populations with equal proportions of zeros. What general conclusion is suggested by these results?

Chapter Nine

CHAPTER CONTENTS

9-1   **Overview**
We identify chapter **objectives**. This chapter presents methods for analyzing the relationship between two variables.

9-2   **Correlation**
We use the **scatter diagram** and **linear correlation coefficient** to determine whether a linear relationship exists between two variables.

9-3   **Regression**
We describe linear relationships between two variables by the equation and graph of the **regression line**.

9-4   **Variation**
We analyze the **variation** between **predicted** and **observed** values.

9-5   **Multiple Regression**
This section presents methods of finding a linear equation that relates three or more variables.

9 Correlation and Regression

You might expect that there is a relationship between the selling price of a home and its size, as measured by the living area. In Table 9-1 we list the actual living areas and selling prices for eight homes recently sold in Dutchess County, New York. The living areas are given in hundreds of square feet, so that the first value of 15 indicates a living area of 1500 sq ft. The selling prices are in thousands of dollars, and so the first value of 145 indicates a selling price of $145,000.

TABLE 9-1								
Living area	15	38	23	16	16	13	20	24
Selling price	145	228	150	130	160	114	142	265

Using the given data, can we conclude that there is a relationship between selling price and living area? If so, what is the relationship? An objective of this chapter is to analyze such relationships. The data of Table 9-1 will be considered in the following sections.

9-1 OVERVIEW

Prior to Chapter 8 we considered data collections that involved only a single variable. A typical example is the list in Chapter 2 of 150 values that represent the single variable of home selling price. We proceeded to use such a data set to estimate the value of a population parameter or to test a hypothesis made about a population parameter. In Chapter 8 we presented methods for dealing with two parameters from two populations. In Chapter 8 we considered paired data arranged in tables similar to Table 9-1, but in this chapter we analyze paired data with a very different objective.

An important principle to be learned in an elementary statistics course is that the arrangement and nature of data can affect the particular method of statistics that should be used. Consider the data in Table 9-2. We will consider two scenarios that lead to different methods of analysis.

TABLE 9-2						
x	110	103	105	98	140	112
y	108	104	101	96	135	107

Scenario 1: In Table 9-2, each x-y pair represents "before and after" weights of subjects who were weighed before and after an experimental diet. An effective diet should result in lower weights. The reasonable and sensible question is this: "Are the after weights *significantly less* than the before weights?" A relevant method of analysis is the test of the hypothesis $\mu_x > \mu_y$ where the two populations are dependent. (See Section 8-3.)

Scenario 2: In Table 9-2, each x-y pair represents the adult weights of a pair of identical twins. A reasonable and sensible question is this: "Is there a *relationship between* the weights of the twins?" The relevant method of analysis will be presented in this chapter.

For scenario 1, we are interested in the differences between the numbers in each pair. For scenario 2, we are interested in the presence or absence of a relationship between the two variables. These are fundamentally different issues, and the statistical methods used to consider them will be very different.

In this chapter we investigate ways of analyzing the relationship between two or more variables. Sample data will be paired as in Tables

9-1 and 9-2. Such paired data is sometimes referred to as **bivariate** data. We begin Section 9-2 by describing the **scatter diagram**, which serves as a graph of the sample data. We then investigate the concept of **correlation**, which is used to decide whether there is a statistically significant relationship between two variables. Section 9-3 investigates **regression analysis** as we attempt to identify the exact nature of the relationship between two variables. Specifically, we show how to determine an equation that relates the two variables. In Section 9-4 we analyze the **variation** between predicted and observed values. In Section 9-5 we use concepts of **multiple regression** to describe the relationship among three or more variables.

Throughout this chapter we deal only with **linear** relationships between two or more variables. (Advanced texts consider more variables and nonlinear relationships.)

As a very simple example, consider Table 9-3.

TABLE 9-3							
x	1	2	4	5	7	8	10
y	3	5	9	11	15	17	21

Note that each y value is one more than twice the corresponding x value. This strongly suggests that there exists a very definite relationship between the two variables. In deciding that there is a relationship, we are dealing with the concept of correlation. For the data of Table 9-3 we can identify exactly what the relationship is. The relationship between x and y can be described by the equation $y = 2x + 1$. We use concepts of regression analysis (discussed in Section 9-3) when we determine such equations and use them to make predictions.

9-2 CORRELATION

As we mentioned in the overview, this section deals with the concept of correlation and scatter diagrams as tools that help us decide whether a linear relationship exists between two variables. Each of the two variables should be normally distributed. That is, for any given value of one variable, the distribution of values of the other variable should be normal. Since these tools are designed to analyze relationships between two variables, the sample data must be collected as paired data. We start with an example.

Using the sample data in Table 9-1, is there a linear relationship between the selling price and living area?

We can often form intuitive and qualitative conclusions about paired data by constructing a scatter diagram similar to the one in Figure 9-1, which represents the data in Table 9-1. The points in the figure seem to follow an upward pattern, so we might conclude that there is a relationship between selling price and living area.

Using other collections of paired data, we may get scatter diagrams similar to the examples illustrated in Figure 9-2. The scatter diagram is easily plotted and doesn't require complex computations. In addition, a large collection of paired data may exhibit a pattern and become meaningful when displayed in this form. However, these advantages are often offset by the subjective and qualitative nature of the conclusions that may be drawn from a scatter diagram.

More precise and objective analyses accompany the computation of the **linear correlation coefficient**, which is denoted by r and is given in Formula 9-1 on page 506.

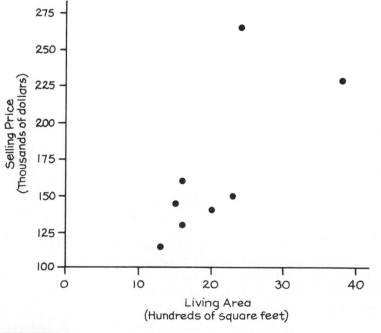

FIGURE 9-1

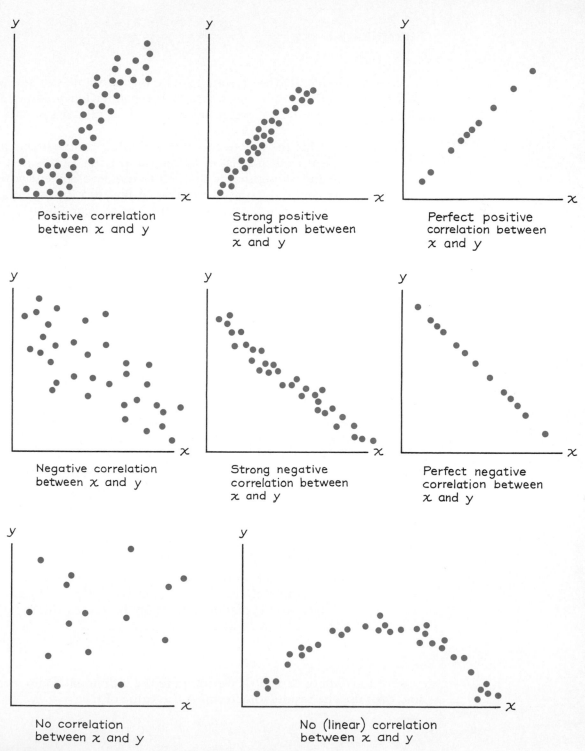

Positive correlation between x and y

Strong positive correlation between x and y

Perfect positive correlation between x and y

Negative correlation between x and y

Strong negative correlation between x and y

Perfect negative correlation between x and y

No correlation between x and y

No (linear) correlation between x and y

FIGURE 9-2

FORMULA 9-1 $\quad r = \dfrac{n\Sigma xy - (\Sigma x)(\Sigma y)}{\sqrt{n(\Sigma x^2) - (\Sigma x)^2}\ \sqrt{n(\Sigma y^2) - (\Sigma y)^2}}$

Since r is calculated using sample data, it is a sample statistic. We might think of r as a point estimate of the population parameter ρ (rho), which is the linear correlation coefficient for all pairs of data in a population.

We now describe the way to compute and interpret the linear correlation coefficient r given a list of paired data. Later in this section we present the underlying theory that led to the development of this formula. Before computing the correlation coefficient r for the data of Table 9-1, we make the following notes relevant to Formula 9-1.

NOTATION
n denotes the **number of pairs** of data present. In Table 9-1, for example, $n = 8$.
Σ denotes the addition of the items indicated.
Σx denotes the sum of all x scores.
Σx^2 indicates that each x score should be squared, and then those squares are added.
$(\Sigma x)^2$ indicates that the x scores should be added and then the total should be squared. It is extremely important to avoid confusion between Σx^2 and $(\Sigma x)^2$.
Σxy indicates that each x score should be multiplied by its corresponding y score. After obtaining all such products, find their sum.
r is the linear correlation coefficient, which measures the strength of the relationship between the paired x and y values in a *sample*. r is a sample statistic.
ρ (rho) is the linear correlation coefficient, which measures the strength of the relationship between all paired x and y values in a *population*. ρ is a population parameter.

For the data of Table 9-1, we compute the individual components and then use the results to determine the value of r.

EXAMPLE

Using the data in Table 9-1, find the value of the linear correlation coefficient r.

Solution

For the sample paired data of Table 9-1 we get $n = 8$ because there are eight pairs of data. The other components required in Formula 9-1 are found from the calculations in the table below. Note how this vertical format makes the calculations easier.

Living Area x	Selling Price y	$x \cdot y$	x^2	y^2
15	145	2175	225	21,025
38	228	8664	1444	51,984
23	150	3450	529	22,500
16	130	2080	256	16,900
16	160	2560	256	25,600
13	114	1482	169	12,996
20	142	2840	400	20,164
24	265	6360	576	70,225
Total: 165	1334	29,611	3855	241,394

$$
\uparrow \qquad \uparrow \qquad \uparrow \qquad \uparrow \qquad \uparrow
$$
$$
\Sigma x \qquad \Sigma y \qquad \Sigma xy \qquad \Sigma x^2 \qquad \Sigma y^2
$$

Using the calculated values, we can now evaluate r as follows.

$$
r = \frac{n(\Sigma xy) - (\Sigma x)(\Sigma y)}{\sqrt{n(\Sigma x^2) - (\Sigma x)^2}\,\sqrt{n(\Sigma y^2) - (\Sigma y)^2}}
$$

$$
= \frac{8(29{,}611) - (165)(1334)}{\sqrt{8(3855) - (165)^2}\sqrt{8(241{,}394) - (1334)^2}}
$$

$$
= \frac{16{,}778}{\sqrt{3615}\sqrt{151{,}596}} = 0.717
$$

After calculating r, how do we interpret the result? Given the way in which Formula 9-1 was derived, it can be shown that the computed value of r must always fall between -1 and $+1$ inclusive. A strong

Student Ratings of Teachers

Many colleges equate high student ratings with good teaching—an equation often fostered by the fact that student evaluations are easy to administer and measure.

However, one study that compared student evaluations of teachers with the amount of material learned found a strong *negative* correlation between the two factors. Teachers rated highly by students seemed to induce less learning.

In a related study, an audience gave a high rating to a lecturer who conveyed very little information but was interesting and entertaining.

positive linear correlation between x and y is reflected by a value of r near $+1$, while a strong negative linear correlation is indicated by a value of r near -1. If r is close to 0, we conclude that there is no significant linear correlation between x and y. We can make this decision process more objective through a formal hypothesis test by using one of two equivalent procedures.

Method 1

Use the Student t distribution with the following.

Null hypothesis: H_0: $\rho = 0$

Test statistic: $t = \dfrac{r - \mu_r}{s_r} = \dfrac{r}{\sqrt{\dfrac{1 - r^2}{n - 2}}}$

Critical value: t (from Table A-3), with $n - 2$ degrees of freedom

In the test statistic given above, we expressed t as $(r - \mu_r)/s_r$ to follow the same format of earlier chapters, but μ_r is actually ρ, which is assumed to be 0.

Method 2

Instead of calculating the test statistic given above, a second method allows us to use the computed value of r as the test statistic. With r as the test statistic, we can find the critical value from Table A-6. The critical values of r in Table A-6 are found by solving

$$t = \frac{r}{\sqrt{\dfrac{1 - r^2}{n - 2}}}$$

for r to get

$$r = \frac{t}{\sqrt{t^2 + n - 2}}$$

where the t value is found from Table A-3 by assuming a two-tailed case with $n - 2$ degrees of freedom. Table A-6 lists the results for selected values of n and α.

EXAMPLE

Use Table A-6 to find the critical values of the linear correlation coefficient r if we have eight pairs of data (as in Table 9-1), the significance level is $\alpha = 0.05$, and the hypothesis test is two-tailed.

Solution

Refer to Table A-6 and locate the critical r value of 0.707 corresponding to $n = 8$ and $\alpha = 0.05$. Since the hypothesis test is two-tailed, we have the critical values of 0.707 and -0.707.

If we reject $H_0: \rho = 0$ and r is positive, we conclude that there is a significant positive linear correlation. If we reject $H_0: \rho = 0$ and r is negative, we conclude that there is a significant negative linear correlation. If we fail to reject $H_0: \rho = 0$, we conclude that there is no significant linear correlation.

See Figure 9-3, which summarizes the two methods we have described. Some instructors prefer the first method because it reenforces concepts introduced in earlier chapters. Others prefer the second method because it eliminates the step of calculating the test statistic in terms of t.

EXAMPLE

From eight pairs of data, r is computed to be 0.717. When testing the claim that $\rho = 0$, what can you conclude at the significance level of $\alpha = 0.05$?

Solution

With $H_0: \rho = 0$ and $H_1: \rho \neq 0$ and $\alpha = 0.05$, we proceed to obtain the value of the test statistic and the critical value.

Method 1: The test statistic is

$$t = \frac{r}{\sqrt{\dfrac{1 - r^2}{n - 2}}} = \frac{0.717}{\sqrt{\dfrac{1 - 0.717^2}{8 - 2}}} = 2.520$$

continued ▶

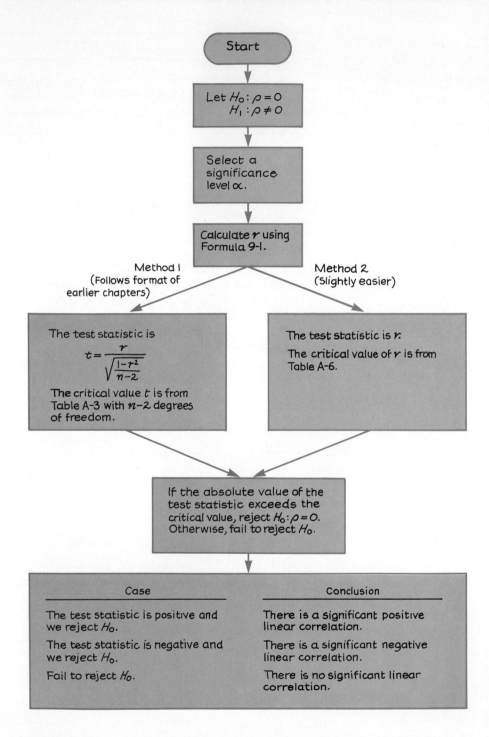

FIGURE 9-3

> **Solution, continued**
>
> The critical value of $t = 2.447$ is found from Table A-3. It corresponds to $n - 2 = 8 - 2 = 6$ degrees of freedom and $\alpha = 0.05$ in two tails. Since the test statistic falls in the critical region, we reject H_0. (See Figure 9-4.) We conclude that there is a significant positive linear correlation.
>
> *Method 2*: The test statistic is $r = 0.717$. The critical value of $r = 0.707$ is found from Table A-6. See Figure 9-5. Since the test statistic falls within the critical region, we reject H_0. We conclude that there is a significant positive linear correlation.
>
> Based on the results of either method, there appears to be a significant positive linear correlation between selling price and living area.

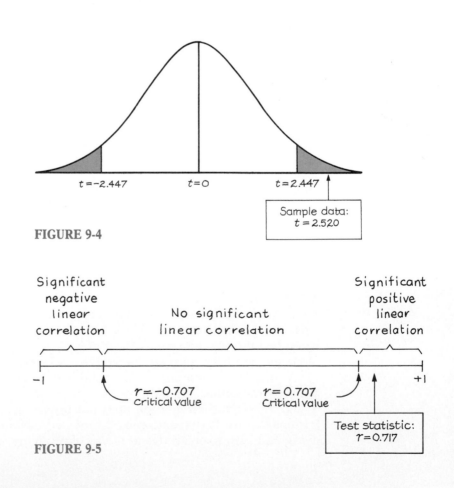

FIGURE 9-4

FIGURE 9-5

The preceding example and Figures 9-4 and 9-5 correspond to a two-tailed hypothesis test. The examples and exercises of this section will generally involve only two-tailed tests. One-tailed tests can occur with a claim of a positive correlation or a claim of a negative correlation. In such cases, the hypotheses will be as shown below.

Left-tailed test	Right-tailed test
$H_0: \rho \geq 0$	$H_0: \rho \leq 0$
$H_1: \rho < 0$	$H_1: \rho > 0$

Method 1 can be handled as in earlier chapters. For method 2, either double the significance level given in Table A-6 or calculate the critical value as in Exercise 9-29.

Common Errors

There are some common errors that are often made with the interpretation of results involving correlation. We will discuss three of these common errors.

1. *We must be careful to avoid the common error of concluding that a significant linear correlation between two variables is proof that there is a cause-and-effect relationship between them.* One study, for example, showed that there is a significant positive linear correlation between teachers' salaries and per capita beer consumption. The significance of the correlation implies that teachers are using their raises to buy more beer, right? Wrong. Perhaps increases in teachers' salaries precipitate higher taxes, which in turn cause taxpayers to drown their sorrows and forget their financial difficulties by drinking more beer. Or perhaps higher teachers' salaries and greater beer consumption are both manifestations of some other factor, such as general improvement in the standard of living. In any event, the techniques in this chapter can be used only to establish a linear relationship. *We cannot establish the existence or absence of any inherent cause-and-effect relationship* between the two variables. The cause-and-effect issue is considered by the professionals in the different fields, such as psychologists, sociologists, biologists, and so on. Mark Twain satirized this lack of causality when he commented on a cold winter by saying: "Cold! If the thermometer had been an inch longer, we'd all have frozen to death."

A medical researcher may establish a significant correlation between the unhealthy habit of smoking and the unhealthy habit of dying. Yet such a correlation does not prove that smoking causes or hastens deaths. Perhaps people become nervous about the prospect of dying and turn to cigarettes as a way of relieving that tension. Maybe dancing puts a strain on the heart that ultimately leads to death and, in the process, creates a biological urge to smoke. Statisticians cannot determine the inherent cause-and-effect relationship, but they can assist and guide the medical researcher in the analysis of the relevant physiological and biological processes.

2. *Another source of potential error arises with data based on rates or averages.* When we use rates or averages for data, we suppress the variation among the individuals, which may easily lead to an inflated correlation coefficient. As an example, one study produced a 0.4 linear correlation coefficient for paired data relating income and education among *individuals*, but the correlation coefficient became 0.7 when regional *averages* were used.

3. *A third misuse of the correlation coefficient involves the concept of linearity.* The linear correlation coefficient r, as discussed in this section, is significant only if the paired data follow a linear or straight-line pattern. Consider the data of Table 9-4 along with the corresponding scatter diagram of Figure 9-6. Table 9-4 and Figure 9-6 represent nine pairs of data obtained from a physical experiment that consists of shooting an object upward and recording the height of the object at different times after its release. For example, the pair of "2 seconds: 192 feet" indicates that the object is at a height of 192 feet exactly 2 seconds after it is shot upward. While Figure 9-6 exhibits a clearly recognizable pattern, the relationship is not linear. The application of

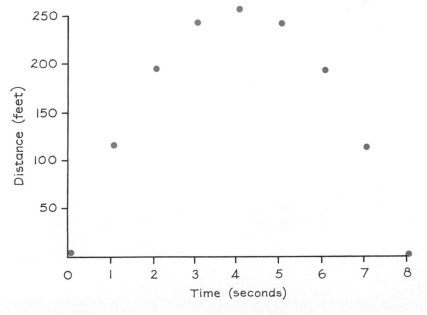

FIGURE 9-6

Formula 9-1 to the data of Table 9-4 reveals that $r = 0$, an indication that there is no *linear* relationship between the two variables. A nonlinear relationship becomes very obvious when we examine Figure 9-6. (The techniques for dealing with these nonlinear cases are beyond the scope of this text.)

TABLE 9-4									
Time (s)	0	1	2	3	4	5	6	7	8
Distance (ft) above ground	0	112	192	240	256	240	192	112	0

So far we have presented the formula for computing the linear correlation coefficient r, but we have given no justification for it. Formula 9-1 is actually a simplified form of the equivalent formula

$$r = \frac{\Sigma(x - \bar{x})(y - \bar{y})}{(n - 1)s_x s_y}$$

While this formula is equivalent to Formula 9-1, we find that Formula 9-1 is generally easier to work with, especially if we use a calculator. (Several inexpensive calculators are designed to compute the linear correlation coefficient r directly. The user simply enters the sample data in pairs and then presses the appropriate key to obtain r.)

Formula 9-1 is a shortcut form of the preceding formula for the linear correlation coefficient, but the following discussion refers to the above formula since its form relates more directly to underlying theory. We will consider the paired data

x	1	1	2	4	7
y	4	5	8	15	23

depicted in the scatter diagram of Figure 9-7. Figure 9-7 includes the point $(\bar{x}, \bar{y})$, which is called the **centroid** of the sample points.

Sometimes r is called **Pearson's product moment**, and that title reflects both the fact that it was first developed by Karl Pearson (1857–1936) and that it is based on the product of the moments $(x - \bar{x})$ and $(y - \bar{y})$. That is, Pearson based the measure of scattering on the statistic $\Sigma(x - \bar{x})(y - \bar{y})$. In any scatter diagram, vertical and horizontal lines through the centroid $(\bar{x}, \bar{y})$ divide the diagram into four quadrants (see

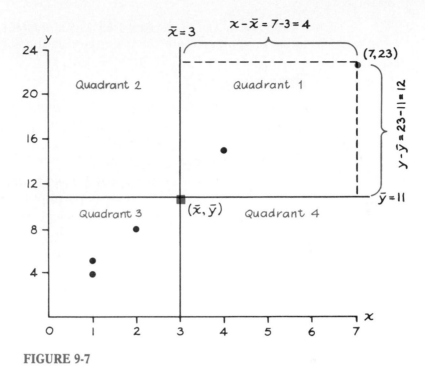

FIGURE 9-7

Figure 9-7). If the points of the scatter diagram tend to approximate an uphill line (as in the figure), individual values of $(x - \bar{x})(y - \bar{y})$ tend to be positive, since the points are predominantly found in the first and third quadrants, where the products of $(x - \bar{x})$ and $(y - \bar{y})$ are positive. If the points of the scatter diagram approximate a downhill line, the points are predominantly in the second and fourth quadrants where $(x - \bar{x})$ and $(y - \bar{y})$ are opposite in sign, so $\Sigma(x - \bar{x})(y - \bar{y})$ tends to be negative. If the points follow no linear pattern, they tend to be scattered among the four quadrants, so $\Sigma(x - \bar{x})(y - \bar{y})$ tends to be close to zero.

The sum $\Sigma(x - \bar{x})(y - \bar{y})$ depends on the magnitude of the numbers used, yet r should not be affected by the particular scale used. For example, r should not change whether heights are measured in meters or centimeters. We can make r independent of the scale used by incorporating the sample standard deviations as follows.

$$r = \frac{\Sigma(x - \bar{x})(y - \bar{y})}{(n - 1)s_x s_y}$$

This expression can be algebraically manipulated into the equivalent form of Formula 9-1 (see Exercise 9-36a).

The format of Formula 9-1 or any equivalent formula leads to the following properties of the linear correlation coefficient *r*.

Properties of *r*

1. *The value of* r *is always between* −1 *and* 1. That is, $-1 \leq r \leq 1$.
2. *The value of* r *does not change if all values of either variable are converted to a different scale.* For example, if the units of *x* are changed from feet to meters, the value of *r* does not change.
3. *The value of* r *is not affected by the choice of* x *or* y. Interchange all *x* and *y* values and the value of *r* will not change.
4. r *measures the strength of a linear relationship.* It is not designed to measure the strength of a relationship that is not linear.

We can use the linear correlation coefficient to decide whether there is a linear relationship between two variables. After deciding that a relationship exists, we can then determine what it is. In the next section we will see how the relationship can be described.

9-2 EXERCISES A

For each part of Exercises 9-1 and 9-2, would you expect positive correlation, negative correlation, or no correlation for each of the given sets of paired data?

9-1   a. The years of education and the incomes of tax payers
      b. People's ages and blood pressures
      c. The amounts of rainfall and vegetation growth
      d. The weights of cars and fuel consumption rates measured in miles per gallon
      e. The hat sizes of adults and their IQ scores

9-2   a. Smoking and lung cancer
      b. Hours spent studying for tests and the resulting test scores
      c. The numbers of absences in a course and the grades in that course
      d. The weights and the heights of children
      e. Annual per capita income for different nations and infant mortality rates for those nations

For each part of Exercises 9-3 and 9-4, a sample of paired data produces a linear correlation coefficient r. What do you conclude in each case? Assume a significance level of $\alpha = 0.05$.

9-3   a. $n = 20, r = 0.5$ 9-4   a. $n = 77, r = 0.35$
       b. $n = 20, r = -0.5$ b. $n = 22, r = 0.37$
       c. $n = 50, r = 0.2$ c. $n = 22, r = 0.40$
       d. $n = 50, r = -0.2$ d. $n = 22, r = -0.5$
       e. $n = 37, r = 0.25$ e. $n = 6, r = -0.8$

In Exercises 9-5 through 9-8, use the given list of paired data.

a. Construct the scatter diagram.   b. Determine n.
c. Find Σx.   d. Find Σx^2.
e. Find $(\Sigma x)^2$.   f. Find Σxy.
g. Find r.

9-5

x	1	1	2	3
y	1	5	4	2

9-6

x	1	2	2	3
y	5	4	3	1

9-7

x	0	1	1	2	5
y	3	3	4	5	6

9-8

x	1	3	3	4	5	5
y	5	3	2	2	0	1

In Exercises 9-9 through 9-28:

a. Construct the scatter diagram.
b. Compute the linear correlation coefficient r.
c. Assume that $\alpha = 0.05$ and find the critical value of r from Table A-6.
d. Based on the results of parts (b) and (c), decide whether there is a significant positive linear correlation, a significant negative linear correlation, or no significant linear correlation. In each case, assume a significance level of $\alpha = 0.05$.
e. Save all your work. The same data will be used in the exercises of the next section.

9-9   The number of drinks consumed and the corresponding blood alcohol concentrations are listed for various subjects with the same body weight (based on data from the Dutchess County STOP-DWI Program).

Number of drinks	2	2	4	5	8
Blood alcohol concentration	0.05	0.06	0.11	0.13	0.22

9-10   The loads (in pounds) on a spring are listed along with the corresponding lengths (in inches) of the spring.

Load	0	1	3	2	6
Length	5	6	8	7	12

9-11   The table below lists the value of exports (in billions of dollars) and the value of imports (in billions of dollars) for several different years (based on data from the U.S. Department of Commerce).

Exports	10	20	43	221	218	218
Imports	9	15	40	245	326	370

9-12   In a study of the factors that affect success in a calculus course, data were collected for 10 different people. Scores on an algebra placement test are given along with calculus achievement scores. (See "Factors Affecting Achievement in the First Course in Calculus" by Edge and Friedberg, *Journal of Experimental Education*, Vol. 52, No. 3.)

Algebra	17	21	11	16	15	11	24	27	19	8
Calculus	73	66	64	61	70	71	90	68	84	52

9-13   For randomly selected states, the table below lists the per capita beer consumption (in gallons) and the per capita wine consumption (in gallons) (based on data from *Statistical Abstract of the United States*).

Beer	32.2	29.4	35.3	34.9	29.9	28.7	26.8	41.4
Wine	3.1	4.4	2.3	1.7	1.4	1.2	1.2	3.0

9-14   When loads were added to a hanging copper wire, the wire stretched. The loads (in Newtons) and increases in length (in centimeters) are given below. (The table is based on data from *College Physics* by Sears, Zemansky, and Young.)

Added load	0	10	20	30	40	50	60	70
Increase in length	0	0.05	0.10	0.15	0.20	0.25	0.30	1.25

9-15   The accompanying table (based on data from the Educational Testing Service) lists the state average SAT verbal score along with the state average SAT math score for several randomly selected states.

Verbal	421	423	429	424	413	437	461	429
Math	476	467	467	470	453	470	515	463

9-16 Randomly selected girls are given the Wide Range Achievement Test. Their ages are listed along with their scores on the reading part of that test (based on data from the National Health Survey, USDHEW publication 72-1011).

Age	6.1	7.2	5.9	6.3	10.5	11.0
Score	17.8	47.4	25.8	24.3	66.6	91.4

9-17 The accompanying table lists the number of registered automatic weapons (in thousands) along with the murder rate (in murders per 100,000) for randomly selected states. (The data are provided by the FBI and the Bureau of Alcohol, Tobacco, and Firearms.)

Automatic weapons	11.6	8.3	3.6	0.6	6.9	2.5	2.4	2.6
Murder rate	13.1	10.6	10.1	4.4	11.5	6.6	3.6	5.3

9-18 The ages (in years) and serum cholesterol levels (in mg/100 ml) are given for randomly selected adult men. (The data are based on results from the National Health Survey, USDHEW publication 78-1652.)

Age	19	27	21	45	46	58	37	42	30
Cholesterol	217.0	221.4	191.3	321.5	196.0	284.6	286.8	194.2	247.7

9-19 Emissions data are given (in grams per meter) for a sample of different vehicles. (See "Determining Statistical Characteristics of a Vehicle Emissions Audit Procedure" by Lorenzen, *Technometrics*, Vol. 22, No. 4.)

HC	0.65	0.55	0.72	0.83	0.57	0.51	0.43	0.37
CO	14.7	12.3	14.6	15.1	5.0	4.1	3.8	4.1

9-20 Two different tests are designed to measure one's understanding of a certain topic. Two tests are given to ten different subjects and the results are listed in the following table.

Test X	75	78	88	92	95	67	55	73	74	80
Test Y	81	73	85	85	89	73	66	81	81	81

9-21 The researchers in a laboratory experiment with a car at different speeds in an attempt to study the fuel consumption rates as measured in miles per gallon (mi/gal). The accompanying data are obtained.

Speed	15	23	30	35	42	45	50	54	60	65
mi/gal	14	17	20	24	26	23	18	15	11	10

9-22 There are many regions where the winter accumulation of snowfall is a primary source of water. Several investigations of snowpack characteristics have used satellite observations from the Landsat series along with measurements taken on earth. Given here are ground measurements of snow depth (in centimeters) along with the corresponding temperatures (in degrees Celsius). (The data are based on information in Kastner's *Space Mathematics*, published by NASA.)

Temperature (°C)	−62	−41	−36	−26	−33	−56	−50	−66
Snow depth (cm)	21	13	12	3	6	22	14	19

9-23 For randomly selected homes recently sold in Dutchess County, New York, the annual tax amounts (in thousands of dollars) are listed along with the selling prices (in thousands of dollars).

Taxes	1.9	3.0	1.4	1.4	1.5	1.8	2.4	4.0
Selling price	145	228	150	130	160	114	142	265

9-24 For randomly selected homes recently sold in Dutchess County, New York, the living areas (in hundreds of square feet) are listed along with the annual tax amounts (in thousands of dollars).

Living area	15	38	23	16	16	13	20	24
Taxes	1.9	3.0	1.4	1.4	1.5	1.8	2.4	4.0

9-25 A manager in a factory randomly selects 15 assembly-line workers and develops scales to measure their dexterity and productivity levels. The results are listed in the following table.

Productivity	63	67	88	44	52	106	99	110	75	58	77	91	101	51	86
Dexterity	2	9	4	5	8	6	9	8	9	7	4	10	7	4	6

9-26 Randomly selected subjects are given a standard IQ test and then tested for their receptivity to hypnosis. The results are listed below.

IQ	103	113	119	107	78	153	114	101	103	111	105	82	110	90	92
Receptivity to hypnosis	55	55	59	64	45	72	42	63	62	46	41	49	57	52	41

9-27 The following table lists per capita cigarette consumption in the United States for various years, along with the percentage of the population admitted to mental institutions as psychiatric cases.

Cigarette consumption	3522	3597	4171	4258	3993	3971	4042	4053
Percentage of psychiatric admissions (in percentage points)	0.20	0.22	0.23	0.29	0.31	0.33	0.33	0.32

9-28   In a study of employee stock ownership plans, data were collected at eight companies on satisfaction with the plan and the amount of organizational commitment. Results are given in the accompanying table, which is based on "Employee Stock Ownership and Employee Attitudes: A Test of Three Models" by Klein, *Journal of Applied Psychology*, Vol. 72, No. 2.

Satisfaction	5.05	4.12	5.39	4.17	4.00	4.49	5.40	4.86
Commitment	5.37	4.49	5.42	4.45	4.24	5.34	5.62	4.90

9-2 EXERCISES B

9-29   Assume that the null hypothesis H_0: $\rho = 0$ is to be tested at the $\alpha = 0.10$ level of significance. Use

$$r = \frac{t}{\sqrt{t^2 + n - 2}}$$

to find the critical values of r corresponding to $n = 4, 5, 10, 20, 30, 40, 50, 100$. Be sure to use $n - 2$ degrees of freedom when referring to Table A-3.

9-30   Use

$$r = \frac{t}{\sqrt{t^2 + n - 2}}$$

to find the critical values of r for the indicated one-tailed cases.
a. H_0: $\rho \geq 0$, $n = 20$, $\alpha = 0.05$
b. H_0: $\rho \leq 0$, $n = 10$, $\alpha = 0.05$
c. H_1: $\rho > 0$, $n = 12$, $\alpha = 0.01$
d. H_1: $\rho < 0$, $n = 25$, $\alpha = 0.10$
e. H_0: $\rho \geq 0$, $n = 16$, $\alpha = 0.10$

9-31   Attempt to compute the linear correlation coefficient r for the data in the table and comment on the results. Also, plot the scatter diagram.

x	0	3	5	5	6
y	2	2	2	2	2

9-32   Do Exercise 9-10 after interchanging each x value with the corresponding y value and then changing all the lengths from inches to centimeters (1 in. = 2.54 cm). How is the value of the linear correlation coefficient affected?

9-33   Do Exercise 9-10 after changing the length of 12 in. to 36 in. How much effect does an extreme value have on the value of the linear correlation coefficient?

9-34 Do Exercise 9-10 after reversing the order of the y values. How is the value of the linear correlation coefficient affected?

9-35 The graph of $y = x^2$ is a parabola, not a straight line, so we might expect that the value of r would not reflect a linear correlation between x and y. Using $y = x^2$, make a table of x and y values for $x = 0, 1, 2, \ldots, 10$ and calculate the value of r. What do you conclude? How do you explain the result?

9-36 a. Show that

$$\frac{\Sigma(x - \bar{x})(y - \bar{y})}{(n - 1)s_x s_y} = \frac{n\Sigma xy - (\Sigma x)(\Sigma y)}{\sqrt{n(\Sigma x^2) - (\Sigma x)^2}\,\sqrt{n(\Sigma y^2) - (\Sigma y)^2}}$$

b. Show that Formula 9-1 is equivalent to

$$r = \frac{(\overline{xy}) - \bar{x} \cdot \bar{y}}{\sqrt{[(\overline{x^2}) - (\bar{x})^2][(\overline{y^2}) - (\bar{y})^2]}}$$

where $\overline{xy} = \Sigma xy/n$ and $(\overline{x^2}) = \Sigma x^2/n$.

9-3 REGRESSION

In Section 9-2 we tested paired data for the presence or absence of the statistical relationship of a linear correlation. In this section we identify that relationship. The relationship is expressed in the form of a linear, or straight-line, equation. Such an equation is often useful in predicting the likely value of one variable given a value of another. The straight line that concisely summarizes the relationship between the two variables is the **regression line**. (It is also called the line of best fit or the least-squares line.)

Sir Francis Galton (1822–1911), a cousin of Charles Darwin, studied the phenomenon of heredity in which certain characteristics regress or revert to more typical values. Galton noted, for example, that children of tall parents tend to be shorter than their parents, while short parents tend to have children taller than themselves (when fully grown, of course). These original studies of regression evolved into a fairly sophisticated branch of mathematics called *regression analysis*, which includes the consideration of linear and nonlinear relationships.

This section is confined to linear relationships for two basic reasons. First, the real relationship between two variables is often a linear relationship, or it can be effectively approximated by a linear relationship. Second, nonlinear or curvilinear regression problems introduce complexities beyond the scope of this introductory text.

Let's reconsider the paired data of Table 9-1, which lists home selling prices along with the sizes of the homes, as measured by their living areas. We saw in Section 9-2 that the value of the linear correlation coefficient is computed to be $r = 0.717$. We also saw that at the 0.05 level of significance, there is a positive linear correlation between home selling price and living area. Knowing that there is a linear correlation between those two variables, we now need to find the equation of the straight line that relates them. We again stipulate that the variable x represents living area, while y represents home selling price. We want an equation of the form $y = Mx + B$, where M and B are the **slope** and **y-intercept** of the true regression line. While we cannot find the true values of the parameters M and B, their point estimates m and b can be found from the paired data by using the Formulas 9-2 and 9-3.

FORMULA 9-2
$$m = \frac{n(\Sigma xy) - (\Sigma x)(\Sigma y)}{n(\Sigma x^2) - (\Sigma x)^2}$$

FORMULA 9-3
$$b = \frac{(\Sigma y)(\Sigma x^2) - (\Sigma x)(\Sigma xy)}{n(\Sigma x^2) - (\Sigma x)^2}$$

Some inexpensive calculators accept entries of paired data and provide the m and b values directly. These formulas appear formidable, but three observations make the required computations easier. First, if the correlation coefficient r has been computed by Formula 9-1, the values of Σx, Σy, Σx^2, $(\Sigma x)^2$, and Σxy have already been computed. These values can now be used again in Formulas 9-2 and 9-3. (Note that the numerator for r in Formula 9-1 is identical to the numerator for m in 9-2.) Second, examine the denominators of the formulas for m and b and note that they are identical. This means that the computation of $n(\Sigma x^2) - (\Sigma x)^2$ need be done only once, and the result can be used in both formulas. Third, the regression line always passes through the centroid $(\bar{x}, \bar{y})$, so the equation $\bar{y} = m\bar{x} + b$ must be true. This implies that $b = \bar{y} - m\bar{x}$, and it is usually easier to evaluate b by computing $\bar{y} - m\bar{x}$ than by using Formula 9-3.

FORMULA 9-4
$$b = \bar{y} - m\bar{x}$$

We now use the home selling price and living area data to find the equation of the regression line. We will use the following statistics already found in the first example of Section 9-2.

$$n = 8 \qquad \Sigma x^2 = 3855$$
$$\Sigma x = 165 \qquad \Sigma y^2 = 241{,}394$$
$$\Sigma y = 1334 \qquad \Sigma xy = 29{,}611$$

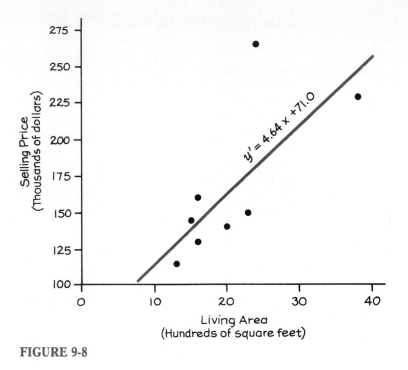

FIGURE 9-8

Having determined the values of the individual components, we can now compute m by using Formula 9-2.

$$m = \frac{n(\Sigma xy) - (\Sigma x)(\Sigma y)}{n(\Sigma x^2) - (\Sigma x)^2} = \frac{8(29,611) - (165)(1334)}{8(3855) - (165)^2}$$

$$= \frac{16,778}{3615} = 4.64$$

We now use Formula 9-4 to find the value of b.

$$b = \bar{y} - m\bar{x} = \frac{1334}{8} - \left(\frac{16,778}{3615}\right)\left(\frac{165}{8}\right) = 71.0$$

We will use y' to represent the **predicted value** of y so that the preceding results allow us to express $y' = mx + b$ as $y' = 4.64x + 71.0$. In Figure 9-8 we show the graph of this line in the scatter diagram of the original sample data.

We should realize that $y' = 4.64x + 71.0$ is an estimate of the true straight-line equation $y = Mx + B$. That estimate is based on one particular set of sample data. Another sample drawn from the same population will probably lead to a slightly different equation. We might think of these equations as point estimates of $y = Mx + B$.

Let's review the notation we are using.

NOTATION		
	Population parameter	**Point estimate**
Linear correlation coefficient	ρ	r
Slope of regression equation	M	m
y-intercept of regression equation	B	b
Equation of the regression line	$y = Mx + B$	$y' = mx + b$

The following example incorporates concepts from this section, along with the linear correlation coefficient from the preceding section.

EXAMPLE

Two different tests are designed to measure the understanding of a certain topic and are administered to ten subjects. The results follow. Find the linear correlation coefficient r, the equation of the regression line, and plot the scatter diagram and regression line on the same graph. Determine whether there is a significant linear correlation and then predict the likely score on test Y for someone receiving a score of 70 on test X. Use a significance level of 0.05.

Test X	75	78	88	92	95	67	55	73	74	80
Test Y	81	73	85	85	89	73	66	81	81	81

Solution

With $H_0: \rho = 0$ and $H_1: \rho \neq 0$, and $\alpha = 0.05$, we proceed to evaluate the test statistic. For the sample data, we determine the following:

$$n = 10 \qquad \Sigma y^2 = 63,629$$
$$\Sigma x = 777 \qquad (\Sigma x)^2 = 603,729$$
$$\Sigma y = 795 \qquad (\Sigma y)^2 = 632,025$$
$$\Sigma x^2 = 61,661 \qquad \Sigma xy = 62,432$$

continued ▶

Solution, continued

We now compute the value of the linear correlation coefficient r.

$$r = \frac{n\Sigma xy - (\Sigma x)(\Sigma y)}{\sqrt{n(\Sigma x^2) - (\Sigma x)^2}\sqrt{n(\Sigma y^2) - (\Sigma y)^2}}$$

$$= \frac{10(62,432) - (777)(795)}{\sqrt{10(61,661) - 603,729}\sqrt{10(63,629) - 632,025}}$$

$$= \frac{6605}{\sqrt{12,881}\sqrt{4265}} = 0.891$$

With $n = 10$, the test statistic of $r = 0.891$ indicates a significant positive linear correlation since it exceeds the critical r value of 0.632 (for $\alpha = 0.05$). That is, at the 0.05 level of significance, we conclude that there is a significant positive linear correlation. Having determined that r is significant, we could proceed to obtain the regression values of m and b.

$$m = \frac{n\Sigma xy - (\Sigma x)(\Sigma y)}{n(\Sigma x^2) - (\Sigma x)^2} = \frac{10(62,432) - (777)(795)}{10(61,661) - 603,729}$$

$$= \frac{6605}{12,881} = 0.51$$

$$b = \bar{y} - m\bar{x} = 79.5 - 0.51(77.7) = 39.87$$

The general equation of the regression line $y' = mx + b$ becomes $y' = 0.51x + 39.87$. The scatter diagram appears in Figure 9-9, along with the graph of the regression line. The line $y' = 0.51x + 39.87$ is plotted by generating the coordinates of two different points on that line. Specifically, when $x = 60$, the predicted value of y becomes $y' = 0.51(60) + 39.87 = 70.47$, so $(60, 70.47)$ is on the regression line and appears as the leftmost asterisk in Figure 9-9. Letting $x = 90$ implies that $y' = 0.51(90) + 39.87 = 85.77$, so that $(90, 85.77)$ represents a second point on the regression line that corresponds to the right asterisk in the figure. Finally, we use the equation of the regression line to find the last item requested. To predict the score on test Y given a score of 70 on test X, we simply substitute 70 for x in the equation of the regression line. If $x = 70$, $y' = 0.51x + 39.87$ becomes $y' = 0.51(70) + 39.87 = 75.57$, which is our predicted point estimate of the score.

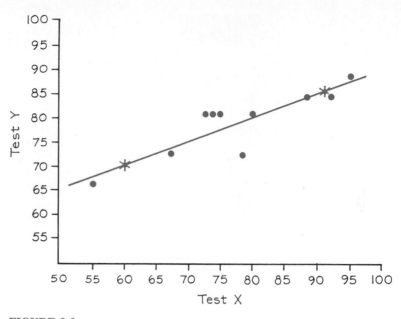

FIGURE 9-9

By examining Figure 9-9 and the computed value of the linear correlation coefficient for the given data, we conclude that the regression line fits the data reasonably well. It is therefore useful in making projections or predictions that do not go far beyond the scope of the available scores. However, if r is close to zero, even though the regression line is the best fitting line, it may not fit the data well enough. It is important to note that the value of r indicates how well the regression line actually fits the available paired data. If $r = 1$ or $r = -1$, then the regression line fits the data perfectly. If r is near $+1$ or -1, the regression line constitutes a very good approximation of the data. But if r is near zero, the regression line fits poorly. When we must estimate the value of one variable given some value of the other, **we should use the equation of the regression line only if r indicates that there is a significant linear correlation. However, in the absence of a significant linear correlation, we should not use the regression equation for projecting or predicting. Instead, our best estimate of the second variable is simply the sample mean of that variable**, regardless of the value assigned to the first variable. See Figure 9-10. To illustrate this concept, let's suppose that we have the two samples of paired data. In both cases we want to estimate y when $x = 5$.

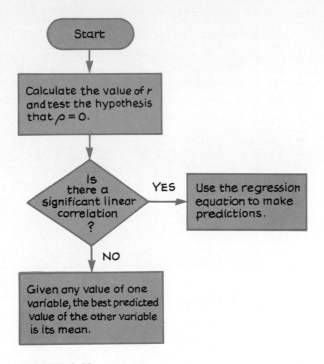

FIGURE 9-10

First collection of paired data	Second collection of paired data
$n = 100$ Regression line: $y' = 2x + 3$ $\bar{y} = 20$ $r = 0.95$	$n = 100$ Regression line: $y' = 2x + 3$ $\bar{y} = 20$ $r = 0.02$
To get the best point estimate of y when $x = 5$, use the equation of the regression line to get $y' = 2(5) + 3 = 13$	To get the best point estimate of y for any value of x, simply select $\bar{y} = 20$.

     In the first case, $r = 0.95$ indicates that the equation $y' = 2x + 3$ will give good results, since the regression line fits the data well. Consequently, the best estimate of y for $x = 5$ can be obtained by substituting 5 for x in the equation of the regression line. The estimated y value of 13 results. However, when we consider the second collection

What SAT Scores Measure

Harvard psychologist David McClellan says that many statistical surveys "have shown that no consistent relationship exists between SAT scores in college students and their actual college accomplishments in social leadership, the arts, science, music, writing, speech, and drama." There does appear to be a significant correlation between SAT scores and income levels of the tested students' families. Higher scores tend to come from high-income families and low scores from low-income families. But some studies show that SAT scores have limited value as predictors of college grades and *no* significant relationship to career success. Motivation seems to be a major success factor not measured by the SAT scores.

of paired sample data, we see that the linear correlation coefficient of 0.02 reflects a poorly fitting regression line that is useless as a predictor. In this case, the best estimate of y is 20 (the value of $\bar{y}$).

As a practical application of this concept, suppose we wanted to predict the height of an adult male with an IQ of 107. Knowing that there is no correlation between height and IQ scores, our best prediction of height would be about 5' 9.5" (the mean). But we do know that there is a positive correlation between the length of a beam as measured in feet and the length of the same beam as measured in inches. The regression equation should reveal, not too surprisingly, that $y' = 12x$ where x is the length in feet and y is the length in inches. To predict the number of inches for a beam 8 ft long, we would use the regression equation $y' = 12x$ and get $y' = 12(8) = 96$ in. We would not be concerned with the mean length of all beams.

As another example, we can use the Table 9-1 data, which lists homes selling prices along with living areas. We have already concluded that there is a significant linear correlation (at the 0.05 level of significance) and we have found the regression equation to be $y' = 4.64x + 71.0$. Recall that the x values are living areas in hundreds of square feet, while the y values are selling prices in thousands of dollars. Given a home with a living area of 2000 sq ft, we substitute $x = 20$ into the regression equation to find the predicted value of y. We get $y' = 4.64(20) + 71.0 = 164$. That is, a home with a living area of 2000 sq ft (or $x = 20$) has a predicted selling price of \$164,000 (from $y' = 164$). These examples should reinforce the important points summarized in Figure 9-10.

The previous discussion refers to one common error in using regression equations to make predictions. We now list that error along with three other common errors.

Common Errors

1. *If there is no significant linear correlation, don't use the regression equation to make predictions.*
2. *When using the regression equation for predictions, stay within the scope of the available sample data.* If you find a regression equation that relates women's heights and shoe sizes, it's absurd to predict the shoe size of a woman who is 10 ft tall.
3. *A regression equation based on old data is not necessarily valid now.* The regression equation relating used car prices and the ages of cars is no longer usable if it's based on data from the 1950s.
4. *Don't make predictions about a population that is different from the population from which the sample data were drawn.* If we collect

sample data from males and develop a regression equation relating SAT math scores and SAT verbal scores, the results don't necessarily apply to females. If we use *state averages* to develop a regression equation relating SAT math scores and SAT verbal scores, the results don't necessarily apply to *individuals*.

We have noted a qualitative relationship between the value of r and how well the regression line fits the data. We can establish a more exact relationship between r and the regression line. For paired data, r is related to m (the slope of the regression line) by

FORMULA 9-5 $$r = \frac{ms_x}{s_y}$$

where s_x is the standard deviation of the x values and s_y is the standard deviation of the y values. In the preceding expression, we can easily solve for m to get $m = rs_y/s_x$, and this expression may simplify the calculation of m when r, s_y, and s_x have already been found.

Formulas 9-2 and 9-3 or 9-4 describe the computations necessary to obtain the regression-line equation $y' = mx + b$, and we now describe the criterion used to arrive at these particular formulas. To be concise, the regression line obtained through these formulas is unique in that **the sum of the squares of the vertical deviations of the sample points from the regression line is the smallest sum possible**. This property is called the **least-squares property** and can be understood by examining Figure 9-11 and the text that follows. In Figure 9-11 we show the paired data contained in the following table.

x	1	2	4	5
y	4	24	8	32

The distances identified as errors in Figure 9-11 are the differences between the actual *observed* y values and the y' values **predicted** by the regression equation. Examination of Figure 9-11 clearly shows that no straight line can pass through all four points, so some error is inevitable, but we want it to be minimized.

Application of the formulas for m and b results in the equation $y' = 4x + 5$, which is graphed in Figure 9-11. The least-squares property refers to the vertical distances by which the original points miss the regression line. In this figure, these vertical deviations are shown to be 5, 11, 13, and 7 for the four pairs of data. The sum of the squares of these vertical errors is $5^2 + 11^2 + 13^2 + 7^2 = 364$, and the regression line $y' = 4x + 5$ is unique in that the value of 364 is the lowest for that particular line. Any other line will yield a sum of squares that exceeds 364, and therefore produces a greater collective error. Thus, using the

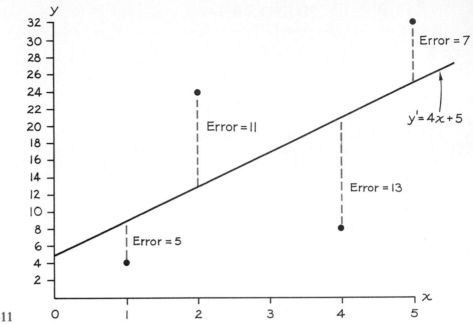

FIGURE 9-11

least-squares property, another line will not fit the data as well as the regression line $y' = 4x + 5$.

For example, the line $y' = 3x + 8$ will produce vertical errors of 7, 10, 12, and 9, so the sum of the squares of those errors is $49 + 100 + 144 + 81 = 374$. The collective error of 374 exceeds a collective error of 364. Consequently, the line $y' = 4x + 5$ provides a better fit to the paired data. Fortunately, we need not deal directly with this least-squares property when we want to obtain the equation of the regression line. Calculus has been used to build the least-squares property into formulas, and our calculations for m and b are a result of that property. Because Formulas 9-2 and 9-3 depend on certain methods of calculus, we have not included their development in this text.

When dealing with larger collections of paired data, the use of a calculator or computer becomes necessary for finding the values of r, m, and b. We have already mentioned that some calculators allow the entry of paired data and provide the values of r, m, and b. Many computer software packages take paired data as input and produce more complete results, such as the STATDISK and Minitab output shown on the next page. The displayed output corresponds to the home selling price and living area data (Table 9-1) introduced in Section 9-1 and used in Sections 9-2 and 9-3. The coefficient of determination and the standard error of estimate will be discussed in the following section.

STATDISK DISPLAY

```
     Linear correlation coefficient ........ r = 0.71671
     Equation of regression line .. Y = 4.64122 X +  71.02490
     Coefficient of determination ............ = 0.51367
     Standard error of estimate .............. = 39.19121
     Level of significance ................... = .05
CONCLUSION: REJECT the claim of no significant linear correlation
     Test statistic is ................... r = 0.71671
     Critical value is ................... r = 0.70737
```

MINITAB DISPLAY

```
          MTB > READ C1 C2
          DATA> 15 145
          DATA> 38 228
          DATA> 23 150
          DATA> 16 130
          DATA> 16 160
          DATA> 13 114
          DATA> 20 142
          DATA> 24 265
          DATA> ENDOFDATA
               8 ROWS READ
          MTB > NAME C1 'AREA' C2 'PRICE'
          MTB > CORRELATION C1 and C2

          Correlation of AREA and PRICE = 0.717

          MTB > REGRESSION C2 1 C1

          The regression equation is
          PRICE = 71.0 + 4.64 AREA
```

9-3 EXERCISES A

In Exercises 9-37 through 9-40, use the data given to obtain the equation of the regression line.

9-37

x	2	3	5
y	4	5	7

9-38

x	2	3	4	6
y	7	9	11	15

9-39

x	1	1	2	3
y	1	5	4	2

9-40

x	1	2	2	3
y	5	4	3	1

In Exercises 9-41 through 9-56, find the equation of the regression line. (The given data is taken from exercises in Section 9-2.)

9-41

Number of drinks	2	2	4	5	8
Blood alcohol concentration	0.05	0.06	0.11	0.13	0.22

9-42

Load	0	1	3	2	6
Length	5	6	8	7	12

9-43

Exports	10	20	43	221	218	218
Imports	9	15	40	245	326	370

9-44

Algebra	17	21	11	16	15	11	24	27	19	8
Calculus	73	66	64	61	70	71	90	68	84	52

9-45

Beer	32.2	29.4	35.3	34.9	29.9	28.7	26.8	41.4
Wine	3.1	4.4	2.3	1.7	1.4	1.2	1.2	3.0

9-46

Added load	0	10	20	30	40	50	60	70
Increase in length	0	0.05	0.10	0.15	0.20	0.25	0.30	1.25

9-47

Verbal	421	423	429	424	413	437	461	429
Math	476	467	467	470	453	470	515	463

9-48

Age	6.1	7.2	5.9	6.3	10.5	11.0
Score	17.8	47.4	25.8	24.3	66.6	91.4

9-49

Automatic weapons	11.6	8.3	3.6	0.6	6.9	2.5	2.4	2.6
Murder rate	13.1	10.6	10.1	4.4	11.5	6.6	3.6	5.3

9-50

Age	19	27	21	45	46	58	37	42	30
Cholesterol	217.0	221.4	191.3	321.5	196.0	284.6	286.8	194.2	247.7

9-51

HC	0.65	0.55	0.72	0.83	0.57	0.51	0.43	0.37
CO	14.7	12.3	14.6	15.1	5.0	4.1	3.8	4.1

9-52

Test X	75	78	88	92	95	67	55	73	74	80
Test Y	81	73	85	85	89	73	66	81	81	81

9-53

Speed	15	23	30	35	42	45	50	54	60	65
mi/gal	14	17	20	24	26	23	18	15	11	10

9-54

Temperature (°C)	−62	−41	−36	−26	−33	−56	−50	−66
Snow depth (cm)	21	13	12	3	6	22	14	19

9-55

Taxes	1.9	3.0	1.4	1.4	1.5	1.8	2.4	4.0
Selling price	145	228	150	130	160	114	142	265

9-56

Living area	15	38	23	16	16	13	20	24
Taxes	1.9	3.0	1.4	1.4	1.5	1.8	2.4	4.0

In Exercises 9-57 through 9-60, use the given paired data.

a. Construct the scatter diagram and sketch the line that appears to fit the data best.

b. Using the graph from part (a), estimate the coordinates of two different points on the estimated regression line and then use these coordinates to approximate the equation of the regression line.

c. Determine the exact equation of the regression line by using the formulas for m and b (Formulas 9-2 and 9-4).

9-57

Productivity	63	67	88	44	52	106	99	110	75	58	77	91	101	51	86
Dexterity	2	9	4	5	8	6	9	8	9	7	4	10	7	4	6

9-58

IQ	103	113	119	107	78	153	114	101	103	111	105	82	110	90	92
Receptivity to hypnosis	55	55	59	64	45	72	42	63	62	46	41	49	57	52	41

9-59

Cigarette consumption	3522	3597	4171	4258	3993	3971	4042	4053
Percentage of psychiatric admissions (in percentage points)	0.20	0.22	0.23	0.29	0.31	0.33	0.33	0.32

9-60

Satisfaction	5.05	4.12	5.39	4.17	4.00	4.49	5.40	4.86
Commitment	5.37	4.49	5.42	4.45	4.24	5.34	5.62	4.90

9-61   Using a collection of paired sample data, the regression equation is found to be $y' = 50.0x + 10.0$ and the sample means are $\bar{x} = 0.30$ and $\bar{y} = 25.0$. In each of the following cases, use the additional information and find the best predicted point estimate of y when $x = 2.0$. Assume a significance level of $\alpha = 0.05$.

     a. $n = 100$; $r = 0.999$ b. $n = 10$; $r = 0.005$

     c. $n = 15$; $r = 0.519$ d. $n = 25$; $r = 0.393$

     e. $n = 22$; $r = 0.567$

9-62   Using a collection of paired sample data, the regression equation is found to be $y' = -20.0x + 50.0$ and the sample means are $\bar{x} = 0.50$ and $\bar{y} = 40.0$. In each of the following cases, use the additional information and find the best predicted point estimate of y when $x = 1.00$. Assume a significance level of $\alpha = 0.05$.

     a. $n = 5$; $r = -0.102$ b. $n = 50$; $r = -0.997$

     c. $n = 20$; $r = -0.403$ d. $n = 20$; $r = -0.449$

     e. $n = 65$; $r = -0.229$

9-63   In each of the following cases, find the best predicted point estimate of y when $x = 5$. The given statistics are summarized from paired sample data. Assume a significance level of $\alpha = 0.01$.

     a. $n = 40$, $\bar{y} = 6$, $r = 0.01$, and the equation of the regression line is $y' = 3x + 2$.

     b. $n = 40$, $\bar{y} = 6$, $r = 0.93$, and the equation of the regression line is $y' = 3x + 2$.

     c. $n = 20$, $\bar{y} = 6$, $r = -0.654$, and the equation of the regression line is $y' = -3x + 2$.

     d. $n = 20$, $\bar{y} = 6$, $r = 0.432$, and the equation of the regression line is $y' = 1.2x + 3.7$.

     e. $n = 100$, $\bar{y} = 6$, $r = -0.175$, and the equation of the regression line is $y' = -2.4x + 16.7$.

9-64   In each of the following cases, find the best predicted point estimate of y when $x = 8.0$. The given statistics are summarized from paired sample data. Assume a significance level of $\alpha = 0.01$.

     a. $n = 10$, $\bar{y} = 8.40$, $r = -0.236$, and the equation of the regression line is $y' = -2.0x + 3.5$.

     b. $n = 10$, $\bar{y} = 8.40$, $r = -0.654$, and the equation of the regression line is $y' = -2.0x + 3.5$.

     c. $n = 10$, $\bar{y} = 8.40$, $r = 0.602$, and the equation of the regression line is $y' = 2.0x + 3.5$.

     d. $n = 32$, $\bar{y} = 8.40$, $r = -0.304$, and the equation of the regression line is $y' = -2.0x + 3.5$.

     e. $n = 75$, $\bar{y} = 8.40$, $r = 0.257$, and the equation of the regression line is $y' = 2.0x + 3.5$.

9-3 EXERCISES B

9-65   Do Exercise 9-43 after changing 218 to 2180. How much of an effect does one exceptional value have on the equation of the regression line?

9-66   Using the paired data in the table below, verify that $r = ms_x/s_y$ where m, s_x, and s_y are as defined in this section and r is computed by using Formula 9-1.

x	1	2	2	3
y	5	4	3	1

9-67   What do you know about s_x and s_y if $r = 0.500$ for paired data having the regression line $y' = \frac{1}{2}x + 7.3$?

9-68   Prove that the point $(\bar{x}, \bar{y})$ will always lie on the regression line.

9-69   Prove that r and m have the same sign.

9-70   Show that

$$\frac{ms_x}{s_y} = \frac{n\Sigma xy - (\Sigma x)(\Sigma y)}{\sqrt{n(\Sigma x^2) - (\Sigma x)^2}\sqrt{n(\Sigma y^2) - (\Sigma y)^2}}$$

9-71   Using the data in Exercise 9-40 and the equation of the regression line, find the sum of the squares of the vertical deviations for the given points. Show that this sum is less than the corresponding sum obtained by replacing the regression line with $y = -x + 6$.

9-72   If the scatter diagram reveals a nonlinear pattern that we recognize as another type of curve, we may be able to apply the methods of this section. For the data given in the table below, find an equation of the form $y' = mx^2 + b$ by using the values of x^2 and y instead of the values of x and y. Use Formulas 9-2 and 9-3 or 9-4, but enter the values of x^2 wherever x occurs.

x	4	5	6	7
y	3	7	11	20

9-4   VARIATION

In the preceding two sections we introduced the fundamental concepts of linear correlation and regression. We saw that the linear correlation coefficient r could be used to determine whether or not there is a significant statistical relationship between two variables. We interpret r in a very limited way when we make one of the following three conclusions:

1. There is a significant positive linear correlation.
2. There is a significant negative linear correlation.
3. There is not a significant linear correlation.

The actual values of r can provide us with more information. We begin with a sample case, which leads up to an important definition.

Let's assume that we have a large collection of paired data, which yields these results:

1. There is a significant positive linear correlation.
2. The equation of the regression line is $y' = 2x + 3$.
3. $\bar{y} = 9$.
4. The scatter diagram contains the point (5, 19), which comes from the original set of observations.

When $x = 5$ we can find the predicted value y' as follows.

$$y' = 2x + 3 = 2(5) + 3 = 13$$

Note that the point (5, 13) is on the regression line, while the point (5, 19) is not. Take the time to examine Figure 9-12 carefully.

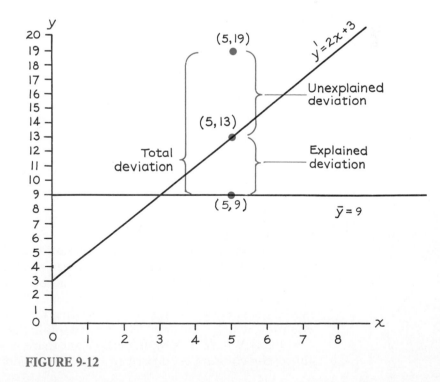

FIGURE 9-12

Unusual Economic Indicators

Forecasting and predicting are important goals of statistics. Investors constantly seek indicators that can be used to forecast stock market trends. Among the more colorful indicators are the hemline index, the Super Bowl omen, and the aspirin count. Developed in 1967, the hemline index is based on the length of women's skirts. According to one theory, rising hemlines precede a rise in the Dow Jones industrial average, while falling hemlines are followed by a drop in that average. According to the Super Bowl omen, a Super Bowl victory by a team with NFL origins is indicative of a year in which the New York Stock Exchange index will rise. A victory by a team with AFL origins means that the market will fall. As of this writing, the Super Bowl indicator has been correct in 19 of the past 21 years. The aspirin count theory also seems to work fairly well. It suggests that bad times cause an increase in aspirin sales, and such an increase is followed by a falling market. A drop in aspirin sales should be followed by rising stock market prices. There are people who observe the pace of elevator traffic at the New York Stock Exchange, the number of limousines on Wall Street, and the price of a seat on the New York Stock Exchange. But consider this: In a 10-year sequence of stock market increases and decreases, there are only 1024 different possible strings of rises and drops. Among the millions of different possible indicators, surely some of them must correspond exactly to the stock market fluctuations, but that does not mean that they will be reliable predictors.

DEFINITION

Given a collection of paired data, the **total deviation** (from the mean) of a particular point (x, y) is $y - \bar{y}$, the **explained deviation** is $y' - \bar{y}$, and the **unexplained deviation** is $y - y'$.

For the specific data under consideration, the total deviation of $(5, 19)$ is $y - \bar{y} = 19 - 9 = 10$.

If we were totally ignorant of correlation and regression concepts and we wanted to predict a value of y given a value of x and a collection of paired (x, y) data, our best guess would be $\bar{y}$. But we are not totally ignorant of correlation and regression concepts: We know that in this case the way to predict the value of y when $x = 5$ is to use the regression equation, which yields $y' = 13$. We can explain the discrepancy between $\bar{y} = 9$ and $y' = 13$ by simply noting that there is a significant positive linear correlation best described by the regression line. Consequently, when $x = 5$, y *should* be 13, and not 9. But while y *should* be 13, *it is* 19, and that discrepancy between 13 and 19 cannot be explained by the regression line and is called an unexplained deviation or a residual. This specific case illustrated in Figure 9-12 can be generalized as follows.

(total deviation) = (explained deviation) + (unexplained deviation)

or $(y - \bar{y})$ = $(y' - \bar{y})$ + $(y - y')$

This last expression can be further generalized and modified to include all of the pairs of sample data as follows. (By popular demand, we omit the intermediate algebraic manipulations.)

FORMULA 9-6

(total variation) = (explained variation) + (unexplained variation)

or $\Sigma(y - \bar{y})^2$ = $\Sigma(y' - \bar{y})^2$ + $\Sigma(y - y')^2$

The components of this last expression are used in the next important definition.

DEFINITION

The amount of the variation in y that is explained by the regression line is indicated by the **coefficient of determination**, which is given by

$$r^2 = \frac{\text{explained variation}}{\text{total variation}}$$

We can compute r^2 by using this definition with Formula 9-6 above, or we can simply square the linear correlation coefficient r, which is found by using the methods given in Section 9-2. To make some sense of this last definition, consider the following example.

EXAMPLE

If $r = 0.8$, then the coefficient of determination is $r^2 = 0.8^2 = 0.64$, which means that 64% of the total variation can be explained by the regression line. Thus 36% of the total variation remains unexplained.

Table 9-5 illustrates that the linear correlation coefficient r can be computed with the total variation and explained variation. This procedure for determining r is not recommended since it is longer, less direct, and more likely to produce errors than the method of Section 9-2. The paired xy data are in the first two columns. The third column of y' values is determined by substituting each value of x into the equation of the regression line ($y' = 2.093x + 1.558$). Also, $\bar{y} = 7.00$. The computed value of $r = 0.996$ agrees with the value found by using Formula 9-1. (If fewer decimal places are used, there would be a small discrepancy due to rounding errors.)

				TABLE 9-5			
x	y	y'	$y - \bar{y}$	$(y - \bar{y})^2$	$y' - \bar{y}$	$(y' - \bar{y})^2$
1	3	3.651	−4	16	−3.349	11.2158
1	4	3.651	−3	9	−3.349	11.2158
2	6	5.744	−1	1	−1.256	1.5775
3	8	7.837	1	1	0.837	0.7006
6	14	14.116	7	49	7.116	50.6375
			Totals:	76		75.3472

$$\uparrow \qquad\qquad\qquad\qquad \uparrow$$
Total variation   Explained variation

$$r^2 = \frac{\text{explained variation}}{\text{total variation}} = \frac{\Sigma(y' - \bar{y})^2}{\Sigma(y - \bar{y})^2} = \frac{75.3472}{76} = 0.9914$$

so that
$$r = \sqrt{0.9914} = 0.996$$

While the computations shown in the table don't seem too involved, we first had to compute the equation of the regression line, which is why this method for finding r is inferior. Also observe that r is not always the *positive* square root of r^2 since we can have negative values for the linear correlation coefficient. Examination of the scatter diagram should reveal whether r is negative or positive. If the pattern of points is predominantly uphill (from left to right), then r should be positive, but if there is a downward pattern, r should be negative.

Suppose we use the regression equation to predict a value of y from a given value of x. For example, we have seen that with a significant positive linear correlation, the regression equation $y' = 2x + 3$ can be used to predict y when $x = 5$; the result is $y' = 13$. Having made that prediction, we should consider how dependable it is. Intuition suggests that if $r = 1$, all sample points lie exactly on the regression line and the regression equation should therefore provide very dependable results, as long as we don't try to project beyond reasonable limits. If r is very close to 1, the sample points are close to the regression line, and again our y' values should be very dependable. What we need is a less subjective measure of the spread of the sample points about the regression line. The next definition provides such a measure.

DEFINITION

The **standard error of estimate** is given by

$$s_e = \sqrt{\frac{\Sigma(y - y')^2}{n - 2}}$$

where y' is the predicted y value.

EXAMPLE

Find the standard error of estimate s_e for the five pairs of data given in the first two columns of Table 9-6.

Solution

We extend the table to include the necessary components.

TABLE 9-6				
x	y	y'	$y - y'$	$(y - y')^2$
1	3	3.651	−0.651	0.4238
1	4	3.651	0.349	0.1218
2	6	5.744	0.256	0.0655
3	8	7.837	0.163	0.0266
6	14	14.116	−0.116	0.0135
				0.6512

↑
unexplained variation

We can now evaluate s_e:

$$s_e = \sqrt{\frac{\Sigma(y - y')^2}{n - 2}} = \sqrt{\frac{0.6512}{3}} = 0.466$$

The development of the standard error of estimate closely parallels that of the ordinary standard deviation introduced in Chapter 2, but this standard error involves deviations away from the regression line instead of the mean. The reasoning behind dividing by $n - 2$ is similar to the reasoning that led to division by $n - 1$ for the ordinary standard deviation, and we will not pursue the complex details. Do remember that smaller values of s_e reflect points that stay close to the regression line, while larger values indicate greater dispersion of points away from the regression line.

Formula 9-7 can also be used to compute the standard error of estimate. It is algebraically equivalent to the expression in the definition, but this form is generally easier to work with since it doesn't require that we compute each of the y' values.

FORMULA   9-7 $$s_e = \sqrt{\frac{\Sigma y^2 - b\Sigma y - m\Sigma xy}{n - 2}}$$

where m and b are the slope and y-intercept of the regression equation.

We can use the standard error of estimate, s_e, to construct interval estimates that will help us to see how dependable our point estimates of y really are. Assume that for each fixed value of x, the corresponding sample values of y are normally distributed about the regression line, and those normal distributions have the same variance. Given a fixed value of x (denoted by x_0), we have the following **prediction interval for an individual y**. (For a confidence interval of the *mean* value of y, see Exercise 9-88.)

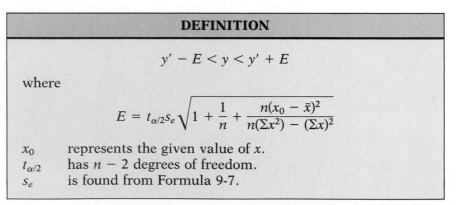

DEFINITION

$$y' - E < y < y' + E$$

where

$$E = t_{\alpha/2} s_e \sqrt{1 + \frac{1}{n} + \frac{n(x_0 - \bar{x})^2}{n(\Sigma x^2) - (\Sigma x)^2}}$$

x_0 represents the given value of x.
$t_{\alpha/2}$ has $n - 2$ degrees of freedom.
s_e is found from Formula 9-7.

The regression equation for the data given in Table 9-6 is $y' = 2.093x + 1.558$, so that when $x = 4$ we find the best point estimate for the predicted value of y to be $y' = 9.930$. Using the same data, we will construct the 95% prediction interval for the individual y value corresponding to the given value of $x = 4$. We have already found that

$s_e = 0.466$. With $n = 5$ we have 3 degrees of freedom, so that $t_{\alpha/2} = 3.182$ for $\alpha = 0.05$ in two tails. With $x_0 = 4$, $\bar{x} = \frac{13}{5} = 2.6$, $\Sigma x^2 = 51$, and $\Sigma x = 13$, we get

$$E = (3.182)(0.466)\sqrt{1 + \frac{1}{5} + \frac{5(4 - 2.6)^2}{5(51) - 169}}$$

$$= (3.182)(0.466)(1.146) = 1.699$$

so that $y' - E < y < y' + E$ becomes $9.930 - 1.699 < y < 9.930 + 1.699$. Given that $x = 4$, we therefore get the following 95% prediction interval for an individual y value.

$$8.231 < y < 11.629$$

EXAMPLE

Refer to the Table 9-1 sample data that lists home selling prices along with living area. In previous sections we have shown that

a. There is a significant linear correlation (at the 0.05 significance level).

b. The regression equation is $y' = 4.64x + 71.0$.

c. When $x = 20$, the predicted y value is 164.

Recall that x is the living area (in hundreds of square feet) and y is the selling price (in thousands of dollars), so that result c shows that for a living area of 2000 square feet, the predicted selling price is $164,000. Construct a 95% prediction interval for an individual selling price, given that the living area is 2000 sq ft.

Solution

We have used the Table 9-1 sample data in previous sections to find the following.

$n = 8$	$\Sigma y = 1334$	$\bar{x} = 20.625$
$\Sigma x = 165$	$\Sigma y^2 = 241,394$	
$\Sigma x^2 = 3855$	$\Sigma xy = 29,611$	

Regression equation (with extra digits carried):

$$m = 4.6412 \quad \text{and} \quad b = 71.025$$

When $x = 20$, the predicted y is $y' = 164$.
Using the preceding values, we first find s_e.

continued ▶

Solution, continued

$$s_e = \sqrt{\frac{\Sigma y^2 - b\Sigma y - m\Sigma xy}{n - 2}}$$

$$= \sqrt{\frac{241{,}394 - (71.025)(1334) - (4.6412)(29{,}611)}{8 - 2}}$$

$$= \sqrt{\frac{9216.1}{6}} = 39.192$$

From Table A-3 we find $t_{\alpha/2} = 2.447$. (We used $8 - 2 = 6$ degrees of freedom with $\alpha = 0.05$ in two tails.) We can now calculate E by letting $x_0 = 20$, since we want the prediction interval of y for $x = 20$.

$$E = t_{\alpha/2}s_e \sqrt{1 + \frac{1}{n} + \frac{n(x_0 - \bar{x})^2}{n(\Sigma x^2) - (\Sigma x)^2}}$$

$$= (2.447)(39.192)\sqrt{1 + \frac{1}{8} + \frac{8(20 - 20.625)^2}{8(3855) - (165)^2}}$$

$$= (2.447)(39.192)(1.0611) = 102 \text{ (rounded off)}$$

With $y' = 164$ and $E = 102$, we get the prediction interval

$$y' - E < y < y' + E$$

$$164 - 102 < y < 164 + 102$$

$$62 < y < 266$$

Recognizing that the selling price y is in thousands of dollars, we see that for a 2000–sq ft house ($x = 20$), the predicted selling price is between \$62,000 and \$266,000. That's quite a range. This interval is so wide mainly because the sample is so small ($n = 8$). The correlation coefficient of 0.717 barely made it into the critical region bounded by the critical value of 0.707. This shows that the regression equation used for the prediction does fit the sample data well, but the fit could be much better.

    In addition to knowing that the predicted selling price is \$164,000, we now have a sense for how reliable that estimate really is. The prediction interval of

$$\$62{,}000 < \text{selling price} < \$266{,}000$$

shows that the \$164,000 estimate can vary substantially.

9-4 EXERCISES A

9-73 If $r = 0.333$, find the coefficient of determination and the percentage of the total variation that can be explained by the regression line.

9-74 If $r = -0.444$, find the coefficient of determination and the percentage of the total variation that can be explained by the regression line.

9-75 If $r = 0.800$, find the coefficient of determination and the percentage of the total variation that can be explained by the regression line.

9-76 If $r = -0.700$, find the coefficient of determination and the percentage of the total variation that can be explained by the regression line.

In Exercises 9-77 through 9-80, find the (a) explained variation, (b) unexplained variation, (c) total variation, (d) coefficient of determination, (e) standard error of estimate s_e.

9-77

x	1	2	3	5	6
y	5	8	11	17	20

(The equation of the regression line is $y' = 3x + 2$.)

9-78

x	1	2	3	5	6
y	5	4	11	13	20

(The equation of the regression line is $y' = 2.9535x + 0.55814$.)

9-79

x	4	7	5	2	3
y	13	22	16	7	9

(The equation of the regression line is $y' = 3.0811x + 0.45946$.)

9-80

x	2	5	7	10	15
y	20	15	14	12	10

(The equation of the regression line is $y' = -0.71660x + 19.789$.)

9-81 Refer to the data given in Exercise 9-77, and assume that the necessary conditions of normality and variance are met.
a. For $x = 4$, find y', the predicted value of y.
b. Find s_e. How does this value of s_e affect the construction of the 95% prediction interval of y for $x = 4$?

9-82 Refer to Exercise 9-78, and assume that the necessary conditions of normality and variance are met.
a. For $x = 4$, find y', the predicted value of y.
b. Find the 99% prediction interval of y for $x = 4$.

9-83 Refer to the data given in Exercise 9-79, and assume that the necessary conditions of normality and variance are met.
a. For $x = 6$, find y', the predicted value of y.
b. Find the 95% prediction interval of y for $x = 6$.

9-84 Refer to the data given in Exercise 9-80, and assume that the necessary conditions of normality and variance are met.
a. For $x = 1$, find y', the predicted value of y.
b. Find the 99% prediction interval of y for $x = 1$.

9-4 EXERCISES B

9-85 a. Find an expression for the unexplained variation in terms of the sample size n and the standard error of estimate s_e.
b. Find an expression for the explained variation in terms of the coefficient of determination r^2 and the unexplained variation.
c. Suppose we have a collection of paired data for which $r^2 = 0.900$ and the regression equation is $y' = -2x + 3$. Find the linear correlation coefficient.

9-86 a. Assuming that a collection of paired data includes at least three pairs of values, what do you know about the linear correlation coefficient if $s_e = 0$?
b. If a collection of paired data is such that the total explained variation is zero, what can be deduced about the slope of the regression line?

9-87 Using the paired data given in Table 9-6, suppose we use each pair of data *twice* instead of once. Is the equation of the regression line affected? Is the value of the standard error of estimate affected? If so, how?

9-88 The formula

$$s_{y'} = s_e \sqrt{1 + \frac{1}{n} + \frac{n(x_0 - \bar{x})^2}{n(\Sigma x^2) - (\Sigma x)^2}}$$

gives the **standard error of the prediction** when predicting for a *single* y given that $x = x_0$. When predicting for the *mean* of all values of y for which $x = x_0$, the point estimate y' is the same, but $s_{y'}$ is described as follows.

$$s_{y'} = s_e \sqrt{\frac{1}{n} + \frac{n(x_0 - \bar{x})^2}{n(\Sigma x^2) - (\Sigma x)^2}}$$

Extend the last example of this section to find the point estimate and a 95% confidence interval for the *mean* selling price of all homes with a living area of 2000 sq ft.

9-5 MULTIPLE REGRESSION

We began this chapter by considering the relationship between home selling price and living area. In Section 9-2 we used the linear correlation coefficient as a measure of the strength of the relationship. Based on the sample data in Table 9-1, we concluded that there is a significant positive linear correlation between the variable of home selling price and the variable of living area. In Section 9-3 we proceeded to describe that relationship by the regression equation $y' = 4.64x + 71.0$. In this equation, x is the living area (in hundreds of square feet) and y is the selling price (in thousands of dollars). For a living area of 2000 sq ft we have a predicted selling price of \$164,000, and we know from Section 9-4 that this estimate is not very reliable. (It can easily vary between \$62,000 and \$266,000.)

It was not too surprising to find a significant linear correlation between home selling price and living area. After all, larger homes cost more to build and that increased cost is reflected in increased value. But we should also recognize that in addition to the living area of a home, there are other important factors that should be considered when trying to predict a home's value and selling price. The number of rooms (or bathrooms!) is an important factor for buyers with children who need their own space. The size of the lot can affect the selling price of a home. There are other less tangible factors such as the type of neighborhood, the state of the economy, the interest rate on mortgages, and so on.

We began with the simple regression equation

$$y' = 4.64x + 71.0$$

$\uparrow \qquad\qquad \uparrow$

selling price (in thousands of dollars)   living area (in hundreds of square feet)

We will now expand Table 9-1 to include the annual tax bill (in thousands of dollars) corresponding to each home. The result is Table 9-7.

TABLE 9-7								
Tax bill (x_2)	1.9	3.0	1.4	1.4	1.5	1.8	2.4	4.0
Living area (x_1)	15	38	23	16	16	13	20	24
Selling price (y)	145	228	150	130	160	114	142	265

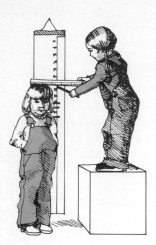

Rising to New Heights

The concept of regression was introduced in 1877 by Sir Francis Galton. His research showed that when tall parents have children, those children tend to have heights that "regress," or revert to the mean height of the population.

Stanford University's Dr. Darrell Wilson reported on a correlation between height and intelligence. Taller children tend to earn higher scores on intelligence tests as well as on achievement tests. The correlation was found to be significant, but the test scores of tall people were not very much greater than those of shorter people.

If we calculate the linear correlation coefficient for the paired data of taxes and selling prices, we get $r = 0.883$, which is significant at the $\alpha = 0.05$ level. This suggests that a predicted selling price should take the living area *and* the tax bill into account. That is, we need a regression equation of this type:

$$y' = b + m_1(\text{living area}) + m_2(\text{tax bill})$$

or

$$y' = b + m_1x_1 + m_2x_2$$

where b, m_1, m_2 are constants and x_1, x_2 are variables representing the living area and the tax bill. Such equations can be found. The result here is

$$y' = 38.4 + 2.04x_1 + 39.7x_2$$

A regression equation that includes more than two variables is an example of a **multiple regression** equation. A multiple regression equation is determined according to the same "least-squares" criterion introduced in Section 9-3. It minimizes the sum of squares $\Sigma(y - y')^2$, where y' represents the predicted value of y that is found through substitution in the regression equation.

The actual process of finding constants b, m_1, m_2 for a multiple regression equation is extremely messy. It involves calculations that are extensive, time consuming, and error-prone. In reality, almost everyone uses a computer software package such as STATDISK, Minitab, SPSS, or SAS. To get some sense for the magnitude of the problem, consider the sample data in Table 9-7. To find the values of the constants b, m_1, m_2 in the multiple regression equation, we must solve these equations.

$$\Sigma y = bn + m_1\Sigma x_1 + m_2\Sigma x_2$$

$$\Sigma x_1y = b\Sigma x_1 + m_1\Sigma x_1^2 + m_2\Sigma x_1x_2$$

$$\Sigma x_2y = b\Sigma x_2 + m_1\Sigma x_1x_2 + m_2\Sigma x_2^2$$

For the data in Table 9-7 we get

$$1334 = 8b + 165m_1 + 17.4\, m_2$$

$$29{,}611 = 165b + 3855m_1 + 388.5m_2$$

$$3197.5 = 17.4b + 388.5m_1 + 43.78m_2$$

Solving the preceding system of equations, we get $b = 38.4$, $m_1 = 2.04$, $m_2 = 39.7$. If our regression equation involves a total of four variables, we need to find the constants b, m_1, m_2, m_3 in the equation

$$y' = b + m_1x_1 + m_2x_2 + m_3x_3$$

and we would get four equations with four unknowns. Get the point? This is *messy* arithmetic and algebra. Now let's proceed on the much more reasonable assumption that we can use a computer. Shown below are the displays from STATDISK and Minitab that result from the data of Table 9-7.

STATDISK DISPLAY

```
        The regression equation is
            y = 38.4 = 2.04 x1 + 39.7 x2
        R-squared = 0.846
```

MINITAB DISPLAY

```
MTB > READ C1 C2 C3
DATA> 1.9 15 145
DATA> 3.0 38 228
DATA> 1.4 23 150
DATA> 1.4 16 130
DATA> 1.5 16 160
DATA> 1.8 13 114
DATA> 2.4 20 142
DATA> 4.0 24 265
DATA> ENDOFDATA
     8 ROWS READ
MTB > NAME C1 'TAXES' C2 'AREA' C3 'PRICE'
MTB > REGRESSION C3 2 C2 C1

The regression equation is
PRICE = 38.4 + 2.04 AREA + 39.7 TAXES

Predictor        Coef       Stdev      t-ratio        P
Constant        38.35      26.86         1.43    0.213
AREA            2.038      1.386         1.47    0.201
TAXES           39.71      12.09         3.28    0.022

s = 24.16     R-sq = 84.6%     R-sq(adj) = 78.4%
```

Once a multiple regression equation has been determined, it can be used to make predictions, as in the following example.

EXAMPLE

Using the sample data from Table 9-7, the multiple regression equation is found to be

$$y' = 38.4 + 2.04x_1 + 39.7x_2$$

where y' is the predicted home selling price (in thousands of dollars)

x_1 is the living area (in hundreds of square feet)

x_2 is the tax bill (in thousands of dollars)

Find the predicted selling price for a home with a living area of 2000 sq ft ($x_1 = 20$) and a tax bill of $2800 ($x_2 = 2.8$).

Solution

Since x_1 is the living area in hundreds of square feet, we represent the 2000-sq ft area by $x_1 = 20$. Since x_2 is the tax bill in thousands of dollars, we represent the $2800 tax bill by $x_2 = 2.8$. We now calculate the predicted value of y.

$$y' = 38.4 + 2.04(20) + 39.7(2.8) = 190$$

The predicted selling price is $190,000.

We will not consider exact procedures for determining how much a predicted value is likely to vary. However, we can use the **multiple coefficient of determination** R^2 as a measure of how well the multiple regression equation fits the available data. A perfect fit would result in $R^2 = 1$. A very good fit results in a value near 1. A very poor fit will result in a value of R^2 close to 0. The actual value of R^2 can be found by calculating

$$R^2 = 1 - \frac{\Sigma(y - y')^2}{\Sigma(y - \bar{y})^2}$$

This is also a messy calculation, but the value of R^2 can usually be found by using the same computer software that gives us the constants in the multiple regression equation. For the data of Table 9-7, $R^2 = 0.846$. (See the STATDISK and Minitab displays.) This value shows that the multiple regression equation $y' = 38.4 + 2.04 x_1 + 39.7x_2$ fits the sample data quite well. It also indicates that 84.6% of the variation in the selling price can be explained by the living area x_1 and the tax bill x_2.

Let's briefly review what we have done so far. Using the sample data of Table 9-1, we began by considering the relationship between home selling price and living area. Having found a significant positive linear correlation, we proceeded to find the regression equation $y' = 4.64x + 71.0$, which relates those two variables. In this section we then proceeded to find the multiple regression equation $y' = 38.4 + 2.04x_1 + 39.7x_2$, which expresses the selling price y in terms of living area x_1 and the annual tax bill x_2.

We could go on to include other relevant factors, such as the size of the lot and the number of rooms. For the homes represented in Table 9-1, the inclusion of the lot size as a third variable x_3 will lead to the following multiple regression equation.

$$y' = 46.6 + 1.21x_1 + 39.0x_2 + 7.22x_3$$

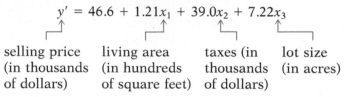

selling price	living area	taxes (in	lot size
(in thousands	(in hundreds	thousands	(in acres)
of dollars)	of square feet)	of dollars)	

However, the inclusion of this third variable of lot size leads to a multiple coefficient of determination given by $R^2 = 0.852$. This is only slightly better than the previous value of 0.846, and it suggests that we really didn't gain much by including the acreage. We might normally expect that larger lot sizes lead to increased values and selling prices, but this is not true here. For the region being considered, some large lots are in rural areas where homes are less expensive. Other large lots are in more populated areas where the prices are higher. These conflicting patterns result in a lower correlation between the lot size and the selling price.

As another example of the use of multiple regression, an Illinois State University study was conducted to identify the factors that affect the academic success of students in the first calculus course. (See "Factors Affecting Achievement in the First Course in Calculus" by Edge and Friedberg, *Journal of Experimental Education*, Vol. 52, No. 3.) The calculus course had high failure and dropout rates. It was hoped that the study would identify the most important characteristics for success so that better placement would improve the situation. The study included factors such as American College Test (ACT) scores, high school rank, high school average, high school algebra grades, scores on an algebra placement test, sex, birth order, family size, and the size of the high school. The best multiple regression equation was

$$y' = 34.8 + 1.21x_1 + 0.23x_2$$

calculus	score on	high school
grade	algebra	rank (percentile)
	placement	
	test	

This multiple regression equation shows that the algebra placement test score and high school rank can be used to predict the calculus grade, and these two variables became factors in placing students. The algebra score appears to measure algebra skills, while the high school rank is a measure of long-term perseverance and competitiveness. The end result is an improved placement procedure that helps students as well as faculty. This is an ideal use of statistics whereby people are helping people. (Does that sound too much like a commercial for General Electric?)

When we discussed regression in Section 9-3, we listed four common errors that should be avoided when using regression equations to make predictions. These same errors should be avoided when using multiple regression equations. We should be especially careful about ascribing causal relationships. In this section we expressed the selling price in terms of the living area and taxes. It's reasonable to claim that a home's selling price will be increased if its living area is increased. But a homeowner would be foolish to think that a higher selling price can be obtained by getting the taxes raised. Here, the annual tax bill may be an important characteristic that's helpful in *predicting* a home's selling price, but a change in taxes will not necessarily *cause* a change in the selling price. Be vary wary of claiming that a variable has a cause-effect relationship with another variable.

9-5 EXERCISES A

In Exercises 9-89 through 9-92, use the regression equation

$$y' = 34.8 + 1.21x_1 + 0.23x_2$$

where y' *is the predicted calculus grade,* x_1 *is the score on an algebra placement test, and* x_2 *is the high school rank expressed as a percentile. (The equation is based on data from "Factors Affecting Achievement in the First Course in Calculus" by Edge and Friedberg,* Journal of Experimental Education, *Vol. 52, No. 3.)*

9-89 Find the predicted calculus grade if the score on the algebra pretest (x_1) is 24 and the high school rank is the 92nd percentile.

9-90 Find the predicted calculus grade if the score on the algebra pretest (x_1) is 12 and the high school rank is the 71st percentile.

9-91 Find the predicted calculus grade if the score on the algebra pretest (x_1) is 18 and the high school rank is the 81st percentile.

9-92 Find the predicted calculus grade if the score on the algebra pretest (x_1) is 31 and the high school rank is the 99th percentile.

In Exercises 9-93 through 9-96, refer to the accompanying Minitab display to answer the given questions.

```
MTB > NAME C1 'X1' C2 'X2' C3 'X3' C4 'X4' C5 'Y'
MTB > REGRESSION C5 4 C1 C2 C3 C4

The regression equation is
Y = - 128 + 1.03 X1 + 12.4 X2 + 2.30 X3 - 1.06 X4

Predictor      Coef      Stdev      t-ratio         p
Constant     -127.7      175.3        -0.73     0.542
X1            1.034      4.181         0.25     0.828
X2           12.434      1.906         6.52     0.023
X3            2.296      1.757         1.31     0.321
X4           -1.059      2.711        -0.39     0.734

s = 10.86     R-sq = 98.5%   R-sq(adj) = 95.6%
```

9-93   What multiple regression equation is suggested by the Minitab display?

9-94   What is the value of the multiple coefficient of determination?

9-95   Find the predicted value of y given that $x_1 = 7$, $x_2 = 20$, $x_3 = 75$, and $x_4 = 42$.

9-96   Find the predicted value of y given that $x_1 = 9$, $x_2 = 29$, $x_3 = 80$, and $x_4 = 40$.

In Exercises 9-97 through 9-100, use Table 9-8 and software such as STATDISK or Minitab to find the indicated multiple regression equation and multiple coefficient of determination. Also, find the predicted y value for $x_1 = 2$, $x_2 = 9$, $x_3 = 1$.

TABLE 9-8					
y	25	13	16	13	10
x_1	5	2	1	3	4
x_2	10	8	7	11	7
x_3	2	3	4	3	1

9-97   Express the variable y in terms of x_1 and x_2.

9-98   Express the variable y in terms of x_1 and x_3.

9-99   Express the variable y in terms of x_2 and x_3.

9-100   Express the variable y in terms of x_1, x_2, and x_3.

In Exercises 9-101 through 9-104, use the sample data given in Table 9-9 and software such as STATDISK or Minitab. The data are based on recent sales of homes in Dutchess County, New York. Selling prices and taxes are in thousands of dollars. The living areas are in hundreds of square feet and the acreage amounts are in acres.

TABLE 9-9								
Selling price	145	228	150	130	160	114	142	265
Living area	15	38	23	16	16	13	20	24
Taxes	1.9	3.0	1.4	1.4	1.5	1.8	2.4	4.0
Acreage	2.0	3.6	1.8	0.53	0.50	0.31	0.75	2.0
Rooms	5	11	9	7	7	7	9	7

9-101   a. Find the multiple regression equation that expresses the tax bill in terms of selling price and living area.
         b. Find the value of the multiple coefficient of determination obtained by using the tax, selling price, and living area data.
         c. Compare the results from parts (a) and (b) to those found by expressing the selling price in terms of living area and tax bill. ($R^2 = 0.846$ and $y' = 38.4 + 2.04x_1 + 39.7x_2$.)

9-102   a. Find the multiple regression equation that expresses the selling price in terms of the living area, acreage, and number of rooms.
         b. Find the value of the multiple coefficient of determination obtained by using the same four variables listed in part (a).

9-103   a. Find the multiple regression equation that expresses the taxes in terms of the living area, acreage, and number of rooms.
         b. Find the value of the multiple coefficient of determination obtained by using the same four variables listed in part (a)

9-104   a. Find the multiple regression equation that expresses the selling price in terms of living area, taxes, acreage, and number of rooms.
         b. Find the value of the multiple coefficient of determination obtained by using the same five variables listed in part (a).

9-5 EXERCISES B

9-105   In some cases, the best-fitting multiple regression equation is of the form $y = b + m_1x + m_2x^2$. The graph of such an equation is a parabola. Let $x_1 = x$ and let $x_2 = x^2$. Use the values of y, x_1, and x_2 to find the multiple regression equation for the parabola that best fits the data given below. Based on the value of the multiple coefficient of determination, how well does this equation fit the given data?

x	1	3	4	7	5
y	5	14	19	42	26

9-106   a. Given the paired data below, find the linear correlation coefficient and the equation of the regression line.

   b. Use the paired data below to find the multiple regression equation

$$y = b + m_1x + m_2x^2$$

   (*Hint:* Let $x_1 = x$ and let $x_2 = x^2$.) Also find the value of the multiple coefficient of determination.

   c. Use the paired data below to find the multiple regression equation

$$y = b + m_1x + m_2x^2 + m_3x^3$$

   (Let $x_1 = x$, $x_2 = x^2$, $x_3 = x^3$.) Also find the value of the multiple coefficient of determination.

   d. Use the paired data below to find the multiple regression equation

$$y = b + m_1x + m_2x^2 + m_3x^3 + m_4x^4$$

   Also find the value of the multiple coefficient of determination.

   e. Based on the preceding results, which equation best fits the given data?

x	−2.0	−1.0	0.0	1.0	2.0	3.0
y	13	4.0	5.0	4.0	13	68

9-107   We noted that with three variables (y, x_1, x_2), solution of the equations

$$\Sigma y = bn + m_1\Sigma x_1 + m_2\Sigma x_2$$

$$\Sigma x_1 y = b\Sigma x_1 + m_1\Sigma x_1^2 + m_2\Sigma x_1 x_2$$

$$\Sigma x_2 y = b\Sigma x_2 + m_1\Sigma x_1 x_2 + m_2\Sigma x_2^2$$

will lead to the values of b, m_1, m_2 in the multiple regression equation

$$y = b + m_1x_1 + m_2x_2$$

Show that if we have only the two variables y and x_1, the solution of the preceding system of equations will lead to the formulas for the slope and y-intercept of the regression line discussed in Section 9-3.

9-108   Using the data in Table 9-7, we found the multiple regression equation $y' = 38.4 + 2.04x_1 + 39.7x_2$, and we also found that the value of the multiple coefficient of determination is $R^2 = 0.846$. Multiply each given value of x_2 by 20 and describe the effect this has on the multiple regression equation and the value of the multiple coefficient of determination.

VOCABULARY LIST

Define and give an example of each term.

bivariate data	y-intercept	unexplained
scatter diagram	least-squares	variation
correlation	property	total variation
linear correlation	predicted value	coefficient of
coefficient	total deviation	determination
centroid	explained deviation	standard error of
Pearson's product	unexplained	estimate
moment	deviation	multiple regression
regression line	explained variation	multiple coefficient
slope		of determination

REVIEW

In this chapter we studied the concepts of **linear correlation** and **regression** so that we could analyze paired sample data. We limited our discussion to linear relationships because consideration of nonlinear relationships requires more advanced mathematics. With correlation, we attempted to decide whether there is a significant linear relationship between the two variables. With regression, we attempted to specify what that relationship is. While a scatter diagram provides a graphic display of the paired data, the linear correlation coefficient r and the equation of the regression line serve as more precise and objective tools for analysis.

Given a list of paired data, we can compute the linear correlation coefficient r by using Formula 9-1. We can use the Student t distribution or Table A-6 to decide whether there is a significant linear relationship. The presence of a significant linear correlation does not necessarily mean that there is a direct cause-and-effect relationship between the two variables.

In Section 9-3 we developed procedures for obtaining the equation of the regression line which, by the least-squares criterion, is the straight line that best fits the paired data. When there is a significant linear correlation, the regression line can be used to predict the value of one variable when given some value of the other variable. The regression line has the form $y' = mx + b$, where the constants m and b can be found by using the formulas given in this section.

In Section 9-4 we introduced the concept of **total variation**, with components of explained and unexplained variation. We defined the coefficient of determination r^2 to be the quotient of explained variation by total variation. We saw that we could measure the amount of spread of the sample points about the regression line by the standard error of estimate, s_e.

In Section 9-5 we considered **multiple regression**, which allows us to investigate relationships among several variables. We discussed procedures for obtaining a multiple regression equation as well as the value of the multiple coefficient of determination R^2. That value gives us an indication of how well the multiple regression equation actually fits the available sample data. Due to the nature of the calculations involved, finding the multiple regression equation and the multiple coefficient of determination usually requires the use of existing computer software.

IMPORTANT FORMULAS

$$r = \frac{n\Sigma xy - (\Sigma x)(\Sigma y)}{\sqrt{n(\Sigma x^2) - (\Sigma x)^2}\sqrt{n(\Sigma y^2) - (\Sigma y)^2}} \qquad \text{or} \qquad r = \frac{ms_x}{s_y}$$

$$m = \frac{n\Sigma xy - (\Sigma x)(\Sigma y)}{n(\Sigma x^2) - (\Sigma x)^2} \qquad \text{or} \qquad m = r\frac{s_y}{s_x}$$

$$b = \frac{(\Sigma y)(\Sigma x^2) - (\Sigma x)(\Sigma xy)}{n(\Sigma x^2) - (\Sigma x)^2} \qquad \text{or} \qquad b = \bar{y} - m\bar{x}$$

$$y' = mx + b$$

$$r^2 = \frac{\text{explained variation}}{\text{total variation}}$$

$$s_e = \sqrt{\frac{\Sigma(y - y')^2}{n - 2}} \qquad \text{or} \qquad \sqrt{\frac{\Sigma y^2 - b\Sigma y - m\Sigma xy}{n - 2}}$$

$$y' - E < y < y' + E$$

$$\text{where } E = t_{\alpha/2}s_e\sqrt{1 + \frac{1}{n} + \frac{n(x_0 - \bar{x})^2}{n(\Sigma x^2) - (\Sigma x)^2}}$$

REVIEW EXERCISES

9-109   In each of the following, determine whether correlation or regression analysis is more appropriate.
  a. Is the value of a car related to its age?
  b. What is the relationship between the age of a car and annual repair costs?
  c. How are Celsius and Fahrenheit temperatures related?
  d. Is the age of a car related to annual repair costs?
  e. Is there a relationship between cigarette smoking and lung cancer?

9-110   a. What should you conclude if 40 pairs of data produce a linear correlation coefficient of $r = -0.508$?
  b. What should you conclude if 10 pairs of data produce a linear correlation coefficient of $r = 0.608$?

In Exercises 9-111 through 9-118, use the given paired data.

  a. Find the value of the linear correlation coefficient r.
  b. Assuming a 5% level of significance, find the critical value of r from Table A-6.
  c. Use the results from parts (a) and (b) to decide whether there is a significant linear correlation.
  d. Find the equation of the regression line.
  e. Plot the regression line on the scatter diagram.

9-111   The following table lists midterm and final exam grades for randomly selected students in a statistics course.

Midterm	82	65	93	70	80
Final	94	77	94	79	91

9-112   The following table lists car rental costs (in dollars) with the corresponding numbers of miles driven.

Cost	40	55	45	80
Mileage	40	100	60	200

9-113   A psychologist suspects that there is a relationship between the scores on a test for motivation and the scores on a personality test. The following sample data is collected.

Motivation test	49	36	62	50	51	42
Personality test	16	14	18	20	17	12

9-114 The table below lists the values (in billions of dollars) of U.S. exports and incomes on foreign investments for various sample years.

Exports	16	20	27	39	56	63
Incomes on foreign investments	2	3	4	7	11	11

9-115 The owner of a bookstore investigates the relationship between outside pedestrian traffic (per hour) and the number of books sold (per hour). The following sample data are obtained.

Traffic	21	37	12	15	21	32	48	25
Books	35	72	25	30	35	63	85	45

9-116 The given paired data lists the heights (in inches) of various males who were measured on their eighth and sixteenth birthdays.

Height (8 yrs)	52	50	48	51	51	48
Height (16 yrs)	69	66	64	67	69	65

9-117 A test designer develops two separate tests, which are intended to measure a person's level of creative thinking. Randomly selected subjects were given both versions and their results follow.

Test A	85	97	100	76	80	116	120	105
Test B	92	109	100	74	85	118	125	90

9-118 A pill designed to lower systolic blood pressure is administered to 10 randomly selected volunteers. The results follow.

Before pill	120	136	160	98	115	110	180	190	138	128
After pill	118	122	143	105	98	98	180	175	105	112

In Exercises 9-119 through 9-123, use Table 9-10, which lists data for eight states in the Northeast region of the country. (The table is based on recent data from USA Today.) *The salary values are reported as the "average" salary of teachers in the state, and they are in thousands of dollars. The cost figures are average costs per pupil for each state, and they are in thousands of dollars. The dropout figures represent the dropout rates (in percent). The SAT values are the average SAT scores for each state. Let* y *represent the SAT scores and let* x_1, x_2, x_3 *represent salary, cost, and dropout rate, respectively. Use software such as STATDISK or Minitab to answer the given questions.*

TABLE 9-10								
State	N.Y.	N.J.	Conn.	Mass.	R.I.	Vt.	Maine	N.H.
Salary (x_1)	32.0	28.7	28.9	28.4	31.1	21.8	21.3	22.0
Cost (x_2)	6.0	5.4	4.7	4.6	4.7	4.0	3.5	3.5
Dropout (x_3)	35.8	22.4	10.2	23.3	32.7	22.4	23.5	26.7
SAT (y)	894	892	912	909	898	914	899	938

9-119   a. Using the salaries (x_1) and costs (x_2), find the value of the linear correlation coefficient r.
  b. Assuming a 0.05 level of significance, find the critical value of r from Table A-6.
  c. Use the results from parts (a) and (b) to decide whether there is a significant linear correlation.
  d. Find the equation of the regression line. (Use x_1 for x and use x_2 for y.)

9-120   a. Using the costs (x_2) and SAT scores (y), find the value of the linear correlation coefficient r.
  b. Assuming a 0.05 level of significance, find the critical value of r from Table A-6.
  c. Use the results from parts (a) and (b) to decide whether there is a significant linear correlation.
  d. Find the equation of the regression line.

9-121 a. Using the costs (x_2) and dropout rates (x_3), find the value of the linear correlation coefficient r.
b. Assuming a 0.05 level of significance, find the critical value of r from Table A-6.
c. Use the results from parts (a) and (b) to decide whether there is a significant linear correlation.
d. Find the equation of the regression line. (Use x_2 for x and use x_3 for y.)

9-122 a. Ignore the dropout rates and use software such as STATDISK or Minitab to find the multiple regression equation of the form

$$y' = b + m_1x_1 + m_2x_2$$

b. Find the multiple coefficient of determination.
c. Based on the results of part (b), how well does the multiple regression equation fit the sample data?

9-123 a. Use software such as STATDISK or Minitab to find the multiple regression equation of the form

$$y' = b + m_1x_1 + m_2x_2 + m_3x_3$$

b. Find the multiple coefficient of determination.
c. Based on the results of part (b), how well does the multiple regression equation fit the sample data?

9-124 A home maintenance firm subcontracts its painting jobs. The painter charges a flat fee for each job, plus an hourly labor rate, plus a fixed amount for each can of paint. The painter will not itemize the component costs, but the following data were collected from several different jobs. (y is the total cost in dollars, x_1 is the number of hours worked, and x_2 is the number of gallons of paint used.)

y	790	702	1030	884	1268	554	716
x_1	30	24	40	33	51	18	27
x_2	5	6	8	7	10	4	4

a. Use software such as STATDISK or Minitab to find the multiple regression equation.
b. What is the flat fee for each job?
c. What is the hourly labor rate?
d. What is the cost of each gallon of paint?
e. What is the predicted cost of a job that will require 40 hours of labor and 7 gallons of paint?

In Exercises 9-125 through 9-128, use the given data set to find (a) explained variation, (b) unexplained variation, (c) total variation, (d) coefficient of determination, and (e) standard error of estimate s_e.

9-125

x	1	2	4	6
y	3	0	1	5

9-126

x	2	5	4	3	7
y	5	11	9	7	15

9-127

x	4	2	1	6
y	2	4	5	0

9-128

x	8	4	2	9	6
y	2	7	7	1	4

COMPUTER PROJECT

1. Develop a computer program that will take a set of paired data as input. Output should consist of the value of the linear correlation coefficient *r* and the values of the slope (*m*) and *y*-intercept (*b*) for the equation of the regression line.

2. Use existing software to obtain the linear correlation coefficient and the equation of the regression line for a set of paired data. Use the paired data given in Exercise 9-9.

CASE STUDY ACTIVITY

Conduct a study to determine whether there is a correlation between two variables. Collect sample paired data consisting of at least 15 different pairs. Identify the population and the method used to obtain the data, and identify any factors suggesting that the sample data are not representative. Calculate the linear correlation coefficient and determine the equation of the regression line. Graph the regression line on the scatter diagram. Assuming a 0.05 level of significance, find the critical value and form a conclusion about the correlation between the two variables.

DATA PROJECT

Refer to Appendix B for the data on homes recently sold in Dutchess County, New York.

a. For each of the 10 different pairs of variables, find the value of the linear correlation coefficient r and enter the results in the table below.

	Selling price	Living area	Acreage	Rooms	Baths
Selling Price	1.000				
Living Area		1.000			
Acreage			1.000		
Rooms				1.000	
Baths					1.000

b. Find the equation of the regression line that relates selling price and living area.

c. Find the multiple regression equation that relates the selling price (y) to the living area (x_1), the acreage (x_2), the number of rooms (x_3), and the number of baths (x_4).

Chapter Ten

CHAPTER CONTENTS

10-1 Overview
We identify chapter **objectives**. This chapter introduces some methods for dealing with data coming from more than two sample groups.

10-2 Multinomial Experiments
This section presents a procedure for testing hypotheses made about more than two proportions from a population. We define and consider **multinomial experiments**.

10-3 Contingency Tables
We test hypotheses that the row and column variables of **contingency tables** are independent.

10-4 Analysis of Variance
In testing the claim that several population means are equal, we use **analysis of variance** techniques to examine the significance of two estimates of the population variance.

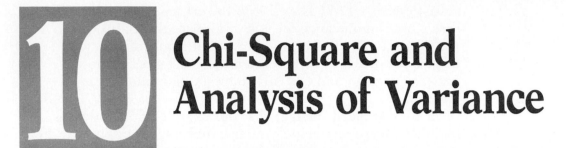

10 Chi-Square and Analysis of Variance

CHAPTER PROBLEM

Many people believe that smoking is unhealthy. In a study of 1000 deaths of males aged 45 to 64, the causes of death are listed along with their smoking habits (see Table 10-1, which is based on data from "Chartbook on Smoking, Tobacco, and Health," USDHEW publication CDC75-7511).

TABLE 10-1			
	Cause of Death		
	Cancer	Heart disease	Other
Smoker	135	310	205
Nonsmoker	55	155	140

Data from the U.S. National Center for Health Statistics show that 45% of males in the 45 to 64 age bracket are smokers. But Table 10-1 shows that among 1000 randomly selected deaths of males aged 45 to 64, a total of 650 deaths (or 65%) were by smokers. If tobacco lobbyists are correct when they claim that smoking doesn't affect health, we would expect 45% of the deaths to be by smokers, not the 65% that the table shows. We could show that this discrepancy is significant by using methods of Chapter 6 to test the claim that the proportion of deaths by smokers is 0.45. The test statistic of $z = 12.71$ is well into the critical region for any reasonable significance level, so we reject the claim that the proportion is 0.45. This shows that smokers are disproportionately represented among the males who die between ages 45 and 64. This suggests that smoking is unhealthy.

If we disregard the fact that our sample of 1000 deaths includes disproportionately many smokers, we might question whether or not the *cause* of death seems to be affected by smoking. In this chapter we will analyze tables such as Table 10-1 as we test the claim that the row variable is independent of the column variable. That is, we will test the claim that the smoking habit is independent of the cause of death. This result can't suggest that smoking is unhealthy, since everyone in our sample died. But it will allow us to investigate whether smoking has an effect on the *cause* of death.

Note that we can use one method from Chapter 6 to investigate one issue (whether smoking is unhealthy) and we can use another method from this chapter to investigate a different issue (whether smoking affects the cause of death). Given sample data, there are often different interesting questions that can be considered by different methods of statistics.

10-1 OVERVIEW

Prior to Chapter 8 we considered statistical analyses involving only one sample. In Chapter 8 we introduced methods for comparing the statistics of two different samples, and in Chapter 9 we introduced ways of analyzing the relationship between two variables. In this chapter we develop methods for comparing more than two samples. We also consider ways of dealing with situations in which we draw a single sample from a single population, but partition the sample into several categories.

In Section 10-2 we begin by considering a method for testing a hypothesis made about several population proportions. Instead of working with the sample proportions, we deal directly with the frequencies with which the events occur. Our objective is to test for the significance of the differences between observed frequencies and the frequencies we would expect in a theoretically ideal experiment. Since we test for how well an observed frequency distribution conforms to

Dangerous to Your Health

In 1965, the first warning labels were put on cigarette packs, and, in 1969, cigarette commercials were banned from television and radio. Reports of the surgeon general include many convincing statistics showing that smoking greatly increases the danger of cancer. The Tobacco Institute contests these reports. Horace Kornegay, the Tobacco Institute's chairman, says that "while many people believe a causal link between smoking and cancer is a given, scientific research has not been able to establish that link." Critics note that not a single case has determined smoking to be the cause of cancer, but the evidence clearly indicates a significant correlation.

(or "fits") some theoretical frequency distribution, this procedure is often referred to as a **goodness-of-fit test**. We introduce the test statistic that measures the differences between observed frequencies and expected frequencies. In repeated large samplings, the test statistic can be approximated by the chi-square distribution discussed in Section 6-6.

In Section 10-3 we analyze tables of frequencies called **contingency tables**, or **two-way tables**. In these tables the rows represent categories of one variable, while the columns represent categories of another variable. We test the hypothesis that the two classification variables are independent. We again use the chi-square distribution to determine whether there is a significant difference between the observed sample frequencies and the frequencies we would expect in a theoretically ideal experiment involving independent variables.

Section 10-4 introduces a technique called **analysis of variance**, or **ANOVA**, to test the claim that several population means are equal. In Section 8-3 we tested the hypothesis $\mu_1 = \mu_2$, but in Section 10-4 we test claims such as $\mu_1 = \mu_2 = \mu_3$. When dealing with more than two samples, we use a fundamentally different approach that incorporates the F distribution instead of the normal or Student t distributions.

10-2 MULTINOMIAL EXPERIMENTS

In Chapter 4 we introduced the binomial probability distribution and indicated that each trial must have all outcomes classified into exactly one of two categories. This feature of two categories is reflected in the prefix *bi*, which begins the term *binomial*. In this section we consider **multinomial experiments** that require each trial to yield outcomes belonging to one of several categories. Except for this difference, binomial and multinomial experiments are essentially the same.

DEFINITION
A **multinomial experiment** is one in which

1. There is a fixed number of trials.
2. The trials are independent.
3. Each trial must have all outcomes classified into exactly one of several categories.
4. The probabilities remain constant for each trial.

We have already discussed methods for testing hypotheses made about one population proportion (Section 6-5) and two population proportions (Section 8-4). However, there is often a need to deal with cases involving a population with more than two proportions. Consider the case involving a study of 147 industrial accidents that required medical attention. The sample data are summarized in Table 10-2, and we will test the claim that accidents occur on the five days with equal frequencies. (The data are based on results from "Counted Data CUSUM's" by Lucas, *Technometrics*, Vol. 27, No. 2.)

TABLE 10-2					
Day	Mon	Tues	Wed	Thurs	Fri
Observed accidents	31	42	18	25	31

If accidents occur with equal frequencies on the five different days, the 147 accidents would average out to 29.4 per day. In Table 10-3 we list the observed frequencies along with the theoretically expected values.

TABLE 10-3					
Day	Mon	Tues	Wed	Thurs	Fri
Observed accidents	31	42	18	25	31
Expected accidents	29.4	29.4	29.4	29.4	29.4

We know that samples deviate from what we theoretically expect, so we now present the key question: Are the differences between the actual *observed* values and the theoretically *expected* values differences that occur just by chance, or are the differences significant? To answer this question we need some way of measuring the significance of the differences between the observed values and the theoretical values. **The expected frequency of an outcome is the product of the probability of that outcome and the total number of trials.** If there are five possible outcomes that are supposed to be equally likely, the probability of each outcome is $\frac{1}{5}$. For 147 trials and 5 equally likely outcomes, the expected frequency of each outcome is $\frac{1}{5} \times 147 = 29.4$.

In testing for the differences among the five sample proportions, one approach might be to compare them two at a time by using the

Did Mendel Fudge His Data?

R. A. Fisher analyzed the results of Mendel's experiments in hybridization. Fisher noted that the data were unusually close to theoretically expected outcomes. He says that "the data have evidently been sophisticated systematically, and after examining various possibilities, I have no doubt that Mendel was deceived by a gardening assistant, who knew only too well what his principal expected from each trial made." Fisher used chi-square tests and concluded that only about a 0.00004 probability exists of such close agreement between expected and reported observations.

methods of Section 8-4. That approach would, however, be very inefficient and would lead to severe problems related to the level of significance. There are 10 different combinations of 2 samples, and if each individual hypothesis test has a 0.95 probability of not leading to a type I error (rejecting a true null hypothesis) and if the tests were independent, then the probability of no type I errors among the 10 tests is only 0.95^{10}, or about 0.599. Instead of pairing off samples and conducting 10 separate tests, we develop one comprehensive test, which is based on a statistic that measures the differences between observed values and the corresponding values that we expect in an ideal case. This statistic will incorporate the observed and expected frequencies instead of the proportions. We introduce the following notation.

NOTATION

O represents the **observed frequency** of an outcome.

E represents the theoretical or **expected frequency** of an outcome.

From Table 10-3 we see that the O values are 31, 42, 18, 25, and 31 and the corresponding values of E are 29.4, 29.4, 29.4, 29.4, and 29.4. The method generally used in testing for agreement between O and E values is based on the following test statistic.

TEST STATISTIC

$$\chi^2 = \sum \frac{(O - E)^2}{E}$$

Simply summing the differences between observed and expected frequencies would not lead to a good measure, since that sum is always zero.

$$\Sigma(O - E) = \Sigma O - \Sigma E = n - n = 0$$

Squaring the $O - E$ values provides a better statistic, which does reflect the differences between observed and expected frequencies, but $\Sigma(O - E)^2$ grows larger as the sample size increases. We compensate for this through division by the expected frequencies. For our accident data, Table 10-3 shows that for Monday, the observed frequency is 31 while the expected frequency is 29.4, so that

$$\frac{(O - E)^2}{E} = \frac{(31 - 29.4)^2}{29.4} = \frac{1.6^2}{29.4} = 0.0871$$

Proceeding in a similar manner with the remaining data, we get

$$\chi^2 = \sum \frac{(O - E)^2}{E}$$

$$= \frac{(31 - 29.4)^2}{29.4} + \frac{(42 - 29.4)^2}{29.4} + \frac{(18 - 29.4)^2}{29.4}$$

$$+ \frac{(25 - 29.4)^2}{29.4} + \frac{(31 - 29.4)^2}{29.4}$$

$$= 0.0871 + 5.4000 + 4.4204 + 0.6585 + 0.0871$$

$$= 10.653$$

The theoretical distribution of $\Sigma(O - E)^2/E$ is a discrete distribution, since there are a limited number of possible values. However, extensive studies have established that, in repeated large samplings, this distribution can be approximated by a chi-square distribution. **This approximation is generally considered to be acceptable, provided that all values of E are at least 5.** In Section 5-5 we saw that the continuous normal probability distribution can reasonably approximate the discrete binomial probability distribution, provided that np and nq are both at least 5. We now see that the continuous chi-square distribution can reasonably approximate the discrete distribution of $\Sigma(O - E)^2/E$, provided that all values of E are at least 5. There are ways of circumventing the problem of an expected frequency that is less than 5, and one procedure requires us to combine categories so that all expected frequencies are at least 5.

When we use the chi-square distribution as an approximation, we obtain the critical test value from Table A-4 after determining the level of significance α and the number of degrees of freedom. In a multinomial experiment with k possible outcomes, the number of degrees of freedom is $k - 1$.

> ### degrees of freedom = $k - 1$

This reflects the fact that, for n trials, the frequencies of $k - 1$ outcomes can be freely varied, but the frequency of the last outcome is determined. Our 147 accidents are distributed among five categories or cells, but we can freely vary the frequencies of only four cells, since the last cell would be 147 minus the total of the first four cell frequencies. In this case, we say that the number of degrees of freedom is 4.

Note that close agreement between observed and expected values will lead to a small value of χ^2. A large value of χ^2 will indicate strong disagreement between observed and expected values. A significantly large value of χ^2 will therefore cause rejection of the null hypothesis of no difference between observed and expected frequencies. Our test is therefore right-tailed since the critical value and critical region are located at the extreme right of the distribution. Unlike previous hypothesis tests where we had to determine whether the test was left-tailed, right-tailed, or two-tailed, these tests are all right-tailed.

For the sample accident data, we have determined the value of the test statistic ($\chi^2 = 10.653$) and the number of degrees of freedom (4). Let's assume a 5% level of significance so that $\alpha = 0.05$. With 4 degrees of freedom and $\alpha = 0.05$, Table A-4 indicates a critical value of 9.488.

Figure 10-1 indicates that our test statistic of 10.653 falls within the critical region, so we reject the null hypothesis that accidents occur on the different weekdays with equal frequencies. From the sample data, it appears that Wednesday has a lower accident rate. However, more specific conclusions such as this would require other methods of analysis.

The first example of this section dealt with the null hypothesis that the frequencies of accidents on the five working days were all equal. However, the theory and methods we present here can also be used in cases where the claimed frequencies are different. The next example illustrates such a case where the claimed frequencies are not all equal.

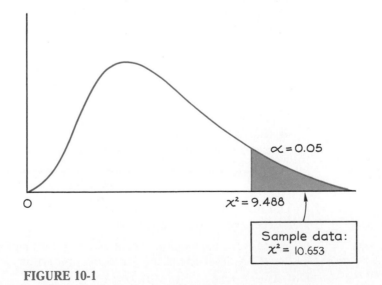

FIGURE 10-1

Cheating Success

Eighteen students from a high school in a depressed area of Los Angeles took an advanced placement exam for college calculus. After all the students passed (seven with the highest possible scores), representatives of the Educational Testing Service challenged the validity of 14 students' results. There was a significant discrepancy between the expected scores and the actual scores. However, a retest of 12 of the students confirmed the original scores; the students knew their calculus! It turned out that Jaime Escalante was a committed teacher who had recruited good students, and they had all worked diligently in special classes before, during, and after regular school hours. What appeared to be cheating was, in fact, the result of hard work by a group of dedicated students and an exceptional teacher.

The results were so dramatic that a movie, *Stand and Deliver*, was made of this exceptional teacher.

EXAMPLE

An ice cream supplier claims that among the four most popular flavors, customers have these preference rates: 62% prefer vanilla, 18% prefer chocolate, 12% prefer neapolitan, and 8% prefer vanilla fudge. (The data are based on results from the International Association of Ice Cream Manufacturers.) A random sample of 200 customers produces these results.

Flavor	Vanilla	Chocolate	Neapolitan	Vanilla fudge
Number prefer	120	40	18	22

At the $\alpha = 0.05$ significance level, test the claim that the percentages given by the supplier are correct.

Solution

The null hypothesis is the claim that the percentages given are correct. We can state this as follows.

H_0: $p_v = 0.62$ and $p_c = 0.18$ and $p_n = 0.12$ and $p_{vf} = 0.08$
H_1: At least one of the preceding proportions is incorrect.

continued ▶

Solution, continued

If the supplier's claim is exactly correct, then the 200 customers would have preferred the flavors with these frequencies:

Flavor	Vanilla	Chocolate	Neapolitan	Vanilla fudge
Number prefer	124	36	24	16

The first table lists actual *observed* values, while this one lists *expected* values. Each expected value is obtained by multiplying the claimed percentage (in decimal form) by the sample size of 200. For example, the expected frequency for vanilla is 124 and it was obtained by multiplying 0.62 by 200. We now compute χ^2 as a measure of the disagreement between the observed and expected values.

$$\chi^2 = \sum \frac{(O - E)^2}{E}$$

$$= \frac{(120 - 124)^2}{124} + \frac{(40 - 36)^2}{36} + \frac{(18 - 24)^2}{24} + \frac{(22 - 16)^2}{16}$$

$$= 0.1290 + 0.4444 + 1.5000 + 2.2500$$

$$= 4.323$$

This is a right-tailed test with $\alpha = 0.05$ and $4 - 1 = 3$ degrees of freedom, so the critical value from Table A-4 is 7.815. Since the test statistic of $\chi^2 = 4.323$ does not fall within the critical region bounded by 7.815, we fail to reject the null hypothesis. There is not sufficient sample evidence to warrant rejection of the claim that the given percentages are correct.

The techniques of this section can be used to test for how well an observed frequency distribution conforms to some theoretical frequency distribution. For the employee accident data considered in this section, we used a goodness-of-fit test to decide whether the observed accidents conformed to a uniform distribution, and we found that the differences were significant. It appears that the observed frequencies do not make a good fit with a uniform distribution. Since many statistical analyses require a normally distributed population, we may use the chi-square test of this section to determine that given samples are drawn from normally distributed populations. (See Exercise 10-24.)

10-2 EXERCISES A

10-1   The following table is obtained from a random sample of 100 absences. At the $\alpha = 0.01$ significance level, test the claim that absences occur on the five days with equal frequency.

Day	Mon	Tues	Wed	Thurs	Fri
Number absent	27	19	22	20	12

a. Find the χ^2 value based on the sample data.
b. Find the critical value of χ^2.
c. What conclusion can you draw?

10-2   In an experiment on perception, 50 subjects are asked to select the most pleasant of five different photographs. The results are as follows. At the $\alpha = 0.05$ significance level, test the hypothesis that the photographs are equally pleasant.

Photograph	A	B	C	D	E
Number selected	0	9	19	10	12

10-3   In the decimal representation of π, the first 100 digits occur with the frequencies described in the table below. At the 0.05 significance level, test the claim that the digits are uniformly distributed.

Digit	0	1	2	3	4	5	6	7	8	9
Frequency	8	8	12	11	10	8	9	8	12	14

10-4   In the decimal representation of 5/71, the first 100 digits occur with the frequencies described in the table below. At the 0.05 significance level, test the claim that the digits are uniformly distributed.

Digit	0	1	2	3	4	5	6	7	8	9
Frequency	17	12	15	9	7	11	12	6	8	3

10-5   When 150 people are randomly selected from each of five different age brackets, the numbers of smokers are found to be the numbers listed in the table below, which is based on data from the National Center for Health Statistics. At the 0.01 significance level, test the claim that the proportions of smokers in the different age brackets are all equal.

Age (years)	20–24	25–34	35–44	45–64	65 and over
Number	36	32	35	31	16

10-6   In a study of fatal car crashes, 216 cases are randomly selected from the pool in which the driver was found to have a blood alcohol content over 0.10. These cases are broken down according to the day of the week, with the results listed in the accompanying table (based on data from the Dutchess County STOP-DWI Program). At the 0.05 significance level, test the claim that such fatal crashes occur on the days of the week with equal frequency.

Day	Sun	Mon	Tues	Wed	Thurs	Fri	Sat
Number	40	24	25	28	29	32	38

10-7   A biochemist conducts an experiment designed to reveal the most effective aspirin. The eight best selling brands of aspirin are tested on 120 random subjects and, for each subject, a best aspirin is selected according to certain technical criteria. The results are as follows. At the $\alpha = 0.05$ significance level, test the claim that the eight brands of aspirin are equally effective.

Brand	A	B	C	D	E	F	G	H
Number of times selected as best	10	14	16	21	17	14	15	13

10-8   A genetics experiment involves 320 mice and is designed to determine whether Mendelian principles hold for a certain list of characteristics. The following table summarizes the actual experimental results and the expected Mendelian results for the five characteristics being considered. At the $\alpha = 0.01$ significance level, test the claim that Mendelian principles hold.

Characteristic	A	B	C	D	E
Observed frequency	30	15	58	83	134
Expected (Mendelian) frequency	20	20	40	120	120

10-9   A marketing specialist claims that among supermarket shoppers who prefer a particular day, these rates apply: 7% prefer Sunday, 5% prefer Monday, 9% prefer Tuesday, 11% prefer Wednesday, 19% prefer Thursday, 24% prefer Friday, and 25% prefer Saturday. (The data are based on results from the Food Marketing Institute.) Sample results are given here. At the 0.05 significance level, test the claim that the given percentages are correct.

Day	Sun	Mon	Tues	Wed	Thurs	Fri	Sat
Number	9	6	10	8	19	23	28

10-10   A psychology course has 10 sections with enrollments listed in the following table. Test the claim that students enroll in the various sections with equal frequencies.

Section	1	2	3	4	5	6	7	8	9	10
Enrollment	22	14	12	25	25	20	17	15	21	19

10-11   In an experiment on extrasensory perception, subjects were asked to identify the month showing on a calendar in the next room. If the results are as shown, test the claim that months were selected with equal frequencies. Assume a significance level of 0.05.

Month	Jan.	Feb.	Mar.	Apr.	May	June	July	Aug.	Sept.	Oct.	Nov.	Dec.
Number selected	8	12	9	15	6	12	4	7	11	11	5	20

10-12   A politician claims that the Republican, Democratic, and Independent mayoral candidates are favored by voters at the rates of 35%, 40%, and 25%, respectively. Test the claim at the 0.05 significance level if a random survey of 30 voters produces the following results.

Candidate	Republican	Democrat	Independent
Number selected	13	11	6

10-13   In a study of the color choices for buyers of compact cars, it is claimed that among the five most frequent choices, these preference rates apply: 22% prefer light red/brown, 22% prefer white, 20% prefer light blue, 18% prefer dark blue, and 18% prefer red. (The data are based on results from the Automotive Information Center.) When 270 compact cars are randomly selected, the following results are found. At the 0.05 level of significance, test the claim that the given percentages are correct.

Color	Lt. red/brown	White	Lt. blue	Dk. blue	Red
Frequency	60	61	43	41	65

10-14   In a study of television viewing, 50 households are surveyed on each night of a week and the numbers of households engaged in watching TV are listed here. (The table is based on data from a survey by A. C. Nielsen Co.)

Day	Mon	Tues	Wed	Thurs	Fri	Sat	Sun
Number	27	24	25	26	23	20	25

At the 0.05 significance level, test the claim that the frequency of television-watching households is the same for all days of the week.

10-15 A television company is told by a consulting firm that its eight leading shows are favored according to the percentages given in the following table. A separate and independent sample is obtained by another consulting firm. Do the figures agree? Assume a significance level of 0.05.

Show	A	B	C	D	E	F	G	H
First consultant	22%	18%	12%	12%	10%	9%	9%	8%
Second consultant (number of respondents favoring show)	29	30	20	16	9	17	10	19

10-16 In a certain county, 80% of the drivers have no accidents in a given year, 16% have one accident, and 4% have more than one accident. A survey of 200 randomly selected teachers from the county produced 172 with no accidents, 23 with one accident, and 5 with more than one accident. At the 5% level of significance, test the claim that the teachers exhibit the same accident rate as the countywide population.

10-17 A college dean expects 40% of the students to register on Monday, 45% on Tuesday, and 15% on Wednesday. Of 1850 students, 1073 register on Monday, 555 register on Tuesday, and 222 register on Wednesday. At the 1% level of significance, test the claim that the observed results are compatible with the dean's expectation.

10-18 For the past several years, the percentage of A's, B's, C's, D's, and F's for a certain statistics course have been 8%, 17%, 45%, 22%, and 8%, respectively. A new testing method was used in the last semester and there were 10 A's, 20 B's, 10 C's, 5 D's, and 5 F's. At the 5% level of significance, test the claim that the grades obtained through the new testing method are in the same proportion as before.

10-19 A pair of dice has 36 possible outcomes that are supposed to be equally likely. If we plan to test a certain pair of dice for fairness using the chi-square distribution, what is the minimum number of rolls necessary if each outcome is to have an expected frequency of at least 5? Suppose we conduct the minimum number of rolls and obtain a χ^2 value of 67.2. What do we conclude at the 0.01 significance level?

10-20 A roulette wheel has 38 possible outcomes that are supposed to be equally likely. If we plan to test a roulette wheel for fairness using the chi-square distribution, what is the minimum number of trials we must make if we are to satisfy the requirement that each expected frequency must be at least 5? Suppose we conduct the minimum number of trials and compute χ^2 to be 49.6. What do we conclude at the 0.01 significance level?

10-2 EXERCISES B

10-21   In this exercise we will show that a hypothesis test involving a multinomial experiment with only two categories is equivalent to a hypothesis test for a proportion (Section 6-5). Assume that a particular multinomial experiment has only two possible outcomes A and B with observed frequencies of f_1 and f_2 respectively.

   a. Find an expression for the χ^2 test statistic and find the critical value for a 0.05 significance level. Assume that we are testing the claim that both categories have the same frequency $(f_1 + f_2)/2$.

   b. The test statistic

$$z = \frac{\hat{p} - p}{\sqrt{\dfrac{pq}{n}}}$$

   is used to test the claim that a population proportion is equal to some value p. With the claim that $p = 0.5$, with $\alpha = 0.05$, and with

$$\hat{p} = \frac{f_1}{f_1 + f_2}$$

   show that z^2 is equivalent to χ^2 (from part (a)). Also show that the square of the critical z score is equal to the critical χ^2 value from part (a).

10-22   An observed frequency distribution is given below.

   a. Assuming a binomial distribution with $n = 3$ and $p = 1/3$, use the binomial probability formula to find the probability corresponding to each category of the table.

   b. Using the probabilities found in part (a), find the expected frequency for each category.

   c. Use a 0.05 level of significance to test the claim that the observed frequencies fit a binomial distribution for which $n = 3$ and $p = 1/3$.

Number of successes	0	1	2	3
Frequency	89	133	52	26

10-23   In conducting a survey of radio listeners, data are collected for different time slots, beginning at 1:00 P.M. The results are listed along with the expected frequencies based on past surveys. We cannot use the chi-square distribution since all expected values are not at least 5. However, we can combine some columns so that all expected values do equal or exceed 5. Use this suggestion to test the claim that the observed and expected frequencies are compatible. Try to combine categories in a meaningful way.

Time slot	1	2	3	4	5	6	7	8	9	10
Observed frequency	2	8	8	9	3	5	3	0	12	3
Expected frequency	4	5	8	7	4	6	5	2	9	3

10-24 An observed frequency distribution of IQ scores is given below.
a. Assuming a normal distribution with $\mu = 100$ and $\sigma = 15$, use the methods of Chapter 5 to find the probability of a randomly selected subject belonging to each class.
b. Using the probabilities found in part (a), find the expected frequency for each category.
c. Use a 0.01 level of significance to test the claim that the IQ scores were randomly selected from a normally distributed population with $\mu = 100$ and $\sigma = 15$.

IQ score	Under 80	80–95	96–110	111–120	Above 120
Frequency	20	20	80	40	40

10-3 CONTINGENCY TABLES

Many people believe that smoking is unhealthy. In a study of 1000 deaths of males aged 45 to 64, the causes of death are listed along with their smoking habits. See Table 10-4. (The data are based on results given in "Chartbook on Smoking, Tobacco, and Health," USDHEW publication CDC75-7511.)

TABLE 10-4		
Cause of Death		
	Cancer	Heart disease	Other
Smoker	135	310	205
Nonsmoker	55	155	140

In the introduction of this problem at the beginning of the chapter, we already noted that smokers are disproportionately represented among the sample of 1000 deaths. That suggests that smoking is unhealthy. We will now consider a different issue: Does smoking affect the cause of death? We will use the sample data of Table 10-4 to test the claim that *the cause of death is independent of smoking*. Having identified the key issue to be considered here, we should also remember that our analysis applies only to the population from which the sample was drawn. That is, we are considering only males who died between the ages of 45 and 64, and we are considering only the three categories of cause of death as given in Table 10-4.

Tables similar to Table 10-4 are generally called contingency tables, or two-way tables. In this context, the word *contingency* refers to dependence, and the contingency table serves as a useful medium for analyzing the dependence of one variable on another. This is only a statistical dependence that cannot be used to establish an inherent cause-and-effect relationship.

We need to test the *null hypothesis that the two variables in question are independent*. That is, we will test the claim that smoking habits and cause of death are independent. Let's select a significance level of $\alpha = 0.05$. We can now compute a test statistic based on the data and then compare that test statistic to the appropriate critical test value. As in the previous section, we use the chi-square distribution where the test statistic is given by χ^2.

TEST STATISTIC

$$\chi^2 = \sum \frac{(O - E)^2}{E}$$

This test statistic allows us to measure the degree of disagreement between the frequencies actually observed and those that we would theoretically expect when the two variables are independent. The reasons underlying the development of the χ^2 statistic in the previous section also apply here. In repeated large samplings, **the distribution of the test statistic χ^2 can be approximated by the chi-square distribution provided that all expected frequencies are at least 5**.

In the preceding section we knew the corresponding probabilities and could easily determine the expected values, but the typical contingency table does not come with the relevant probabilities. Consequently, we need to devise a method for obtaining the corresponding

If You Can Read This . . .

In the *New York Times* article headlined "Illiteracy Statistics: A Numbers Game," Jonathan Kozol noted gross inconsistencies in reported illiteracy statistics, which ranged from just over 1 million Americans (0.5%) to 21 million (9%) and possibly as high as 40 million Americans (17%). Kozol wrote that "every month, a new statistical report appears that purports to present the 'latest truth' about illiteracy in the United States." He suggested that when reporting about the various studies of illiteracy, journalists take the time to "present a story devoid of statistical narcotics that induce a false euphoria, which inevitably is followed by equally exaggerated anguish."

expected values. We will first describe the procedure for finding the values of the expected frequencies, and we will then proceed to justify that procedure. For each cell in the frequency table, the expected frequency E can be calculated by using the following.

EXPECTED FREQUENCY

$$\text{expected frequency } E = \frac{(\text{row total}) \cdot (\text{column total})}{(\text{grand total})}$$

---

where *grand total* refers to the total number of observations in the table. For example, in the lower right cell of Table 10-4, we see the observed frequency of 140. The total of all frequencies for that row is 350, the total of the column frequencies is 345, and the total of all frequencies in the table is 1000, so we get an expected frequency of

$$E = \frac{(350)(345)}{1000} = 120.75$$

In the lower right cell, the *observed* frequency is 140, while the *expected* frequency is 120.75. Table 10-5 reproduces Table 10-4 with the expected frequencies inserted in parentheses. As in Section 10-2, we require that all expected frequencies be at least 5 before we can conclude that the chi-square distribution serves as a suitable approximation to the distribution of χ^2 values.

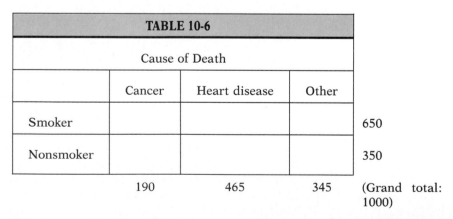

TABLE 10-5

	Cause of Death		
	Cancer	Heart disease	Other
Smoker	135 (123.50)	310 (302.25)	205 (224.25)	650
Nonsmoker	55 (66.50)	155 (162.75)	140 (120.75)	350
	190	465	345	(Grand total: 1000)

To better understand the rationale for this procedure, let's pretend that we know only the row and column totals and that we must fill in the cell frequencies by assuming that there is no relationship between the two variables involved (see Table 10-6).

TABLE 10-6

	Cause of Death		
	Cancer	Heart disease	Other
Smoker				650
Nonsmoker				350
	190	465	345	(Grand total: 1000)

We begin with the cell in the upper left corner that corresponds to smokers who died of cancer. Since 650 of the 1000 subjects are smokers, we have P(smoker) = 650/1000. Similarly, 190 of the 1000 subjects died from cancer, so that P(death by cancer) = 190/1000. Since we assume that smoking and cause of death are independent, we conclude that

$$P(\text{smoker and death by cancer}) = \frac{650}{1000} \times \frac{190}{1000}$$

This follows from the multiplication rule of probability whereby $P(A \text{ and } B) = P(A) \times P(B)$ if A and B are independent events. To obtain the expected value for the upper left cell, we simply multiply the probability for that cell by the total number of subjects available to get

$$\frac{650}{1000} \times \frac{190}{1000} \times 1000 = 123.50$$

The form of this product suggests a general way to obtain the expected frequency of a cell:

$$\text{expected frequency } E = \frac{(\text{row total})}{(\text{grand total})} \cdot \frac{(\text{column total})}{(\text{grand total})} \cdot (\text{grand total})$$

This expression can be simplified to

$$E = \frac{(\text{row total}) \cdot (\text{column total})}{(\text{grand total})}$$

Using the observed and expected frequencies shown in Table 10-5, we can now compute the χ^2 test statistic based on the sample data.

$$
\begin{aligned}
\chi^2 &= \sum \frac{(O - E)^2}{E} \\
&= \frac{(135 - 123.50)^2}{123.50} + \frac{(310 - 302.25)^2}{302.25} + \frac{(205 - 224.25)^2}{224.25} \\
&\quad + \frac{(55 - 66.50)^2}{66.50} + \frac{(155 - 162.75)^2}{162.75} + \frac{(140 - 120.75)^2}{120.75} \\
&= 1.0709 + 0.1987 + 1.6525 \\
&\quad + 1.9887 + 0.3690 + 3.0688 \\
&= 8.349
\end{aligned}
$$

With $\alpha = 0.05$ we proceed to Table A-4 to obtain the critical value of χ^2, but we must first know where the critical region lies and the number of degrees of freedom. **Tests of independence with contingency tables involve only right-tailed critical** regions since small values of χ^2 support the claimed independence of the two variables. That is, χ^2 is small if observed and expected frequencies are close. Large values of χ^2 are to the right of the chi-square distribution, and they reflect significant differences between observed and expected frequencies.

In a contingency table with r rows and c columns, the number of degrees of freedom is given by

degrees of freedom $(r - 1)(c - 1)$

Thus Table 10-4 has $(2 - 1)(3 - 1) = 2$ degrees of freedom. In a right-tailed test with $\alpha = 0.05$ and with 2 degrees of freedom we refer to Table A-4 to get a critical χ^2 value of 5.991. Since the calculated χ^2 value of 8.349 falls in the critical region bounded by $\chi^2 = 5.991$, we reject the null hypothesis of independence between the two variables. See Figure 10.2. It appears that smoking and cause of death are dependent.

We now summarize the important components of this hypothesis test. We used the sample data of Table 10-4 with these hypotheses.

H_0: Smoking is independent of the cause of death (cancer, heart disease, other).

H_1: Smoking and the cause of death are dependent.

Significance level: $\alpha = 0.05$

Test statistic: $\chi^2 = 8.349$

Critical value: $\chi^2 = 5.991$

Conclusion: Reject the null hypothesis. There is sufficient sample evidence to warrant rejection of the claim that smoking and cause of death are independent. It appears that they are dependent.

Note that this hypothesis test doesn't show that smoking is unhealthy; we discussed that at the beginning of the chapter. Instead, this hypothesis test suggests that smoking and cause of death are dependent. We should always be careful about jumping to a conclusion that there is a cause-effect relationship. We have not established that smoking actually affects the cause of death. We have simply shown that they appear to be dependent and we have not identified the true causal factors.

We might note that the original study involved a sample size of 43,221 deaths of males aged 45 to 64, so the results are even more dramatic than for the sample of 1000 included here. We might also comment that smoking really *is* unhealthy. No ifs, ands, or . . .

Because of the nature of the calculations, it is often helpful to use a statistics software package for tests of the type discussed in this

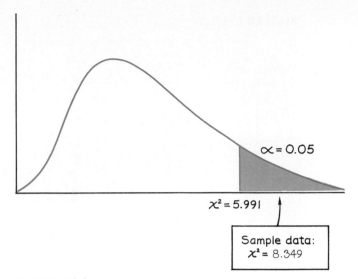

FIGURE 10-2

section. The STATDISK and Minitab displays show the results obtained for the data of Table 10-4.

STATDISK DISPLAY

```
        A  [ 135 ] [ 310 ] [ 205 ]
           (123.5) (302.3) (224.3)


        B  [  55 ] [ 155 ] [ 140 ]
           (66.5 ) (162.8) (120.8)


     Test statistic . . Chi Square = 8.3
     Critical value . . Chi Square = 5.9
     P-value . . . . . . . . . . . = 0.0156547
     Significance level  . . . . . = .05
     Degrees of freedom  . . . . . = 2

     CONCLUSION: REJECT the null hypothesis
   that row and column variables are independent.
```

MINITAB DISPLAY

```
MTB > READ C1 C2 C3
DATA> 135 310 205
DATA>  55 155 140
DATA> ENDOFDATA
     2 ROWS READ
MTB > CHISQUARE analysis for C1 C2 C3

Expected counts are printed below observed counts

            C1        C2        C3    Total
    1      135       310       205      650
        123.50    302.25    224.25

    2       55       155       140      350
         66.50    162.75    120.75

Total      190       465       345     1000

ChiSq = 1.071 +   0.199 +   1.652 +
        1.989 +   0.369 +   3.069 = 8.349
df = 2
```

10-3 EXERCISES A

10-25   A survey on a local gun control proposal produced the data summarized in the following table. At the $\alpha = 0.05$ significance level, test the claim that voting on the bill is independent of the political party of the voter.

a. Assume independence of party and voter opinion and determine the expected frequency for each of the four cells.

b. Compute the χ^2 test statistic based on the sample data.

c. Determine the critical value of χ^2.

d. What conclusion can you draw?

	In favor	Opposed
Democrat	20	30
Republican	30	10

10-26   Many people believe that criminals who plead guilty tend to get lighter sentences than those who are convicted by trial. In the table below we summarize sample data for San Francisco defendants in burglary cases. All of the subjects had prior prison sentences. (See "Does It Pay to Plead Guilty? Differential Sentencing and the Functioning of the Criminal Courts" by Brereton and Casper, *Law and Society Review*, Vol. 16, No. 1.) At the 0.05 significance level, test the claim that the sentence (sent to prison or not sent to prison) is independent of the plea.

	Guilty plea	Plea of not guilty
Sent to prison	392	58
Not sent to prison	564	14

10-27   In the judicial case of *United States v. City of Chicago*, fair employment practices were challenged. A minority group (group A) and a majority group (group B) took the Fire Captain Examination. At the 0.05 significance level, use the given results and test the claim that success on the test is independent of the group.

	Pass	Fail
Group A	10	14
Group B	417	145

10-28   Researchers studied 90 jurors to find the effect of sex on punitive attitudes. They administered psychological tests and classified each subject as having liberal, fair, or vindictive tendencies. The results follow. At the $\alpha = 0.05$ level of significance, test the claim that the presence of these tendencies is independent of the sex of the juror.

	Liberal	Fair	Vindictive
Females	21	24	15
Males	5	18	7

10-29   A study of medical practices involved subjects from different groups. Sample data is given in the accompanying table, which is based on data from "A Randomized Controlled Trial of Academic Group Practice" by Goldberg and others, *Journal of the American Medical Association*, Vol. 257, No. 15. At the 0.05 level of significance, test the claim that the payer is independent of the group.

		Control group	Experimental group
	Medicare	38	43
Payer	Medicaid	16	17
	County welfare	11	10

10-30   At the 0.05 significance level, test the claim that the type of college is independent of the dropout rate. (The sample data are based on results from the American College Testing Program.)

	Four-year public	Four-year private	Two-year public	Two-year private
Freshmen dropouts	10	9	15	9
Freshmen who stay	26	28	18	27

10-31   In a study of car accidents and drivers who use cellular phones, the following sample data are obtained. (The data are based on results from AT&T and the Automobile Association of America.) At the 0.05 level of significance, test the claim that the occurrence of accidents is independent of the use of cellular phones.

	Had accident in last year	Had no accident in last year
Cellular phone user	23	282
Not cellular phone user	46	407

10-32 The following table summarizes results from randomly selected drivers convicted of motor vehicle violations in New York State. (The data are based on results from the New York State Department of Motor Vehicles.) At the 0.05 significance level, test the claim that the county is independent of the type of violation. (DWI is driving while intoxicated and DTD is driving with a disabled traffic device.)

	Albany	Monroe	Orange	Westchester
DWI	6	19	6	10
Speeding	103	261	160	226
DTD	60	152	25	174

10-33 The manager of an assembly operation wants to determine whether the number of defective parts is dependent on the day of the week. She develops the following sample data. Test the claim of the union representative that the day of the week makes no difference in the number of defects.

	Mon	Tues	Wed	Thurs	Fri
Acceptable products	80	100	95	93	82
Defective products	15	5	5	7	12

10-34 In a study of seat belt use in taxi cabs, sample data is collected and summarized in the following table (based on "The Phantom Taxi Seat Belt" by Welkon and Reisinger, *American Journal of Public Health*, Vol. 67, No. 11). At the 0.05 level of significance, test the claim that the occurrence of usable seat belts is independent of the city.

		New York	Chicago	Pittsburgh
Taxi has usable seat belt?	Yes	3	42	2
	No	74	87	70

10-35   For a random sample of deaths of U.S. Army veterans, a medical panel established the cause of death by reviewing records. Among deaths considered to be alcohol-related, the cause of death is listed and the results are included in the accompanying table. (See "Underreporting of Alcohol-Related Mortality on Death Certificates of Young U.S. Army Veterans" by Pollock and others, *Journal of the American Medical Association*, Vol. 258, No. 3.) At the 0.05 significance level, test the claim that the cause of death is independent of the determining source.

	Natural	Motor vehicle	Suicide/homicide
Original death certificate	9	7	4
Medical panel	30	53	35

10-36   Many people believe that criminals who plead guilty tend to get lighter sentences than those who are convicted by trial. In the following table we summarize sample data for 434 San Francisco defendents in robbery cases. All of the subjects had prior prison sentences. (See "Does It Pay to Plead Guilty? Differential Sentencing and the Functioning of Criminal Courts" by Brereton and Casper, *Law and Society Review*, Vol. 16, No. 1.) At the 0.01 level of significance, test the claim that the sentence (sent to prison or not sent to prison) is independent of the plea.

	Guilty plea	Plea of not guilty
Sent to prison	191	64
Not sent to prison	169	10

10-37   A study is conducted to determine the rate of smoking among people from different age groups. Sample data are summarized in the accompanying table, which is based on data from the National Center for Health Statistics. At the 0.05 significance level, test the claim that smoking is independent of the four listed age groups.

	Age (years)			
	20–24	25–34	35–44	45–64
Smoke	18	15	17	15
Don't smoke	32	35	33	35

10-38   In a study of Reye's syndrome, the stage of the disease and the time of year of its onset are listed below for 227 subjects. The stages (0 for mild, 1 for severe) follow uniform standards. (See "Reye's Syndrome" by Holtzhauer and others, *American Journal of Diseases of Children*, Vol. 140.) At the 0.05 significance level, test the claim that the stage is independent of the time of onset.

	Time of year of onset		
	Jan.–Mar.	Apr.–June	July–Dec.
Stage 0	55	6	6
Stage 1 or greater	122	17	21

10-39   Data drawn from randomly selected students is summarized in the table below. At the 0.10 level of significance, test the claim that the subject and grade are independent.

Course	Grade				
	A	B	C	D	F
Math	22	24	28	10	16
Art	40	64	44	9	22
English	65	150	160	38	62

10-40   A sociologist conducts a study of attitudes and obtains results for the first question in a survey. Those results are listed in the following table. At the 0.01 level of significance, test the claim that age and response are independent.

Age	Response				
	Strongly agree	Agree	Undecided	Disagree	Strongly disagree
Under 21	20	27	10	15	15
21–30	40	52	30	20	10
31–40	8	8	12	16	15
Over 40	10	12	15	20	8

10-3 EXERCISES B

10-41   The only sample data available on attitudes about capital punishment are summarized as follows. We wish to test the independence of geographic region and opinion, but each cell does not have an expected value of at least 5. Combine rows in a reasonable way so that the expected value of each cell is at least 5, and then complete the test using a significance level of $\alpha = 0.05$.

	In favor	Opposed	No opinion
Northeast	4	8	3
Southeast	7	8	6
North Central	2	4	4
South Central	9	3	5
Northwest	12	20	4
Southwest	2	14	3

10-42   The chi-square distribution is continuous while the test statistic used in this section is actually discrete. Some statisticians use **Yates' correction for continuity** in cells with an expected frequency less than 10 or in all cells of a contingency table with two rows and two columns. With Yates' correction, we replace

$$\sum \frac{(O - E)^2}{E} \quad \text{with} \quad \sum \frac{(|O - E| - 0.5)^2}{E}$$

Given the contingency table below, find the value of the χ^2 test statistic with and without Yates' correction. In general, what effect does Yates' correction have on the value of the test statistic?

	X	Y
A	5	25
B	65	5

10-43   If each observed frequency in a contingency table is multiplied by a positive integer K (where $K \geq 2$), how is the value of the test statistic affected?

10-44　a. For the contingency table given below, verify that the test statistic becomes

$$\chi^2 = \frac{(a + b + c + d)(ad - bc)^2}{(a + b)(c + d)(b + d)(a + c)}$$

	Column	
	1	2
Row 1	a	b
Row 2	c	d

b. Let $\hat{p}_1 = \dfrac{a}{a + c}$ and let $\hat{p}_2 = \dfrac{b}{b + d}$ and show that the test statistic

$$z = \frac{(\hat{p}_1 - \hat{p}_2) - 0}{\sqrt{\bar{p}\bar{q}\left(\dfrac{1}{n_1} + \dfrac{1}{n_2}\right)}}$$

is such that $z^2 = \chi^2$ (the same result from part (a)). This shows that the chi-square test involving a 2×2 table is equivalent to the test for the difference between two proportions, as described in Section 8-4.

10-4 ANALYSIS OF VARIANCE

In Section 10-2 we developed a procedure for testing the hypothesis that the differences among several sample proportions are due to chance. In this section we develop a procedure for testing the hypothesis that differences among several sample **means** are due to chance. We now require that the populations under consideration have distributions that are essentially normal. Also, the variances of the populations must be equal, and the samples must be independent. The requirements of normality and equal variances are somewhat loose since the methods in this section work reasonably well unless there is a very nonnormal distribution or unless the population variances differ by very large amounts. (If the samples are independent but the distributions are not normal, we can use the Kruskal-Wallis test presented in Section 11-5.) The method we will describe in this section is called analysis of variance (often referred to as ANOVA) because it is based on a comparison of two different estimates of the variance that is common to the different populations. We will begin with cases involving samples having the same numbers of scores. We will then proceed to consider cases in which the sample sizes are not all equal.

Equal Sample Sizes

Let's assume that we have k different populations and that a sample of size n is drawn from each population. From the sample data we compute the sample means $\bar{x}_1, \bar{x}_2, \ldots, \bar{x}_k$ and the sample standard deviations $s_1, s_2, \ldots, s_k$. Our method of testing the null hypothesis $\mu_1 = \mu_2 = \ldots = \mu_k$ requires that we obtain two different estimates of the common population variance σ^2 and then compare those estimates by using the F distribution. Let's consider a specific example.

EXAMPLE

A pilot does extensive bad weather flying and decides to buy a battery-powered radio as an independent backup for her regular radios, which depend on the airplane's electrical system. She has a choice of three brands of rechargeable batteries that vary in cost. She obtains the sample data in the following table. She randomly selects five batteries for each brand, and tests them for the operating time (in hours) before recharging is necessary. Do the three brands have the same mean usable time before recharging is required?

Brand X	Brand Y	Brand Z
26.0	29.0	30.0
28.5	28.8	26.3
27.3	27.6	29.2
25.9	28.1	27.1
28.2	27.0	29.8
$\bar{x}_1 = 27.18$	$\bar{x}_2 = 28.10$	$\bar{x}_3 = 28.48$
$s_1^2 = 1.46$	$s_2^2 = 0.69$	$s_3^2 = 2.81$

Solution

$H_0: \mu_x = \mu_y = \mu_z$
$H_1:$ The preceding means are not all equal.

Using these sample results we develop two separate estimates of σ^2, where σ^2 is assumed to be the population variance common

continued ▶

Solution, continued

to all three brands. We employ the F distribution in the comparison of the two separate estimates. In Section 8-2, we discussed the use of the F distribution in testing for equality of two variances. We saw that when two independent samples were drawn from populations that were approximately normally distributed, the sampling distribution of $F = s_1^2/s_2^2$ was the F distribution. Here we again use the F distribution, but we replace the two sample variances by two different estimates of the population variance common to all of the populations under consideration. Specifically, we use

TEST STATISTIC

For cases with equal sample sizes:

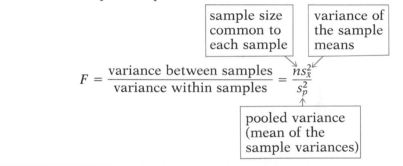

$$F = \frac{\text{variance between samples}}{\text{variance within samples}} = \frac{ns_{\bar{x}}^2}{s_p^2}$$

sample size common to each sample

variance of the sample means

pooled variance (mean of the sample variances)

The variance *between* samples is obtained by finding the product of n and $s_{\bar{x}}^2$; n is the number of scores in each sample and $s_{\bar{x}}^2$ is the variance of the sample means. For the data in the table, $n = 5$ since each sample consists of 5 scores, and $s_{\bar{x}}^2 = 0.45$ is found by computing the variance of the sample means (27.18, 28.10, 28.48). (Use Formula 2-5.) Thus $ns_{\bar{x}}^2 = (5)(0.45) = 2.25$. The expression $ns_{\bar{x}}^2$ is justified as an estimate of σ^2 since $\sigma_{\bar{x}} = \sigma/\sqrt{n}$. Squaring both sides of this last expression and solving for σ^2, we get $\sigma^2 = n\sigma_{\bar{x}}^2$, which indicates that σ^2 can be estimated by $ns_{\bar{x}}^2$.

The variance *within* samples is obtained by computing s_p^2, which denotes the mean of the sample variances. Since we have three sets of samples, s_p^2 is the pooled variance found by computing

$$\frac{s_1^2 + s_2^2 + s_3^2}{3} = \frac{1.46 + 0.69 + 2.81}{3} = 1.65$$

continued ▶

Solution, continued

This approach seems reasonable since we assume that equal population variances imply that representative samples yield sample variances which, when pooled, provide a good estimate of the value of the common population variance.

We can now proceed to include both estimates of σ^2 in the determination of the value of the F statistic based on the data.

$$F = \frac{\text{variance between samples}}{\text{variance within samples}} = \frac{ns_{\bar{x}}^2}{s_p^2} = \frac{2.25}{1.65} = 1.36$$

If the two estimates of variance are close, the calculated value of F will be close to 1 and we could conclude that there are no significant differences among the sample means. But if the value of F is excessively *large*, then we would reject the claim of equal means. The estimate of variance in the denominator (s_p^2) depends only on the sample variances and is not affected by differences among the sample means. However, if there are extreme differences among the sample means, the numerator will be larger so the value of F will be larger. In general, as the sample means move farther apart, the value of $ns_{\bar{x}}^2$ grows larger and the value of F itself grows larger. Since excessively large values of F reflect unequal means, the test is right tailed.

The critical value of F that separates excessive values from acceptable values is found in Table A-5, where α is again the level of significance and the numbers of degrees of freedom are as follows (assuming that there are k sets of separate samples with n scores in each set).

For cases with equal sample sizes:

numerator degrees of freedom = $k - 1$
denominator degrees of freedom = $k(n - 1)$

Our battery example involves $k = 3$ separate sets of samples and there are $n = 5$ scores in each set, so

numerator degrees of freedom = $k - 1 = 3 - 1 = 2$
denominator degrees of freedom = $k(n - 1) = 3(5 - 1) = 12$

With $\alpha = 0.05$, the critical F value corresponding to these degrees of freedom is 3.8853. The computed test statistic of $F = 1.36$ does not fall within the right-tailed critical region bounded by $F = 3.8853$ (see Figure 10-3), so we fail to reject the null hypothesis of equal population means. That is, there are no significant differences among the means of the three given sets of samples. The pilot should purchase the least expensive battery.

94022

Zip Codes Reveal Much

The Claritas Corporation has developed a way of obtaining much information about people from their zip codes. Zip code data were extracted from a variety of mailing lists, purchase orders, warranty cards, census data, and market research surveys. With people of the same social and economic levels tending to live in the same areas, it became possible to match zip codes with such factors as purchase patterns, types of cars driven, foods preferred, leisure time activities, and television viewing choices. The company helps clients target their advertising efforts to regions that are most likely to accept particular products.

The preceding analysis led to a right-tailed critical region, and every other similar situation will also involve a right-tailed critical region. Since the individual components of n, $s_{\bar{x}}^2$, and s_p^2 are all positive, F is always positive, and we know that values of F near 1 correspond to relatively close sample means. The only values of the F test statistic that are indicative of significant differences among the sample means are those F values that exceed 1 beyond the critical value obtained from Table A-5.

It may seem strange that we are testing equality of several means by analyzing only variances, but the sample means directly affect the value of the test statistic F. To illustrate this important point, we add 10 to each score listed under brand X, but we leave the brand Y and brand Z values unchanged. The revised sample statistics follow.

Brand X	Brand Y	Brand Z
$\bar{x}_1 = 37.18$	$\bar{x}_2 = 28.10$	$\bar{x}_3 = 28.48$
$s_1^2 = 1.46$	$s_2^2 = 0.69$	$s_3^2 = 2.81$

We again assume a significance level of $\alpha = 0.05$ and test the claim that the three brands have the same population mean. In this way, we can see the effect of increasing the brand X scores by 10. The F statistic based on the revised sample data follows.

$$F = \frac{\text{variance between samples}}{\text{variance within samples}} = \frac{ns_{\bar{x}}^2}{s_p^2} = \frac{(5)(26.38)}{(1.46 + 0.69 + 2.81)/3}$$

$$= \frac{131.90}{1.65} = 79.94$$

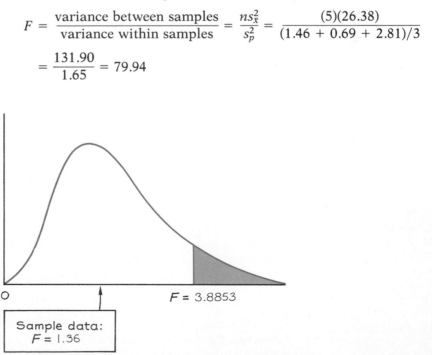

F = 3.8853

Sample data:
F = 1.36

FIGURE 10-3

Statistics and Baseball Strategy

Statisticians are using computers to develop very sophisticated measures of baseball performance and strategy. They have found, for example, that sacrifice bunts, sacrifice fly balls, and stolen bases rarely help win games and that it is seldom wise to have a pitcher intentionally walk the batter. They can identify the ballparks that favor pitchers and those that favor batters. Instead of simply comparing the batting averages of two different players, they can develop better measures of offensive strength by taking into account such factors as pitchers faced, position in the lineup, ballparks played, and weather conditions.

We get $n = 5$ since each brand yields 5 sample values. The value of $s_{\bar{x}}^2 = 26.38$ is obtained by calculating the variance of the sample means 37.18, 28.10, 28.48. We find s_p^2 to be 1.65 by computing the mean of the three sample variances. With three samples of five scores each, we have $3 - 1$, or 2, degrees of freedom for the numerator and $3(5 - 1)$, or 12, degrees of freedom for the denominator. With $\alpha = 0.05$ we therefore obtain a critical F value of 3.8853. Since the computed F statistic of 79.94 far exceeds the critical F value of 3.8853, we reject the null hypothesis of equal population means. Further examination of the data reveals that brand X appears to last significantly longer than brands Y and Z, which don't appear to be significantly different.

Before adding 10 to each score for brand X, we obtained a test statistic of $F = 1.36$, but the increased brand X values cause F to become 79.94. Note that, in both cases, the brand X variance is 1.46, so the difference in the F test statistic is attributable only to the change in $\bar{x}_1$. By changing $\bar{x}_1$ from 27.18 to 37.18 and retaining the same values of $\bar{x}_2$, $\bar{x}_3$, s_1^2, s_2^2, and s_3^2, we find that the F test statistic changes from 1.36 to 79.94. This illustrates that the F statistic is very sensitive to sample *means* even though it is obtained through two different estimates of the common population *variance*.

Unequal Sample Sizes

Up to this point, our discussion of analysis of variance has involved only examples with the same number of values in each sample. We will now proceed to consider the treatment of samples with unequal sizes, and we will introduce some of the common terminology and notation used in analysis of variance. We will again proceed by analyzing the significance of the difference between **variation due to treatment** (variance between samples) and **variation due to error** (variation within samples). When we refer to "treatments," we are actually referring to the different samples as somehow being the results of different factors that we are analyzing. An agriculture study might involve treatments consisting of different fertilizer mixtures. An athlete might analyze treatments consisting of different training techniques. In each case, we obtain sample data from the different treatments, and we proceed to analyze the results in order to decide whether the effects of the treatments are different.

As with the case of equal sample sizes, we again use the test statistic

$$F = \frac{\text{variance between samples}}{\text{variance within samples}}$$

However, for unequal sample sizes we weight the two estimates of variance to account for the different sample sizes. We get the following.

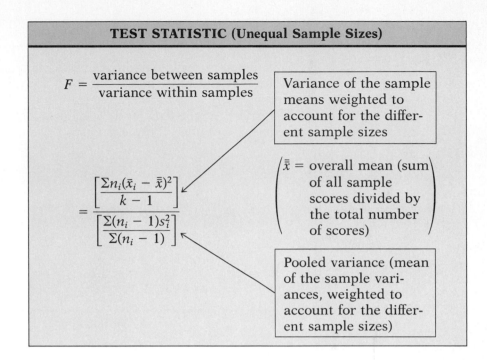

TEST STATISTIC (Unequal Sample Sizes)

$$F = \frac{\text{variance between samples}}{\text{variance within samples}}$$

Variance of the sample means weighted to account for the different sample sizes

$$= \frac{\left[\dfrac{\Sigma n_i(\bar{x}_i - \bar{\bar{x}})^2}{k - 1}\right]}{\left[\dfrac{\Sigma(n_i - 1)s_i^2}{\Sigma(n_i - 1)}\right]}$$

$\left(\begin{array}{l}\bar{\bar{x}} = \text{overall mean (sum} \\ \quad \text{of all sample} \\ \quad \text{scores divided by} \\ \quad \text{the total number} \\ \quad \text{of scores)}\end{array}\right)$

Pooled variance (mean of the sample variances, weighted to account for the different sample sizes)

Note that the numerator is really a form of the formula

$$s^2 = \frac{\Sigma(x - \bar{x})^2}{n - 1}$$

for variance that was given in Chapter 2. The factor of n_i is included so that larger samples carry more weight. The denominator of the test statistic is simply the mean of the sample variances, but it is a weighted mean with the weights corresponding to the sample sizes. (If all samples have the same number of scores, the test statistic simplifies to $F = ns_{\bar{x}}^2/s_p^2$, as given earlier.)

Assuming that we have sample data for which the value of the above test statistic has been found, we need to find the critical value. We refer to Table A-5 and use the following degrees of freedom.

degrees of freedom (numerator) = $k - 1$ where k is the number of samples

degrees of freedom (denominator) = $N - k$ where N is the total number of scores in all samples combined

---

EXAMPLE

Three different computer programming languages (A, B, and C) are used by different students to solve a problem; the times (in hours) required for the solution are listed below each language designation. At the 0.05 level of significance, test the claim that the mean times for all three languages are the same.

A	B	C
7	9	2
4	5	3
4	7	5
3		3
		8

$n_1 = 4$	$n_2 = 3$	$n_3 = 5$
$\bar{x}_1 = 4.5$	$\bar{x}_2 = 7.0$	$\bar{x}_3 = 4.2$
$s_1^2 = 3.0$	$s_2^2 = 4.0$	$s_3^2 = 5.7$

Solution

The null hypothesis is the claim that the samples come from populations with the same mean.

We have

$$H_0: \mu_A = \mu_B = \mu_C \qquad H_1: \text{The preceding means are not all equal.}$$

With $\alpha = 0.05$, we proceed to calculate the test statistic as follows.

Number of samples: $k = 3$ $\quad \bar{\bar{x}} = $ mean of all scores $= \dfrac{60}{12} = 5.0$

← variance between samples (numerator of F) =

$$\frac{\Sigma n_i (\bar{x}_i - \bar{\bar{x}})^2}{k - 1} = \frac{4(4.5 - 5.0)^2 + 3(7.0 - 5.0)^2 + 5(4.2 - 5.0)^2}{3 - 1}$$

$$= \frac{16.2}{2} = 8.1$$

← variance within samples (denominator of F) =

$$\frac{\Sigma(n_i - 1)s_i^2}{\Sigma(n_i - 1)} = \frac{(4 - 1)(3.0) + (3 - 1)(4.0) + (5 - 1)(5.7)}{(4 - 1) + (3 - 1) + (5 - 1)}$$

$$= \frac{39.8}{9} = 4.4222$$

$$\to F = \frac{\text{variance between samples}}{\text{variance within samples}} = \frac{8.1}{4.4222} = 1.8317$$

continued ▶

Solution, continued

With a test statistic of $F = 1.8317$ we now proceed to obtain the critical value. The significance level indicates that $\alpha = 0.05$. The degrees of freedom are as follows.

degrees of freedom (numerator) $= k - 1 = 3 - 1 = 2$

degrees of freedom (denominator) $= N - k = 12 - 3 = 9$

With 2 degrees of freedom for the numerator and 9 degrees of freedom for the denominator, we refer to Table A-5 and find the critical value of $F = 4.2565$. Since the test statistic of $F = 1.8317$ does not exceed the critical value of $F = 4.2565$, we fail to reject the null hypothesis that the means are equal. There is not sufficient sample evidence to warrant rejection of the claim that the means are equal.

The method illustrated in the preceding example is based on sample statistics $(n_i, \bar{x}_i, s_i^2)$ that would probably be found in the early stages of research. That method is a generalization of the same method used for the case of equal sample sizes; it reinforces earlier concepts and follows patterns already established. However, it often leads to larger errors from rounding than some other equivalent methods. In the table on the next page we identify the components used in another method that produces the same results, but usually with less error due to rounding. Using this second method, we list key results for the data of the preceding example.

$$SS \text{ (total)} = 56.0$$

$$SS(\text{treatment}) = 16.2 \qquad df(\text{treatment}) = 2 \qquad MS(\text{treatment}) = 8.1$$

$$SS(\text{error}) = 39.8 \qquad df(\text{error}) = 9 \qquad MS(\text{error}) = 4.4222$$

$$F = \frac{MS(\text{treatment})}{MS(\text{error})} = \frac{8.1}{4.4222} = 1.8317$$

These results show that the same test statistic is obtained with this second method. The first method is superior if you want to better understand the underlying concepts, and the second method is generally better if you want easier calculations. However, the "easier" calculations of the second method continue to be messy enough so that the use of computer software should be seriously considered for most real problems. Shown on page 603 are the STATDISK and Minitab displays resulting from the data of the preceding problem.

k = number of different samples (or the number of columns of data)

n_i = number of values in the ith sample

c_i = total of the values in the ith sample

$N = \Sigma n_i$ = total number of values in all samples combined. (N has been used previously to represent the size of a population.)

$\Sigma x = \Sigma c_i$ = sum of the scores of all samples combined

Σx^2 = sum of the squares of all scores from all samples combined

$SS(\text{total}) = \Sigma x^2 - \dfrac{(\Sigma x)^2}{N}$	Total sum of squares. This expression is algebraically the same as $\Sigma(x - \bar{x})^2$.
$SS(\text{treatment}) = \left(\Sigma \dfrac{c_i^2}{n_i}\right) - \dfrac{(\Sigma c_i)^2}{N}$	Sum of squares that represents variation among the different samples (sometimes called *explained variation*).
$SS(\text{error}) = SS(\text{total}) - SS(\text{treatment})$	Sum of squares that represents the variation within samples that is due to chance (sometimes called *unexplained variation*.)
$df(\text{treatment}) = k - 1$	Degrees of freedom associated with the different treatments.
$df(\text{error}) = N - k$	Degrees of freedom associated with the errors within samples.
$MS(\text{treatment}) = \dfrac{SS(\text{treatment})}{df(\text{treatment})}$	Mean square or variance estimate explained by the different treatments.
$MS(\text{error}) = \dfrac{SS(\text{error})}{df(\text{error})}$	Mean square or variance estimate that is unexplained (due to chance).
$F = \dfrac{MS(\text{treatment})}{MS(\text{error})}$	Test statistic representing the ratio of two estimates of variance.

ANOVA Table				
Source of variation	Sum of squares SS	Degrees of freedom	Mean square MS	Test statistic
Treatments	$SS(\text{treatment})$	$k - 1$	$MS(\text{treatment}) = \dfrac{SS(\text{treatment})}{df(\text{treatment})}$	$F = \dfrac{MS(\text{treatment})}{MS(\text{error})}$
Error	$SS(\text{error})$	$N - k$	$MS(\text{error}) = \dfrac{SS(\text{error})}{df(\text{error})}$	
Total	$SS(\text{total})$	$N - 1$		

STATDISK DISPLAY

```
D.F.    (TREATMENT) = 2              D.F.    (ERROR) = 9

Sum sq. (TREATMENT) = 16.2000        Sum sq. (ERROR) = 39.8000

            Total sum of squares = 56.0000

Mean sq.(TREATMENT) = 8.1000         Mean sq.(ERROR) = 4.4222

            Significance level . = .05

            F . . . . . . . . . = 1.8317

            P-value . . . . . . = 0.2151

    CONCLUSION: FAIL TO REJECT the null hypothesis of equal means
```

MINITAB DISPLAY

```
MTB > SET C1
DATA> 7 4 4 3
DATA> ENDOFDATA
MTB > SET C2
DATA> 9 5 7
DATA> ENDOFDATA
MTB > SET C3
DATA> 2 3 5 3 8
DATA> ENDOFDATA
MTB > AOVONEWAY C1 C2 C3

ANALYSIS OF VARIANCE
SOURCE    DF        SS        MS        F         P
FACTOR     2      16.20      8.10      1.83     0.215
ERROR      9      39.80      4.42
TOTAL     11      56.00
```

As efficient and reliable as such computer programs may be, they are totally worthless if we don't understand the relevant concepts. We should recognize that the methods of this section are used to test the claim that several samples come from populations with the same mean. **These methods require normally distributed populations with the same variance, and the samples must be independent.** We reject or fail to reject the null hypothesis of equal means by analyzing the two estimates of variance. The MS(treatment) is an estimate of the variation between samples, while the MS(error) is an estimate of the variation within samples. If MS(treatment) is significantly greater than MS(error), then we reject the claim of equal means, otherwise we fail to reject that claim.

The procedure of analysis of variance, as presented in this section, can be used to decide whether differences among sample means are significant or attributable to chance. This procedure is referred to as **one-way analysis of variance** or **single-factor analysis of variance** to indicate that the data are classified into groups according to a single criterion. In the last example, the three samples were categorized according to the single factor of programming language used. A more complex analysis might involve samples with the times classified according to more than one variable, such as experience of the programmer, education of the programmer, and other relevant factors. Methods for dealing with any number of classification variables have been developed, but this text considers only cases involving a single factor.

When we use the single-factor analysis of variance technique and conclude that the differences among the means are significant, we cannot necessarily conclude that the given factor is responsible for those differences. In the last example, if the differences were significant, we would reject the null hypothesis of equal means. However, that rejection would not necessarily imply that the differences were due to the programming language used. Perhaps all the newer programmers used the same language, but their inexperience made them slower, so that significant differences are the results of experience rather than the programming language used. One way to reduce the effect of extraneous factors (such as experience of the programmer) is to design the experiment so that it has a **completely randomized design**. That is, each element is given the same chance of belonging to the different categories or treatments. Another way to reduce the effect of extraneous factors is to use a **rigorously controlled design** in which all other factors are forced to be constant. To eliminate the effect of experience and education, for example, we might use only programmers with the same experience and education. Obviously, this approach is often difficult or impossible to implement. In any event, we should carefully scrutinize the way in which the sample data were collected so that we avoid serious errors in attributing significant differences among means to factors that might not be responsible for those differences. The design of the experiment is critically important, and no statistical calisthenics can salvage a poor design.

10-4 EXERCISES A

10-45   The dean of a college wants to compare grade-point averages of resident, commuting, and part-time students. A random sample of each group is selected, and the results are as follows. At the $\alpha = 0.05$ level of significance, test the claim that the three populations have equal means.

Residents	Commuters	Part-time
$n = 15$	$n = 15$	$n = 15$
$\bar{x} = 2.60$	$\bar{x} = 2.55$	$\bar{x} = 2.30$
$s^2 = 0.30$	$s^2 = 0.25$	$s^2 = 0.16$

10-46   Do Exercise 10-45 after changing the sample mean grade-point average for resident students from 2.60 to 3.00.

10-47   Five socioeconomic classes are being studied by a sociologist, and members from each class are rated for their adjustments to society. The sample data are summarized as follows. At the $\alpha = 0.05$ level of significance, test the claim that the five populations have equal means.

A	B	C	D	E
$n_1 = 10$	$n_2 = 10$	$n_3 = 10$	$n_4 = 10$	$n_5 = 10$
$\bar{x}_1 = 103$	$\bar{x}_2 = 97$	$\bar{x}_3 = 102$	$\bar{x}_4 = 100$	$\bar{x}_5 = 110$
$s_1^2 = 230$	$s_2^2 = 75$	$s_3^2 = 200$	$s_4^2 = 150$	$s_5^2 = 100$

10-48   A unit on elementary algebra is taught to five different classes of randomly selected students with the same academic backgrounds. A different method of teaching is used in each class, and the final averages of the 20 students in each class are compiled. The results yield the following data. At the $\alpha = 0.05$ level of significance, test the claim that the five population means are equal.

Traditional	Programmed	Audio	Audiovisual	Visual
$n = 20$	$n = 20$	$n = 20$	$n = 20$	$n = 20$
$\bar{x} = 76$	$\bar{x} = 74$	$\bar{x} = 70$	$\bar{x} = 75$	$\bar{x} = 74$
$s^2 = 60$	$s^2 = 50$	$s^2 = 100$	$s^2 = 36$	$s^2 = 40$

10-49   Five car models are studied in a test that involves four of each model. For each of the four cars in each of the five samples, exactly 1 gallon of gas is placed in the tank and the car is driven until the gas is used up. The results follow. At the $\alpha = 0.05$ significance level, test the claim that the five population means are all equal.

Distance traveled in miles				
A	B	C	D	E
16	18	18	19	15
22	23	18	21	16
17	15	20	22	20
17	20	20	22	17

10-50   A sociologist randomly selects subjects from three types of family structure: stable families, divorced families, and families in transition. The selected subjects are interviewed and rated for their social adjustment. The sample results are given at left. At the $\alpha = 0.05$ level of significance, test the claim that the three population means are equal.

Stable	Divorced	Transition
110	115	90
105	105	120
100	110	125
95	130	100
120	105	105

10-51   Do Exercise 10-50 after adding 30 to each score in the stable group.

10-52   Readability studies are conducted to determine the clarity of four different texts, and the sample scores follow. At the $\alpha = 0.05$ level of significance, test the claim that the four texts produce the same mean readability score.

Text A	Text B	Text C	Text D
50	59	48	60
51	60	51	65
53	58	47	62
58	57	49	68
53	61	50	70

10-53   An introductory calculus course is taken by students with varying high school records. The sample results from each of three groups follow. The values given are the final numerical averages in the calculus course. At the $\alpha = 0.05$ level of significance, test the claim that the mean scores are equal in the three groups.

Good high school record	Fair high school record	Poor high school record
90	80	60
86	70	60
88	61	55
93	52	62
80	73	50
	65	70
	83	

10-54   A preliminary study is conducted to determine whether there is any relationship between education and income. The sample results are as follows. The figures represent, in thousands of dollars, the lifetime incomes of randomly selected workers from each category. At the $\alpha = 0.05$ level of significance, test the claim that the samples come from populations with equal means.

Years of education				
8 years or less	9–11 years	12 years	13–15 years	16 or more years
300	270	400	420	570
210	330	430	480	640
260	380	370	510	590
330	310	390	390	700
		420	470	620
				660

A	B	C
16	14	20
20	16	21
18	16	19
18	17	22
	15	18
	18	24
		18
		20

10-55   Three car models are studied in tests that involve several cars of each model. In each case, the car is run on exactly 1 gallon of gas until the fuel supply is exhausted, and the distances traveled are on the left. Test the claim that the population means are equal.

A	B	C
0.11	0.08	0.04
0.10	0.09	0.04
0.09	0.07	0.05
0.09	0.07	0.05
0.10	0.06	0.06
		0.04
		0.05

10-56   Three groups of adult men were selected for an experiment designed to measure their blood alcohol levels after consuming five drinks. Members of group A were tested after one hour, members of group B were tested after two hours, and members of group C were tested after four hours, and the results are given at left. At the 0.05 level of significance, test the claim that the three groups have the same mean level.

A	B	C
6	9	11
8	8	9
12	7	10
9	6	8
7	9	11
2		9
		12
		14

10-57   Medical researchers use three different treatments in an experiment involving diseased rabbits, and the recovery times (in days) are given at left. Using a 0.01 significance level, test the claim that the different treatments result in the same means.

10-58   A dental research team investigating a new tooth filling material experiments with four different hardening methods, and the sample results are given below. At the 0.01 significance level, test the claim that the four methods yield the same mean index of hardness.

A	B	C	D
8.2	6.9	8.2	8.0
7.9	7.3	8.5	7.2
8.4	7.5	8.9	7.3
8.0	8.2	8.7	7.1
8.0	6.3	8.6	7.9
	6.8	8.4	7.3
	6.7	8.8	7.1
			7.4

10-59   The numbers of program errors are recorded for four different programmers on randomly selected days. Test the claim that they produce the same mean number of errors. Use a 0.01 level of significance.

1	2	3	4
14	3	17	16
16	5	20	18
18	12	22	20
14	8	24	17
22	7	26	21
9	6	18	
	6	9	
	4	11	
	7		
	16		

10-60   Five different machines are used to produce floppy disks and the number of defects are recorded for batches randomly selected on different days. Use a 0.01 significance level to test the claim that the machines produce the same mean number of defects.

1	2	3	4	5
8	11	14	32	10
9	11	13	33	8
6	8	9	26	11
10	10	10	15	14
12	13	12	18	22
	12		25	
			31	
			40	

Zone	SP	LA	Acres	Taxes
1	147	20	0.50	1.9
1	160	18	1.00	2.4
1	128	27	1.05	1.5
1	162	17	0.42	1.6
1	135	18	0.84	1.6
1	132	13	0.33	1.5
1	181	24	0.90	1.7
1	138	15	0.83	2.2
1	145	17	2.00	1.6
1	165	16	0.78	1.4
4	160	18	0.55	2.8
4	140	20	0.46	1.8
4	173	19	0.94	3.2
4	113	12	0.29	2.1
4	85	9	0.26	1.4
4	120	18	0.33	2.1
4	285	28	1.70	4.2
4	117	10	0.50	1.7
4	133	15	0.43	1.8
4	119	12	0.25	1.6
7	215	21	3.04	2.7
7	127	16	1.09	1.9
7	98	14	0.23	1.3
7	147	23	1.00	1.7
7	184	17	6.20	2.2
7	109	17	0.46	2.0
7	169	20	3.20	2.2
7	110	14	0.77	1.6
7	68	12	1.40	2.5
7	160	18	4.00	1.8

In Exercises 10-61 through 10-64 use the data given at the left. The data represent 30 different homes recently sold in Dutchess County, New York. The zones (1, 4, 7) correspond to different geographical regions of the county. The values of SP are the selling prices in thousands of dollars. The values of LA are the living areas in hundreds of square feet. The Acres values are the lot sizes in acres, and the Taxes values are the annual tax bills in thousands of dollars. For example, the first home is in zone 1, it sold for $147,000, it has a living area of 2000 square feet, it is on a 0.50-acre lot, and the annual taxes are $1900.

10-61 At the 0.05 significance level, test the claim that the means of the selling prices are the same in zones 1, 4, and 7.

10-62 At the 0.05 significance level, test the claim that the means of the living areas are the same in zones 1, 4, and 7.

10-63 At the 0.05 significance level, test the claim that the means of the lot sizes (in acres) are the same in zones 1, 4, and 7.

10-64 At the 0.05 significance level, test the claim that the means of the tax amounts are the same in zones 1, 4, and 7.

In Exercises 10-65 through 10-68, in addition to the given sample statistics, the value of $\bar{\bar{x}}$ may be needed. Recall that $\bar{\bar{x}}$ denotes the overall mean of all sample scores combined. It can be found from the given sample statistics by computing a weighted mean as follows.

$$\bar{\bar{x}} = \frac{\Sigma n_i \bar{x}_i}{\Sigma n_i}$$

10-65 Do Exercise 10-45 after changing the sample size for residents to $n = 30$.

10-66 Do Exercise 10-47 after changing the sample size for group E from 10 to 20.

10-67 Do Exercise 10-47 after changing the sample sizes for groups D and E to 20 and 50, respectively.

10-68 Do Exercise 10-48 after changing the sample size for the audiovisual class from 20 to 60.

10-4 EXERCISES B

Precinct 1

$n_1 = 50$
$\bar{x}_1 = 170$ s
$s_1 = 18$ s

Precinct 2

$n_2 = 50$
$\bar{x}_2 = 202$ s
$s_2 = 20$ s

Precinct 3

$n_3 = 50$
$\bar{x}_3 = 165$ s
$s_3 = 23$ s

10-69 A study is made of three police precincts to determine the time (in seconds) required for a police car to be dispatched after a crime is reported. Sample results are given at left.

a. At the 5% level of significance, test the claim that $\mu_1 = \mu_2$. Use the methods discussed in Chapter 8.

b. At the 5% level of significance, test the claim that $\mu_2 = \mu_3$. Use the methods discussed in Chapter 8.

c. At the 5% level of significance, test the claim that $\mu_1 = \mu_3$. Use the methods discussed in Chapter 8.

d. At the 5% level of significance, test the claim that $\mu_1 = \mu_2 = \mu_3$. Use analysis of variance.

e. Compare the methods and results of parts (a), (b), and (c) to part (d).

10-70 Five independent samples of 50 scores are randomly drawn from populations that are normally distributed with equal variances. We wish to test the claim that $\mu_1 = \mu_2 = \mu_3 = \mu_4 = \mu_5$.

a. If we use only the methods of Chapter 8, we would test the individual claims $\mu_1 = \mu_2, \mu_1 = \mu_3, \ldots, \mu_4 = \mu_5$. What is the number of claims? That is, how many ways can we pair off five means?

b. Assume that for each test of equality between two means, there is a 0.95 probability of not making a type I error. If all possible pairs of means are tested for equality, what is the probability of making no type I errors? (Although the tests are not actually independent, assume that they are.)

c. If we use analysis of variance to test the claim $\mu_1 = \mu_2 = \mu_3 = \mu_4 = \mu_5$ at the 5% level of significance, what is the probability of not making a type I error?

d. Compare the results of parts (b) and (c).

10-71 Five independent samples of 50 scores are randomly drawn from populations that are normally distributed with equal variances, and the values of n, $\bar{x}$, and s are obtained in each case. Analysis of variance is then used to test the claim that $\mu_1 = \mu_2 = \mu_3 = \mu_4 = \mu_5$.

a. If a constant is added to each of the five sample means, how is the value of the test statistic affected?

b. If each of the five means is multiplied by a constant, how is the value of the test statistic affected?

10-72 Each of six different people was given four different tests. In testing the hypothesis that the tests produce the same mean scores, the test statistic is found to be $F = 4.000$ and the estimate of variation within samples is calculated as 60. Construct the corresponding ANOVA table.

VOCABULARY LIST

Define and give an example of each term.

goodness-of-fit test
contingency table
two-way table
analysis of variance
ANOVA
multinomial
   experiment
observed frequency

expected frequency
variance between
   samples
variance within
   samples
variation due to
   treatment
variation due to
   error

one-way analysis of
   variance
single-factor analysis
   of variance
completely random-
   ized design
rigorously controlled
   design

REVIEW

We began this chapter by developing methods for testing hypotheses made about more than two population proportions. For **multinomial experiments** we tested for goodness-of-fit or agreement between observed and expected frequencies by using the chi-square test statistic given in the table on the next page. In repeated large samplings, the distribution of the χ^2 test statistic can be approximated by the chi-square distribution. This approximation is generally considered acceptable as long as all expected frequencies are at least 5. In a multinomial experiment with k cells or categories, the number of degrees of freedom is $k - 1$.

In Section 10-3 we used the sample χ^2 test statistic to measure disagreement between observed and expected frequencies in **contingency tables**. A contingency table contains frequencies; the rows correspond to categories of one variable while the columns correspond to categories of another variable. With contingency tables, we test the hypothesis that the two variables of classification are independent. In this test of independence, we can again approximate the sampling distribution of that statistic by the chi-square distribution as long as all expected frequencies are at least 5. In a contingency table with r rows and c columns, the number of degrees of freedom is $(r - 1)(c - 1)$.

In Section 10-4 we used **analysis of variance** to determine whether differences among three or more sample means are due to chance fluctuations or whether the differences are significant. This method requires normally distributed populations with equal variances. Our comparison of sample means is based on two different estimates of the common population variance. In repeated samplings, the distribution of the F test statistic can be approximated by the F distribution, which has critical values given in Table A-5.

		IMPORTANT FORMULAS		
Application	Applicable distribution	Test statistic	Degrees of freedom	Table of critical values
Multinomial	chi-square	$\chi^2 = \sum \dfrac{(O - E)^2}{E}$	$k - 1$	Table A-4
Contingency table	chi-square	$\chi^2 = \sum \dfrac{(O - E)^2}{E}$ where $E = \dfrac{\text{(row total)(column total)}}{\text{(grand total)}}$	$(r - 1)(c - 1)$	Table A-4
Analysis of variance (equal sample sizes only)	F	$F = \dfrac{ns_{\bar{x}}^2}{s_p^2}$	num: $k - 1$ den: $k(n - 1)$	Table A-5
(all cases)	F	$F = \dfrac{\left[\dfrac{\sum n_i(\bar{x}_i - \bar{\bar{x}})^2}{k - 1}\right]}{\left[\dfrac{\sum(n_i - 1)s_i^2}{\sum(n_i - 1)}\right]}$	num: $k - 1$ den: $N - k$	Table A-5
		or $F = \dfrac{MS(\text{treatment})}{MS(\text{error})}$ $\vert$ (see below) $\downarrow$	num: $k - 1$ den: $N - k$	Table A-5

For analysis of variance:

k = number of samples N = total number of values

n_i = number of values in the ith sample c_i = total of values in the ith sample

Σx = sum of all sample values Σx^2 = sum of the squares of all sample values

$$SS(\text{total}) = \Sigma x^2 - \frac{(\Sigma x)^2}{N}$$

$$SS(\text{treatment}) = \left(\sum \frac{c_i^2}{n_i} \right) - \frac{(\Sigma c_i)^2}{N}$$

$$SS(\text{error}) = SS(\text{total}) - SS(\text{treatment})$$

$$df(\text{treatment}) = k - 1$$

$$df(\text{error}) = N - k$$

$$MS(\text{treatment}) = \frac{SS(\text{treatment})}{df(\text{treatment})}$$

$$MS(\text{error}) = \frac{SS(\text{error})}{df(\text{error})}$$

$$F = \frac{MS(\text{treatment})}{MS(\text{error})}$$

REVIEW EXERCISES

10-73   A candidate for national office wants to know whether there are regional differences in his popularity and conducts a survey to find out. At the $\alpha = 0.05$ level of significance, test the claim that the candidate's popularity is independent of geographic region. The sample results are given in the following table.

	In favor	Against	No opinion
Northeast	10	20	5
South	12	36	8

10-74   A random number generator produces the outcomes listed in the following table. At the 0.05 significance level, test the claim that the values of 1, 2, 3, 4, 5, 6, are equally likely.

Outcome	1	2	3	4	5	6
Frequency	16	13	8	9	6	8

10-75   A lawyer is studying punishments for a certain crime and wants to compare the sentences imposed by three different judges. Randomly selected results follow. At the $\alpha = 0.05$ level of significance, test the claim that the three judges impose sentences that have the same mean.

Judge A	Judge B	Judge C
$n = 36$	$n = 36$	$n = 36$
$\bar{x} = 5.2$ years	$\bar{x} = 4.1$ years	$\bar{x} = 5.5$ years
$s = 1.4$ years	$s = 1.1$ years	$s = 1.5$ years

10-76   In a certain region, a survey is made of companies that officially declared bankruptcy during the past year. Of the 120 small bankrupt businesses, 72 advertised in weekly newspapers. Of the 65 medium-sized bankrupt businesses, 25 advertised in weekly newspapers, while 8 of the 15 large bankrupt businesses did so. At the 5% level of significance, test the claim that the three proportions of bankrupt businesses that used weekly newspaper ads are equal.

10-77   The owner of a new grocery store records the number of customers arriving on the different days for one week. The results are as follows. At the $\alpha = 0.05$ significance level, test the claim that customers arrive on the different days with equal frequencies.

Sun	Mon	Tues	Wed	Thurs	Fri	Sat
97	72	55	68	70	88	110

10-78   Two different research teams attempt to develop paint mixtures with more durability than the current product. Samples are tested and the resulting measures of durability are given below. At the 0.05 level of significance, test the claim that the three different mixtures have the same mean index of durability.

Current Product	Team I Product	Team II Product
12.5	12.1	14.0
12.3	12.6	14.2
11.8	12.9	12.6
12.4	13.5	14.8
12.9	12.7	15.1
	12.7	13.9
		14.3
		14.4

10-79   The following table summarizes sample grades for two subjects. At the $\alpha = 0.05$ level of significance, test the claim that grade distribution is independent of the subject.

	Grades				
	A	B	C	D	F
Math	13	16	13	8	9
English	6	15	15	5	6

10-80   A psychologist conducted studies on the relationship between forgetfulness and IQ scores. The results are in the following table. At the 0.05 level of significance, test the claim that IQ scores and levels of forgetfulness are independent.

	Forgets infrequently	Forgets occasionally	Forgets often
Low IQ	15	10	5
Medium IQ	20	30	10
High IQ	15	25	15

10-81 In testing the effectiveness of four different diets, subjects with the same overweight characteristics are randomly selected for each diet and the weight losses are listed below. At the 0.01 level of significance, test the claim that the diets produce the same mean weight loss.

Diet 1	Diet 2	Diet 3	Diet 4	
8	10	6	21	35
12	14	24	23	12
14	14	12	16	19
16	21	10	19	15
3	5	10	27	40

10-82 Three teaching methods are used with three groups of randomly selected students and the results follow. At the 0.05 level of significance, test the claim that the samples came from populations with equal means.

Method A	Method B	Method C
$n = 20$	$n = 20$	$n = 20$
$\bar{x} = 72.0$	$\bar{x} = 76.0$	$\bar{x} = 71.0$
$s = 9.0$	$s = 10.0$	$s = 12.0$

10-83 At the 0.01 level of significance, test the claim that four separate and distinct genetic characteristics are equally likely. Sample data consist of 100 randomly selected subjects, and the four characteristics occurred with the frequencies of 20, 30, 15, and 35, respectively.

10-84 A marketing study is conducted in order to determine if a product's appeal is affected by geographical region. Given the sample data in the following table, use a 0.01 significance level to test the claim that the consumer's opinion is independent of region.

	Like	Dislike	Uncertain
Northeast	30	15	15
Southeast	10	30	20
West	40	60	15

COMPUTER PROJECT

1. Use existing software to solve Exercise 10-29. Repeat that exercise by adding 10 to each frequency in the table, and note the effect of that change. Finally, repeat Exercise 10-29 a third time after multiplying each frequency by 10, and note the effect of that change.

2. Use existing software to employ analysis of variance in solving Exercise 10-55. Then repeat that exercise after adding 10 to each score and note the effect of that change. Finally, repeat Exercise 10-55 a third time after multiplying each score by 10, and note the effect of that change.

CASE STUDY ACTIVITY

Conduct a survey by asking the question "Do you favor or oppose the death penalty for people convicted of murder?" Record each response (yes, no, undecided) along with the sex (female, male) of the respondent. Include in your data set only responses of "yes" or "no." At the 0.05 level of significance, test the claim that the opinion is independent of the sex of the respondent. Be sure to survey enough people so that the expected frequency of each cell in the resulting contingency table is at least 5. Identify any factors suggesting that your sample is not representative of the people in your region.

DATA PROJECT

Refer to the data sets in Appendix B.

1. a. Using the student data set, find the column that consists of three-digit random numbers selected by the respondents. With three such digits for each of 92 respondents, there are 276 digits. Use that set of 276 digits to complete the table below.

Digit	0	1	2	3	4	5	6	7	8	9
Frequency										

   If the digits are randomly selected, the expected frequencies are equal. Test the claim that these digits are randomly selected.
   b. Repeat part (a) for the set of 276 digits that came from social security numbers.
   c. Compare the results from parts (a) and (b).   *(continued)*

2. Using the data for homes sold, enter the frequencies in the table below.

	Number of baths	
	1 or 1.5	2 or more
Under 2000 sq ft		
2000 sq ft or more		

Use this data to test the claim that the number of baths is independent of the living area.

3. Using the data for homes sold, enter the selling prices in the appropriate category below. Then test the claim that the samples in all three categories come from populations having the same mean.

Number of baths

1 or 1.5	2 or 2.5	3 or more

Chapter Eleven

CHAPTER CONTENTS

11-1 Overview
We present the general nature and the advantages and disadvantages of **nonparametric** methods, introduce the concept of **ranked data**, and identify chapter **objectives**.

11-2 Sign Test
The sign test is a nonparametric method that can be used to test the claim that two sets of dependent data have the same median.

11-3 Wilcoxon Signed-Ranks Test for Two Dependent Samples
The Wilcoxon signed-ranks test can be used to test the claim that two sets of dependent data come from identical populations. This test takes into account the magnitudes of the numbers.

11-4 Wilcoxon Rank-Sum Test for Two Independent Samples
The Wilcoxon rank-sum test can be used to test the claim that two independent samples come from identical populations.

11-5 Kruskal-Wallis Test
The Kruskal-Wallis test can be used to test the claim that several independent samples come from identical populations.

11-6 Rank Correlation
The rank correlation coefficient can be used to test for an association between two sets of paired data.

11-7 Runs Test for Randomness
The runs test can be used to test for randomness in the way data are selected.

11 Nonparametric Statistics

CHAPTER PROBLEM

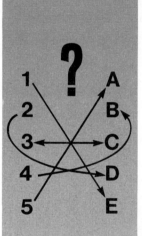

When eight states are randomly selected and ranked according to average teacher's salary and average SAT scores, the following results (based on data from the U.S. Department of Education) are obtained.

State	N.Y.	Calif.	Fla.	N.J.	Tex.	N.C.	Md.	Ore.
Teacher's salary (rank)	1	2	7	4	6	8	3	5
SAT score (rank)	4	3	5	6	7	8	2	1

In analyzing these data, an important and relevant consideration would be the presence or absence of a correlation between the two sets of ranks. Is there a relationship between money spent on teacher's salaries and SAT scores of students? Because these values are ranks, we cannot use the linear correlation coefficient discussed in Chapter 9. The linear correlation coefficient from Chapter 9 requires that the two variables be normally distributed, and ranks do not satisfy that requirement. We cannot use the methods of Chapter 9, but we will present an alternative method that can be used with data sets of the type given here. This particular set of data will be considered later in this chapter.

11-1   OVERVIEW

Most of the methods of inferential statistics covered before Chapter 10 can be called **parametric methods**, because their validity is based on sampling from a population with particular parameters such as the mean, standard deviation, or proportion. Parametric methods can usually be applied only to circumstances in which some fairly strict requirements are met. One typical requirement is that the sample data must come from a normally distributed population. What do we do when the necessary requirements are not satisfied? It is very possible that there may be an alternative approach among the many methods that are classified as **nonparametric**. In addition to being an alternative to parametric methods, nonparametric techniques are frequently valuable in their own right.

In this chapter, we introduce six of the more popular nonparametric methods that are in current use. These methods have advantages and disadvantages.

Advantages of Nonparametric Methods

1. Nonparametric methods can be applied to a wider variety of situations, because they do not have the more rigid requirements of their parametric counterparts. In particular, nonparametric methods do not require normally distributed populations. For this reason, nonparametric tests of hypotheses are often called **distribution-free** tests.

2. Unlike the parametric methods, nonparametric methods can often be applied to nominal data that lack exact numerical values.

3. Nonparametric methods usually involve computations that are simpler than the corresponding parametric methods.

4. Since nonparametric methods tend to require simpler computations, they tend to be easier to understand.

If all of these terrific advantages could be accrued without any significant disadvantages, we could ignore the parametric methods and enjoy much less-complicated procedures. Unfortunately, there are some disadvantages.

Disadvantages of Nonparametric Methods

1. Nonparametric methods tend to waste information, since exact numerical data are often reduced to a qualitative form.

2. Nonparametric methods are generally less sensitive than the corresponding parametric methods. This means that we need stronger evidence before we reject a null hypothesis.

As an example of the way that information is wasted, there is one nonparametric method in which weight losses by dieters are recorded simply as negative signs. With this particular method, a weight loss of only 1 pound receives the same representation as a weight loss of 50 pounds. This would not thrill dieters.

Although nonparametric tests are less sensitive than their parametric counterparts, this can be compensated by an increased sample size. The **efficiency** of a nonparametric method is one concrete measure of its sensitivity. Section 11-6 deals with a concept called the *rank correlation coefficient*, which has an efficiency rating of 0.91 when compared to the linear correlation coefficient of Chapter 9. This means that with all things being equal, this nonparametric approach would require 100 sample observations to achieve the same results as 91 sample observations analyzed through the parametric approach, assuming the stricter requirements for using the parametric method are met. Not bad! The point, though, is that an increased sample size can overcome lower sensitivity. Table 11-1 lists nonparametric methods covered in this chapter, along with the corresponding parametric approach and efficiency rating. You can see from this table that the lower efficiency might not be a critical factor.

TABLE 11-1			
Application	Parametric test	Nonparametric test	Efficiency of nonparametric test with normal population
Two dependent samples	t test or z test	Sign test Wilcoxon signed-ranks	0.63 0.95
Two independent samples	t test or z test	Wilcoxon rank-sum	0.95
Several independent samples	Analysis of variance (F test)	Kruskal-Wallis test	0.95
Correlation	Linear correlation	Rank correlation	0.91
Randomness	No parametric test	Runs test	No basis for comparison

In choosing between a parametric method and a nonparametric method, the key factors that should govern our decision are cost, time, efficiency, amount of data available, type of data available, method of sampling, nature of the population, and probabilities (α and β) of making errors. In one experiment we might have abundant data with strong assurances that all of the requirements of a parametric test are satisfied, and we would probably be wise to choose that parametric test. But given another experiment with relatively few cases drawn from some mysterious population, we would probably fare better with a nonparametric test. Sometimes we don't really have a choice. If we want to test data to see if they are randomly selected, the only available test happens to be nonparametric. As another example, only nonparametric methods can be used on data consisting of observations that can only be ranked. Some methods included in this chapter are based on ranks. Instead of describing ranks in each section or making some sections dependent on others, we will now discuss ranks so that we will be prepared to use them wherever they are required.

Data are **ranked** when they are arranged according to some criterion, such as smallest to largest or best to worst. The first item in the arrangement is given a rank of 1, the second item is given a rank of 2, and so on. For example, the numbers 5, 3, 40, 10, and 12 can be arranged from lowest to highest as 3, 5, 10, 12, and 40, so that 3 has rank 1, 5 has rank 2, 10 has rank 3, 12 has rank 4, and 40 has rank 5. If a tie in ranks should occur, the usual procedure is to find the mean of the ranks involved and then assign that mean rank to each of the tied items. This might sound complicated, but it's really quite simple. For example, the numbers 3, 5, 5, 10, and 12 would be given ranks of 1, 2.5, 2.5, 4, and 5, respectively. In this case, there is a tie for ranks 2 and 3, so we find the mean of 2 and 3 (which is 2.5) and assign it to the scores that created the tie. As another example, the scores 3, 5, 5, 7, 10, 10, 10, and 15 would be ranked 1, 2.5, 2.5, 4, 6, 6, 6, and 8, respectively. From these examples we can see how to convert numbers to ranks, but there are many situations in which the original data consist of ranks. If a judge ranks five piano contestants, we get ranks of 1, 2, 3, 4, 5 corresponding to five names; it's this type of data that precludes the use of parametric methods and enhances the importance of nonparametric methods.

Scores	5	3	40	10	12
Scores in order	3	5	10	12	40
	↑	↑	↑	↑	↑
Ranks	1	2	3	4	5

Scores	3	5	5	10	12
	↑	↑	↑	↑	↑
Ranks	1	2.5	2.5	4	5

(2 and 3 are tied)

11-2 SIGN TEST

The **sign test** is one of the easiest nonparametric tests to use, and it is applicable to a few different types of situations. One application is to the paired data that form the basis for hypothesis tests involving two dependent samples. We considered such cases by using parametric tests in Section 8-3. One example from Section 8-3 refers to a study conducted to investigate the effectiveness of hypnotism in reducing pain. The sample results for randomly selected subjects are given in Table 11-2. (The values are before and after hypnosis. The measurements are in centimeters on the mean visual analogue scale, and the data are based on "An Analysis of Factors That Contribute to the Efficacy of Hypnotic Analgesia" by Price and Barber, *Journal of Abnormal Psychology*, Vol. 96, No. 1.) We will test the claim that the sensory measurements are lower after hypnotism. Note that Table 11-2 also includes the *signs* of the changes from the "before" values to the "after" values.

In Section 8-3 we used the parametric Student *t* test, but in this section we apply the nonparametric sign test, which can be used to test for equality between two medians. The key concept underlying the sign test is this: **If the two sets of data have equal medians, the number of positive signs should be approximately equal to the number of negative signs.** For the data in Table 11-2, we can conclude that hypnotism is effective if there is an excess of negative signs and a deficiency of positive signs.

TABLE 11-2								
Subject	A	B	C	D	E	F	G	H
Before	6.6	6.5	9.0	10.3	11.3	8.1	6.3	11.6
After	6.8	2.4	7.4	8.5	8.1	6.1	3.4	2.0
Sign of change from before to after	+	−	−	−	−	−	−	−

In our sign test procedure, we exclude any ties (represented by zeros). We now have this specific question: Do the seven negative signs in Table 11-2 *significantly* outnumber the single positive sign? Or, to put it another way, is the number of positive signs small enough? The answer to this question depends on the level of significance, so let's use $\alpha = 0.05$ as we did in Section 8-3. When we assume the null

Seat Belts Save Lives

Statistical analysis of data sometimes evolves into changes in public policy. The car seat belt issue is one such example. The Highway Users Federation recently estimated that if everyone in the United States used seat belts, the number of highway deaths would drop by 12,000 each year, and there would be 330,000 fewer disabling injuries each year. Similar statistics have convinced some state legislatures to pass laws mandating the use of seat belts.

One study of 1126 accidents showed that riders wearing seat belts had 86% fewer life-threatening injuries. Another study of 28,780 accident victims showed that riders not wearing seat belts died in crashes involving speeds as low as 12 miles per hour, but no one wearing a seat belt and shoulder harness was killed at a speed under 60 miles per hour. It is estimated that accident victims wearing seat belts are half as likely to be killed when compared to beltless accident victims.

Some people question the use of seat belts because they know of cases where serious injury was avoided when an unbelted rider was thrown clear of a wreck. There are cases where the seat belt had a negative effect, but such cases are far outnumbered by incidents where the seat belt was helpful. Clearly, the wisest strategy is to buckle up.

hypothesis of no decrease in sensory measurements, we assume that positive signs and negative signs occur with equal frequency, so $P(\text{positive sign}) = P(\text{negative sign}) = 0.5$. (The null hypothesis of no decrease also includes the possibility of an increase, but we continue to assume that positive signs and negative signs are equally likely.)

H_0: There is no decrease in sensory measurements.
H_1: There is a decrease in sensory measurements.

Since our results fall into two categories (positive or negative) and we have a fixed number of independent cases, we could use the binomial probability distribution (Section 4-4) to determine the likelihood of getting one or no positive signs among the eight subjects. Instead, we have used the binomial probability formula to construct a separate table (Table A-7) that lists critical values for the sign test. For consistency and ease, we will stipulate that

> **The test statistic x is the number of times the less frequent sign occurs.**

With seven negative signs and one positive sign, the test statistic x is the lesser of 7 and 1, so $x = 1$. We have 8 sample cases so that $n = 8$. Our test is one-tailed with $\alpha = 0.05$, and Table A-7 indicates that the critical value is 1. We should therefore reject the null hypothesis only if the test statistic is less than or equal to 1. With a test statistic of $x = 1$ we do reject the null hypothesis of no decrease. It appears that hypnotism does result in lower sensory measurements.

Because of the way that we are determining the value of the test statistic, we should check to ensure that our conclusion is consistent with the circumstances. It is only when the sense of the sample data is *against* the null hypothesis that we should even consider rejecting it. If the sense of the data supports the null hypothesis, we should fail to reject it regardless of the test statistic and critical value. Figure 11-1 summarizes the procedure for the sign test and includes this check for consistency of results.

In the preceding example we arrived at the same conclusion obtained in Section 8-3. However, consider the new data set given in Table 11-3.

TABLE 11-3								
Subject	A	B	C	D	E	F	G	H
Before	6.6	6.5	9.0	10.3	11.3	8.1	6.3	11.6
After	6.8	6.6	2.4	3.1	2.9	2.8	2.4	2.0
Difference	0.2	0.1	−6.6	−7.2	−8.4	−5.3	−3.9	−9.6
Sign of change from before to after	+	+	−	−	−	−	−	−

For this data set there are two positive signs and six negative signs. Using the sign test, the test statistic is $x = 2$ and we fail to reject the null hypothesis of no decrease. But if we use the Student t test from Section 8-3, we get

$$t = \frac{\bar{d} - 0}{s_d/\sqrt{n}} = \frac{-5.0875}{3.6725/\sqrt{8}} = -3.918$$

which causes *rejection* of the null hypothesis because the test statistic of $t = -3.918$ is in the critical region bounded by the critical score of $t = -1.895$. An intuitive analysis of Table 11-3 suggests that the "after" scores are significantly lower, but the sign test is blind to the *magnitude*

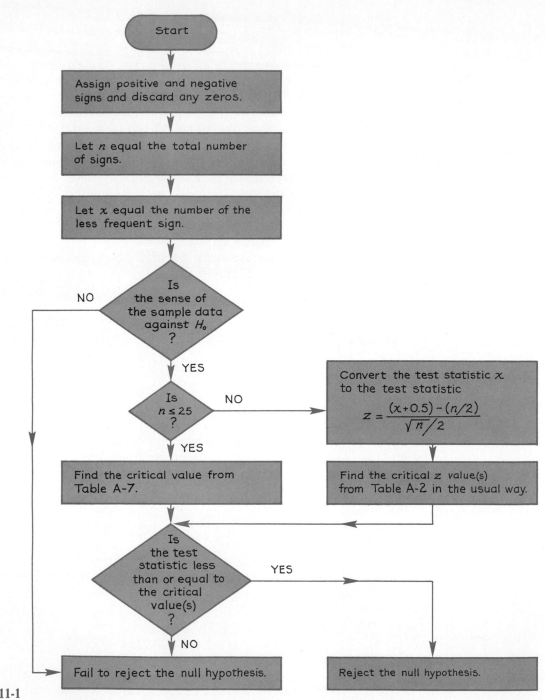

FIGURE 11-1
Sign test

Air Is Healthier Than Tobacco

A consumer testing group studied the tar and nicotine levels of various brands of cigarettes. Now cigarettes were advertised as being lowest in tar and nicotine, but they seemed to burn more quickly than other brands. Samples of Now and Winston were randomly selected and weighed. Both brands were products of the R. J. Reynolds Tobacco Company, and Winston was their best seller at the time of the comparison. Results showed that the average weight of the tobacco in a Now was about two-thirds the average weight of the tobacco in a Winston. With one-third less tobacco, it isn't difficult to get lower tar and nicotine levels. The study also noted that smokers tend to consume more cigarettes when they are of the low tar and nicotine variety. The net effect for Now was that more cigarettes could be sold at lower production costs.

of the changes. This illustrates the previous assertion that nonparametric tests lack the sensitivity of parametric tests, with the resulting tendency that stronger evidence is required before a null hypothesis is rejected.

An examination of Figure 11-1 shows that when $n \leq 25$, we should use Table A-7 to find the critical value, but for $n > 25$, we use a normal approximation to obtain the critical values.

Claims Involving Nominal Data

The next example illustrates the fact that nonparametric methods can be used with nominal data. Also, since it involves a sample size greater than 25, the normal approximation is used.

EXAMPLE

A company claims that its hiring practices are fair, it does not discriminate on the basis of sex, and the fact that 40 of the last 50 new employees are men is just a fluke. The company acknowledges that applicants are about half men and half women. Test the null hypothesis that men and women are equal in their ability to be employed by this company. Use a significance level of 0.05.

continued ▶

Example, continued

Solution

H_0: $p_1 = p_2$ (the proportions of men and women are equal)
H_1: $p_1 \neq p_2$

If we denote hired women by $+$ and hired men by $-$, we have 10 positive signs and 40 negative signs. Refer now to the flowchart of Figure 11-1. The test statistic x is the smaller of 10 and 40, so $x = 10$. This test involves two tails since a disproportionately low number of either sex will cause us to reject the claim of equality. The sense of the sample data is against the null hypothesis because 10 and 40 aren't exactly equal. Continuing with the procedure of Figure 11-1, we note that the value of $n = 50$ is above 25, so the test statistic x is converted to the test statistic z as follows.

$$z = \frac{(x + 0.5) - (n/2)}{\sqrt{n}/2}$$

$$= \frac{(10 + 0.5) - (50/2)}{\sqrt{50}/2}$$

$$= -4.10$$

With $\alpha = 0.05$ in a two-tailed test, the critical values are $z = -1.96$ and 1.96. The test statistic $z = -4.10$ is less than these critical values (see Figure 11-2), so we reject the null hypothesis of equality. There is sufficient sample evidence to warrant rejection of the claim that the hiring practices are fair.

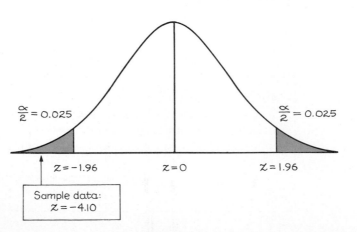

FIGURE 11-2

When $n > 25$, the test statistic z is based on a normal approxima-tion to the binomial probability distribution with $p = q = \frac{1}{2}$. In Section 5-5 we saw that the normal approximation to the binomial distribution is acceptable when both $np \geq 5$ and $nq \geq 5$. Also, in Section 4-5 we saw that $\mu = np$ and $\sigma = \sqrt{n \cdot p \cdot q}$ for binomial experiments. Since this sign test assumes that $p = q = \frac{1}{2}$, we meet the $np \geq 5$ and $nq \geq 5$ prerequisites whenever $n \geq 10$; we have a table of critical values (Table A-7) for n up to 25, so that we need the normal approximation only for values of n above 25. Also, with the assumption that $p = q = \frac{1}{2}$, we get $\mu = np = n/2$ and $\sigma = \sqrt{n \cdot p \cdot q} = \sqrt{n/4} = \sqrt{n}/2$, so that $z = (x - \mu)/\sigma$ becomes

$$z = \frac{x - \left(\dfrac{n}{2}\right)}{\dfrac{\sqrt{n}}{2}}$$

Finally, we replace x by $x + 0.5$ as a correction for continuity. That is, the values of x are discrete, but since we are using a continuous prob-ability distribution, a discrete value such as 10 is actually represented by the interval from 9.5 to 10.5. Because x represents the less frequent sign, we need to concern ourselves only with $x + 0.5$; we thus get the test statistic z as given above and in Figure 11-1.

Claims About a Median

The previous examples involved application of the sign test to a com-parison of *two* sets of data, but we can also use the sign test to inves-tigate a claim made about the median of one set of data, as the next example shows.

EXAMPLE

Use the sign test to test the claim that the median IQ of pilots is at least 100 if a sample of 50 pilots contained exactly 22 members with IQs of 100 or higher.

Solution

The null hypothesis is the claim that the median is equal to or greater than 100; the alternative hypothesis is the claim that the median is less than 100.

continued ▶

Solution, continued

H_0: Median is at least 100. (Median $\geq$ 100)
H_1: Median is less than 100. (Median $<$ 100)

We select a significance level of 0.05, and we use + to denote each IQ score that is at least 100. We therefore have 22 positive signs and 28 negative signs. We can now determine the significance of getting 22 positive signs out of a possible 50. Referring to Figure 11-1, we note that $n = 50$ and $x = 22$ (the smaller of 22 and 28). The sense of the data is against the null hypothesis, since a median of at least 100 would require at least 25 (half of 50) scores of 100 or higher. The value of n exceeds 25, so we convert the test statistic x to the test statistic z.

$$z = \frac{(x + 0.5) - (n/2)}{\sqrt{n}/2}$$

$$= \frac{(22 + 0.5) - (50/2)}{\sqrt{50}/2}$$

$$= \frac{22.5 - 25}{\sqrt{50}/2} = -0.71$$

In this one-tailed test with $\alpha = 0.05$, we use Table A-2 to get the critical z value of -1.645. From Figure 11-3, we can see that the computed value of -0.71 does not fall within the critical region. We therefore fail to reject the null hypothesis. Based on the available sample evidence, we cannot reject the claim that the median IQ is at least 100. A corresponding parametric test may or may not lead to the same conclusion, depending on the specific values of the 50 sample scores.

We have shown that the sign test wastes information because it uses only information about the direction of the differences between pairs of data, while the magnitudes of those differences are ignored. The next section introduces the Wilcoxon signed-ranks test, which largely overcomes that disadvantage.

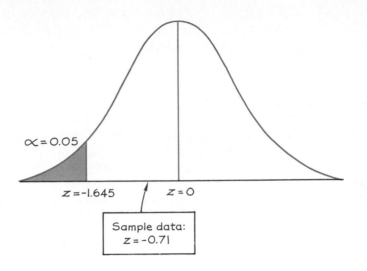

FIGURE 11-3

11-2 EXERCISES A

In the following exercises, use the sign test.

11-1 In a study of techniques used to measure lung volumes, physiological data were collected for 10 subjects. The values given in the table are in liters and they represent the measured forced vital capacities of the 10 subjects in a sitting position and in a supine (lying) position. (See "Validation of Esophageal Balloon Technique at Different Lung Volumes and Postures," by Baydur, Cha, and Sassoon, *Journal of Applied Physiology*, Vol. 62, No. 1.) At the 0.05 significance level, test the claim that there is no significant difference between the measurements from the two positions.

Sitting	4.66	5.70	5.37	3.34	3.77	7.43	4.15	6.21	5.90	5.77
Supine	4.63	6.34	5.72	3.23	3.60	6.96	3.66	5.81	5.61	5.33

11-2 In a study of techniques used to measure lung volumes, physiological data were collected for 10 subjects. The values given in the table are in liters and they represent the measured functional residual capacities of the 10 subjects in a sitting position and in a supine (lying) position. (See "Validation of Esophageal Balloon Technique at Different Lung Volumes and Postures," by Baydur, Cha, and Sassoon, *Journal of Applied Physiology*, Vol. 62, No. 1.) At the 0.05 significance level, test the claim that there is no significant difference between the measurements from the two positions.

Sitting	2.96	4.65	3.27	2.50	2.59	5.97	1.74	3.51	4.37	4.02
Supine	1.97	3.05	2.29	1.68	1.58	4.43	1.53	2.81	2.70	2.70

11-3   At the 0.05 level of significance, test the claim that the sample x and y values come from the same population. Assume that the x and y values are paired as shown.

x	1	2	2	3	5	6	8	4	6	7	2	5	3	2
y	2	0	1	4	4	4	9	3	5	7	2	6	2	1

11-4   Two different firms design their own IQ tests, and a psychologist administers both tests to randomly selected subjects with the results given below. At the 0.05 level of significance, test the claim that there is no significant difference between the two tests.

Subject	A	B	C	D	E	F	G	H	I	J
Test I	98	94	111	102	108	105	92	88	100	99
Test II	105	103	113	98	112	109	97	95	107	103

11-5   A pill designed to lower systolic blood pressure is administered to ten randomly selected volunteers. The results follow. At the $\alpha = 0.05$ significance level, test the claim that systolic blood pressure is not affected by the pill. That is, test the claim that the before and after values are equal.

Before pill	120	136	160	98	115	110	180	190	138	128
After pill	118	122	143	105	98	98	180	175	105	112

11-6   A test of driving ability is given to a random sample of ten student drivers before and after they completed a formal driver education course. The results follow. At the $\alpha = 0.05$ significance level, test the claim that the course does not affect scores.

Before course	100	121	93	146	101	109	149	130	127	120
After course	136	129	125	150	110	138	136	130	125	129

11-7   A student hears that fish is a "brain food" that helps to make people more intelligent. She participates in an experiment involving 12 randomly selected volunteers who take an IQ test and then begin an intensive fish diet. A second IQ test is given at the end of the experiment, with the results given in the table below. At the 0.05 significance level, test the claim that the fish diet has no effect on IQ scores.

IQ score before diet	98	110	105	121	100	88	112	92	99	109	103	104
IQ score after diet	98	112	106	118	102	97	115	90	99	110	105	109

11-8   A course is designed to increase readers' speed and comprehension. To evaluate the effectiveness of this course, a test is given both before and after the course, and sample results follow. At the 0.05 significance level, test the claim that the scores are higher after the course.

Before	100	110	135	167	200	118	127	95	112	116
After	136	160	120	169	200	140	163	101	138	129

11-9   The following chart lists a random sampling of the ages of married couples. The age of each husband is listed above the age of his wife. At the 0.01 significance level, test the claim that there is no difference between the ages of husbands and wives.

Husband	28.1	33.0	29.8	53.1	56.7	41.6	50.6	21.4	62.0	19.7
Wife	28.4	27.6	32.7	52.0	58.1	41.2	50.7	20.6	61.1	18.1

11-10   Ten randomly selected volunteers test a new diet, with the following results. At the 0.01 level of significance, test the claim that the diet is effective—that is, that weights (in kilograms) are lower after the diet.

Subject	A	B	C	D	E	F	G	H	I	J
Weight before diet	68	54	59	60	57	62	62	65	88	76
Weight after diet	65	52	52	60	58	59	60	63	78	75

11-11   A study was conducted to investigate the effectiveness of hypnotism in reducing pain. At the 0.05 significance level, test the claim that the affective responses to pain are the same before and after hypnosis. Results for randomly selected subjects are given below. (The data are based on results from "An Analysis of Factors That Contribute to the Efficacy of Hypnotic Analgesia," by Price and Barber, *Journal of Abnormal Psychology*, Vol. 96, No. 1.)

Before	-5.5	-5.0	-6.6	-9.7	-4.0	-7.0	-7.0	-8.4
After	-1.4	-0.5	0.7	1.0	2.0	0.0	-0.6	-1.8

11-12   A study was conducted to investigate some effects of physical training. Sample data are listed in the accompanying table (based on data from "Effect of Endurance Training on Possible Determinants of VO_2 During Heavy Exercise," by Casaburi and others, *Journal of Applied Physiology*, Vol. 62, No. 1). At the 0.05 level of significance, test the claim that pretraining weights equal posttraining weights (in kilograms).

Pretraining	99	57	62	69	74	77	59	92	70	85
Posttraining	94	57	62	69	66	76	58	88	70	84

11-13 When 20 students are asked if they understand the purpose of their student senate, 15 respond affirmatively and 5 respond negatively. At the 0.05 level of significance, test the claim that most (more than half) students feel that they understand the purpose of their student senate.

11-14 A standardized aptitude test yields a mathematics score M and a verbal score V for each person. Among 15 male subjects, $M - V$ is positive in 12 cases, negative in 2 cases, and zero in 1 case. At the 0.05 level of significance, use the sign test to test the claim that males do better on the mathematics portion of the test.

11-15 A political party preference poll is taken among 20 randomly selected voters. If 7 prefer the Republican Party while 13 prefer the Democratic Party, apply the sign test to test the claim that both parties are preferred equally. Use a 0.05 level of significance.

11-16 A television commercial advertises that seven out of ten dentists surveyed prefer Covariant toothpaste over the leading competitor. Assume that ten dentists are surveyed and seven do prefer Covariant, while three favor the other brand. Is this a reasonable basis for making the claim that most (more than half) dentists favor Covariant toothpaste? Use the sign test with a significance level of 0.05.

11-17 Use the sign test to test the claim that the median life of a battery is at least 40 hours if a random sample of 75 includes exactly 32 that last 40 hours or more. Assume a significance level of 0.05.

11-18 A college aptitude test is given to 100 randomly selected high school seniors. After a period of intensive training, another similar test is given to the same students, and 59 receive higher grades, 36 receive lower grades, and 5 students receive the same grades. At the 0.05 level of significance, use the sign test to test the claim that the training is effective.

11-19 A company is experimenting with a new fertilizer at 50 different locations. In 32 of the locations there is an increase in production, while in 18 locations there is a decrease. At the 0.05 level of significance, use the sign·test to test the claim that production is increased by the new fertilizer.

11-20 A new diet is designed to lower cholesterol levels. In six months, 36 of the 60 subjects on the diet have lower cholesterol levels, 22 have slightly higher levels, and 2 register no change. At the 0.01 level of significance, use the sign test to test the claim that the diet produces no change in cholesterol levels.

11-21 After 30 drivers are tested for reaction times, they are given two drinks and tested again, with the result that 22 have slower reaction times, 6 have faster reaction times, and 2 receive the same scores as before the drinks. At the 0.01 significance level, use the sign test to test the claim that the drinks had no effect on the reaction times.

11-22   In target practice, 40 police academy students use two different pistols. Analysis of the scores shows that 24 students get higher scores with the more expensive pistol, while 16 students get better scores with the less expensive pistol. At the 0.05 level of significance, use the sign test to test the claim that both pistols are equally effective.

11-23   Of 50 voters surveyed, 28 favor a tax revision bill before Congress, while all the others are opposed. At the 0.10 level of significance, use the sign test to test the claim that the majority (more than half) of voters favor the bill.

11-24   Use the sign test to test the claim that the median IQ score of Philadelphians is at least 100 if a sample of 200 Philadelphians contains 86 with IQs of 100 or more.

11-2 EXERCISES B

11-25   Given n sample scores sorted in ascending order $(x_1, x_2, \ldots, x_n)$, if we wish to find the approximate $1 - \alpha$ confidence interval for the population median M, we get

$$x_{k+1} < M < x_{n-k}$$

where k is the critical value (Table A-7) for the number of signs in a two-tailed hypothesis test conducted at the significance level of α. Find the approximate 95% confidence interval for the sample scores listed below.

3, 8, 6, 2, 1, 7, 9, 11, 17, 23, 25, 10, 14, 8, 30

11-26   Of the voters surveyed, 57 favor passage of a certain bill, and they constitute a majority. At the 0.05 significance level, we apply the sign test and reject the claim that voters are equally split on the bill. Given the preceding information, what is the largest sample size possible?

11-27   Of n subjects tested for high blood pressure, a majority of exactly 50 provided negative results. (That is, their blood pressure is not high.) This is sufficient for us to apply the sign test and reject (at the 0.01 level of significance) the claim that the median blood pressure level is high. Find the largest value n can assume.

11-28   Table A-7 lists critical values for limited choices of α. Use Table A-1 to add a new column in Table A-7 that would represent a significance level of 0.03 in one tail or 0.06 in two tails. For any particular n we use $p = 0.5$, since the sign test requires the assumption that

$$P(\text{positive sign}) = P(\text{negative sign}) = 0.5$$

The probability of x or fewer like signs is the sum of the probabilities up to and including x.

11-3 WILCOXON SIGNED-RANKS TEST FOR TWO DEPENDENT SAMPLES

In the preceding section, we used the sign test to analyze the differences between paired data. The sign test used only the signs of the differences, while ignoring their actual magnitudes. In this section we introduce the **Wilcoxon signed-ranks test**, which takes the magnitudes into account. Because this test incorporates and uses more information than the ordinary sign test, it tends to yield better results than the sign test. However, the Wilcoxon signed-ranks test requires that the two sets of data come from populations with a common distribution. Unlike the *t* test for paired data (see Section 8-3), the Wilcoxon signed-ranks test does *not* require normal distributions.

Consider the data given in Table 11-4. The 13 subjects are given a test for logical thinking. They are then given a tranquilizer and retested. We will use the Wilcoxon signed-ranks test to test the claim that the tranquilizer has no effect, so that there is no significant difference between before and after scores. We will assume a 0.05 level of significance.

H_0: The tranquilizer has no effect on logical thinking.
H_1: The tranquilizer has an effect on logical thinking.

TABLE 11-4					
Subject	Before	After	Difference	Ranks of differences	Signed-ranks
A	67	68	−1	1	−1
B	78	81	−3	2	−2
C	81	85	−4	3	−3
D	72	60	+12	10	+10
E	75	75	0	—	—
F	92	81	+11	8.5	+8.5
G	84	73	+11	8.5	+8.5
H	83	78	+5	4	+4
I	77	84	−7	5	−5
J	65	56	+9	6	+6
K	71	61	+10	7	+7
L	79	64	+15	11	+11
M	80	63	+17	12	+12

In general, the null hypothesis will be the claim that both samples come from the same population distribution. We summarize here the procedure for using the Wilcoxon signed-ranks test with paired data.

PROCEDURE

1. For each pair of data, find the difference d by subtracting the second score from the first. Retain signs, but discard any pairs for which $d = 0$.

2. Ignoring the signs of those differences, rank them from lowest to highest. When differences have the same numerical value, assign to them the mean of the ranks involved in the tie.

3. Assign to each rank the sign of the difference from which it came.

4. Find the sum of the absolute values of the negative ranks. Also find the sum of the positive ranks.

5. Let T be the smaller of the two sums found in step 4.

6. Let n be the number of pairs of data for which the difference d is not zero.

7. If $n \leq 30$, use Table A-8 to find the critical value of T. Reject the null hypothesis if the sample data yield a value of T less than or equal to the value in Table A-8. Otherwise, fail to reject the null hypothesis. If $n > 30$, compute the test statistic z by using Formula 11-1.

$$\text{FORMULA 11-1} \quad z = \frac{T - \dfrac{n(n+1)}{4}}{\sqrt{\dfrac{n(n+1)(2n+1)}{24}}}$$

When Formula 11-1 is used, the critical z values are found from Table A-2 in the usual way. Again reject the null hypothesis if the test statistic z is less than or equal to the critical value(s) of z. Otherwise, fail to reject the null hypothesis.

We will now use this procedure to test the null hypothesis that the tranquilizer has no effect on logical thinking.

Step 1. In Table 11-4, the column of differences is obtained by subtracting each "after" score from the corresponding "before" score. Differences of zero are discarded.

Step 2. Ignoring their signs, the differences are then ranked from lowest to highest, with ties being treated in the manner described at the end of Section 11-1. See the fifth column of Table 11-4.

Step 3. The signed-ranks column is then created by applying to each rank the sign of the corresponding difference. The results are listed in the last column of Table 11-4. If the tranquilizer really has no effect, we would expect the number of positive ranks to be approximately equal to the number of negative ranks. If the tranquilizer tends to lower scores, then ranks of positive sign would tend to outnumber ranks of negative sign. If the tranquilizer tends to raise scores, then ranks of negative sign would tend to outnumber ranks of positive sign. We can detect a domination by either sign through analysis of the rank sums.

Step 4. Now find the sum of the absolute values of the negative ranks and the sum of the positive ranks. For the data of Table 11-4, we get

sum of absolute values of negative ranks = 1 + 2 + 3 + 5 = 11

sum of positive ranks = 10 + 8.5 + 8.5 + 4 + 6 + 7 + 11 + 12 = 67

Step 5. We will base our test on the smaller of those two sums, denoted by T. For the given data we have $T = 11$. We use the following notation.

NOTATION

T Smaller of these two sums:
   1. The sum of the absolute values of the negative ranks.
   2. The sum of the positive ranks.

n Number of *pairs* of data after excluding any pairs in which both values are the same.

Step 6. $n = 12$ since there are 12 pairs of data with nonzero differences.

Step 7. Whenever $n \leq 30$, we use Table A-8 to find the critical values. If $n > 30$ we can use a normal approximation with the test statistic given in Formula 11-1 and critical values given in Table A-2. As in

Section 11-2, we would be justified in using the normal approximation whenever $n \geq 10$, but we have a table of critical values for values of n up to 30 so that we really need the normal approximation only for $n > 30$. Because the data of Table 11-4 yield $n = 12$, we use Table A-8 to get the critical value of 14. ($\alpha = 0.05$ and the test is two-tailed since our null hypothesis is the claim that the scores have not changed significantly.) Since $T = 11$ is less than or equal to the critical value of 14, we reject the null hypothesis. It appears that the drug does affect scores.

In this last example, the unsigned ranks go from 1 through 12 and the sum of those 12 integers is 78. If the two sets of data have no significant differences, each of the two signed-rank totals should be in the neighborhood of $78 \div 2$, or 39. However, for the given sample data, the negative ranks were the smaller values, while the positive ranks tended to the larger values. We got 11 for one total and 67 for the other; this 11-67 split was a significant departure from the 39-39 split expected with a true null hypothesis. The table of critical values shows that at the 0.05 level of significance with 12 pairs of data, a 14-64 split represents a significant departure from the null hypothesis, and any split farther apart (such as 13-65 or 12-66) will also represent a significant departure from the null hypothesis. Conversely, splits like 15-63, 16-62, or 38-40 do not represent significant departures away from a 39-39 split, and they would not be a basis for rejecting the null hypothesis. See Figure 11-4. The Wilcoxon signed-ranks test is based on the lower rank total, so that instead of analyzing both numbers that constitute the split, it is necessary to analyze only the lower number.

In general, the sum $1 + 2 + 3 + \cdots + n$ is equal to $n(n + 1)/2$; if this is a rank sum to be divided equally between two categories (positive and negative), each of the two totals should be near $n(n + 1)/4$, which is $n(n + 1)/2$ after it is halved. Recognition of this principle forms a basis for understanding the rationale behind Formula 11-1. The denominator in that formula represents a standard deviation of T and is based on the principle that $1^2 + 2^2 + 3^2 + \cdots + n^2 = n(n + 1)(2n + 1)/6$.

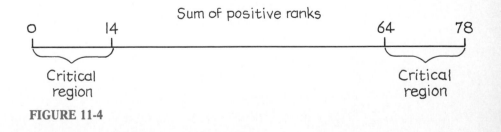

Sum of positive ranks

0   14   64   78

Critical region   Critical region

FIGURE 11-4

Technology Clouds Television Ratings

In the past it was a relatively simple task to provide television advertisers with accurate data about the numbers of viewers who saw their commercials. The simple task, which consisted of rating three major television networks, is now complicated by the presence of local television stations, cable TV, satellite receiving dishes, videocassette recorders, and remote control devices. One Nielsen study found that when shows are recorded on home VCRs for future viewing, only 80% of those shows are actually played back at a later time. In addition, viewers avoid most commericals when playing back shows. Large amounts of money are at stake since one ratings point equates to about $55 million in annual advertising per year for each network.

If we were to apply the ordinary sign test (Section 11-2) to the example given in this section, we would fail to reject the null hypothesis of no change in before and after scores. This is not the conclusion reached through the Wilcoxon signed-ranks test, which is more sensitive to the magnitudes of the differences and is therefore more likely to be correct.

This section can be used for paired data only, but the next section involves a rank-sum test that can be applied to two sets of data that are not paired.

11-3 EXERCISES A

In Exercises 11-29 through 11-36, first arrange the given data in order of lowest to highest and then find the rank of each entry.

11-29	5, 8, 12, 15, 10
11-30	1, 3, 6, 8, 99
11-31	150, 600, 200, 100, 50, 400
11-32	47, 53, 46, 57, 82, 63, 90, 55
11-33	6, 8, 8, 9, 12, 20
11-34	6, 8, 8, 8, 9, 12, 20
11-35	16, 13, 16, 13, 13, 14, 15, 18, 12
11-36	36, 27, 27, 27, 41, 39, 58, 63, 63

In Exercises 11-37 through 11-40, use the given before and after test scores in the Wilcoxon signed-ranks test procedure to:

a. Find the differences d.
b. Rank the differences while ignoring their signs.
c. Find the signed ranks.　　　d. Find T.

11-37

Before	103	98	112	94	118	99	90	101
After	100	105	114	98	119	99	100	116

11-38

Before	66	58	59	58	63	52	54	60
After	58	51	56	53	53	51	45	64

11-39

Before	83	76	91	59	62	75	80	66	73
After	82	77	89	62	68	70	90	86	73

11-40

Before	52	49	37	45	50	48	39	49	55	42	40
After	44	46	40	35	41	43	41	34	35	35	40

In Exercises 11-41 through 11-44, assume a 0.05 level of significance in a two-tailed hypothesis test. Use the given statistics to find the critical score from Table A-8, then form a conclusion about the null hypothesis H_0.

11-41	a. $T = 24, n = 15$	b. $T = 25, n = 15$
11-42	a. $T = 26, n = 15$	b. $T = 81, n = 24$
11-43	a. $T = 25, n = 17$	b. $T = 8, n = 10$
11-44	a. $T = 5, n = 10$	b. $T = 15, n = 12$

In Exercises 11-45 through 11-56, use the Wilcoxon signed-ranks test.

11-45 A psychologist wants to test the claim that two different IQ tests produce the same results. Both tests are given to a sample of nine randomly selected students, with the results given below. At the 0.05 level of significance, test the claim that both tests produce the same results.

Test A	100	111	93	92	99	85	117	110	98
Test B	106	112	95	90	107	100	126	105	110

11-46 A biomedical researcher wants to test the effectiveness of a synthetic antitoxin. The 12 randomly selected subjects are tested for resistance to a particular poison. They are retested after receiving the antitoxin, with the results given below. At the 0.05 level of significance, test the claim that the antitoxin is not effective and produces no change.

Before	18.2	21.6	23.5	22.9	16.3	19.2	21.6	21.8	20.3	19.5	18.9	20.3
After	18.4	20.3	21.5	20.2	17.6	18.5	21.7	22.3	19.4	18.6	20.1	19.7

11-47 In a study of techniques used to measure lung volumes, physiological data were collected for 10 subjects. The values given in the table are in liters and they represent the measured forced vital capacities of the 10 subjects in a sitting position and in a supine (lying) position. (See "Validation of Esophageal Balloon Technique at Different Lung Volumes and Postures," by Baydur, Cha, and Sassoon, *Journal of Applied Physiology*, Vol. 62, No. 1.) At the 0.05 significance level, test the claim that both positions have the same distribution.

Sitting	4.66	5.70	5.37	3.34	3.77	7.43	4.15	6.21	5.90	5.77
Supine	4.63	6.34	5.72	3.23	3.60	6.96	3.66	5.81	5.61	5.33

11-48 In a study of techniques used to measure lung volumes, physiological data were collected for 10 subjects. The values given in the table are in liters and they represent the measured functional residual capacities of the 10 subjects in a sitting position and in a supine (lying) position. (See "Validation of Esophageal Balloon Technique at Different Lung Volumes and Postures," by Baydur, Cha, and Sassoon, *Journal of Applied Physiology*, Vol. 62, No. 1.) At the 0.05 significance level, test the claim that both positions have the same distribution.

Sitting	2.96	4.65	3.27	2.50	2.59	5.97	1.74	3.51	4.37	4.02
Supine	1.97	3.05	2.29	1.68	1.58	4.43	1.53	2.81	2.70	2.70

11-49 A researcher devises a test of depth perception while the subject has one eye covered. The test is repeated with the other eye covered. Results are given below for eight randomly selected subjects. At the 0.05 level of significance, test the claim that depth perception is the same for both eyes.

Right eye	14.7	16.3	12.4	8.1	21.6	13.9	14.2	15.8
Left eye	15.2	16.7	12.6	10.4	24.1	17.2	11.9	18.4

11-50 To test the effect of smoking on pulse rate, a researcher compiled data consisting of pulse rate before and after smoking. The results are given below. At the 0.05 level of significance, test the claim that smoking does not affect pulse rate.

Before smoking	68	72	69	70	70	74	66	71
After smoking	69	76	68	73	72	76	66	71

11-51 An anxiety-level index is invented for third grade students, and results for each of 14 randomly selected students are obtained in a classroom setting and a recess situation. The results follow. At the 0.05 level of significance, test the claim that anxiety levels are the same for both situations.

Classroom	7.2	7.8	6.0	5.1	3.9	4.7	8.2	9.1	8.7	4.1	3.8	6.7	5.8	4.9
Recess	8.7	7.1	5.8	4.2	4.0	3.6	7.0	8.7	7.9	2.5	2.4	5.0	4.0	3.4

11-52 Two types of cooling systems are being tested in preparation for construction of a nuclear power plant. Eight different standard experimental situations yield the temperature in degrees Fahrenheit of water expelled by each of the cooling systems, and the results follow. At the 0.05 level of significance, test the claim that both cooling systems produce the same results.

Type A	72	78	81	77	84	76	79	74
Type B	75	71	71	72	73	74	70	74

11-53 Randomly selected voters are given two different tests designed to measure their attitudes about conservatism. Both tests supposedly use the same rating scale and the same criteria. At the 0.05 level of significance, test the claim that both tests produce the same results. The sample data follow.

Test A	237	215	312	190	217	250	341	380	270	245
Test B	217	190	307	192	220	233	314	367	249	238

11-54 Randomly selected executives are surveyed in an attempt to measure their attitudes toward two different minority groups, and the sample results follow. At the 0.05 level of significance, test the claim that there is no difference in their attitudes toward the two groups.

Group A	420	490	380	570	630	710	425	576	550	610	580	575
Group B	520	510	450	530	600	705	415	600	625	730	500	530

11-55 A study was conducted to investigate the effectiveness of hypnotism in reducing pain. At the 0.05 significance level, test the claim that the distribution of affective responses to pain is the same before and after hypnosis. Results for randomly selected subjects are given in the accompanying table which is based on data from "An Analysis of Factors That Contribute to the Efficacy of Hypnotic Analgesia," by Price and Barber, *Journal of Abnormal Psychology*, Vol. 96, No. 1.

Before	-5.5	-5.0	-6.6	-9.7	-4.0	-7.0	-7.0	-8.4
After	-1.4	-0.5	0.7	1.0	2.0	0.0	-0.6	-1.8

11-56   A study was conducted to investigate some effects of physical training. Sample data are listed in the following table (based on data from "Effect of Endurance Training on Possible Determinants of VO_2 During Heavy Exercise," by Casaburi and others, *Journal of Applied Physiology*, Vol. 62, No. 1). At the 0.05 level of significance, test the claim that the pretraining weights and the posttraining weights have the same distribution. All weights are given in kilograms.

Pretraining	99	57	62	69	74	77	59	92	70	85
Posttraining	94	57	62	69	66	76	58	88	70	84

11-3 EXERCISES B

11-57   a. Two checkout systems are being tested at a department store. One system uses an optical scanner to record prices while the other system has prices manually entered by the clerk. Randomly selected customers are paid to use both checkout systems, and their processing times are recorded. Listed here are the differences (in seconds) obtained when the times for the scanner system are subtracted from the corresponding times for the manual system. At the 0.01 significance level, use the Wilcoxon signed-ranks test to test the claim that both systems require the same times.

30	33	27	0	−5	−3	18	10	16	12	3	52	14	−8
−27	0	42	26	19	35	72	14	5	1	12	−6	23	52
47	33	19	16	0	−12	44	40	29	59	38			

   b. Do part (a) after subtracting 20 from each value listed. How does that change affect the results?

11-58   a. With $n = 8$ pairs of data, find the lowest and highest possible values of T.

   b. With $n = 10$ pairs of data, find the lowest and highest possible values of T.

   c. With $n = 50$ pairs of data, find the lowest and highest possible values of T.

11-59   Use Formula 11-1 to find the critical value of T for a two-tailed hypothesis test with a significance level of 0.05. Assume that there are $n = 100$ pairs of data with no differences of zero.

11-60   The Wilcoxon signed-ranks test can be used to test the claim that a sample comes from a population with a specified median. This use of the Wilcoxon signed-ranks test requires that the population be approximately symmetrical. That is, when the population distribution is separated in the middle, the left half approximates a mirror image of the right half. The procedure for testing hypotheses is the same as the one described in this section, except that the differences (step 1) are obtained by subtracting the value of the median from each score. At the 0.05 level of significance, test the claim that the values below are drawn from a population with a median of 10,000 lb. The scores are the weights (in pounds) of 50 different loads handled by a moving company in Dutchess County, New York.

8,090	9,110	17,810	12,350	3,670	14,800	10,100	26,580	17,330	15,970
8,800	11,860	7,770	8,450	12,430	10,780	13,260	5,030	10,220	11,430
13,490	11,600	13,520	7,470	4,510	14,310	14,760	13,410	4,480	7,450
7,540	3,250	10,630	6,400	10,330	8,160	10,510	9,310	12,700	9,900
7,200	6,170	12,010	16,200	11,450	8,770	9,140	6,820	7,280	6,390

11-4 WILCOXON RANK-SUM TEST FOR TWO INDEPENDENT SAMPLES

While Section 11-3 uses ranks to analyze dependent or paired data, this section introduces the **Wilcoxon rank-sum test**, which can be applied to situations involving two samples that are *independent* and not paired. (This test is equivalent to the **Mann-Whitney U test** found in some other books. See Exercise 11-78.)

We will test the null hypothesis that two independent samples come from populations with the same distribution. The alternative hypothesis is the claim that the two distributions are different in some way. **This procedure requires that each sample size be greater than 10.** For cases involving samples with 10 or fewer values, special tables are available in other reference books. Being typical of nonparametric tests, this procedure does not require that any distribution must be normal. The type of problem we consider will closely resemble many of the problems given in Section 8-3, where we tested hypotheses made about the means of two populations. However, the methods of Section 8-3 required that the samples come from normally distributed populations. The Wilcoxon rank-sum test of this section is a very reasonable alternative to those situations where the prerequisite of a normally distributed population is in question. Also, the Wilcoxon rank-sum test can be used with data at the ordinal level of measurement, such as data that consist of ranks.

In Section 11-1, we noted that this test has a 0.95 efficiency rating when compared with the parametric *t* test or *z* test. Because this test has such a high efficiency rating and involves easier calculations, it is

Class Attendance *Does* Help

In a study of 424 undergraduates at the University of Michigan, it was found that students with the worst attendance records tended to get the lowest grades. (Is anybody surprised?) Those who were absent less than 10% of the time tended to receive grades of B or above. The study also showed that students who sit in the front of the class tend to get significantly better grades.

often preferred over the Section 8-3 parametric tests, even when the condition of normality is satisfied.

We will illustrate the procedure of the Wilcoxon rank-sum test with the following example. The basis of the procedure is the principle that if two samples are drawn from identical populations and the individual scores are all ranked as one combined collection of values, then the high and low ranks should be dispersed evenly between the two samples. If we find that the low (or high) ranks are found predominantly in one of the samples, we suspect that the two populations are not identical. While this is called a rank-sum test, it is actually the means of the ranks that will affect the results.

EXAMPLE

Random samples of teachers' salaries from New York State and Florida are as follows. (The data are based on results from the U.S. Department of Education.) At the 0.05 level of significance, test the claim that the salaries of teachers are the same in both states.

New York		Florida		
$36,200	(28)	$22,200	(8)	
32,500	(25)	22,900	(10)	
31,000	(23)	21,500	(5.5)	←
31,100	(24)	22,700	(9)	
33,700	(26)	24,000	(11)	Tie between
28,400	(20)	20,500	(1)	5 and 6
29,100	(21)	21,000	(4)	
28,000	(18)	24,100	(12)	
27,900	(17)	21,500	(5.5)	←
33,900	(27)	22,000	(7)	
39,900	(29)	25,700	(13)	
		28,100	(19)	
		20,900	(3)	
		20,800	(2)	
		26,400	(14)	
		29,900	(22)	
		26,600	(15)	
		26,700	(16)	

$$n_1 = 11 \qquad n_2 = 18$$

$$R_1 = 258 \qquad R_2 = 177$$

$$\frac{R_1}{n_1} = 23.45 \qquad \frac{R_2}{n_2} = 9.83$$

continued ▶

Example, continued

Solution

H_0: The populations of salaries are identical.
H_1: The populations are not identical.

We may be tempted to use the Student t test to compare the means of two independent samples (as in Section 8-3), but we cannot meet the prerequisite of having normally distributed populations, since salaries are not normally distributed. We therefore require a nonparametric method; Wilcoxon's rank-sum test is appropriate here.

We rank all 29 salaries, beginning with a rank of 1 (assigned to the lowest salary of $20,500), a rank of 2 (assigned to the second lowest salary of $20,800), and so forth. The ranks corresponding to the various salaries are shown in parentheses in the preceding table. Note that there is a tie between the fifth and sixth scores. We treat ties in the manner described at the end of Section 11-1, by computing the mean of the ranks involved in the tie and assigning that mean rank to each of the tying values. In the case of our tie between the fifth and sixth scores, we assign the rank of 5.5 to each of those two salaries since 5.5 is the mean of 5 and 6. We denote by R the sum of the ranks for one of the two samples. If we choose the New York salaries we get

$$R = 28 + 25 + 23 + 24 + 26 + 20 + 21 + 18 + 17 + 27 + 29$$

$$= 258$$

With a null hypothesis of identical populations and with both sample sizes greater than 10, the sampling distribution of R is approximately normal. Denoting the mean of the sample R values by μ_R and the standard deviation of the sample R values by σ_R, we are able to use the test statistic

FORMULA 11-2
$$z = \frac{R - \mu_R}{\sigma_R}$$

where $\mu_R = \dfrac{n_1(n_1 + n_2 + 1)}{2}$

$$\sigma_R = \sqrt{\frac{n_1 n_2(n_1 + n_2 + 1)}{12}}$$

n_1 = size of the sample from which the rank sum R is found
n_2 = size of the other sample
R = sum of ranks of the sample with size n_1

continued ▶

Solution, continued

The expression for μ_R is a variation of a result of mathematical induction, which states that the sum of the first n positive integers is given by $1 + 2 + 3 + \cdots + n = n(n + 1)/2$, and the expression for σ_R is a variation of a result that states that the integers 1, 2, 3, . . . , n have standard deviation $\sqrt{(n^2 - 1)/12}$.

For the salary data given in the table we have already found that the rank sum R for the New York salaries is 258. Since there are 11 New York salaries, we have $n_1 = 11$. Also, $n_2 = 18$ since there are 18 Florida salaries. We can now determine the values of μ_R, σ_R, and z.

$$\mu_R = \frac{n_1(n_1 + n_2 + 1)}{2} = \frac{11(11 + 18 + 1)}{2} = 165.00$$

$$\sigma_R = \sqrt{\frac{n_1 n_2(n_1 + n_2 + 1)}{12}} = \sqrt{\frac{(11)(18)(11 + 18 + 1)}{12}} = 22.25$$

$$z = \frac{R - \mu_R}{\sigma_R} = \frac{258 - 165.00}{22.25} = 4.18$$

A large positive value of z indicates that the higher ranks are disproportionately found in the New York salaries, while a large negative value of z would indicate that New York has a disproportionate share of lower ranks. In either case, we would have strong evidence against the claim that the New York and Florida salaries are identical. The test is therefore two-tailed.

The significance of the test statistic z can now be treated in the same manner as in previous chapters. We are now testing (with $\alpha = 0.05$) the hypothesis that the two populations are the same, so we have a two-tailed test with critical z values of 1.96 and -1.96. The test statistic of $z = 4.18$ falls within the critical region and we therefore reject the null hypothesis that the salaries are the same in both states. New York appears to have significantly higher teacher salaries than Florida.

We can verify that if we interchange the two sets of salaries, we will find that $R = 177$, $\mu_R = 270.0$, $\sigma_R = 22.25$, and $z = -4.18$, so that the same conclusion will be reached.

Like the Wilcoxon signed-ranks test, this test also considers the relative magnitudes of the sample data, whereas the sign test does not. In the sign test, a weight loss of 1 lb or 50 lb receives the same sign, so the actual magnitude of the loss is ignored. While rank-sum tests do not directly involve quantitative differences between data from two samples, changes in magnitude do cause changes in rank, and these in turn affect the value of the test statistic. For example, if we change the Florida salary of $22,200 to $30,000, then the value of the rank-sum R will change, and the value of the z test statistic will also change.

11-4 EXERCISES A

11-61 The following scores were randomly selected from last year's college entrance examination scores. Use the Wilcoxon rank-sum test to test the claim that the performance of New Yorkers equals that of Californians. Assume a significance level of 0.05.

New York	520	490	571	398	602	475	557	621	737	403	511	598
California	508	563	385	617	704	401	409	527	393	478	521	536

11-62 The following scores represent the reaction times (in seconds) of randomly selected subjects from two age groups. Use the Wilcoxon rank-sum approach at the 0.05 level of significance to test the claim that both groups have the same reaction times.

18 years old	1.96	0.94	0.96	1.51	1.36	1.41	1.03	1.12
	2.12	0.86	0.79	1.17	1.13	1.00	1.01	
50 years old	1.03	1.42	1.75	2.01	0.93	1.92	2.00	1.87
	2.09	1.73	1.49	1.82				

11-63 A class of statistics students rated the president's performance using several different criteria, and the composite scores follow. Use the Wilcoxon rank-sum test at the 0.05 level of significance to test the claim that the ratings of both sexes are the same.

Males	27	36	42	57	88	92	60	60	43	29	76	79
Females	21	43	38	40	40	60	87	72	73	10	12	

11-64 Job applicants from two cultural backgrounds are tested to determine the number of trials they need to learn a certain task. The results are as follows. Use the Wilcoxon rank-sum test at the 0.05 level of significance to test the claim that the number of trials is the same for both groups.

Group A	7	12	18	15	13	14	22	9	11	10	10	10
Group B	21	19	17	8	16	16	20	24	6	19	23	

11-65   An auto parts supplier must send many shipments from the central warehouse to the city in which the assembly takes place, and she wants to determine the faster of two railroad routes. A search of past records provides the following data. (The shipment times are in hours.) At the 0.05 level of significance, use the Wilcoxon rank-sum test to determine whether or not there is a significant difference between the routes.

Route A	98	102	83	117	128	92	112	108	108	100	93	72	95	91
Route B	96	132	121	87	106	102	116	95	99	76	97	104	115	114

11-66   A large city police department offers a refresher course on arrest procedures. The effectiveness of this course is examined by testing 15 randomly selected officers who have recently completed the course. The same test is given to 15 randomly selected officers who have not had the refresher course. The results are as follows. At the 0.05 level of significance, test the claim that the course has no effect on the test grades by using the Wilcoxon rank-sum approach.

Group completing the course	173	141	219	157	163	165	178	200
	154	189	192	201	157	168	181	
Group without the course	159	124	170	148	135	133	137	189
	181	111	144	127	138	151	162	

11-67   A study is conducted to determine whether a drug affects eye movements. A standardized scale is developed and the drug is administered to one group, while a control group is given a placebo that produces no effects. The eye movement ratings of subjects are as follows. At the 0.01 level of significance, test the claim that the drug has no effect on eye movements. Use the Wilcoxon rank-sum test.

Drugged group	652	512	711	621	508	603	787	747	516	624	627	777	729
Control group	674	676	821	830	565	821	837	652	549	668	772	563	703
	789	800	711	598									

11-68   Two coffee-vending machines are studied to determine whether they distribute the same amounts. Samples are obtained and the contents (in liters) are as follows. At the 0.05 level of significance, use the Wilcoxon rank-sum test to test the claim that the machines distribute the same amount.

Machine A	0.210	0.213	0.206	0.195	0.180	0.250	0.212	0.217
	0.213	0.222	0.201	0.205	0.209			
Machine B	0.229	0.224	0.221	0.247	0.270	0.233	0.237	0.235
	0.238	0.200	0.198	0.216	0.241	0.273	0.205	

11-69 Groups of randomly selected men and women are given questionnaires designed to measure their attitudes toward capital punishment, and the results are as follows. At the 0.05 level of significance, test the claim that there is no difference between the attitudes of men and women concerning capital punishment. Use the Wilcoxon rank-sum test.

Men	67	72	48	30	92	15	5	87	91	54	66	72	98	97	75	74
Women	20	40	37	42	51	15	68	35	12	31	85					

11-70 In a study of longevity, two groups of adult males are randomly selected; their longevity data (in years) are summarized below. Use the Wilcoxon rank-sum test to test the claim that there is no difference between the two groups.

Group A	65	66	73	78	54	39	47	59	67	67
	69	71	74	77	62	73	75	76	68	67
Group B	64	67	73	70	58	69	72	71	63	64
	63	63	55	43	35	50	74	61	62	69

11-71 In a study of crop yields, two different fertilizer treatments are tested on parcels with the same area and soil conditions. Listed below are the yields (in bushels of corn) for sample plots. Use a 0.05 significance level and apply the Wilcoxon rank-sum test to determine whether there is a difference between the two treatments.

Treatment A	132	137	129	142	160	139	143	147	145	140	131	136
Treatment B	162	180	149	157	159	159	152	167	163	165	180	156
	158	151										

11-72 A consumer investigator obtains prices from mail order companies and computer stores. Listed below are the prices (in dollars) quoted for boxes of ten floppy disks from various manufacturers. Use a 0.05 level of significance to test the claim that there is no difference between mail order and store prices. Use the Wilcoxon rank-sum test.

Mail order	23.00	26.00	27.99	31.50	32.75	27.00
	27.98	24.50	24.75	28.15	29.99	29.99
Computer store	30.99	33.98	37.75	38.99	35.79	33.99
	34.79	32.99	29.99	33.00	32.00	

11-73   The effectiveness of mental training was tested in a military training program. In an antiaircraft artillary examination, scores for an experimental group and a control group were recorded. Use the Wilcoxon rank-sum approach to test the claim that both groups come from populations with the same scores. Use a 0.05 significance level. (See "Routinization of Mental Training in Organizations: Effects on Performance and Well-Being" by Larsson, *Journal of Applied Psychology*, Vol. 72, No. 1.)

Experimental				Control			
60.83	117.80	44.71	75.38	122.80	70.02	119.89	138.27
73.46	34.26	82.25	59.77	118.43	54.22	118.58	74.61
69.95	21.37	59.78	92.72	121.70	70.70	99.08	120.76
72.14	57.29	64.05	44.09	104.06	94.23	111.26	121.67
80.03	76.59	74.27	66.87				

11-74   Sample data were collected in a study of calcium supplements and the effects on blood pressure. A placebo group and a calcium group began the study with measures of blood pressures. At the 0.05 significance level, use the Wilcoxon rank-sum approach to test the claim that the two sample groups come from populations with the same blood pressure levels. (The data are based on "Blood Pressure and Metabolic Effects of Calcium Supplementation in Normotensive White and Black Men" by Lyle and others, *Journal of the American Medical Association*, Vol. 257, No. 13.)

Placebo				Calcium			
124.6	104.8	96.5	116.3	129.1	123.4	102.7	118.1
106.1	128.8	107.2	123.1	114.7	120.9	104.4	116.3
118.1	108.5	120.4	122.5	109.6	127.7	108.0	124.3
113.6				106.6	121.4	113.2	

11-75   In a study involving motivation and test scores, data were obtained for males and females. Use the following data and the Wilcoxon rank-sum test to test the claim that both samples come from populations with the same scores. Use a 0.05 significance level. (See "Relationships Between Achievement-Related Motives, Extrinsic Conditions, and Task Performance" by Schroth, *Journal of Social Psychology*, Vol. 127, No. 1.)

Male				Female			
12.27	39.53	32.56	23.93	31.13	18.71	14.34	23.90
19.54	25.73	32.20	19.84	13.96	13.88	29.85	20.15
20.20	23.01	25.63	17.98	6.66	19.20	15.89	
22.99	22.12	12.63	18.06				

11-76 The arrangement of test items was studied for its effect on anxiety. Sample results are given below. At the 0.05 level of significance, use the Wilcoxon rank-sum test to test the claim that the two samples come from populations with the same scores. (The data are based on "Item Arrangement, Cognitive Entry Characteristics, Sex and Test Anxiety as Predictors of Achievement Examination Performance" by Klimko, *Journal of Experimental Education*, Vol. 52, No. 4.)

Easy to difficult				Difficult to easy			
24.64	39.29	16.32	32.83	33.62	34.02	26.63	30.26
28.02	33.31	20.60	21.13	35.91	26.68	29.49	35.32
26.69	28.90	26.43	24.23	27.24	32.34	29.34	33.53
7.10	32.86	21.06	28.89	27.62	42.91	30.20	32.54
28.71	31.73	30.02	21.96				
25.49	38.81	27.85	30.29				
30.72							

11-4 EXERCISES B

11-77 a. The *ranks* for group A are 1, 2, . . . , 15 and the *ranks* for group B are 16, 17, . . . , 30. At the 0.05 level of significance, use the Wilcoxon rank-sum test to test the claim that both groups come from the same population.

b. The *ranks* for group A are 1, 3, 5, 7, . . . , 29 and the *ranks* for group B are 2, 4, 6, . . . , 30. At the 0.05 level of significance, use the Wilcoxon rank-sum test to test the claim that both groups come from the same population.

c. Compare parts (a) and (b).

d. What changes occur when the rankings of the two groups in part (a) are interchanged?

e. Use the two groups in part (a) and interchange the ranks of 1 and 30 and then note the changes that occur.

11-78 The Mann-Whitney U test is equivalent to the Wilcoxon rank-sum test for independent samples in the sense that they both apply to the same situations and they always lead to the same conclusions. In the Mann-Whitney U test we calculate

$$z = \frac{U - \dfrac{n_1 n_2}{2}}{\sqrt{\dfrac{n_1 n_2 (n_1 + n_2 + 1)}{12}}}$$

where

$$U = n_1 n_2 + \frac{n_1(n_1 + 1)}{2} + R$$

(continued)

Show that if the expression for U is substituted into the preceding expression for z, we get the same test statistic (with opposite sign) used in the Wilcoxon rank-sum test for two independent samples.

11-79 Assume that we have two treatments (A and B) that produce measurable results, and we have only two observations for treatment A and two observations for treatment B. We cannot use Formula 11-2 because both sample sizes do not exceed 10.

a. Complete the table below by listing the other five rows corresponding to the other five cases, and enter the corresponding rank sums for treatment A.

1	Rank 2	3	4	(Rank sum for treatment A) R
A	A	B	B	3
	.			.
	.			.
	.			.

b. List the possible values of R along with their corresponding probabilities. (Assume that the rows of the table from part (a) are equally likely.)

c. Is it possible, at the 0.10 significance level, to reject the null hypothesis that there is no difference between treatments A and B? Explain.

11-80 Do Exercise 11-79 for the case involving a sample of size three for treatment A and a sample of size three for treatment B.

11-5 KRUSKAL-WALLIS TEST

In Section 10-4 we used analysis of variance to test hypotheses that differences among several sample means are due to chance. That parametric F test required that all the involved populations possess normal distributions with variances that are approximately equal. In this section we introduce the **Kruskal-Wallis test** as a nonparametric alternative that does not require normal distributions. In using the Kruskal-Wallis test (also called the **H test**), we test the null hypothesis that independent and random samples come from the same or identical populations. We compute the test statistic **H, which has a distribution that can be approximated by the chi-square distribution as long as each sample has at least five observations**. (For cases involving samples with fewer than five observations, we must refer to more advanced books for special tables of critical values.) When we use the chi-square distribution in this context, the number of degrees of freedom is $k - 1$, where k is the number of samples.

$$\textbf{degrees of freedom} = k - 1$$

FORMULA 11-3 $H = \dfrac{12}{N(N + 1)}\left(\dfrac{R_1^2}{n_1} + \dfrac{R_2^2}{n_2} + \cdots + \dfrac{R_k^2}{n_k}\right) - 3(N + 1)$

where N = total number of observations in all samples combined

   R_1 = sum of ranks for the first sample

   R_2 = sum of ranks for the second sample

   R_k = sum of ranks for the kth sample

   k = the number of samples

   In using the Kruskal-Wallis test, we replace the original scores by their corresponding ranks. We then proceed to calculate the test statistic H, which is basically a measure of the variance of the rank sums $R_1, R_2, \ldots, R_k$. If the ranks are distributed evenly among the sample groups, then H should be a relatively small number. If the samples are very different, then the ranks will be excessively low in some groups and high in others, with the net effect that H will be large. Consequently, only large values of H lead to rejection of the null hypothesis that the samples come from identical populations. **The Kruskal-Wallis test is therefore a right-tailed test.**

   Begin by considering all observations together, and then assign a rank to each one. We rank from lowest to highest, and we also treat ties as we did in the previous sections of this chapter—the mean value of the ranks is assigned to each of the tied observations. Then take each individual sample and find the sum of the ranks and the corresponding sample size. We will illustrate the Kruskal-Wallis test by considering the following example. The Kruskal-Wallis test is especially appropriate here because there may be some doubt that home selling prices are normally distributed. We will see similarities between the Kruskal-Wallis test and Wilcoxon's rank-sum test since both are based on rank sums.

EXAMPLE

A real estate investor randomly selects homes recently sold in three different zones of the same county. The selling prices (based on data from homes recently sold in Dutchess County, New York) are listed here. At the 0.05 significance level, test the claim that selling prices are the same in all three zones.

continued ▶

Example, continued

Selling prices (dollars)

Zone 1		Zone 4		Zone 7	
$147,000	(15.5)	$160,000	(17.5)	$215,000	(24)
160,000	(17.5)	140,000	(14)	127,000	(8)
128,000	(9)	173,000	(21)	98,000	(2)
162,000	(19)	113,000	(4)	147,000	(15.5)
135,000	(12)	85,000	(1)	184,000	(23)
132,000	(10)	120,000	(7)	109,000	(3)
181,000	(22)	285,000	(25)	169,000	(20)
138,000	(13)	117,000	(5)		
		133,000	(11)		
		119,000	(6)		

$$n_1 = 8 \qquad n_2 = 10 \qquad n_3 = 7$$

$$R_1 = 118 \qquad R_2 = 111.5 \qquad R_3 = 95.5$$

Solution

H_0: The populations are identical.
H_1: The populations are not identical.

Step 1. Rank the combined samples from lowest to highest. Begin with the lowest observation of $85,000, which is assigned a rank of 1. Ranks are listed with the original data in the preceding list.

Step 2. For each individual sample, find the number of observations and the sum of the ranks. The first sample (Zone 1) has eight observations, so $n_1 = 8$. Also, $R_1 = 15.5 + 17.5 + \cdots + 13 = 118$. The values of $n_2, R_2, n_3,$ and R_3 are shown above. Since the total number of observations is 25, we have $N = 25$.

Step 3. Compute the value of the test statistic H. Using Formula 11-3 for the given data, we get

$$H = \frac{12}{N(N+1)}\left(\frac{R_1^2}{n_1} + \frac{R_2^2}{n_2} + \frac{R_3^2}{n_3}\right) - 3(N+1)$$

$$= \frac{12}{25(26)}\left(\frac{118^2}{8} + \frac{111.5^2}{10} + \frac{95.5^2}{7}\right) - 3(26)$$

$$= \frac{12}{650}(1740.5 + 1243.225 + 1302.893) - 78$$

$$= 1.138$$

continued ▶

> **Solution, continued**
>
> *Step 4.* Since each sample has at least five observations, the distribution of H is approximately a chi-square distribution with $k - 1$ degrees of freedom. The number of samples is $k = 3$, so we get $3 - 1$, or 2, degrees of freedom. Refer to Table A-4 to find the critical value of 5.991, which corresponds to 2 degrees of freedom and a significance level of $\alpha = 0.05$. (This use of the chi-square distribution is always right-tailed, since only large values of H reflect disparity in the distribution of ranks among the samples.) If any sample has fewer than five observations, use the special tables found in texts devoted exclusively to nonparametric statistics. Such cases are not included in this text.
>
> *Step 5.* The test statistic $H = 1.138$ is less than the critical value of 5.991, so we fail to reject the null hypothesis of identical populations. We reject the null hypothesis of identical populations only when H exceeds the critical value. The three zones appear to have selling prices that are not significantly different.

The test statistic H, as described in Formula 11-3, is the rank version of the test statistic F used in the analysis of variance discussed in Section 10-4. When dealing with ranks R instead of raw scores x, many components are predetermined. For example, the sum of all ranks can be expressed as $N(N + 1)/2$, where N is the total number of scores in all samples combined. The expression

$$H = \frac{12}{N(N + 1)} \Sigma n_i(\bar{R}_i - \bar{\bar{R}})^2$$

combines weighted variances of ranks in a test statistic that is algebraically equivalent to Formula 11-1, which is easier to work with. (In this expression, $\bar{R}_i = R_i/n_i$ and $\bar{\bar{R}} = (\Sigma R_i)/(\Sigma n_i)$.)

In comparing the procedures of the parametric F test for analysis of variance and the nonparametric Kruskal-Wallis test, we see that the Kruskal-Wallis test is much simpler to apply. We need not compute the sample variances and sample means. We do not require normal population distributions. Life becomes so much easier. However, the Kruskal-Wallis test is not as efficient as the F test, and it may require more dramatic differences for the null hypothesis to be rejected.

11-5 EXERCISES A

In Exercises 11-81 through 11-96, use the Kruskal-Wallis test.

11-81   An experiment involves raising samples of corn under identical conditions except for the type of fertilizer used. The yields are obtained for three different fertilizers, and those values are ranked with the results shown below. Find the value of the test statistic H, where H is given in Formula 11-3.

	Treatment	
A	B	C
1	2	3
6	4	5
7	8	9
12	11	10
14	15	13

11-82   Do Exercise 11-81 after replacing the given ranks with those listed below.

	Treatment	
A	B	C
1	6	11
2	7	12
3	8	13
4	9	14
5	10	15

11-83   A sociologist randomly selects subjects from three different types of family structure: stable families, divorced families, and families in transition. The selected subjects are interviewed and rated for their social adjustment, and the sample results are given below. (The numbers in parentheses are the ranks.) At the 0.05 level of significance, test the claim that the samples come from identical populations.

Stable		Divorced		Transition	
108	(8.5)	113	(10)	92	(1)
104	(4)	106	(7)	123	(13)
103	(3)	108	(8.5)	126	(14)
97	(2)	127	(15)	105	(5.5)
118	(12)	114	(11)	105	(5.5)

11-84 Readability studies are conducted to determine the clarity of four different texts, and the sample scores follow. (The numbers in parentheses are the ranks.) At the 0.05 level of significance, test the claim that the four texts have the same readability level.

Text A	Text B	Text C	Text D
50 (3.5)	59 (10.5)	45 (1)	62 (13)
50 (3.5)	60 (12)	48 (2)	64 (15)
53 (6)	63 (14)	51 (5)	68 (18)
58 (9)	65 (16)	54 (7)	70 (19)
59 (10.5)	67 (17)	55 (8)	72 (20)

11-85 An introductory calculus course is taken by students with varying high school records. The sample results of six students from each of the three groups follow. The values given are the final numerical averages in the calculus course. At the 0.05 level of significance, test the claim that the three groups come from identical populations.

Good high school record	Fair high school record	Poor high school record
90	80	60
86	70	60
88	61	55
93	52	62
80	73	50
96	65	70

11-86 A store owner records the gross receipts for days randomly selected from times during which she used only newspaper advertising, only radio advertising, or no advertising. The results are listed below. At the 0.05 level of significance, test the claim that the receipts are the same, regardless of advertising.

Newspaper	Radio	None
845	811	612
907	782	574
639	749	539
883	863	641
806	872	666

11-87 Three methods of instruction are used to train air traffic controllers. With method A, an experienced controller is assigned to a trainee for practical field experience. With method B, trainees are given extensive classroom instruction and are then placed without supervision.

(*continued*)

Method C requires a moderate amount of classroom training and then several trainees are supervised on the job by an experienced instructor. A standardized test is used to measure levels of competency, and sample results are given below. At the 0.01 level of significance, test the claim that the three methods are equally effective.

Method A	Method B	Method C
195	187	193
198	210	212
223	222	215
240	238	231
251	256	252
		260
		267

11-88   Do Exercise 11-87 after adding 100 to each value of method A.

Zone	LA	Acres	Taxes
1	20	0.50	1.9
1	18	1.00	2.4
1	27	1.05	1.5
1	17	0.42	1.6
1	18	0.84	1.6
1	13	0.33	1.5
1	24	0.90	1.7
1	15	0.83	2.2
4	18	0.55	2.8
4	20	0.46	1.8
4	19	0.94	3.2
4	12	0.29	2.1
4	9	0.26	1.4
4	18	0.33	2.1
4	28	1.70	4.2
4	10	0.50	1.7
4	15	0.43	1.8
4	12	0.25	1.6
7	21	3.04	2.7
7	16	1.09	1.9
7	14	0.23	1.3
7	23	1.00	1.7
7	17	6.20	2.2
7	17	0.46	2.0
7	20	3.20	2.2

11-89   The accompanying data represent 25 different homes sold in Dutchess County, New York. The zones (1, 4, 7) correspond to different geographic regions of the county. The values of LA are living areas in hundreds of square feet. The Acres values are lot sizes in acres, and the Taxes values are the annual tax bills in thousands of dollars. At the 0.05 significance level, test the claim that living areas are the same in all three zones.

11-90   Use the data from Exercise 11-89. At the 0.05 level of significance, test the claim that lot sizes (as measured in acres) are the same in all three zones.

11-91   Use the data from Exercise 11-89. At the 0.05 level of significance, test the claim that taxes are the same in all three zones.

11-92   Refer to the data from Exercise 11-89 and change the zone 7 Tax values to the following: 3.0, 1.8, 1.1, 1.4, 2.3, 2.0, 2.4. Note that the mean Tax amount does not change, but these values vary more. Now repeat Exercise 11-91 with this modified data set and note the effect of increasing the spread of the values in one of the samples.

11-93 A unit on basic consumer economics is taught to five different classes of randomly selected students. A different teaching method is used for each group, and sample final test data are given below. The scores represent the final averages of the individual students. Test the claim that the five methods are equally effective.

Traditional	Programmed	Audio	Audiovisual	Visual
76.2	85.2	67.3	75.8	50.5
78.3	74.3	60.1	81.6	70.2
85.1	76.5	55.4	90.3	88.8
63.7	80.3	72.3	78.0	67.1
91.6	67.4	40.0	67.8	77.7
87.2	67.9		57.6	73.9
	72.1			
	60.4			

11-94 A study is made of the response time of police cars after dispatching occurs. Sample results for three precincts are given below in seconds. At the 0.05 level of significance, test the claim that the three precincts have the same response time.

Precinct 1	Precinct 2	Precinct 3
160	165	162
172	174	175
176	180	177
176	181	179
176	184	187
178	186	195
	190	210
	200	215
		216
		220

11-95 Given the measurements randomly obtained for the five samples below, can we conclude that the samples come from identical populations? Use a 0.05 level of significance. The given values are levels of algae in different ponds located in the same county.

Sample A	Sample B	Sample C	Sample D	Sample E
67.2	67.7	70.2	72.6	72.2
67.4	68.3	70.3	73.0	72.3
67.9	68.5	70.5	73.8	72.5
69.3	68.7	71.0	74.0	72.8
69.5	68.8	71.6	75.0	73.4
69.8	68.9	71.7	75.6	73.7
		71.9	75.7	74.6
		72.0	75.9	74.9

11-96 a. If 20 is added to each observed sample value of Exercise 11-95, how is the value of the test statistic affected?

b. If each observed sample value in Exercise 11-95 is multiplied by 5, how is the value of the test statistic affected?

11-5 EXERCISES B

11-97 Simplify Formula 11-3 for the special case of eight samples, all consisting of exactly six observations each.

11-98 For three samples, each of size five, what are the largest and smallest possible values of H?

11-99 In using the Kruskal-Wallis test, there is a correction factor that should be applied whenever there are many ties: Divide H by

$$1 - \frac{\Sigma T}{N^3 - N}$$

where $T = t^3 - t$. For each group of tied scores, find the number of observations that are tied and represent this number by t. Then compute $t^3 - t$ to find the value of T. Repeat this procedure for all cases of ties and find the total of the T values, which is ΣT. For the example presented in this section, use this procedure to find the corrected value of H.

11-100 Show that for the case of two samples, the Kruskal-Wallis test is equivalent to the Wilcoxon rank-sum test. This can be done by showing that for the case of two samples, the test statistic H equals the square of the test statistic z used in the Wilcoxon rank-sum test. Also note that with one degree of freedom, the critical values of χ^2 correspond to the square of the critical z score.

11-6 RANK CORRELATION

In Chapter 9 we considered the concept of correlation, and we introduced the *linear correlation coefficient* as a measure of the strength of the association between two variables. In this section we will study rank correlation, the nonparametric counterpart of that parametric measure. In Chapter 9 we computed values for the linear correlation coefficient r, but in this section we will be computing values for the **rank correlation coefficient**. The method of this section has some distinct advantages over the parametric methods discussed in Chapter 1.

Does Television Watching Cause Violence?

A study of 1650 British youngsters yielded a strong correlation between the commission of violent acts and the number of hours of television viewed. A separate California study showed a very strong negative correlation between academic skills and television viewing; more television watching tended to correspond to lower skill levels. More than 6000 research studies on the effects of television on children have not led to much agreement, but the following conclusions seem to emerge from many of those studies.

- Prolonged television watching by children tends to correspond to aggressive behavior.
- Prolonged television watching by children tends to correspond to lower vocabulary and reading abilities.
- Children who view television extensively tend to view the world as being more violent than it really is.

Studies may show a correlation between television watching and aggressive behavior, but that doesn't imply that television is the cause of the aggressive behavior. Perhaps aggressive children watch television because it complements their existing behavior. Perhaps some other factor encourages aggressive behavior and television viewing.

Advantages

1. One major advantage of this nonparametric approach is that *it allows us to analyze some types of data that can be ranked but not measured*; yet such data could not be considered with the parametric linear correlation coefficient *r* of Chapter 9.

2. A second major advantage of rank correlation is that *it can be used to detect some relationships that are not linear*. An example illustrating this will be given later in this section.

3. Another possible advantage of this nonparametric approach is that *the computations are much simpler* than those for the linear correlation coefficient *r*. This can be readily seen by comparing Formula 11-4 (given later in this section) to Formula 9-1. With many calculators, you can get the value of *r* easily, but if you do not have a calculator or computer, you would probably find that the rank correlation coefficient is easier to compute.

4. A fourth advantage of the rank correlation approach is that *it can be used when some of the more restrictive requirements of the linear correlation approach are not met*. That is, the nonparametric

approach can be used in a wider variety of circumstances than can the parametric method. For example, the parametric approach requires that the involved populations have normal distributions; the nonparametric approach does not require normality.

As an example, observe the data in Table 11-5. Seven teachers are being ranked for promotion by two separate committees.

TABLE 11-5		
Teacher	Faculty rank	Administrator's rank
Aquino	1	2
Bennet	7	7
Cohen	2	6
Davis	5	3
Ellis	4	5
Fujimoto	3	1
Gallo	6	4

Even though we lack any of the real numerical data that led to the above table, we can still use the given ranks to obtain a rank correlation coefficient by using Formula 11-4.

FORMULA 11-4 $$r_s = 1 - \frac{6\Sigma d^2}{n(n^2 - 1)}$$

where r_s = rank correlation coefficient

    n = number of *pairs* of data

    d = difference between ranks for the two observations within a pair

We use the notation r_s for the *rank* correlation coefficient so that we don't confuse it with the *linear* correlation coefficient r. The subscript s is commonly used in honor of Charles Spearman (1863–1945), who originated the rank correlation approach. In fact, r_s is often called **Spearman's rank correlation coefficient.** The subscript s has nothing to do with standard deviation, and for that we should be thankful. Just as r is a sample statistic that can be considered an estimate of the population parameter ρ, we can also consider r_s to be an estimate of ρ_s.

If the data lead to ties in ranks, we proceed as in previous sections: Calculate the mean of the ranks involved in the tie and then assign that mean rank to each of the tied items. Formula 11-4 leads to an exact value of r_s only if no ties occur. With a relatively low number of ties, Formula 11-4 results in a good approximation to r_s. (With ties, we can get an exact value of r_s by using the parametric Formula 9-1; after using Formula 9-1 to evaluate r_s, we should continue with the procedures of this section.)

We now calculate r_s for the data given in Table 11-5. Since there are seven pairs of data, $n = 7$. We obtain the differences d for each pair by subtracting the lower rank from the higher rank.

Faculty rank	Administrator's rank	d (difference)	d^2
1	2	1	1
7	7	0	0
2	6	4	16
5	3	2	4
4	5	1	1
3	1	2	4
6	4	2	4
		Total: $\Sigma d^2 = 30$	

With $n = 7$ and $\Sigma d^2 = 30$, we get

$$r_s = 1 - \frac{6\Sigma d^2}{n(n^2 - 1)}$$

$$= 1 - \frac{6(30)}{7(7^2 - 1)}$$

$$= 1 - 0.536$$

$$= 0.464$$

For practical reasons, we are omitting the theoretical derivation of Formula 11-4, but we can gain some insight by considering the following three cases. If we intuitively examine Formula 11-4, we can see that strong agreement between the two sets of ranks will lead to values of d near zero, so that r_s will be close to 1 (see Case I). Conversely, when the ranks of one set tend to be at opposite extremes when compared to the ranks of the second set, then the values of d tend to be high, which will cause r_s to be near -1 (see Case II). If there is no relationship between the two sets of ranks, then the values of d will be neither high nor low and r_s will tend to be near zero (see Case III).

Case I:
Perfect Positive
Correlation

Rank x	Rank y	Difference d	d^2
1	1	0	0
3	3	0	0
5	5	0	0
4	4	0	0
2	2	0	0
		$\Sigma d^2 = 0$	

$$r_s = 1 - \frac{6(0)}{5(5^2 - 1)} = 1$$

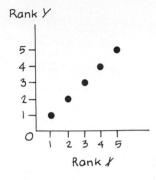

Case II:
Perfect Negative
Correlation

Rank x	Rank y	Difference d	d^2
1	5	4	16
2	4	2	4
3	3	0	0
4	2	2	4
5	1	4	16
		$\Sigma d^2 = 40$	

$$r_s = 1 - \frac{6(40)}{5(5^2 - 1)} = -1$$

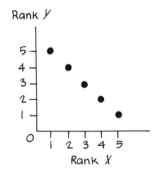

Case III:
No Correlation

Rank x	Rank y	Difference d	d^2
1	2	1	1
2	5	3	9
3	3	0	0
4	1	3	9
5	4	1	1
		$\Sigma d^2 = 20$	

$$r_s = 1 - \frac{6(20)}{5(5^2 - 1)} = 0$$

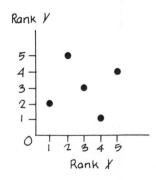

Cases I and II illustrate the most extreme cases, so that the following property applies.

$$-1 \le r_s \le 1$$

Recall that this same property applies to the linear correlation coefficient r described in Chapter 9. We can formulate conclusions about correlation by comparing the computed rank correlation coefficient r_s to a critical value. Suppose we have somehow established that at the 0.05 level of significance, the critical value of r_s for 12 pairs of ranks is 0.591. Figure 11-5 shows us how to interpret the computed rank correlation coefficient r_s for the critical value $r_s = 0.591$. Note that this interpretation follows the precedent set by the interpretation of the linear correlation coefficient r in Chapter 9. We have a table of critical values of r_s in Table A-9, which stops at $n = 30$ pairs of ranks. (Be careful to avoid confusion between Table A-6 for the linear correlation coefficient r and Table A-9 for Spearman's rank correlation coefficient r_s.) When the number of pairs of ranks, n, exceeds 30, the sampling distribution of r_s is approximately a normal distribution with mean zero and standard deviation $1/\sqrt{n-1}$. We therefore get

$$z = \frac{r_s - 0}{\dfrac{1}{\sqrt{n-1}}} = r_s\sqrt{n-1}$$

In a two-tailed case we would use the positive and negative z values. Solving for r_s, we then get the critical values by evaluating

FORMULA 11-5 $\quad r_s = \dfrac{\pm z}{\sqrt{n-1}} \qquad$ (when n > 30)

The value of z would correspond to the significance level.

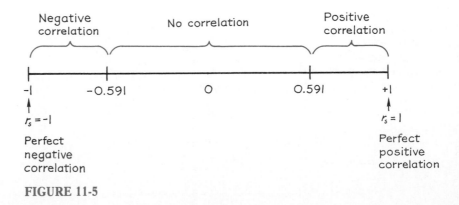

FIGURE 11-5

Study Criticized as Misleading

A study sponsored by the National Association of Elementary School Principals and the Kettering Foundation suggested that there was a *correlation* between children with one parent and children who were low achievers. The media coverage of the final report implied that the *cause* of lower achievement was the absence of one of the parents. Critics charged that the media confused correlation with cause and effect and that the study itself was misleading since it emphasized negative factors. Also, some conclusions were based on samples too small to be statistically significant.

EXAMPLE

Find the critical values of Spearman's rank correlation coefficient r_s when the data consist of 40 pairs of ranks. Assume a two-tailed case with a 0.05 significance level.

Solution

Since there are 40 pairs of data, $n = 40$. Because n exceeds 30, we use Formula 11-5 instead of Table A-9. With $\alpha = 0.05$ in two tails, we let $z = 1.96$ to get

$$r_s = \frac{\pm 1.96}{\sqrt{40 - 1}} = \pm 0.314$$

In Figure 11-6 we summarize the procedure to be followed when using Spearman's nonparametric approach in testing for correlation. This next example illustrates that procedure.

EXAMPLE

Ten different students study for a test; the table below lists the number of hours studied (x) and the corresponding number of correct answers (y). At the 0.05 level of significance, use Spearman's rank correlation approach to determine if there is a correlation between hours studied and the number of correct answers.

x	5	9	17	1	2	21	3	29	7	100
y	6	16	18	1	3	21	7	20	15	22

Solution

$H_0: \rho_s = 0$
$H_1: \rho_s \neq 0$

Refer to Figure 11-6, which we will follow in this solution. The given data are not ranks, so we convert the table into ranks, as follows. (Section 11-1 describes the procedure for converting scores into ranks.)

continued ▶

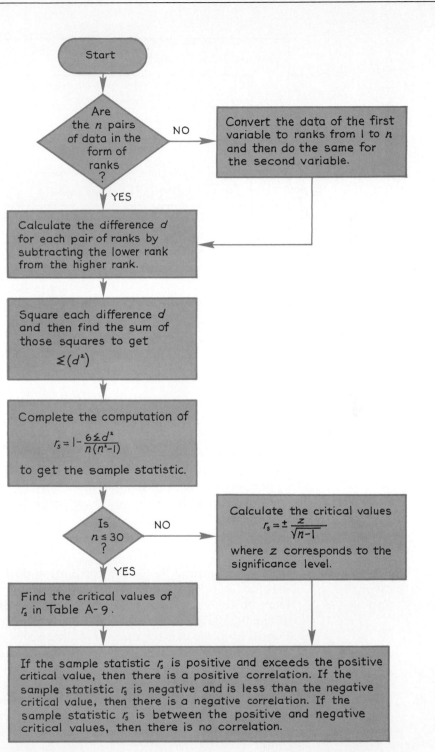

FIGURE 11-6
Spearman's rank correlation

Solution, continued

x	4	6	7	1	2	8	3	9	5	10
y	3	6	7	1	2	9	4	8	5	10
d	1	0	0	0	0	1	1	1	0	0
d^2	1	0	0	0	0	1	1	1	0	0

After expressing all data as ranks, we next calculate the differences, d, and then we square them. The sum of the d^2 values is 4. We now calculate

$$r_s = 1 - \frac{6\Sigma d^2}{n(n^2 - 1)}$$

$$= 1 - \frac{6(4)}{10(10^2 - 1)}$$

$$= 1 - 0.024$$

$$= 0.976$$

Proceeding with Figure 11-6, $n = 10$, so we answer yes when asked if $n \leq 30$. We use Table A-9 to get the critical values of -0.648 and 0.648. Finally, the sample statistic of 0.976 exceeds 0.648, so we conclude that there is significant positive correlation. More hours of study appear to be associated with higher grades. (You didn't really think we would suggest otherwise, did you?)

If we compute the linear correlation coefficient r (using Formula 9-1) for the original data in this last example, we get $r = 0.629$, which leads to the conclusion that there is no significant *linear* correlation at the 0.05 level of significance. If we examine the scatter diagram of Figure 11-7, we can see that there does seem to be a relationship, but it's not linear. This last example is intended to illustrate two advantages of the nonparametric approach over the parametric approach. We have already noted the advantage of detecting some relationships that are not linear. This last example also illustrates this additional advantage: *Spearman's rank correlation coefficient* r_s *is less sensitive to a value that is very far out of line*, such as the 100 hours in the preceding data.

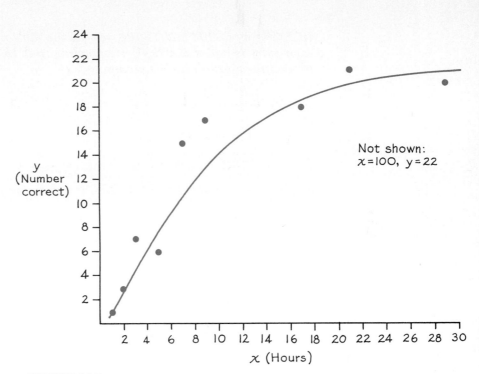

FIGURE 11-7

Like other nonparametric methods, the rank correlation approach does have the disadvantage of being less efficient. In Section 11-1 we noted that rank correlation has a 0.91 efficiency rating when compared to the parametric linear correlation method from Chapter 9. If the conditions required for the parametric approach are satisfied, the nonparametric rank correlation approach would require 100 sample observations to achieve the same result as 91 sample observations analyzed through the parametric linear correlation approach of Chapter 9.

11-6 EXERCISES A

In Exercises 11-101 and 11-102, find the critical value for r_s by using Table A-9 or Formula 11-5, as appropriate. Assume two-tailed cases where α represents the level of significance and n *represents the number of pairs of data.*

11-101
a. $n = 20, \alpha = 0.05$
b. $n = 50, \alpha = 0.05$
c. $n = 40, \alpha = 0.02$
d. $n = 25, \alpha = 0.02$
e. $n = 37, \alpha = 0.04$

11-102
a. $n = 82, \alpha = 0.04$
b. $n = 15, \alpha = 0.01$
c. $n = 50, \alpha = 0.01$
d. $n = 37, \alpha = 0.01$
e. $n = 43, \alpha = 0.05$

In Exercises 11-103 and 11-104, rewrite the table of paired data so that each entry is replaced by its corresponding rank in accordance with the procedure described in this section.

11-103

x	2	5	8	6	10
y	30	25	20	15	5

11-104

x	15	14	10	9	3
y	3	7	10	11	20

In Exercises 11-105 through 11-108, compute the sample statistic r_s by using Formula 11-4.

11-105

x	1	2	3	4
y	4	2	3	1

11-106

x	10	20	30	40
y	40	20	30	10

11-107

x	63	68	71	55	70	75
y	43	44	39	30	28	20

11-108

x	28	28	35	37	40
y	16	17	12	19	20

In Exercises 11-109 through 11-124:

a. Compute the rank correlation coefficient r_s for the given sample data.
b. Assume that $\alpha = 0.05$ and find the critical value of r_s from Table A-9 or Formula 11-5.
c. Based on the results of parts (a) and (b), decide whether there is significant positive correlation, significant negative correlation, or no significant correlation. In each case, assume a significance level of 0.05.

11-109 The following table ranks eight states according to teacher's salaries and SAT scores. (The table is based on data from the U.S. Department of Education.)

	N.Y.	Calif.	Fla.	N.J.	Tex.	N.C.	Md.	Ore.
Teacher's salary	1	2	7	4	6	8	3	5
SAT score	4	3	5	6	7	8	2	1

11-110 The following table (based on data from the U.S. Department of Education) ranks eight states according to teacher's salaries and cost per student.

	N.Y.	Calif.	Fla.	N.J.	Tex.	N.C.	Md.	Ore.
Teacher's salary	1	2	7	4	6	8	3	5
Cost per student	1	5	6	2	7	8	3	4

11-111 The following table ranks eight states according to SAT scores and monetary aid to families with dependent children. (The table is based on data from the U.S. Department of Education.)

	N.Y.	Calif.	Fla.	N.J.	Tex.	N.C.	Md.	Ore.
SAT score	4	3	5	6	7	8	2	1
Aid per family with dependent children	2	1	7	3	8	6	5	4

11-112 The following table (based on data from the U.S. Department of Education) ranks eight states according to SAT scores and cost per student.

	N.Y.	Calif.	Fla.	N.J.	Tex.	N.C.	Md.	Ore.
SAT score	4	3	5	6	7	8	2	1
Cost per student	1	5	6	2	7	8	3	4

11-113 The accompanying table lists the value of exports (in billions of dollars) and the value of imports (in billions of dollars) for several different years. (The data are provided by the U.S. Department of Commerce.)

Exports	10	20	43	221	218	218
Imports	9	15	40	245	326	370

11-114   For randomly selected states, the following table lists the per capita beer consumption (in gallons) and the per capita wine consumption (in gallons). (The table is based on data from *Statistical Abstract of the United States*.)

Beer	32.2	29.4	35.3	34.9	29.9	28.7	26.8	41.4
Wine	3.1	4.4	2.3	1.7	1.4	1.2	1.2	3.0

11-115   When loads were added to a hanging copper wire, the wire is stretched. The loads (in Newtons) and increases in length (in centimeters) are given here. (The table is based on data from *College Physics* by Sears, Zemansky, and Young.)

Added load	0	10	20	30	40	50	60	70
Increase in length	0	0.05	0.10	0.15	0.20	0.25	0.30	1.25

11-116   The following table lists the number of registered automatic weapons (in thousands) along with the murder rate (in murders per 100,000) for randomly selected states. (The data are from the FBI and the Bureau of Alcohol, Tobacco, and Firearms.)

Automatic weapons	11.6	8.3	3.6	0.6	6.9	2.5	2.4	2.6
Murder rate	13.1	10.6	10.1	4.4	11.5	6.6	3.6	5.3

11-117   Two of the judges in a gymnastics competition rank 10 athletes, with the results given below.

	A	B	C	D	E	F	G	H	I	J
Judge Allen	8	6	1	3	5	10	9	7	2	4
Judge Zeller	6	7	1	4	3	10	9	8	5	2

11-118   A company ranks its sales personnel after giving them a test designed to measure aggressiveness. They are also ranked in order of sales for the last year. All these results are summarized below for 12 randomly selected employees.

	A	B	C	D	E	F	G	H	I	J	K	L
Aggressiveness	6	4	7	3	12	5	10	2	11	1	8	9
Sales	2	1	8	9	11	10	3	4	12	5	6	7

11-119 Several cities were ranked according to the number of hotel rooms and the amount of office space. The results are as follows.

City	NY	Ch	SF	Ph	LA	At	Mi	KC	NO	Da	Ba	Bo	Se	Ho	SL
Office rank	1	3	2	7	6	10	14	15	11	9	12	4	8	5	13
Hotel rank	1	2	3	8	6	7	14	15	4	10	13	5	9	11	12

11-120 A medical researcher tests the effects of a drug on the time it takes a patient to perform a standard manual task. The drug amounts in milligrams and corresponding times in seconds for randomly selected patients are given in the table.

Drug amount	15	20	25	30	35	40	45	50	55	60	65	70	75
Time	48	46	55	54	60	58	73	74	82	90	105	130	200

11-121 There are many regions where the winter accumulation of snowfall is a primary source of water. Several investigations of snowpack characteristics have used satellite observations from the Landsat series along with measurements taken on earth. Given in the following table are ground measurements of snow depth (in centimeters) along with the corresponding temperatures (in degrees Celsius) (data based on information in Kastner's *Space Mathematics*, published by NASA).

Temperature (°C)	−62	−41	−36	−26	−33	−56	−50	−66
Snow depth (cm)	21	13	12	3	6	22	14	19

11-122 For randomly selected homes recently sold in Dutchess County, New York, the annual tax amounts (in thousands of dollars) are listed along with the selling prices (in thousands of dollars).

Taxes	1.9	3.0	1.4	1.4	1.5	1.8	2.4	4.0
Selling price	145	228	150	130	160	114	142	265

11-123 For randomly selected homes recently sold in Dutchess County, New York, the living areas (in hundreds of square feet) are listed along with the annual tax amounts (in thousands of dollars).

Living area	15	38	23	16	16	13	20	24
Taxes	1.9	3.0	1.4	1.4	1.5	1.8	2.4	4.0

11-124 In a study of employee stock ownership plans, data was collected at eight companies on satisfaction with the plan and the amount of organizational commitment. Results are given below. (See "Employee Stock Ownership and Employee Attitudes: A Test of Three Models" by Klein, *Journal of Applied Psychology*, Vol. 72, No. 2.)

Satisfaction	5.05	4.12	5.39	4.17	4.00	4.49	5.40	4.86
Commitment	5.37	4.49	5.42	4.45	4.24	5.34	5.62	4.90

11-6 EXERCISES B

11-125 Two judges each rank three contestants, and the ranks for the first judge are 1, 2, and 3, respectively.
a. List all possible ways that the second judge can rank the same three contestants. (No ties allowed.)
b. Compute r_s for each of the cases found in part (a).
c. Assuming that all of the cases from part (a) are equally likely, find the probability that the sample statistic, r_s, is greater than 0.9.

11-126 One alternative to using Table A-9 involves an approximation of critical values for r_s given as

$$r_s = \pm \sqrt{\frac{t^2}{t^2 + n - 2}}$$

where t is the t score from Table A-3 corresponding to the significance level and $n - 2$ degrees of freedom. Apply this approximation to find critical values of r_s for the following cases.
a. $n = 8$, $\alpha = 0.05$
b. $n = 15$, $\alpha = 0.05$
c. $n = 30$, $\alpha = 0.05$
d. $n = 30$, $\alpha = 0.01$
e. $n = 8$, $\alpha = 0.01$

11-127 a. Given the bivariate data depicted in the scatter diagram, which would be more likely to detect the relationship between x and y: the linear correlation coefficient r or the rank correlation coefficient r_s? Explain.
b. How is r_s affected if one variable is ranked from low to high while the other variable is ranked from high to low?

11-128 Assume that a set of paired data has been converted to ranks according to the procedure described in this section, and also assume that no ties occur. Show that Formula 11-4 (for r_s) and Formula 9-1 (for r) will provide the same result when the ranks are used as the x and y values.

11-7 RUNS TEST FOR RANDOMNESS

A classic example of the use of statistics involves a manufacturer of pantyhose who obtained sample data describing the production output and later became convinced of employee stealing when fewer pairs of pantyhose were produced. As it turned out, the problem was not employee theft. Instead, the initial sampling was done with newly serviced machinery, which produced a finer mesh and more items than the same machinery used for many hours. The initial sampling was not random, and it led to misleading results and embarassment for the poor president who proclaimed that pantyhose were being pilfered.

In many of the examples and exercises already discussed in this book, it was explicitly assumed that data were randomly selected. But how can we really check for randomness? If a pollster reports back with a sequence of 20 interviews with females followed by a string of 20 interviews with males, we might strongly suspect a lack of randomness. But where do we draw the line? We need a systematic and standard procedure for testing for the randomness of data. We can use the **runs test** as a test for randomness. We will work with sequences of data that fall into two distinct categories.

DEFINITION

A **run** is a sequence of data that exhibit the same characteristic; the sequence is preceded and followed by different data or no data at all.

As an example, consider the political party of the sequence of ten voters interviewed by a pollster. We let D denote a Democrat, while R indicates a Republican. The following sequence contains exactly four runs.

EXAMPLE 1

$$D, D, D, D, \qquad R, R, \qquad D, D, D, \qquad R$$

1st run　　　2nd run　　3rd run　　4th run

The All-Male Jury Finds Dr. Spock . . .

Guilty. Dr. Benjamin Spock was convicted of conspiracy as a result of activities that encouraged resistance to the military draft during the Vietnam War. The defense claimed that Dr. Spock was handicapped by the fact that all 12 jurors were men. Women would tend to be more inclined to think of Spock more favorably, since they were generally more opposed to the war; also, Dr. Spock was a well-known baby doctor who wrote popular books used by many mothers.

A statistician testified that the presiding judge had a consistently lower proportion of women jurors than the other six judges serving the same judicial district. Spock's conviction was overturned for other reasons, but we now randomly select federal court jurors so that sex bias will not occur.

We would use the runs test in this situation to test for the randomness with which Democrats and Republicans occur. Let's use common sense to see how runs relate to randomness. Examine the sequence in Example 2 and then stop to consider how randomly Democrats and Republicans occur. Also count the number of runs.

EXAMPLE 2

D, D, D, D, D, D, D, D, D, D, R, R, R, R, R, R, R, R, R, R

In Example 2, it is reasonable to conclude that Democrats and Republicans occur in a sequence that is *NOT* random. Note that in the sequence of 20 data, there are only two runs. This example might suggest that if the number of runs is very low, randomness may be lacking. Now consider the sequence of 20 data given in Example 3. Try again to form your own conclusion about randomness before you continue reading.

EXAMPLE 3

D, R, D, R, D, R, D, R, D, R, D, R, D, R, D, R, D, R, D, R

In Example 3, it should be apparent that the sequence of Democrats and Republicans is again *NOT* random, since there is a distinct, predictable pattern. In this case, the number of runs is 20; this example suggests that randomness is lacking when the number of runs is too high.

It is important to note that this test for randomness is based on the *order* in which the data occur. This test is *NOT* based on the *frequency* of the data. For example, a particular sequence containing 3 Democrats and 20 Republicans might lead to the conclusion that the sequence is random. The issue of whether or not 3 Democrats and 20 Republicans is a *biased* sample is another issue not addressed by the runs test.

The sequences in Examples 2 and 3 are obvious in their lack of randomness, but most sequences are not so obvious; we therefore need more sophisticated techniques for analysis. We begin by introducing some notation.

NOTATION

n_1 Number of elements in the sequence that have the same characteristic

n_2 Number of elements in the sequence that have the other characteristic

G Number of runs

We use G to represent the number of runs because n and r have already been used for other statistics, and G is a relatively innocuous letter that deserves more attention.

EXAMPLE

In the sequence $D, D, D, D, R, R, D, D, D, R, R, R, R, D$, we obtain the following values for n_1, n_2, and G.

$$n_1 = 8 \quad \text{since there are eight Democrats}$$
$$n_2 = 6 \quad \text{since there are six Republicans}$$
$$G = 5 \quad \text{since there are five runs}$$

Uncle Sam Wants You, if You're Randomly Selected

A considerable amount of controversy was created when a lottery was used to determine who would be drafted into the U.S. Army. In 1970, the lottery approach was instituted in an attempt to make the selection process random, but many claimed that the outcome was unfair because men born later in the year had a better chance of being drafted.

The 1970 lottery involved 366 capsules corresponding to the dates in a leap year. (The first dates selected would be the birthdays of the first men drafted.) First, the 31 January capsules were placed in a box. The 29 February capsules were added and the two months were mixed. Then the 31 March capsules were added

and the three months were mixed. This process continued; one result was that January capsules were mixed 11 times, while December was mixed only once. Later arguments claimed that this process tended to place early dates closer to the bottom while dates later in the year tended to be near the top. The first ten dates selected, in order of priority, were September 14, April 24, December 30, February 14, October 18, September 6, October 26, September 7, November 22, and December 6. Although the runs test indicated randomness, other statistical tests did not. There was enough criticism of the process to cause a revised procedure the following year.

In 1971, statisticians used two drums of cap-

sules. One drum contained capsules with dates, which were deposited in a random order according to a table of random numbers. A second drum contained the draft priority numbers, which were also deposited randomly according to a table of random numbers. Both drums were rotated for an hour before the selection process was begun. September 16 was drawn from one drum and 139 was drawn from the other. This meant that men born on September 16 had draft priority number 139. This process continued until all dates had a priority number. The randomness of this procedure was significantly less controversial.

We can now revert to our standard procedure for hypothesis testing. **The null hypothesis H_0 will be the claim that the sequence is random;** the alternative hypothesis H_1 will be the claim that the sequence is *NOT* random. The flowchart in Figure 11-8 summarizes the mechanics of the procedure. That flowchart directs us to a table (Table A-10) of critical G values when the following three conditions are all met:

1. $\alpha = 0.05$, and
2. $n_1 \leq 20$, and
3. $n_2 \leq 20$.

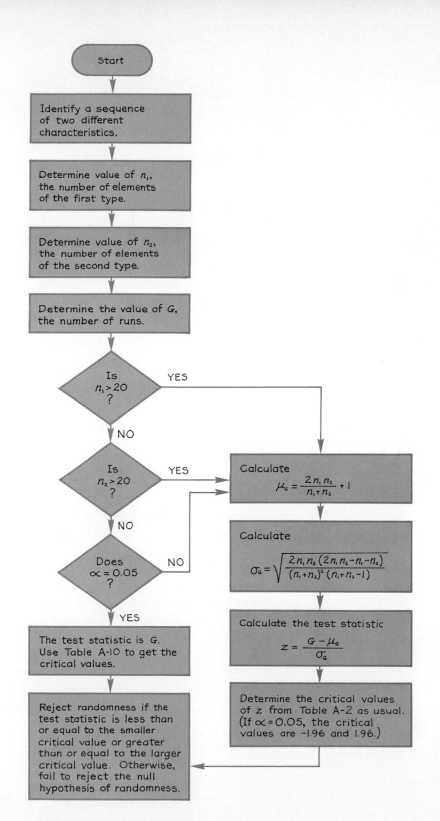

FIGURE 11-8
The runs test of randomness

If all these conditions are not satisfied, we use the fact that G has a distribution that is approximately normal with mean and standard deviation as follows.

FORMULA 11-6

$$\mu_G = \frac{2n_1 n_2}{n_1 + n_2} + 1$$

FORMULA 11-7

$$\sigma_G = \sqrt{\frac{(2n_1 n_2)(2n_1 n_2 - n_1 - n_2)}{(n_1 + n_2)^2 (n_1 + n_2 - 1)}}$$

When this normal approximation is used, the test statistic is

FORMULA 11-8

$$z = \frac{G - \mu_G}{\sigma_G}$$

and the critical values are found by using the procedures introduced in Chapter 6. This normal approximation is quite good. If the entire table of critical values (Table A-10) had been computed using this normal approximation, no critical value would be off by more than one unit.

We now illustrate the use of the runs test for randomness by presenting examples of complete tests of hypotheses.

EXAMPLE

The president of an investment firm has observed that men and women have been hired in the following sequence: $M, M, M, W,$ M, M, M, M, W, W, W, M. At the 0.05 level of significance, test the personnel officer's claim that the sequence of men and women is random. (Note that we are not testing for a *bias* in favor of one sex over the other. There are eight men and four women, but we are testing only for the *randomness* in the way they appear in the given sequence.)

Solution

The null hypothesis is the claim of randomness so we get

H_0: The eight men and four women have been hired in a random sequence.

H_1: The sequence is not random.

continued ▶

Solution, continued

The significance level is $\alpha = 0.05$. Figure 11-8 summarizes the procedure for the runs test, so we refer to that flowchart. We now determine the values of n_1, n_2, and G for the given sequence.

$$n_1 = \text{number of men} = 8$$

$$n_2 = \text{number of women} = 4$$

$$G = \text{number of runs} = 5$$

Continuing with Figure 11-8, we answer no when asked if $n_1 > 20$ (since $n_1 = 8$), no when asked if $n_2 > 20$ (since $n_2 = 4$), and yes when asked if $\alpha = 0.05$. The test statistic is $G = 5$, and we refer to Table A-10 to find the critical values of 3 and 10. Figure 11-9 shows that the test statistic $G = 5$ does not fall in the critical region. We therefore fail to reject the null hypothesis that the given sequence of men and women is random.

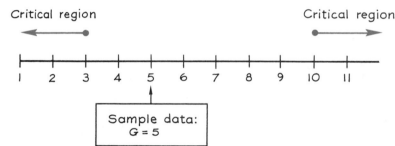

FIGURE 11-9

The next example will illustrate the procedure to be followed when Table A-10 cannot be used and the normal approximation must be used instead.

EXAMPLE

A machine produces defective (D) items and acceptable (A) items in the following sequence.

$D, A, A, A, A, D, A, A, A, A, D, A, A, A, A,$

$D, A, A, A, A, D, A, A, A, A, D, A, A, A, A$

At the 0.05 level of significance, test the claim that the sequence is random.

H_0: The sequence is random. H_1: The sequence is not random.

continued ▶

Example, continued

Here $\alpha = 0.05$, and the statistic relevant to this test is G. Referring to the flowchart in Figure 11-8, we determine the values of n_1, n_2, and G as follows:

$$n_1 = \text{number of defective items} = 6$$

$$n_2 = \text{number of acceptable items} = 24$$

$$G = \text{number of runs} = 12$$

Continuing with the flowchart, we answer no when asked if $n_1 > 20$ (since $n_1 = 6$) and yes when asked if $n_2 > 20$ (since $n_2 = 24$). The flowchart now directs us to calculate μ_G. Because of the sample sizes involved, we are on our way to using a normal approximation.

$$\mu_G = \frac{2n_1n_2}{n_1 + n_2} + 1$$

$$= \frac{2(6)(24)}{6 + 24} + 1$$

$$= \frac{288}{30} + 1 = 9.6 + 1 = 10.6$$

The next step requires the computation of σ_G.

$$\sigma_G = \sqrt{\frac{(2n_1n_2)(2n_1n_2 - n_1 - n_2)}{(n_1 + n_2)^2(n_1 + n_2 - 1)}}$$

$$= \sqrt{\frac{2(6)(24)[2(6)(24) - 6 - 24]}{(6 + 24)^2(6 + 24 - 1)}}$$

$$= \sqrt{\frac{288(258)}{900(29)}} = \sqrt{\frac{74,304}{26,100}}$$

$$= \sqrt{2.847} = 1.69$$

We now calculate the test statistic z.

$$z = \frac{G - \mu_G}{\sigma_G} = \frac{12 - 10.6}{1.69} = 0.83$$

At the 0.05 level of significance, the critical values of z are -1.96 and 1.96. The test statistic $z = 0.83$ does not fall in the critical region, so we fail to reject the null hypothesis that the sequence is random. This is a good example of why we do not say "accept the null hypothesis" because examination of the data reveals that there is a consistent pattern that is clearly not random.

Randomness Above and Below the Mean or Median

In each of the preceding examples, the data clearly fit into two categories, but we can also test for randomness in the way numerical data fluctuate above or below a mean or median. That is, in addition to analyzing sequences of nominal data of the type already discussed, we can also analyze sequences of data at the interval or ratio levels of measurement. See the following example.

EXAMPLE

An investor wants to analyze long-term trends in the stock market. The annual high points of the Dow Jones Industrial Average are given below for a recent sequence of 10 years.

   908, 898, 1000, 1024, 1071, 1287, 1287, 1553, 1956, 2722

At the 0.05 significance level, test for randomness above and below the median.

Solution

For the given data, the median is 1179. Let B denote a value *below* the median of 1179 and let A represent a value *above* 1179. We can rewrite the given sequence as follows. (If a value is equal to the median, we simply delete it from the sequence.)

$$
\begin{array}{cccccccccc}
B & B & B & B & B & A & A & A & A & A \\
\downarrow & \downarrow & \downarrow & \downarrow & \downarrow & \downarrow & \downarrow & \downarrow & \downarrow & \downarrow \\
908 & 898 & 1000 & 1024 & 1071 & 1287 & 1287 & 1553 & 1956 & 2722
\end{array}
$$

It is helpful to write the A's and B's directly above the numbers they represent. This makes checking easier and it also reduces the chance of having the wrong number of letters. After finding the sequence of letters, we can proceed to apply the runs test in the usual way. We get

$$n_1 = \text{number of } B\text{'s} = 5$$

$$n_2 = \text{number of } A\text{'s} = 5$$

$$G = \text{number of runs} = 2$$

From Table A-10, the critical values of G are found to be 2 and 10. At the 0.05 level of significance, we reject a null hypothesis

continued ▶

> ### Solution, continued
>
> of randomness above and below the median if the number of runs is 2 or lower, or 10 or more. Since $G = 2$, we reject the null hypothesis of randomness above and below the median. It isn't necessary to be a financial wizard in order to recognize that the yearly high levels of the Dow Jones Industrial Average appear to be following an upward trend. This example illustrates one way to recognize such trends, especially those that are not so obvious.

We could also test for randomness above and below the *mean* by following the same procedure (after we have deleted from the sequence any values that are equal to the mean).

Economists use the runs test for randomness above and below the median in an attempt to identify trends or cycles. An upward economic trend would contain a predominance of B's in the beginning and A's at the end, so that the number of runs would be small, as in the preceding example. A downward trend would have A's dominating the beginning and B's at the end, with a low number of runs. A cyclical pattern would yield a sequence that systematically changes, so that the number of runs would tend to be large.

11-7 EXERCISES A

In Exercises 11-129 through 11-136, use the given sequence to determine the values of n_1, n_2, the number of runs, G, and the appropriate critical values from Table A-10. (Assume a 0.05 significance level.)

11-129 A, B, B, B, A, A, A, A

11-130 $A, A, A, B, A, B, A, A, B, B$

11-131 $A, A, A, A, B, B, B, B, B, B$

11-132 $A, A, B, B, A, B, A, B, A, B, A$

11-133 $A, A, A, A, A, A, A, B, B$

11-134 $A, A, A, B, A, A, B, B, A, B, B, B, B$

11-135 $A, B, B, B, A, B, B, B, B, B, A, B, A, A, A, A, A, B, A, A$

11-136 $A, A, A, B, B, B, B, A, A, A, A, A, A, B, B, B, B, B, A, A, A, A, B, B, B, B, B,$
B, A, A, A, A, A

In Exercises 11-137 through 11-140, use the given sequence to answer the following.

a. Find the mean.
b. Let *B* represent a value below the mean and let *A* represent a value above the mean. Rewrite the given numerical sequence as a sequence of *A*'s and *B*'s.
c. Find the values of n_1, n_2, and *G*.
d. Assuming a 0.05 level of significance, use Table A-10 to find the appropriate critical values of *G*.
e. What do you conclude about the randomness of the values above and below the mean?

11-137   3, 8, 7, 7, 9, 12, 10, 16, 20, 18

11-138   2, 2, 3, 2, 4, 4, 5, 8, 4, 6, 7, 9, 9, 12

11-139   15, 12, 12, 10, 17, 11, 8, 7, 7, 5, 6, 5, 5, 9

11-140   3, 3, 4, 4, 4, 8, 8, 8, 9, 7, 10, 4, 4, 3, 2, 4, 4, 10, 10, 12, 9, 10, 10, 11, 5, 4, 3, 3, 4, 5

In Exercises 11-141 through 11-144, use the given sequence to answer the following.

a. Find the median.
b. Let *B* represent a value below the median and let *A* represent a value above the median. Rewrite the given numerical sequence as a sequence of *A*'s and *B*'s.
c. Find the values of n_1, n_2, and *G*.
d. Assuming a 0.05 level of significance, use Table A-10 to find the appropriate critical values of *G*.
e. What do you conclude about the randomness of the values above and below the median?

11-141   3, 8, 7, 7, 9, 12, 10, 16, 20, 18

11-142   2, 2, 3, 2, 4, 4, 5, 8, 4, 6, 7, 9, 9, 12

11-143   3, 3, 4, 5, 8, 8, 8, 9, 9, 9, 10, 15, 14, 13, 16, 18, 20

11-144   19, 17, 3, 2, 2, 15, 4, 12, 14, 13, 9, 7, 7, 7, 8, 8, 8

11-145   A pollster is hired to determine the sex and income of the heads of households. The sexes, in order of occurrence, are given below. At the 0.05 level of significance, test the claim that males and females occur randomly.

F, F, F, F, F, F, M, M, M, F, F, F, M, M, M, M, M, M, M, M, M, M, M, M, M, M, M

11-146   At the 0.05 level of significance, test for randomness of odd (O) and even (E) numbers on a roulette wheel that yielded the results given in the sequence below.

$O, E, O, O, O, O, E, E, O, O, E, O, O, O, O, O, O, E, E, E, O, O, O, E, O, O, O$

11-147   A machine produces defective (D) items and acceptable (A) items in the sequence given below. At the 0.05 level of significance, test the claim that the sequence is random.

$A, A, A, D, A, A, A, A, D, A, A, A, A, A, D, A, A, A, A, A, A, D, A, A, A, A, A, A, A,$
$D, D, D, A, D, D, D, D, A, A, A, A, D, D, D, D$

11-148   The Apgar rating scale is used to rate the health of newborn babies. A sequence of births results in the following values.

$9, 6, 9, 8, 7, 9, 5, 8, 9, 10, 8, 9, 7, 5, 9, 8, 8$

At the 0.05 level of significance, test the claim that the ratings above and below the sample mean are random.

11-149   The number of housing starts for the most recent years in one particular county are listed below. At the 0.05 level of significance, test the claim of randomness of those values above and below the *median*.

973, 1067, 856, 971, 1456, 903, 899, 905, 812, 630, 720, 676, 731, 655, 598, 617

11-150   The gold reserves of the United States (in millions of fine troy ounces) are given here for a recent 12-year sequence (based on data from the International Monetary Fund). At the 0.05 significance level, test for randomness above (A) and below (B) the median. (The data are arranged in chronological order by rows.)

274.71	274.68	277.55	276.41	264.60	264.32
264.11	264.03	263.39	262.79	262.65	262.04

11-151 An elementary school nurse records the grade level of students that develop chicken pox; the sequential list is given here. Consider children in grades 1, 2, and 3 to be younger, while those in grades 4, 5, and 6 are older. At the 0.05 level of significance, test the claim that the sequence is random with respect to the categories of younger and older.

3, 3, 3, 2, 3, 1, 4, 3, 6, 5, 3, 3, 2, 1, 4, 4, 4, 4, 4, 4, 4, 4, 4, 5, 5, 5, 5, 4, 3, 2, 1, 6, 6, 6, 6, 1, 3, 4, 2, 5, 6, 6, 6, 6, 6

11-152 A teacher develops a true-false test with these answers:

T, F, F, F, F, T, T, T, T, T, T, F, F, F, F, T, T, T, T, T, T, T, F, F, F, F, F, F, F, F, F, T, T, T, T, T, T, F, F, F

At the 0.05 level of significance, test the claim that the sequence of answers is random.

11-153 Listed below are the daily lottery numbers drawn in 40 consecutive days in New York State. At the 0.05 level of significance, test the claim that the occurrence of even and odd numbers in this sequence of 40 numbers is random.

968	816	305	164	931	671	256	421	872	205	658
238	510	988	922	090	756	856	787	334	507	753
458	329	783	860	530	468	716	273	492	926	758
282	139	539	974	746	275	696				

11-154 The following values are the numbers of nerve impulses per unit time produced by a visual neuron located in the lateral geniculate nucleus of a macaque monkey. (The data are provided by Robert Shapley of the Rockefeller Institute.) At the 0.05 level of significance, test the claim that the sequence of even (E) and odd (O) numbers is random. The scores are arranged in order by row.

```
3 4 3 3 3 2 3 5 1 1 3 2 3 3 3 3 2 2 4 4 2 1 3 1 2 3 2
3 1 3 2 2 0 5 3 1 5 2 2 2 3 2 1 2 2 2 2 4 2 5 4 4 3 4
3 4 2 4 1 1 3 4 3 4 1 3 4 3 3 1 3 5 1 1 2 5 3 0 2 5 1
1 1 3 4 1 4 2 3 2 5 1 4 2 1 3 3 6 4 5 2 4 3 3 2 2 3 2
2 5 1 4 1 4 2 1 4 3 3 2 1 4 3 2 4 3 2 5
```

11-155   A *New York Times* article about the calculation of decimal places of π noted that "mathematicians are pretty sure that the digits of π are indistinguishable from any random sequence." Given below are the first 100 decimal places of π. At the 0.05 level of significance, test for randomness of odd (*O*) and even (*E*) digits.

1 4 1 5 9 2 6 5 3 5 8 9 7 9 3 2 3 8 4 6 2 6 4 3 3 8
3 2 7 9 5 0 2 8 8 4 1 9 7 1 6 9 3 9 9 3 7 5 1 0 5 8
2 0 9 7 4 9 4 4 5 9 2 3 0 7 8 1 6 4 0 6 2 8 6 2 0 8
9 9 8 6 2 8 0 3 4 8 2 5 3 4 2 1 1 7 0 6 7 9

11-156   Use the 100 decimal digits for π given in Exercise 11-155. At the 0.05 level of significance, test for randomness above (*A*) and below (*B*) the value of 4.5.

11-157   A manufacturer of ball bearings analyzes the sequence of output on a particular machine. Observation of the sequence of the first 60 bearings results in 30 below the median and 30 above. The number of runs is found to be eight. At the 0.05 level of significance, test for randomness above and below the median.

11-158   A sequence of 50 altimeters produced by a manufacturer yields 30 altimeters that read too low and 20 that read too high. The number of runs is determined to be 18. At the 0.05 level of significance, test for randomness of high and low readings in the sequence of 50 altimeters.

11-159   An economist analyzes a sequence of 80 consumer price indices. In testing for randomness above and below the median, six runs were found. What do you conclude? Assume a 0.05 level of significance.

11-160   A sequence of dates is selected through a process that has the outward appearance of being random. At the 0.05 level of significance, test the resulting sequence for randomness before and after the middle of the year.

Nov. 27, July 7, Aug. 3, Oct. 19, Dec. 19, Sept. 21, Apr. 1, Mar. 5, June 10, May 21, June 27, Jan. 5

11-7 EXERCISES B

11-161   Using A, A, B, B, what is the minimum number of possible runs that can be arranged? What is the maximum number of runs? Now refer to Table A-10 to find the critical G values for $n_1 = n_2 = 2$. What do you conclude about this case?

11-162   Let $z = 1.96$ and $n_1 = n_2 = 20$, and then compute μ_G and σ_G. Use those values in Formula 11-8 to solve for G. What is the importance of this result? How does this result compare to the corresponding value found in Table A-10? How do you explain any discrepancy?

11-163   a. Using all of the elements $A, A, A, B, B, B, B, B, B$, list the 84 different possible sequences.
b. Find the number of runs for each of the 84 sequences.
c. Assuming that each sequence has a probability of 1/84, then find $P(2 \text{ runs})$, $P(3 \text{ runs})$, $P(4 \text{ runs})$, and so on.
d. Use the results of part (c) to establish your own critical G values in a two-tailed test with 0.025 in each tail.
e. Compare your results to those given in Table A-10.

11-164   Using all of the elements $A, A, A, B, B, B, B, B, B, B, B, B$, it is possible to arrange 220 different sequences.
a. List all of those sequences having exactly three runs.
b. Using your result from part (a), find $P(3 \text{ runs})$.
c. Using your answer to part (b), should $G = 3$ be in the critical region?
d. Find the lower critical value from Table A-10.
e. Find the lower critical value by using the normal approximation.

VOCABULARY LIST

Define and give an example of each term.

parametric methods
nonparametric
  methods
efficiency
rank
sign test

Wilcoxon signed-
  ranks test
Wilcoxon rank-sum
  test
Kruskal-Wallis test
H test
rank correlation
  coefficient

Spearman's rank
  correlation
  coefficient
runs test
run

REVIEW

In this chapter we examined six different **nonparametric** methods for analyzing statistics. Besides excluding involvement with population parameters like μ and σ, nonparametric methods are not encumbered by many of the restrictions placed on parametric methods, such as the requirement that data must come from normally distributed populations. While nonparametric methods generally lack the sensitivity of their parametric counterparts, they can be used in a wider variety of circumstances and can accommodate more types of data. Also, the computations required in nonparametric tests are generally much simpler than the computations required in the corresponding parametric tests.

In Table 11-6, we list the nonparametric tests of this chapter, along with their functions. The table also lists the corresponding parametric tests. Following Table 11-6 is a listing of the important formulas from this chapter.

TABLE 11-6		
Nonparametric test	Function	Parametric test
Sign test (Section 11-2)	Test for claimed value of average with one sample	t test or z test (Sections 6-2, 6-3, 6-4)
	Test for a difference between two dependent samples	t test or z test (Section 8-3)
Wilcoxon signed-ranks test (Section 11-3)	Test for difference between two dependent samples	t test or z test (Section 8-3)
Wilcoxon rank-sum test (Section 11-4)	Test for difference between two independent samples	t test or z test (Section 8-3)
Kruskal-Wallis test (Section 11-5)	Test for more than two independent samples coming from identical populations	Analysis of variance (Section 10-4)
Rank correlation (Section 11-6)	Test for a relationship between two variables	Linear correlation (Section 9-2)
Runs test (Section 11-7)	Test for randomness of sample data	(No parametric test)

IMPORTANT FORMULAS		
Test	Test statistic	Distribution
Sign test	x = number of times the less frequent sign occurs Test statistic when $n > 25$: $$z = \frac{(x + 0.5) - \left(\frac{n}{2}\right)}{\frac{\sqrt{n}}{2}}$$	If $n \le 25$, see Table A-7. If $n > 25$, use the normal distribution (Table A-2).
Wilcoxon signed-ranks test for two dependent samples	T = smaller of the rank sums Test statistic when $n > 30$: $$z = \frac{T - \frac{n(n + 1)}{4}}{\sqrt{\frac{n(n + 1)(2n + 1)}{24}}}$$	If $n \le 30$, see Table A-8. If $n > 30$, use the normal distribution (Table A-2). *(continued)*

	IMPORTANT FORMULAS (*continued*)	
Test	Test statistic	Distribution
Wilcoxon rank-sum test for two independent samples	$$z = \frac{R - \mu_R}{\sigma_R}$$ where R = sum of ranks of the sample with size n_1 $$\mu_R = \frac{n_1(n_1 + n_2 + 1)}{2}$$ $$\sigma_R = \sqrt{\frac{n_1 n_2(n_1 + n_2 + 1)}{12}}$$	Normal distribution (Table A-2) (Requires that $n_1 > 10$ and $n_2 > 10$.)
Kruskal-Wallis test	$$H = \frac{12}{N(N + 1)}\left(\frac{R_1^2}{n_1} + \frac{R_2^2}{n_2} + \cdots + \frac{R_k^2}{n_k}\right) - 3(N + 1)$$ where N = total number of sample values R_k = sum of ranks for the kth sample k = the number of samples	Chi-square with $k - 1$ degrees of freedom (Requires that each sample has at least five values.)
Rank correlation	$$r_s = 1 - \frac{6\Sigma d^2}{n(n^2 - 1)}$$	If $n \le 30$, use Table A-9. If $n > 30$, critical values are $$r_s = \frac{\pm z}{\sqrt{n - 1}}$$ where z is from the normal distribution.
Runs test for randomness	Test statistic when $n > 20$: $$z = \frac{G - \mu_G}{\sigma_G}$$ where $\mu_G = \frac{2n_1 n_2}{n_1 + n_2} + 1$ $$\sigma_G = \sqrt{\frac{(2n_1 n_2)(2n_1 n_2 - n_1 - n_2)}{(n_1 + n_2)^2(n_1 + n_2 - 1)}}$$	If $n \le 20$, use Table A-10. If $n > 20$, use the normal distribution (Table A-2).

REVIEW EXERCISES

11-165 An index of civic awareness is obtained for 10 randomly selected subjects before and after they take a short course. The results follow. Use the Wilcoxon signed-ranks test to test the claim that the course does not affect the index. Use a 0.05 level of significance.

Before course	63	65	67	70	71	77	78	80	85	91
After course	73	64	72	74	77	80	79	82	96	91

11-166 Randomly selected married couples were given questionnaires about family attitudes, and their scores are summarized below. Use the sign test to test the claim that husbands and wives produce the same results.

Husband	62	68	54	73	72	83	76	74	73	74
Wife	54	65	69	55	61	71	79	79	73	81

11-167 Sample entrance exam scores are given below for applicants randomly selected from three counties. Test the claim that the Kings County and Queens County populations are identical.

Kings County	76	52	39	27	88	73	75	92	99	83	79		
Queens County	67	48	52	40	53	91	30	20	18	23	61	63	62
Bronx County	78	76	52	63	85	77	68						

11-168 Refer to the sample data given in Exercise 11-167. Test the claim that the three county populations are identical.

11-169 A college dean randomly selects several faculty who earn money for extra duties and lists those earnings (in dollars) along with their corresponding ages (in years). Is there a correlation between age and extra service income? Use the rank correlation coefficient at the 0.05 level of significance.

Age	27	42	29	30	32	55	57	33
Income	3600	120	2800	1200	1200	50	140	900

11-170 Final exam grades are listed below according to the order in which they were completed. At the 0.05 level of significance, test for randomness above and below the mean.

45, 50, 92, 87, 79, 89, 93, 75, 76, 74, 76, 73, 65, 68, 69, 70, 78, 60, 60, 60, 55, 100

11-171   A police academy gives an entrance exam; sample results for applicants from two different counties follow. Use the Wilcoxon rank-sum test to test the claim that there is no difference between the scores from the two counties. Assume a significance level of 0.05.

Orange County	63	39	26	14	75	60	62	79	86	70	66	
Westchester County	54	35	39	27	40	78	17	7	5	10	48	50	49

11-172   An annual art award was won by a woman in 30 out of 40 presentations. At the 0.05 significance level, use the sign test to test the claim that men and women are equal in their abilities to win this award.

11-173   Three judges grade entries in a science fair. The grades for eight randomly selected contestants are given below. Use the data for judge A and judge B to test the claim that there is no difference between their scoring. Use the Wilcoxon signed-ranks test at the 0.05 level of significance.

Judge A	6.3	7.2	6.6	8.5	9.7	7.0	7.3	8.8
Judge B	7.1	6.5	8.2	8.6	9.0	6.1	6.3	8.8
Judge C	7.0	6.4	8.3	8.5	8.9	6.0	6.2	8.7

11-174   Use the data from all three judges listed in Exercise 11-173 and test the claim that they score the same way. Use a 0.05 level of significance.

11-175   Refer to the data for judges A and B as given in Exercise 11-173. Use the rank correlation coefficient to determine if there is a significant positive correlation, significant negative correlation, or no significant correlation between those two judges. Use a significance level of 0.05.

11-176   A telephone solicitor writes a report listing N for unanswered calls and Y for answered calls. At the 0.05 level of significance, test the claim that the sample listed below is random.

$N, N, N, N, N, N, N, N, Y, Y, N, N, N, N, N, Y, Y, Y, Y, Y, Y, N, N, N, N, N,$
N, N, N, N, N, N, N

11-177   Test the claim that the median age of a teacher in Montana is 38 years. The ages of a sample group of teachers from Montana are 35, 31, 27, 42, 38, 39, 56, 64, 61, 33, 35, 24, 25, 28, 37, 36, 40, 43, 54, and 50. Use a 0.05 significance level.

11-178 A dose of the drug captopril, designed to lower systolic blood pressure, is administered to 10 randomly selected volunteers. The results follow. Use the Wilcoxon signed-ranks test at the 0.05 significance level to test the claim that systolic blood pressure is not affected by the pill.

Before pill	120	136	160	98	115	110	180	190	138	128
After pill	118	122	143	105	98	98	180	175	105	112

11-179 Three machines are programmed to produce keychains, and the daily outputs are listed below for randomly selected days. Test the claim that the machines produce the same amount. Use a 0.05 level of significance.

Machine A:	660	690	690	672	683
Machine B:	590	588	560	570	592
Machine C:	520	572	578	553	564

11-180 A study was conducted of drivers on similar sections of an interstate highway in Iowa and Nevada, and randomly selected speeds are listed below. At the 0.05 level of significance, use the Wilcoxon rank-sum test to test the claim that there is no difference between the speeds in the two states.

Iowa	56	58	62	54	47	68	70	55	56	52	53	51
Nevada	63	60	59	57	67	64	65	72	50	55	55	49
	48	46										

11-181 A medical researcher selects blood samples and tests for the presence (*P*) or absence (*A*) of a certain virus. Test the claim that the following sequence of results is random.

$$A, A, A, A, A, A, P, P, A, A, A, A, A, A, A, A, A, P, P, P, A, A, A, A, A$$

11-182 A dentist records the ages (in years) and total annual bills (in dollars) for several randomly selected patients, and the results are listed below. Is there a correlation between age and the amount billed?

Age	7	9	12	36	42	15	16	17	27	53	55	8
Dental bill	60	72	95	312	108	220	90	40	65	120	30	430

11-183   Three groups of adult men were selected for an experiment designed to measure their blood alcohol levels after consuming five drinks. Members of group A were tested after one hour, members of group B were tested after two hours, and members of group C were tested after four hours, and the results are given below. At the 0.05 level of significance, use the Kruskal-Wallis test to test the claim that the three groups come from identical populations.

A	B	C
0.11	0.08	0.04
0.10	0.09	0.04
0.09	0.07	0.05
0.09	0.07	0.05
0.10	0.06	0.06
		0.04
		0.05

11-184   Samples of equal amounts of two brands of a food substance were randomly selected and the amounts of carbohydrates were measured (in grams). The results are listed below. Use the Wilcoxon rank-sum test to test the claim that both brands are the same when compared on the basis of carbohydrate content. Use a 0.05 level of significance.

Brand x	20.3	21.2	19.3	19.2	19.1	19.0	22.6	23.6	22.9	20.7	20.7
Brand y	18.9	18.8	19.1	21.0	20.0	18.6	20.4	23.3	20.1	17.9	17.7

11-185   A pollster is hired to collect data from 30 randomly selected adults. As the data are turned in, the sex of the interviewed subjects is noted and the sequence below is obtained. At the 0.05 level of significance, test the claim that the sequence is random.

$M, M, M, M, M, M, M, M, F, M, M, M, M, F, F, F, F, F, F, F, F, F, M, M,$
M, M, M, M, M, M

11-186   Randomly selected cars are tested for fuel consumption and then retested after a tune-up. The measures of fuel consumption follow. At the 0.05 significance level, use rank correlation to test for a relationship between the before and after values.

Before tune-up	16	23	12	13	7	31	27	18	19	19	19	11	9	15
After tune-up	18	23	16	17	8	29	31	21	19	20	24	13	14	18

11-187 Refer to the data given in Exercise 11-186. At the 0.05 level of significance, use the sign test to test the claim that the tune-up has no effect on fuel consumption.

11-188 Refer to the data given in Exercise 11-186. At the 0.05 level of significance, use the Wilcoxon signed-ranks test to test the claim that the tune-up has no effect on fuel consumption.

COMPUTER PROJECT

Most computer systems have a random number generator. Use such a system to generate 100 numbers and then use the runs test to test for randomness above and below the mean.

CASE STUDY ACTIVITY

Obtain a collection of sample data listed in the order of selection, and test for randomness. You might refer to recent summaries of state lottery numbers, or recent stock market changes, or priority numbers used in the draft (see *U. S. News and World Report*, p. 34 of the December 15, 1969, issue, or p. 27 of the July 13, 1970, issue).

DATA PROJECT

Refer to the data in Appendix B.

1. Use the sign test to test the claim that the median home selling price is $150,000.

2. Use the Wilcoxon rank-sum test to test the claim that homes with fewer than two baths have the same selling price as those with two or more baths.

3. Use the Wilcoxon rank-sum test to test the claim that men and women have equal pulse rates.

4. Use the runs test to test for randomness of pulse rates above and below the median.

Chapter Twelve

CHAPTER CONTENTS

12-1   **Designing the Experiment**
The complete plan for data collection must be defined before the collection process is started.

12-2   **Sampling and Collecting Data**
We present different methods of sampling: random sampling, stratified sampling, systematic sampling, cluster sampling, and convenience sampling.

12-3   **Analyzing Data and Drawing Conclusions**
The conclusions must be developed according to the previously developed design of the experiment.

12-4   **Writing the Report**
We present a suggested outline for a written report.

12-5   **A Do-It-Yourself Project**
We present suggestions for a student project.

12

Design, Sampling, and Report Writing

IDENTIFYING OBJECTIVES

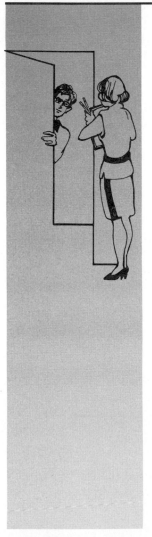

In setting up a statistical experiment, we must begin by determining exactly what question we want answered. Beginning researchers are often overcome by enthusiasm as they set out to collect facts without considering *precisely* what they are investigating and which facts are truly relevant. The original statement of the problem is often too vague or too broad. When this happens, too many different directions are sometimes pursued.

As an example, suppose some group wants to survey public opinion about abortion. What is the population we will sample? All Americans? Clearly, the opinion of a two-year-old child on this topic is not very relevant. Do we restrict our population to adults? Over 16, over 18, or over 21? Men and women? After determining the exact population of interest, we need to determine the number of people that will be surveyed. After considering factors such as the degree of confidence and the amount of error we can allow, we can use methods from Chapter 7 to determine the sample size. How do we select the actual subjects? What options do we have? Do we mail a questionnaire, contact them by telephone, or interview them in person? How do we word the questions? In an article on polls, *Money* magazine presented these two questions:

1. Should there be a constitutional amendment prohibiting abortions?

2. Should there be an amendment protecting the life of the unborn child?

It was noted that a majority of those surveyed answered no to the first version, while 20% switched their answers when presented with the second version. Obviously, the wording of the question can dramatically affect the responses. (Can you explain why such different results occurred?) After selecting and analyzing the sample data, how might we organize a report of our findings?

Many of the preceding questions will be discussed in this chapter.

12-1 DESIGNING THE EXPERIMENT

In order to obtain meaningful data in an efficient way, we must develop a complete plan for collecting data *before* the collection is actually begun. Researchers are often frustrated and discouraged if they learn that the method of collection or the data themselves cannot be used to answer their questions. Will the experiment be conducted on the entire population (in which case we will use descriptive statistics) or will a sample be drawn from the population (in which case we will use inferential statistics)? The population size usually requires us to make inferences on the basis of sample data. We need to determine the size of the sample and the method of sampling at the very beginning.

Some aspects of sample size determination were discussed in Chapter 7. In Chapter 7 we saw that samples with sizes between 1000 and 2000 people are often quite good for accurately gauging public opinion, even when the population consists of millions of people. It is this realization that has encouraged the growth and use of survey results. Surveys allow us to measure opinions and other characteristics in a way that is relatively quick and inexpensive.

Later in this chapter we discuss some types of sampling. In addition to determining sample size and sampling plan, we need to obtain a source from which to draw our samples. This source may be voter registration files, a class roster, a car registration list, computer-generated telephone numbers, and so on.

Finally, we should describe how we intend to analyze the data after sampling is completed. We should note which statistics are to be computed and which formulas will be used in making estimates and conducting tests of hypotheses.

12-2 SAMPLING AND COLLECTING DATA

Sampling and data collection usually require the most time, effort, and money. Careful planning will minimize the expenditure of those precious resources. Take care that the sampling is done according to plan and the data are recorded in a complete and accurate manner.

If you are obtaining measurements of some characteristic from people, realize that you will get better results if you can do the measuring instead of asking the subject for the value. Analysis of the weights included in Appendix B shows a disproportionate number of weights that end in 0 or 5. This suggests that the subjects often rounded their results, and there is a tendency to round down (often *way* down). As a result, the weights are distorted and the sample is flawed.

When conducting a survey, consider the medium to be used. **Mail** surveys, **telephone** surveys, and **personal interviews** are most common, although other media are used. Mail surveys tend to get lower response rates. Personal interviews are obviously more time consuming and expensive, but they may be necessary if detailed and complex data are required. Telephone interviews are relatively efficient and inexpensive.

Be especially careful with the method of sampling. Some different methods of sampling are described below.

Random Sampling

In **random sampling**, each member of the population has an equal chance of being selected. Random sampling is also called representative or proportionate sampling since all groups of the population should be proportionately represented in the sample.

Random sampling is not the same as haphazard or unsystematic sampling. Much effort and planning must be invested in order to avoid any bias. For example, if a list of all elements from the population is available, the names of those elements can be placed in capsules and put in a bowl; those capsules can then be mixed and samples selected.

Another approach involves numbering the list and using a computer or table of random numbers to determine which specific elements are to be selected.

A major problem with these approaches is the difficulty of finding a complete list of *all* elements in the population. Even when a complete list seems to be available, we must be sure that it is not biased and the resulting sample is not biased. For example, if we decide that voters

A Professional Speaks About Sampling Error

Daniel Yankelovich, in an essay for *Time*, commented on the sampling error often reported along with poll results. He stated that sampling error refers only to the inaccuracy created by using random sample data to make an inference about a population; the sampling error does not address issues of poorly stated, biased, or emotional questions. He says that "most important of all, warning labels about sampling error say nothing about whether or not the public is conflict-ridden or has given a subject much thought. This is the most serious source of opinion poll misinterpretation."

are to be sampled by selecting numbers from a telephone directory, we automatically eliminate all unlisted numbers and our sample may be biased. People who choose to have unlisted phone numbers may constitute a special interest group when certain issues are raised. This can dramatically affect some results because there are regions with high proportions of unlisted numbers. In Los Angeles, 42.5% of the telephone numbers are unlisted. The percentages for Chicago and San Francisco are 40.9% and 39.5%, respectively. (The data are from Survey Sampling, Inc.) Pollsters commonly circumvent this problem by using computers to generate phone numbers so that all of them become accessible. However, we still run the risk of having a biased sample if we simply ignore those who are unavailable or refuse to comment. Humphrey Taylor, president of the Harris polling company, states that the refusal rate for telephone interviews generally is at least 20%. Ignore those people who initially refuse and you run a real risk of having a biased sample.

A recent example of a nonrandom sample is the book *Women and Love: A Cultural Revolution in Progress* by Shere Hite. She based conclusions on 4500 responses from 100,000 questionnaires distributed to women. Such results are not random, and they may well reflect the opinions of respondents who have strong feelings about the survey topics. This particular book was widely criticized for its obvious bias and lack of sound statistical methodology.

Stratified Sampling

After classifying the population into at least two different **strata** (or classes) that share the same characteristics, we draw a sample from each stratum. In surveying views on an Equal Rights Amendment to the Constitution, we might use sex as a basis for creating two strata. After obtaining a list of men and a list of women, we use some suitable method (such as random sampling) to select a certain number of people from each list. If it should happen that some strata are not represented in the proper proportion, then the results can be adjusted or weighted accordingly. Stratified sampling is often the most efficient of the various types.

Systematic Sampling

In **systematic sampling** we select some starting point and then select every kth element. For example, we use a telephone directory of 10,000 names as our population, and we must choose 200 of those names. We can randomly select one of the first 50 names and then choose every 50th name after that. This method is simple and is used frequently.

Polls

Each year, about one-third of all Americans are polled by organizations that *appear* to be research oriented. About 45% of these calls are really attempts to sell products or services, but the remaining 55% are from the 150 legitimate research firms. These firms charge clients about $2 billion for their services.

Many of the polls involve telephone surveys. Ideally, telephone numbers are randomly selected so that the 16% of American homes with unlisted numbers will be included. Here is a typical case in which the Gallup organization did research for Zenith: The United States was first partitioned into four categories, according to community populations. These four strata were further broken down by location. Using this process, sample locations were drawn, and in each location about eight telephone starts (the first five digits) were drawn from a working bank of telephone numbers. These starts determined the list of people to be called. For example, the start 987-65?? would suggest 100 telephone numbers, beginning with 987-6500 and ending with 987-6599. Each number was called until an interview was completed, except that no number was called more than four times.

Cluster Sampling

In **cluster sampling** we first divide the population area into sections and then randomly select a few of those sections. We can then choose all the members from those sections. For example, in conducting a preelection poll, we could randomly select 30 election precincts and then survey 50 people from each of those chosen precincts. This would be much more efficient than selecting one person from each of the 1500 precincts. The results can be adjusted or weighted to correct for any disproportionate representations of groups. Cluster sampling is used extensively by the government and by private research organizations.

Convenience Sampling

In **convenience sampling** we simply use results that are readily available (or convenient). In some cases, results may be quite good, while they may be seriously biased in other cases. In investigating the proportion of left-handed people, it would be convenient for a teacher to survey students. Even though the sample is not random, it will tend to be unbiased because left-handedness is not the type of characteristic that would be related to presence in class. But if the same teacher questions the same students about their opinions on federal aid to education, the results will be clearly biased and not representative of the general population. Be very wary of convenience sampling.

Realize that the preceding descriptions of different sampling methods are intended to be brief and general. If you want to thoroughly understand those different methods so that you can successfully use them, you must undertake much more extensive study than is practical in a single introductory course of this type.

Importance of Sampling

Even experienced and reputable research organizations sometimes get erroneous results due to biased sampling or poor methodology. In 1948 the Gallup poll was wrong when it predicted Truman would lose to Dewey. That mistake led to a revision of Gallup's polling methods. A quota system had been used to obtain the opinions of a proportionate number of men, women, rich, poor, Catholics, Protestants, Jews, and so on. After the 1948 error, Gallup abandoned the quota system and instituted random sampling based on clusters of interviews in several hundred areas throughout the nation. This example illustrates the importance of the sample.

12-3 ANALYZING DATA AND DRAWING CONCLUSIONS

The data must be analyzed and the conclusions drawn according to the methods specified when the experiment was designed. After the data have been analyzed and the conclusions have been formed, it is important to note the level of confidence used in any hypothesis test or estimation.

In analyzing results, it is often helpful to discuss two sources of errors, namely, **sampling errors** and **nonsampling errors**. Sampling errors result from the actual sampling process. They include such factors as the smallness of the sample size. Nonsampling errors arise from other external factors not related to sampling, such as a measuring instrument that is defective. Sampling errors should be described in a clear and understandable way. For example, we might state that for a reported percentage, "there is a 95% chance that the error is no more than three percentage points."

Nonsampling errors can't be described as objectively as sampling errors, but we should note any factors that might significantly affect the results. We might comment on a large number of response refusals or missing values. We might comment on errors in coding or recoding data. We might comment on a discovered bias in the sample.

Invisible Ink Deceives Subscribers

Newspapers and magazines routinely survey subscribers so that they can better serve advertisers and readers. The *National Observer* once hired a firm to conduct a confidential mail survey. The *National Observer* editor, Henry Gemmill, wrote in a letter accompanying the survey: "Each individual reply will be kept confidential, of course, but when your reply is combined with others from all over this land, we'll have a composite picture of our subscribers." The questionnaire did not ask for name or address, but one very clever subscriber used an ultraviolet light to detect a code written with invisible ink. This code could be used to identify the person who answered the "confidential survey." Gemmill was unaware that this procedure was used, and he publicly apologized to all subscribers. In this situation, confidentiality was observed as promised, but anonymity was not directly promised and it was certainly not observed.

12-4 WRITING THE REPORT

In writing the final report, the author should consider the people who will read the report. Statisticians will expect specific and detailed results accompanied by fairly complete descriptions of methodology. However, a lay audience would not benefit from this type of report, so a different approach is necessary. A suggested outline for the written report follows.

1. Front matter
   a. Title page
   b. Table of contents
   c. List of tables and illustrations
   d. Preface
   e. Summary of results and conclusions

2. Body of report
   a. Statement of objectives
   b. Description of procedure and methods
   c. Analysis of data
   d. Conclusions drawn from data

3. Supplementary material
   a. Appendixes
   b. Bibliography
   c. Glossary of terms requiring definition
   d. Index

Ethics in Experiments

Sample data can often be obtained by simply observing or surveying members selected from the population. Many other situations require that we somehow manipulate circumstances to obtain sample data. In both cases, ethical questions may arise. Researchers in Tuskegee, Alabama, withheld the effective penicillin treatment to syphilis victims so that the disease could be studied. This continued for a period of 27 years!

12-5 A DO-IT-YOURSELF PROJECT

An ideal way to gain insight into statistical methods is to conduct a statistical experiment from beginning to end. It might be advantageous to relate the subject of such an experiment to another course or discipline. The project should involve the actual collection of raw data and should conclude with a typewritten report that generally follows the outline given in Section 12-4. A title page should list your name, the course, the instructor's name, and the title of the report. An example of a typical title is "Statistical Analysis of the Hours Worked by Full-Time Students." A suggested outline and format for the table of contents is listed below.

Table of Contents

Summary 2

Objectives 3

Procedure 3

Analysis of Data 4

Conclusion 4

Appendix of Raw Data 5–6

Be sure to identify the type of sampling (such as random or systematic). Identify the features of the sampling process that tend to make your sample representative of the population. Also be sure to identify the population from which the sample was drawn.

Caution: It may be necessary to gain approval and permission before collecting certain types of data. Reference to published statistical almanacs and abstracts can normally be made at your own discretion, but there may be regulations governing the process of polling people in person or by telephone. We now know that some past studies have adversely affected the people being studied. In one experiment, a psychologist had subjects play a game in which it was impossible to win without cheating. Some of the subjects underwent some behavior modification when they decided to cheat and, in the process, altered their attitudes about cheating in general. As another example, if we were to give people an in-depth survey on suicide, we might run the risk of pushing a potential suicide victim over the edge. Any survey should be seriously and carefully constructed with consideration for the sensitivities and emotions of those surveyed. If our data source is people, we should take every necessary precaution and obtain any necessary approval.

Code of Ethics for Survey Research

The American Association for Public Opinion Research developed a voluntary code of ethics to be used in news reports of survey results. This code requires that the following items be included in reports.

1. Identification of the person, group, or organization that sponsored the survey
2. Date the survey was conducted
3. Size of the sample
4. Nature of the population sampled
5. Type of survey used
6. Exact wording of survey questions

Surveys funded by the U.S. government are subject to a pre-screening that assesses the risk to those surveyed, the scientific merit of the survey, and the guarantee of the subject's consent to participate.

In the future, you may be required to conduct statistical experiments as part of job or course requirements. It is natural and common to have feelings of apprehension (or even outright fear or dread) because you might sense that the subject matter of this course was not mastered to the extent necessary. To keep things in a proper perspective, you should recognize that nobody expects an introductory statistics course to make you an expert statistician. It is also very important to recognize that you can get help from those who have much more extensive training and experience with statistical methods. Even when not mastered fully, this course will help you open a dialogue with a professional statistician who can be of help. At the very least, this course gives you a literacy so necessary in this era of data proliferation.

EXERCISES A

12-1 Describe each of the following types of sampling.
   a. Random sampling
   b. Stratified sampling
   c. Systematic sampling
   d. Cluster sampling
   e. Convenience sampling

12-2 In addition to telephone interviews, identify two more common media used for surveys.

12-3   Assume that you are employed by a car manufacturer to collect data on the waist sizes of drivers. Why is it better to obtain direct measurements than to ask people the sizes of their waists?

12-4   Distinguish between sampling error and nonsampling error.

12-5   Identify the type of sampling used in each of the following:
a. A teacher selects every fifth student in the class for a test.
b. A teacher writes the name of each student on a card, shuffles the cards, and then draws five names.
c. A teacher selects five students from each of 12 classes.
d. A teacher selects five men and five women from each of four classes.
e. A teacher surveys all students in order to study public opinion.

12-6   Public opinion is sometimes measured by asking television viewers to call a "900" number. The cost of such a call is usually around 50¢. Identify two different factors that cause the resulting sample to be biased.

12-7   You plan to estimate the mean weight of all passenger cars used in the United States. Is there universal agreement about what a "passenger car" is? Are there any factors that might lead to regional differences among the weights of passenger cars? How can you obtain a sample?

12-8   Two categories of survey questions are *open* and *closed*. An open question allows a free response, while a closed question allows only fixed responses. Here are examples based on Gallup surveys.

Open question: What do you think can be done to reduce crime?
Closed question: Which of the following approaches would be most effective in reducing crime?

- Hire more police officers.
- Get parents to discipline children more.
- Correct social and economic conditions in slums.
- Improve rehabilitation efforts in jails.
- Give convicted criminals tougher sentences.
- Reform courts.

What are the advantages and disadvantages of open questions? What are the advantages and disadvantages of closed questions? Which type is easier to analyze with formal statistical procedures?

EXERCISES B

12-9   Find an article that deals with some hypothesis test in a professional journal. Use the given information to write a report that follows, as closely as possible, the outline for a written report. Include the name of the journal, the author, and the title of the article.

12-10   Conduct a complete statistical experiment following the outline given in Section 12-5. Collect sample data consisting of single numbers. Include a frequency table, histogram, point estimate of the population mean, point estimate of the population standard deviation, and interval estimate of the population mean; then state and test some appropriate null hypothesis.

12-11   Conduct a complete statistical experiment following the outline given in Section 12-5. Collect sample data consisting of pairs of numbers. Include a scatter diagram, the correlation coefficient, the regression equation, and the coefficient of determination. Form a conclusion about the relationship between the two variables that produced the pairs of data. (See Chapter 9.)

12-12   Conduct a complete statistical experiment following the outline given in Section 12-5. Collect sample data that can be summarized as a contingency table. Form a conclusion about the independence of the two variables used in classifying the table. (See Section 10-3.)

12-13   Conduct a complete statistical experiment following the outline given in Section 12-5. Collect sample data that can be used to establish some population proportion or percentage p. Include the point estimate of p and an interval estimate of p; then state and test some appropriate hypothesis.

12-14   Conduct a complete statistical experiment following the outline given in Section 12-5. Select a topic of your own, or use one of the following general topics for suggestions.
   a. The use of videocassette recorders
   b. Television viewing habits
   c. Voting habits
   d. Spending habits
   e. Career goals of college students
   f.   Health status of college students
   g. Smoking and/or drinking habits of college students
   h. Analysis of the college registration process

VOCABULARY LIST

Define and give an example of each term.

random sampling	cluster sampling	sampling errors
stratified sampling	convenience	nonsampling errors
systematic sampling	sampling	

Appendix A: Tables

TABLE A-1	(0+ represents a positive probability less than 0.0005)

Binomial Probabilities

n	x	.01	.05	.10	.20	.30	.40	p .50	.60	.70	.80	.90	.95	.99	x
2	0	980	902	810	640	490	360	250	160	090	040	010	002	0+	0
	1	020	095	180	320	420	480	500	480	420	320	180	095	020	1
	2	0+	002	010	040	090	160	250	360	490	640	810	902	980	2
3	0	970	857	729	512	343	216	125	064	027	008	001	0+	0+	0
	1	029	135	243	384	441	432	375	288	189	096	027	007	0+	1
	2	0+	007	027	096	189	288	375	432	441	384	243	135	029	2
	3	0+	0+	001	008	027	064	125	216	343	512	729	857	970	3
4	0	961	815	656	410	240	130	062	026	008	002	0+	0+	0+	0
	1	039	171	292	410	412	346	250	154	076	026	004	0+	0+	1
	2	001	014	049	154	265	346	375	346	265	154	049	014	001	2
	3	0+	0+	004	026	076	154	250	346	412	410	292	171	039	3
	4	0+	0+	0+	002	008	026	062	130	240	410	656	815	961	4
5	0	951	774	590	328	168	078	031	010	002	0+	0+	0+	0+	0
	1	048	204	328	410	360	259	156	077	028	006	0+	0+	0+	1
	2	001	021	073	205	309	346	312	230	132	051	008	001	0+	2
	3	0+	001	008	051	132	230	312	346	309	205	073	021	001	3
	4	0+	0+	0+	006	028	077	156	259	360	410	328	204	048	4
	5	0+	0+	0+	0+	002	010	031	078	168	328	590	774	951	5
6	0	941	735	531	262	118	047	016	004	001	0+	0+	0+	0+	0
	1	057	232	354	393	303	187	094	037	010	002	0+	0+	0+	1
	2	001	031	098	246	324	311	234	138	060	015	001	0+	0+	2
	3	0+	002	015	082	185	276	312	276	185	082	015	002	0+	3
	4	0+	0+	001	015	060	138	234	311	324	246	098	031	001	4
	5	0+	0+	0+	002	010	037	094	187	303	393	354	232	057	5
	6	0+	0+	0+	0+	001	004	016	047	118	262	531	735	941	6
7	0	932	698	478	210	082	028	008	002	0+	0+	0+	0+	0+	0
	1	066	257	372	367	247	131	055	017	004	0+	0+	0+	0+	1
	2	002	041	124	275	318	261	164	077	025	004	0+	0+	0+	2
	3	0+	004	023	115	227	290	273	194	097	029	003	0+	0+	3
	4	0+	0+	003	029	097	194	273	290	227	115	023	004	0+	4
	5	0+	0+	0+	004	025	077	164	261	318	275	124	041	002	5
	6	0+	0+	0+	0+	004	017	055	131	247	367	372	257	066	6
	7	0+	0+	0+	0+	0+	002	008	028	082	210	478	698	932	7
8	0	923	663	430	168	058	017	004	001	0+	0+	0+	0+	0+	0
	1	075	279	383	336	198	090	031	008	001	0+	0+	0+	0+	1
	2	003	051	149	294	296	209	109	041	010	001	0+	0+	0+	2
	3	0+	005	033	147	254	279	219	124	047	009	0+	0+	0+	3
	4	0+	0+	005	046	136	232	273	232	136	046	005	0+	0+	4
	5	0+	0+	0+	009	047	124	219	279	254	147	033	005	0+	5
	6	0+	0+	0+	001	010	041	109	209	296	294	149	051	003	6
	7	0+	0+	0+	0+	001	008	031	090	198	336	383	279	075	7
	8	0+	0+	0+	0+	0+	001	004	017	058	168	430	663	923	8

(continued)

Binomial Probabilities

n	x	.01	.05	.10	.20	.30	.40	p .50	.60	.70	.80	.90	.95	.99	x
9	0	914	630	387	134	040	010	002	0+	0+	0+	0+	0+	0+	0
	1	083	299	387	302	156	060	018	004	0+	0+	0+	0+	0+	1
	2	003	063	172	302	267	161	070	021	004	0+	0+	0+	0+	2
	3	0+	008	045	176	267	251	164	074	021	003	0+	0+	0+	3
	4	0+	001	007	066	172	251	246	167	074	017	001	0+	0+	4
	5	0+	0+	001	017	074	167	246	251	172	066	007	001	0+	5
	6	0+	0+	0+	003	021	074	164	251	267	176	045	008	0+	6
	7	0+	0+	0+	0+	004	021	070	161	267	302	172	063	003	7
	8	0+	0+	0+	0+	0+	004	018	060	156	302	387	299	083	8
	9	0+	0+	0+	0+	0+	0+	002	010	040	134	387	630	914	9
10	0	904	599	349	107	028	006	001	0+	0+	0+	0+	0+	0+	0
	1	091	315	387	268	121	040	010	002	0+	0+	0+	0+	0+	1
	2	004	075	194	302	233	121	044	011	001	0+	0+	0+	0+	2
	3	0+	010	057	201	267	215	117	042	009	001	0+	0+	0+	3
	4	0+	001	011	088	200	251	205	111	037	006	0+	0+	0+	4
	5	0+	0+	001	026	103	201	246	201	103	026	001	0+	0+	5
	6	0+	0+	0+	006	037	111	205	251	200	088	011	001	0+	6
	7	0+	0+	0+	001	009	042	117	215	267	201	057	010	0+	7
	8	0+	0+	0+	0+	001	011	044	121	233	302	194	075	004	8
	9	0+	0+	0+	0+	0+	002	010	040	121	268	387	315	091	9
	10	0+	0+	0+	0+	0+	0+	001	006	028	107	349	599	904	10
11	0	895	569	314	086	020	004	0+	0+	0+	0+	0+	0+	0+	0
	1	099	329	384	236	093	027	005	001	0+	0+	0+	0+	0+	1
	2	005	087	213	295	200	089	027	005	001	0+	0+	0+	0+	2
	3	0+	014	071	221	257	177	081	023	004	0+	0+	0+	0+	3
	4	0+	001	016	111	220	236	161	070	017	002	0+	0+	0+	4
	5	0+	0+	002	039	132	221	226	147	057	010	0+	0+	0+	5
	6	0+	0+	0+	010	057	147	226	221	132	039	002	0+	0+	6
	7	0+	0+	0+	002	017	070	161	236	220	111	016	001	0+	7
	8	0+	0+	0+	0+	004	023	081	177	257	221	071	014	0+	8
	9	0+	0+	0+	0+	001	005	027	089	200	295	213	087	005	9
	10	0+	0+	0+	0+	0+	001	005	027	093	236	384	329	099	10
	11	0+	0+	0+	0+	0+	0+	0+	004	020	086	314	569	895	11
12	0	886	540	282	069	014	002	0+	0+	0+	0+	0+	0+	0+	0
	1	107	341	377	206	071	017	003	0+	0+	0+	0+	0+	0+	1
	2	006	099	230	283	168	064	016	002	0+	0+	0+	0+	0+	2
	3	0+	017	085	236	240	142	054	012	001	0+	0+	0+	0+	3
	4	0+	002	021	133	231	213	121	042	008	001	0+	0+	0+	4
	5	0+	0+	004	053	158	227	193	101	029	003	0+	0+	0+	5
	6	0+	0+	0+	016	079	177	226	177	079	016	0+	0+	0+	6
	7	0+	0+	0+	003	029	101	193	227	158	053	004	0+	0+	7
	8	0+	0+	0+	001	008	042	121	213	231	133	021	002	0+	8
	9	0+	0+	0+	0+	001	012	054	142	240	236	085	017	0+	9
	10	0+	0+	0+	0+	0+	002	016	064	168	283	230	099	006	10
	11	0+	0+	0+	0+	0+	0+	003	017	071	206	377	341	107	11
	12	0+	0+	0+	0+	0+	0+	0+	002	014	069	282	540	886	12

NOTE: 0+ represents a positive probability less than 0.0005.

(continued)

TABLE A-1 (continued)

Binomial Probabilities

n	x	.01	.05	.10	.20	.30	.40	p .50	.60	.70	.80	.90	.95	.99	x
13	0	878	513	254	055	010	001	0+	0+	0+	0+	0+	0+	0+	0
	1	115	351	367	179	054	011	002	0+	0+	0+	0+	0+	0+	1
	2	007	111	245	268	139	045	010	001	0+	0+	0+	0+	0+	2
	3	0+	021	100	246	218	111	035	006	001	0+	0+	0+	0+	3
	4	0+	003	028	154	234	184	087	024	003	0+	0+	0+	0+	4
	5	0+	0+	006	069	180	221	157	066	014	001	0+	0+	0+	5
	6	0+	0+	001	023	103	197	209	131	044	006	0+	0+	0+	6
	7	0+	0+	0+	006	044	131	209	197	103	023	001	0+	0+	7
	8	0+	0+	0+	001	014	066	157	221	180	069	006	0+	0+	8
	9	0+	0+	0+	0+	003	024	087	184	234	154	028	003	0+	9
	10	0+	0+	0+	0+	001	006	035	111	218	246	100	021	0+	10
	11	0+	0+	0+	0+	0+	001	010	045	139	268	245	111	007	11
	12	0+	0+	0+	0+	0+	0+	002	011	054	179	367	351	115	12
	13	0+	0+	0+	0+	0+	0+	0+	001	010	055	254	513	878	13
14	0	869	488	229	044	007	001	0+	0+	0+	0+	0+	0+	0+	0
	1	123	359	356	154	041	007	001	0+	0+	0+	0+	0+	0+	1
	2	008	123	257	250	113	032	006	001	0+	0+	0+	0+	0+	2
	3	0+	026	114	250	194	085	022	003	0+	0+	0+	0+	0+	3
	4	0+	004	035	172	229	155	061	014	001	0+	0+	0+	0+	4
	5	0+	0+	008	086	196	207	122	041	007	0+	0+	0+	0+	5
	6	0+	0+	001	032	126	207	183	092	023	002	0+	0+	0+	6
	7	0+	0+	0+	009	062	157	209	157	062	009	0+	0+	0+	7
	8	0+	0+	0+	002	023	092	183	207	126	032	001	0+	0+	8
	9	0+	0+	0+	0+	007	041	122	207	196	086	008	0+	0+	9
	10	0+	0+	0+	0+	001	014	061	155	229	172	035	004	0+	10
	11	0+	0+	0+	0+	0+	003	022	085	194	250	114	026	0+	11
	12	0+	0+	0+	0+	0+	001	006	032	113	250	257	123	008	12
	13	0+	0+	0+	0+	0+	0+	001	007	041	154	356	359	123	13
	14	0+	0+	0+	0+	0+	0+	0+	001	007	044	229	488	869	14
15	0	860	463	206	035	005	0+	0+	0+	0+	0+	0+	0+	0+	0
	1	130	366	343	132	031	005	0+	0+	0+	0+	0+	0+	0+	1
	2	009	135	267	231	092	022	003	0+	0+	0+	0+	0+	0+	2
	3	0+	031	129	250	170	063	014	002	0+	0+	0+	0+	0+	3
	4	0+	005	043	188	219	127	042	007	001	0+	0+	0+	0+	4
	5	0+	001	010	103	206	186	092	024	003	0+	0+	0+	0+	5
	6	0+	0+	002	043	147	207	153	061	012	001	0+	0+	0+	6
	7	0+	0+	0+	014	081	177	196	118	035	003	0+	0+	0+	7
	8	0+	0+	0+	003	035	118	196	177	081	014	0+	0+	0+	8
	9	0+	0+	0+	001	012	061	153	207	147	043	002	0+	0+	9
	10	0+	0+	0+	0+	003	024	092	186	206	103	010	001	0+	10
	11	0+	0+	0+	0+	001	007	042	127	219	188	043	005	0+	11
	12	0+	0+	0+	0+	0+	002	014	063	170	250	129	031	0+	12
	13	0+	0+	0+	0+	0+	0+	003	022	092	231	267	135	009	13
	14	0+	0+	0+	0+	0+	0+	0+	005	031	132	343	366	130	14
	15	0+	0+	0+	0+	0+	0+	0+	0+	005	035	206	463	860	15

NOTE: 0+ represents a positive probability less than 0.0005.

(continued)

Binomial Probabilities

n	x	.01	.05	.10	.20	.30	.40	p .50	.60	.70	.80	.90	.95	.99	x
16	0	851	440	185	028	003	0+	0+	0+	0+	0+	0+	0+	0+	0
	1	138	371	329	113	023	003	0+	0+	0+	0+	0+	0+	0+	1
	2	010	146	275	211	073	015	002	0+	0+	0+	0+	0+	0+	2
	3	0+	036	142	246	146	047	009	001	0+	0+	0+	0+	0+	3
	4	0+	006	051	200	204	101	028	004	0+	0+	0+	0+	0+	4
	5	0+	001	014	120	210	162	067	014	001	0+	0+	0+	0+	5
	6	0+	0+	003	055	165	198	122	039	006	0+	0+	0+	0+	6
	7	0+	0+	0+	020	101	189	175	084	019	001	0+	0+	0+	7
	8	0+	0+	0+	006	049	142	196	142	049	006	0+	0+	0+	8
	9	0+	0+	0+	001	019	084	175	189	101	020	0+	0+	0+	9
	10	0+	0+	0+	0+	006	039	122	198	165	055	003	0+	0+	10
	11	0+	0+	0+	0+	001	014	067	162	210	120	014	001	0+	11
	12	0+	0+	0+	0+	0+	004	028	101	204	200	051	006	0+	12
	13	0+	0+	0+	0+	0+	001	009	047	146	246	142	036	0+	13
	14	0+	0+	0+	0+	0+	0+	002	015	073	211	275	146	010	14
	15	0+	0+	0+	0+	0+	0+	0+	003	023	113	329	371	138	15
	16	0+	0+	0+	0+	0+	0+	0+	0+	003	028	185	440	851	16
17	0	843	418	167	023	002	0+	0+	0+	0+	0+	0+	0+	0+	0
	1	145	374	315	096	017	002	0+	0+	0+	0+	0+	0+	0+	1
	2	012	158	280	191	058	010	001	0+	0+	0+	0+	0+	0+	2
	3	001	041	156	239	125	034	005	0+	0+	0+	0+	0+	0+	3
	4	0+	008	060	209	187	080	018	002	0+	0+	0+	0+	0+	4
	5	0+	001	017	136	208	138	047	008	001	0+	0+	0+	0+	5
	6	0+	0+	004	068	178	184	094	024	003	0+	0+	0+	0+	6
	7	0+	0+	001	027	120	193	148	057	009	0+	0+	0+	0+	7
	8	0+	0+	0+	008	064	161	185	107	028	002	0+	0+	0+	8
	9	0+	0+	0+	002	028	107	185	161	064	008	0+	0+	0+	9
	10	0+	0+	0+	0+	009	057	148	193	120	027	001	0+	0+	10
	11	0+	0+	0+	0+	003	024	094	184	178	068	004	0+	0+	11
	12	0+	0+	0+	0+	001	008	047	138	208	136	017	001	0+	12
	13	0+	0+	0+	0+	0+	002	018	080	187	209	060	008	0+	13
	14	0+	0+	0+	0+	0+	0+	005	034	125	239	156	041	001	14
	15	0+	0+	0+	0+	0+	0+	001	010	058	191	280	158	012	15
	16	0+	0+	0+	0+	0+	0+	0+	002	017	096	315	374	145	16
	17	0+	0+	0+	0+	0+	0+	0+	0+	002	023	167	418	843	17
18	0	835	397	150	018	002	0+	0+	0+	0+	0+	0+	0+	0+	0
	1	152	376	300	081	013	001	0+	0+	0+	0+	0+	0+	0+	1
	2	013	168	284	172	046	007	001	0+	0+	0+	0+	0+	0+	2
	3	001	047	168	230	105	025	003	0+	0+	0+	0+	0+	0+	3
	4	0+	009	070	215	168	061	012	001	0+	0+	0+	0+	0+	4
	5	0+	001	022	151	202	115	033	004	0+	0+	0+	0+	0+	5
	6	0+	0+	005	082	187	166	071	015	001	0+	0+	0+	0+	6
	7	0+	0+	001	035	138	189	121	037	005	0+	0+	0+	0+	7
	8	0+	0+	0+	012	081	173	167	077	015	001	0+	0+	0+	8
	9	0+	0+	0+	003	039	128	185	128	039	003	0+	0+	0+	9

NOTE: 0+ represents a positive probability less than 0.0005.

(*continued*)

TABLE A-1 (*continued*)

Binomial Probabilities

n	x	.01	.05	.10	.20	.30	.40	.50	.60	.70	.80	.90	.95	.99	x
	10	0+	0+	0+	001	015	077	167	173	081	012	0+	0+	0+	10
	11	0+	0+	0+	0+	005	037	121	189	138	035	001	0+	0+	11
	12	0+	0+	0+	0+	001	015	071	166	187	082	005	0+	0+	12
	13	0+	0+	0+	0+	0+	004	033	115	202	151	022	001	0+	13
	14	0+	0+	0+	0+	0+	001	012	061	168	215	070	009	0+	14
	15	0+	0+	0+	0+	0+	0+	003	025	105	230	168	047	001	15
	16	0+	0+	0+	0+	0+	0+	001	007	046	172	284	168	013	16
	17	0+	0+	0+	0+	0+	0+	0+	001	013	081	300	376	152	17
	18	0+	0+	0+	0+	0+	0+	0+	0+	002	018	150	397	835	18
19	0	826	377	135	014	001	0+	0+	0+	0+	0+	0+	0+	0+	0
	1	159	377	285	068	009	001	0+	0+	0+	0+	0+	0+	0+	1
	2	014	179	285	154	036	005	0+	0+	0+	0+	0+	0+	0+	2
	3	001	053	180	218	087	017	002	0+	0+	0+	0+	0+	0+	3
	4	0+	011	080	218	149	047	007	001	0+	0+	0+	0+	0+	4
	5	0+	002	027	164	192	093	022	002	0+	0+	0+	0+	0+	5
	6	0+	0+	007	095	192	145	052	008	001	0+	0+	0+	0+	6
	7	0+	0+	001	044	153	180	096	024	002	0+	0+	0+	0+	7
	8	0+	0+	0+	017	098	180	144	053	008	0+	0+	0+	0+	8
	9	0+	0+	0+	005	051	146	176	098	022	001	0+	0+	0+	9
	10	0+	0+	0+	001	022	098	176	146	051	005	0+	0+	0+	10
	11	0+	0+	0+	0+	008	053	144	180	098	017	0+	0+	0+	11
	12	0+	0+	0+	0+	002	024	096	180	153	044	001	0+	0+	12
	13	0+	0+	0+	0+	001	008	052	145	192	095	007	0+	0+	13
	14	0+	0+	0+	0+	0+	002	022	093	192	164	027	002	0+	14
	15	0+	0+	0+	0+	0+	001	007	047	149	218	080	011	0+	15
	16	0+	0+	0+	0+	0+	0+	002	017	087	218	180	053	001	16
	17	0+	0+	0+	0+	0+	0+	0+	005	036	154	285	179	014	17
	18	0+	0+	0+	0+	0+	0+	0+	001	009	068	285	377	159	18
	19	0+	0+	0+	0+	0+	0+	0+	0+	001	014	135	377	826	19
20	0	818	358	122	012	001	0+	0+	0+	0+	0+	0+	0+	0+	0
	1	165	377	270	058	007	0+	0+	0+	0+	0+	0+	0+	0+	1
	2	016	189	285	137	028	003	0+	0+	0+	0+	0+	0+	0+	2
	3	001	060	190	205	072	012	001	0+	0+	0+	0+	0+	0+	3
	4	0+	013	090	218	130	035	005	0+	0+	0+	0+	0+	0+	4
	5	0+	002	032	175	179	075	015	001	0+	0+	0+	0+	0+	5
	6	0+	0+	009	109	192	124	037	005	0+	0+	0+	0+	0+	6
	7	0+	0+	002	055	164	166	074	015	001	0+	0+	0+	0+	7
	8	0+	0+	0+	022	114	180	120	035	004	0+	0+	0+	0+	8
	9	0+	0+	0+	007	065	160	160	071	012	0+	0+	0+	0+	9
	10	0+	0+	0+	002	031	117	176	117	031	002	0+	0+	0+	10
	11	0+	0+	0+	0+	012	071	160	160	065	007	0+	0+	0+	11
	12	0+	0+	0+	0+	004	035	120	180	114	022	0+	0+	0+	12
	13	0+	0+	0+	0+	001	015	074	166	164	055	002	0+	0+	13
	14	0+	0+	0+	0+	0+	005	037	124	192	109	009	0+	0+	14

NOTE: 0+ represents a positive probability less than 0.0005.

(*continued*)

Binomial Probabilities

n	x	.01	.05	.10	.20	.30	.40	p .50	.60	.70	.80	.90	.95	.99	x
	15	0+	0+	0+	0+	0+	001	015	075	179	175	032	002	0+	15
	16	0+	0+	0+	0+	0+	0+	005	035	130	218	090	013	0+	16
	17	0+	0+	0+	0+	0+	0+	001	012	072	205	190	060	001	17
	18	0+	0+	0+	0+	0+	0+	0+	003	028	137	285	189	016	18
	19	0+	0+	0+	0+	0+	0+	0+	0+	007	058	270	377	165	19
	20	0+	0+	0+	0+	0+	0+	0+	0+	001	012	122	358	818	20
21	0	810	341	109	009	001	0+	0+	0+	0+	0+	0+	0+	0+	0
	1	172	376	255	048	005	0+	0+	0+	0+	0+	0+	0+	0+	1
	2	017	198	284	121	022	002	0+	0+	0+	0+	0+	0+	0+	2
	3	001	066	200	192	058	009	001	0+	0+	0+	0+	0+	0+	3
	4	0+	016	100	216	113	026	003	0+	0+	0+	0+	0+	0+	4
	5	0+	003	038	183	164	059	010	001	0+	0+	0+	0+	0+	5
	6	0+	0+	011	122	188	105	026	003	0+	0+	0+	0+	0+	6
	7	0+	0+	003	065	172	149	055	009	0+	0+	0+	0+	0+	7
	8	0+	0+	001	029	129	174	097	023	002	0+	0+	0+	0+	8
	9	0+	0+	0+	010	080	168	140	050	006	0+	0+	0+	0+	9
	10	0+	0+	0+	003	041	134	168	089	018	001	0+	0+	0+	10
	11	0+	0+	0+	001	018	089	168	134	041	003	0+	0+	0+	11
	12	0+	0+	0+	0+	006	050	140	168	080	010	0+	0+	0+	12
	13	0+	0+	0+	0+	002	023	097	174	129	029	001	0+	0+	13
	14	0+	0+	0+	0+	0+	009	055	149	172	065	003	0+	0+	14
	15	0+	0+	0+	0+	0+	003	026	105	188	122	011	0+	0+	15
	16	0+	0+	0+	0+	0+	001	010	059	164	183	038	003	0+	16
	17	0+	0+	0+	0+	0+	0+	003	026	113	216	100	016	0+	17
	18	0+	0+	0+	0+	0+	0+	001	009	058	192	200	066	001	18
	19	0+	0+	0+	0+	0+	0+	0+	002	022	121	284	198	017	19
	20	0+	0+	0+	0+	0+	0+	0+	0+	005	048	255	376	172	20
	21	0+	0+	0+	0+	0+	0+	0+	0+	001	009	109	341	810	21
22	0	802	324	098	007	0+	0+	0+	0+	0+	0+	0+	0+	0+	0
	1	178	375	241	041	004	0+	0+	0+	0+	0+	0+	0+	0+	1
	2	019	207	281	107	017	001	0+	0+	0+	0+	0+	0+	0+	2
	3	001	073	208	178	047	006	0+	0+	0+	0+	0+	0+	0+	3
	4	0+	018	110	211	096	019	002	0+	0+	0+	0+	0+	0+	4
	5	0+	003	044	190	149	046	006	0+	0+	0+	0+	0+	0+	5
	6	0+	001	014	134	181	086	018	001	0+	0+	0+	0+	0+	6
	7	0+	0+	004	077	177	131	041	005	0+	0+	0+	0+	0+	7
	8	0+	0+	001	036	142	164	076	014	001	0+	0+	0+	0+	8
	9	0+	0+	0+	014	095	170	119	034	003	0+	0+	0+	0+	9
	10	0+	0+	0+	005	053	148	154	066	010	0+	0+	0+	0+	10
	11	0+	0+	0+	001	025	107	168	107	025	001	0+	0+	0+	11
	12	0+	0 1	0+	0+	010	066	154	148	053	005	0+	0+	0+	12
	13	0+	0+	0+	0+	003	034	119	170	095	014	0+	0+	0+	13
	14	0+	0+	0+	0+	001	014	076	164	142	036	001	0+	0+	14

NOTE: 0+ represents a positive probability less than 0.0005.

(*continued*)

Binomial Probabilities

n	x	.01	.05	.10	.20	.30	.40	p .50	.60	.70	.80	.90	.95	.99	x
	15	0+	0+	0+	0+	0+	005	041	131	177	077	004	0+	0+	15
	16	0+	0+	0+	0+	0+	001	018	086	181	134	014	001	0+	16
	17	0+	0+	0+	0+	0+	0+	006	046	149	190	044	003	0+	17
	18	0+	0+	0+	0+	0+	0+	002	019	096	211	110	018	0+	18
	19	0+	0+	0+	0+	0+	0+	0+	006	047	178	208	073	001	19
	20	0+	0+	0+	0+	0+	0+	0+	001	017	107	281	207	019	20
	21	0+	0+	0+	0+	0+	0+	0+	0+	004	041	241	375	178	21
	22	0+	0+	0+	0+	0+	0+	0+	0+	0+	007	098	324	802	22
23	0	794	307	089	006	0+	0+	0+	0+	0+	0+	0+	0+	0+	0
	1	184	372	226	034	003	0+	0+	0+	0+	0+	0+	0+	0+	1
	2	020	215	277	093	013	001	0+	0+	0+	0+	0+	0+	0+	2
	3	001	079	215	163	038	004	0+	0+	0+	0+	0+	0+	0+	3
	4	0+	021	120	204	082	014	001	0+	0+	0+	0+	0+	0+	4
	5	0+	004	051	194	133	035	004	0+	0+	0+	0+	0+	0+	5
	6	0+	001	017	145	171	070	012	001	0+	0+	0+	0+	0+	6
	7	0+	0+	005	088	178	113	029	003	0+	0+	0+	0+	0+	7
	8	0+	0+	001	044	153	151	058	009	0+	0+	0+	0+	0+	8
	9	0+	0+	0+	018	109	168	097	022	002	0+	0+	0+	0+	9
	10	0+	0+	0+	006	065	157	136	046	005	0+	0+	0+	0+	10
	11	0+	0+	0+	002	033	123	161	082	014	0+	0+	0+	0+	11
	12	0+	0+	0+	0+	014	082	161	123	033	002	0+	0+	0+	12
	13	0+	0+	0+	0+	005	046	136	157	065	006	0+	0+	0+	13
	14	0+	0+	0+	0+	002	022	097	168	109	018	0+	0+	0+	14
	15	0+	0+	0+	0+	0+	009	058	151	153	044	001	0+	0+	15
	16	0+	0+	0+	0+	0+	003	029	113	178	088	005	0+	0+	16
	17	0+	0+	0+	0+	0+	001	012	070	171	145	017	001	0+	17
	18	0+	0+	0+	0+	0+	0+	004	035	133	194	051	004	0+	18
	19	0+	0+	0+	0+	0+	0+	001	014	082	204	120	021	0+	19
	20	0+	0+	0+	0+	0+	0+	0+	004	038	163	215	079	001	20
	21	0+	0+	0+	0+	0+	0+	0+	001	013	093	277	215	020	21
	22	0+	0+	0+	0+	0+	0+	0+	0+	003	034	226	372	184	22
	23	0+	0+	0+	0+	0+	0+	0+	0+	0+	006	089	307	794	23
24	0	786	292	080	005	0+	0+	0+	0+	0+	0+	0+	0+	0+	0
	1	190	369	213	028	002	0+	0+	0+	0+	0+	0+	0+	0+	1
	2	022	223	272	081	010	001	0+	0+	0+	0+	0+	0+	0+	2
	3	002	086	221	149	031	003	0+	0+	0+	0+	0+	0+	0+	3
	4	0+	024	129	196	069	010	001	0+	0+	0+	0+	0+	0+	4
	5	0+	005	057	196	118	027	003	0+	0+	0+	0+	0+	0+	5
	6	0+	001	020	155	160	056	008	0+	0+	0+	0+	0+	0+	6
	7	0+	0+	006	100	176	096	021	002	0+	0+	0+	0+	0+	7
	8	0+	0+	001	053	160	136	044	005	0+	0+	0+	0+	0+	8
	9	0+	0+	0+	024	122	161	078	014	001	0+	0+	0+	0+	9

NOTE: 0+ represents a positive probability less than 0.0005.

(*continued*)

TABLE A-1 (continued)

Binomial Probabilities

n	x	.01	.05	.10	.20	.30	.40	p .50	.60	.70	.80	.90	.95	.99	x
	10	0+	0+	0+	009	079	161	117	032	003	0+	0+	0+	0+	10
	11	0+	0+	0+	003	043	137	149	061	008	0+	0+	0+	0+	11
	12	0+	0+	0+	001	020	099	161	099	020	001	0+	0+	0+	12
	13	0+	0+	0+	0+	008	061	149	137	043	003	0+	0+	0+	13
	14	0+	0+	0+	0+	003	032	117	161	079	009	0+	0+	0+	14
	15	0+	0+	0+	0+	001	014	078	161	122	024	0+	0+	0+	15
	16	0+	0+	0+	0+	0+	005	044	136	160	053	001	0+	0+	16
	17	0+	0+	0+	0+	0+	002	021	096	176	100	006	0+	0+	17
	18	0+	0+	0+	0+	0+	0+	008	056	160	155	020	001	0+	18
	19	0+	0+	0+	0+	0+	0+	003	027	118	196	057	005	0+	19
	20	0+	0+	0+	0+	0+	0+	001	010	069	196	129	024	0+	20
	21	0+	0+	0+	0+	0+	0+	0+	003	031	149	221	086	002	21
	22	0+	0+	0+	0+	0+	0+	0+	001	010	081	272	223	022	22
	23	0+	0+	0+	0+	0+	0+	0+	0+	002	028	213	369	190	23
	24	0+	0+	0+	0+	0+	0+	0+	0+	0+	005	080	292	786	24
25	0	778	277	072	004	0+	0+	0+	0+	0+	0+	0+	0+	0+	0
	1	196	365	199	024	001	0+	0+	0+	0+	0+	0+	0+	0+	1
	2	024	231	266	071	007	0+	0+	0+	0+	0+	0+	0+	0+	2
	3	002	093	226	136	024	002	0+	0+	0+	0+	0+	0+	0+	3
	4	0+	027	138	187	057	007	0+	0+	0+	0+	0+	0+	0+	4
	5	0+	006	065	196	103	020	002	0+	0+	0+	0+	0+	0+	5
	6	0+	001	024	163	147	044	005	0+	0+	0+	0+	0+	0+	6
	7	0+	0+	007	111	171	080	014	001	0+	0+	0+	0+	0+	7
	8	0+	0+	002	062	165	120	032	003	0+	0+	0+	0+	0+	8
	9	0+	0+	0+	029	134	151	061	009	0+	0+	0+	0+	0+	9
	10	0+	0+	0+	012	092	161	097	021	001	0+	0+	0+	0+	10
	11	0+	0+	0+	004	054	147	133	043	004	0+	0+	0+	0+	11
	12	0+	0+	0+	001	027	114	155	076	011	0+	0+	0+	0+	12
	13	0+	0+	0+	0+	011	076	155	114	027	001	0+	0+	0+	13
	14	0+	0+	0+	0+	004	043	133	147	054	004	0+	0+	0+	14
	15	0+	0+	0+	0+	001	021	097	161	092	012	0+	0+	0+	15
	16	0+	0+	0+	0+	0+	009	061	151	134	029	0+	0+	0+	16
	17	0+	0+	0+	0+	0+	003	032	120	165	062	002	0+	0+	17
	18	0+	0+	0+	0+	0+	001	014	080	171	111	007	0+	0+	18
	19	0+	0+	0+	0+	0+	0+	005	044	147	163	024	001	0+	19
	20	0+	0+	0+	0+	0+	0+	002	020	103	196	065	006	0+	20
	21	0+	0+	0+	0+	0+	0+	0+	007	057	187	138	027	0+	21
	22	0+	0+	0+	0+	0+	0+	0+	002	024	136	226	093	002	22
	23	0+	0+	0+	0+	0+	0+	0+	0+	007	071	266	231	024	23
	24	0+	0+	0+	0+	0+	0+	0+	0+	001	024	199	365	196	24
	25	0+	0+	0+	0+	0+	0+	0+	0+	0+	004	072	277	778	25

NOTE: 0+ represents a positive probability less than 0.0005.

Frederick Mosteller, Robert E. K. Rourke, and George B. Thomas, Jr., *Probability with Statistical Applications*, 2nd ed. (Reading, Mass.: Addison-Wesley, 1961 and 1970). Reprinted with permission of the publisher.

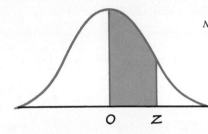

Notes: 1. For values of *z* above 3.09, use 0.4999 for the area.

*2. Use these common values that result from interpolation:

z score	area
1.645	0.4500
2.575	0.4950

TABLE A-2

The Standard Normal (*z*) Distribution

z	.00	.01	.02	.03	.04	.05	.06	.07	.08	.09
0.0	.0000	.0040	.0080	.0120	.0160	.0199	.0239	.0279	.0319	.0359
0.1	.0398	.0438	.0478	.0517	.0557	.0596	.0636	.0675	.0714	.0753
0.2	.0793	.0832	.0871	.0910	.0948	.0987	.1026	.1064	.1103	.1141
0.3	.1179	.1217	.1255	.1293	.1331	.1368	.1406	.1443	.1480	.1517
0.4	.1554	.1591	.1628	.1664	.1700	.1736	.1772	.1808	.1844	.1879
0.5	.1915	.1950	.1985	.2019	.2054	.2088	.2123	.2157	.2190	.2224
0.6	.2257	.2291	.2324	.2357	.2389	.2422	.2454	.2486	.2517	.2549
0.7	.2580	.2611	.2642	.2673	.2704	.2734	.2764	.2794	.2823	.2852
0.8	.2881	.2910	.2939	.2967	.2995	.3023	.3051	.3078	.3106	.3133
0.9	.3159	.3186	.3212	.3238	.3264	.3289	.3315	.3340	.3365	.3389
1.0	.3413	.3438	.3461	.3485	.3508	.3531	.3554	.3577	.3599	.3621
1.1	.3643	.3665	.3686	.3708	.3729	.3749	.3770	.3790	.3810	.3830
1.2	.3849	.3869	.3888	.3907	.3925	.3944	.3962	.3980	.3997	.4015
1.3	.4032	.4049	.4066	.4082	.4099	.4115	.4131	.4147	.4162	.4177
1.4	.4192	.4207	.4222	.4236	.4251	.4265	.4279	.4292	.4306	.4319
1.5	.4332	.4345	.4357	.4370	.4382	.4394	.4406	.4418	.4429	.4441
1.6	.4452	.4463	.4474	.4484	.4495	* .4505	.4515	.4525	.4535	.4545
1.7	.4554	.4564	.4573	.4582	.4591	.4599	.4608	.4616	.4625	.4633
1.8	.4641	.4649	.4656	.4664	.4671	.4678	.4686	.4693	.4699	.4706
1.9	.4713	.4719	.4726	.4732	.4738	.4744	.4750	.4756	.4761	.4767
2.0	.4772	.4778	.4783	.4788	.4793	.4798	.4803	.4808	.4812	.4817
2.1	.4821	.4826	.4830	.4834	.4838	.4842	.4846	.4850	.4854	.4857
2.2	.4861	.4864	.4868	.4871	.4875	.4878	.4881	.4884	.4887	.4890
2.3	.4893	.4896	.4898	.4901	.4904	.4906	.4909	.4911	.4913	.4916
2.4	.4918	.4920	.4922	.4925	.4927	.4929	.4931	.4932	.4934	.4936
2.5	.4938	.4940	.4941	.4943	.4945	.4946	.4948	.4949	* .4951	.4952
2.6	.4953	.4955	.4956	.4957	.4959	.4960	.4961	.4962	.4963	.4964
2.7	.4965	.4966	.4967	.4968	.4969	.4970	.4971	.4972	.4973	.4974
2.8	.4974	.4975	.4976	.4977	.4977	.4978	.4979	.4979	.4980	.4981
2.9	.4981	.4982	.4982	.4983	.4984	.4984	.4985	.4985	.4986	.4986
3.0	.4987	.4987	.4987	.4988	.4988	.4989	.4989	.4989	.4990	.4990

Frederick Mosteller and Robert E. K. Rourke, *Sturdy Statistics* Table A-1 (Reading, Mass.: Addison-Wesley, 1973). Reprinted with permission.

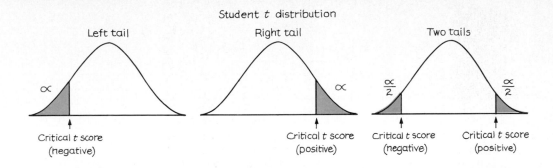

Student t distribution

Left tail	Right tail	Two tails
α	α	$\frac{\alpha}{2}$ $\frac{\alpha}{2}$
Critical t score (negative)	Critical t score (positive)	Critical t score (negative) / Critical t score (positive)

TABLE A-3

t Distribution

	α					
Degrees of freedom	.005 (one tail) .01 (two tails)	.01 (one tail) .02 (two tails)	.025 (one tail) .05 (two tails)	.05 (one tail) .10 (two tails)	.10 (one tail) .20 (two tails)	.25 (one tail) .50 (two tails)
1	63.657	31.821	12.706	6.314	3.078	1.000
2	9.925	6.965	4.303	2.920	1.886	.816
3	5.841	4.541	3.182	2.353	1.638	.765
4	4.604	3.747	2.776	2.132	1.533	.741
5	4.032	3.365	2.571	2.015	1.476	.727
6	3.707	3.143	2.447	1.943	1.440	.718
7	3.500	2.998	2.365	1.895	1.415	.711
8	3.355	2.896	2.306	1.860	1.397	.706
9	3.250	2.821	2.262	1.833	1.383	.703
10	3.169	2.764	2.228	1.812	1.372	.700
11	3.106	2.718	2.201	1.796	1.363	.697
12	3.054	2.681	2.179	1.782	1.356	.696
13	3.012	2.650	2.160	1.771	1.350	.694
14	2.977	2.625	2.145	1.761	1.345	.692
15	2.947	2.602	2.132	1.753	1.341	.691
16	2.921	2.584	2.120	1.746	1.337	.690
17	2.898	2.567	2.110	1.740	1.333	.689
18	2.878	2.552	2.101	1.734	1.330	.688
19	2.861	2.540	2.093	1.729	1.328	.688
20	2.845	2.528	2.086	1.725	1.325	.687
21	2.831	2.518	2.080	1.721	1.323	.686
22	2.819	2.508	2.074	1.717	1.321	.686
23	2.807	2.500	2.069	1.714	1.320	.685
24	2.797	2.492	2.064	1.711	1.318	.685
25	2.787	2.485	2.060	1.708	1.316	.684
26	2.779	2.479	2.056	1.706	1.315	.684
27	2.771	2.473	2.052	1.703	1.314	.684
28	2.763	2.467	2.048	1.701	1.313	.683
29	2.756	2.462	2.045	1.699	1.311	.683
Large (z)	2.575	2.327	1.960	1.645	1.282	.675

TABLE A-4

The Chi-Square (χ^2) Distribution

Area to the Right of the Critical Value

Degrees of freedom	0.995	0.99	0.975	0.95	0.90	0.10	0.05	0.025	0.01	0.005
1	—	—	0.001	0.004	0.016	2.706	3.841	5.024	6.635	7.879
2	0.010	0.020	0.051	0.103	0.211	4.605	5.991	7.378	9.210	10.597
3	0.072	0.115	0.216	0.352	0.584	6.251	7.815	9.348	11.345	12.838
4	0.207	0.297	0.484	0.711	1.064	7.779	9.488	11.143	13.277	14.860
5	0.412	0.554	0.831	1.145	1.610	9.236	11.071	12.833	15.086	16.750
6	0.676	0.872	1.237	1.635	2.204	10.645	12.592	14.449	16.812	18.548
7	0.989	1.239	1.690	2.167	2.833	12.017	14.067	16.013	18.475	20.278
8	1.344	1.646	2.180	2.733	3.490	13.362	15.507	17.535	20.090	21.955
9	1.735	2.088	2.700	3.325	4.168	14.684	16.919	19.023	21.666	23.589
10	2.156	2.558	3.247	3.940	4.865	15.987	18.307	20.483	23.209	25.188
11	2.603	3.053	3.816	4.575	5.578	17.275	19.675	21.920	24.725	26.757
12	3.074	3.571	4.404	5.226	6.304	18.549	21.026	23.337	26.217	28.299
13	3.565	4.107	5.009	5.892	7.042	19.812	22.362	24.736	27.688	29.819
14	4.075	4.660	5.629	6.571	7.790	21.064	23.685	26.119	29.141	31.319
15	4.601	5.229	6.262	7.261	8.547	22.307	24.996	27.488	30.578	32.801
16	5.142	5.812	6.908	7.962	9.312	23.542	26.296	28.845	32.000	34.267
17	5.697	6.408	7.564	8.672	10.085	24.769	27.587	30.191	33.409	35.718
18	6.265	7.015	8.231	9.390	10.865	25.989	28.869	31.526	34.805	37.156
19	6.844	7.633	8.907	10.117	11.651	27.204	30.144	32.852	36.191	38.582
20	7.434	8.260	9.591	10.851	12.443	28.412	31.410	34.170	37.566	39.997
21	8.034	8.897	10.283	11.591	13.240	29.615	32.671	35.479	38.932	41.401
22	8.643	9.542	10.982	12.338	14.042	30.813	33.924	36.781	40.289	42.796
23	9.260	10.196	11.689	13.091	14.848	32.007	35.172	38.076	41.638	44.181
24	9.886	10.856	12.401	13.848	15.659	33.196	36.415	39.364	42.980	45.559
25	10.520	11.524	13.120	14.611	16.473	34.382	37.652	40.646	44.314	46.928
26	11.160	12.198	13.844	15.379	17.292	35.563	38.885	41.923	45.642	48.290
27	11.808	12.879	14.573	16.151	18.114	36.741	40.113	43.194	46.963	49.645
28	12.461	13.565	15.308	16.928	18.939	37.916	41.337	44.461	48.278	50.993
29	13.121	14.257	16.047	17.708	19.768	39.087	42.557	45.772	49.588	52.336
30	13.787	14.954	16.791	18.493	20.599	40.256	43.773	46.979	50.892	53.672
40	20.707	22.164	24.433	26.509	29.051	51.805	55.758	59.342	63.691	66.766
50	27.991	29.707	32.357	34.764	37.689	63.167	67.505	71.420	76.154	79.490
60	35.534	37.485	40.482	43.188	46.459	74.397	79.082	83.298	88.379	91.952
70	43.275	45.442	48.758	51.739	55.329	85.527	90.531	95.023	100.425	104.215
80	51.172	53.540	57.153	60.391	64.278	96.578	101.879	106.629	112.329	116.321
90	59.196	61.754	65.647	69.126	73.291	107.565	113.145	118.136	124.116	128.299
100	67.328	70.065	74.222	77.929	82.358	118.498	124.342	129.561	135.807	140.169

Donald B. Owen, *Handbook of Statistical Tables*, U.S. Department of Energy (Reading, Mass.: Addison-Wesley, 1962). Reprinted with permission of the publisher.

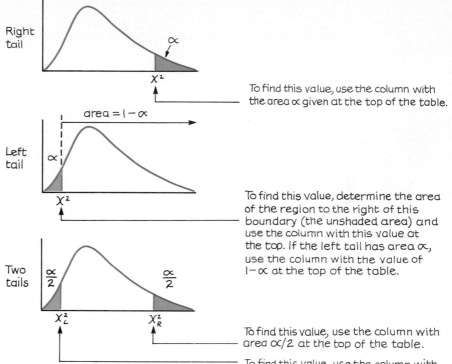

Right tail

α

χ^2

To find this value, use the column with the area α given at the top of the table.

Left tail

area $= 1 - \alpha$

α

χ^2

To find this value, determine the area of the region to the right of this boundary (the unshaded area) and use the column with this value at the top. If the left tail has area α, use the column with the value of $1 - \alpha$ at the top of the table.

Two tails

$\frac{\alpha}{2}$

$\frac{\alpha}{2}$

χ^2_L

χ^2_R

To find this value, use the column with area $\alpha/2$ at the top of the table.

To find this value, use the column with area $1 - \alpha/2$ at the top of the table.

TABLE A-5

F Distribution ($\alpha = 0.01$ in the right tail)

df_2 \ df_1	1	2	3	4	5	6	7	8	9
1	4052.2	4999.5	5403.4	5624.6	5763.6	5859.0	5928.4	5981.1	6022.5
2	98.503	99.000	99.166	99.249	99.299	99.333	99.356	99.374	99.388
3	34.116	30.817	29.457	28.710	28.237	27.911	27.672	27.489	27.345
4	21.198	18.000	16.694	15.977	15.522	15.207	14.976	14.799	14.659
5	16.258	13.274	12.060	11.392	10.967	10.672	10.456	10.289	10.158
6	13.745	10.925	9.7795	9.1483	8.7459	8.4661	8.2600	8.1017	7.9761
7	12.246	9.5466	8.4513	7.8466	7.4604	7.1914	6.9928	6.8400	6.7188
8	11.259	8.6491	7.5910	7.0061	6.6318	6.3707	6.1776	6.0289	5.9106
9	10.561	8.0215	6.9919	6.4221	6.0569	5.8018	5.6129	5.4671	5.3511
10	10.044	7.5594	6.5523	5.9943	5.6363	5.3858	5.2001	5.0567	4.9424
11	9.6460	7.2057	6.2167	5.6683	5.3160	5.0692	4.8861	4.7445	4.6315
12	9.3302	6.9266	5.9525	5.4120	5.0643	4.8206	4.6395	4.4994	4.3875
13	9.0738	6.7010	5.7394	5.2053	4.8616	4.6204	4.4410	4.3021	4.1911
14	8.8616	6.5149	5.5639	5.0354	4.6950	4.4558	4.2779	4.1399	4.0297
15	8.6831	6.3589	5.4170	4.8932	4.5556	4.3183	4.1415	4.0045	3.8948
16	8.5310	6.2262	5.2922	4.7726	4.4374	4.2016	4.0259	3.8896	3.7804
17	8.3997	6.1121	5.1850	4.6690	4.3359	4.1015	3.9267	3.7910	3.6822
18	8.2854	6.0129	5.0919	4.5790	4.2479	4.0146	3.8406	3.7054	3.5971
19	8.1849	5.9259	5.0103	4.5003	4.1708	3.9386	3.7653	3.6305	3.5225
20	8.0960	5.8489	4.9382	4.4307	4.1027	3.8714	3.6987	3.5644	3.4567
21	8.0166	5.7804	4.8740	4.3688	4.0421	3.8117	3.6396	3.5056	3.3981
22	7.9454	5.7190	4.8166	4.3134	3.9880	3.7583	3.5867	3.4530	3.3458
23	7.8811	5.6637	4.7649	4.2636	3.9392	3.7102	3.5390	3.4057	3.2986
24	7.8229	5.6136	4.7181	4.2184	3.8951	3.6667	3.4959	3.3629	3.2560
25	7.7698	5.5680	4.6755	4.1774	3.8550	3.6272	3.4568	3.3239	3.2172
26	7.7213	5.5263	4.6366	4.1400	3.8183	3.5911	3.4210	3.2884	3.1818
27	7.6767	5.4881	4.6009	4.1056	3.7848	3.5580	3.3882	3.2558	3.1494
28	7.6356	5.4529	4.5681	4.0740	3.7539	3.5276	3.3581	3.2259	3.1195
29	7.5977	5.4204	4.5378	4.0449	3.7254	3.4995	3.3303	3.1982	3.0920
30	7.5625	5.3903	4.5097	4.0179	3.6990	3.4735	3.3045	3.1726	3.0665
40	7.3141	5.1785	4.3126	3.8283	3.5138	3.2910	3.1238	2.9930	2.8876
60	7.0771	4.9774	4.1259	3.6490	3.3389	3.1187	2.9530	2.8233	2.7185
120	6.8509	4.7865	3.9491	3.4795	3.1735	2.9559	2.7918	2.6629	2.5586
∞	6.6349	4.6052	3.7816	3.3192	3.0173	2.8020	2.6393	2.5113	2.4073

Numerator degrees of freedom

Denominator degrees of freedom

(continued)

From Maxine Merrington and Catherine M. Thompson, "Tables of Percentage Points of the Inverted Beta (F) Distribution," Biometrika 33 (1943): 80–84. Reproduced by permission of Professor E. S. Pearson.

Numerator degrees of freedom

df_1 df_2	10	12	15	20	24	30	40	60	120	∞
1	6055.8	6106.3	6157.3	6208.7	6234.6	6260.6	6286.8	6313.0	6339.4	6365.9
2	99.399	99.416	99.433	99.449	99.458	99.466	99.474	99.482	99.491	99.499
3	27.229	27.052	26.872	26.690	26.598	26.505	26.411	26.316	26.221	26.125
4	14.546	14.374	14.198	14.020	13.929	13.838	13.745	13.652	13.558	13.463
5	10.051	9.8883	9.7222	9.5526	9.4665	9.3793	9.2912	9.2020	9.1118	9.0204
6	7.8741	7.7183	7.5590	7.3958	7.3127	7.2285	7.1432	7.0567	6.9690	6.8800
7	6.6201	6.4691	6.3143	6.1554	6.0743	5.9920	5.9084	5.8236	5.7373	5.6495
8	5.8143	5.6667	5.5151	5.3591	5.2793	5.1981	5.1156	5.0316	4.9461	4.8588
9	5.2565	5.1114	4.9621	4.8080	4.7290	4.6486	4.5666	4.4831	4.3978	4.3105
10	4.8491	4.7059	4.5581	4.4054	4.3269	4.2469	4.1653	4.0819	3.9965	3.9090
11	4.5393	4.3974	4.2509	4.0990	4.0209	3.9411	3.8596	3.7761	3.6904	3.6024
12	4.2961	4.1553	4.0096	3.8584	3.7805	3.7008	3.6192	3.5355	3.4494	3.3608
13	4.1003	3.9603	3.8154	3.6646	3.5868	3.5070	3.4253	3.3413	3.2548	3.1654
14	3.9394	3.8001	3.6557	3.5052	3.4274	3.3476	3.2656	3.1813	3.0942	3.0040
15	3.8049	3.6662	3.5222	3.3719	3.2940	3.2141	3.1319	3.0471	2.9595	2.8684
16	3.6909	3.5527	3.4089	3.2587	3.1808	3.1007	3.0182	2.9330	2.8447	2.7528
17	3.5931	3.4552	3.3117	3.1615	3.0835	3.0032	2.9205	2.8348	2.7459	2.6530
18	3.5082	3.3706	3.2273	3.0771	2.9990	2.9185	2.8354	2.7493	2.6597	2.5660
19	3.4338	3.2965	3.1533	3.0031	2.9249	2.8442	2.7608	2.6742	2.5839	2.4893
20	3.3682	3.2311	3.0880	2.9377	2.8594	2.7785	2.6947	2.6077	2.5168	2.4212
21	3.3098	3.1730	3.0300	2.8796	2.8010	2.7200	2.6359	2.5484	2.4568	2.3603
22	3.2576	3.1209	2.9779	2.8274	2.7488	2.6675	2.5831	2.4951	2.4029	2.3055
23	3.2106	3.0740	2.9311	2.7805	2.7017	2.6202	2.5355	2.4471	2.3542	2.2558
24	3.1681	3.0316	2.8887	2.7380	2.6591	2.5773	2.4923	2.4035	2.3100	2.2107
25	3.1294	2.9931	2.8502	2.6993	2.6203	2.5383	2.4530	2.3637	2.2696	2.1694
26	3.0941	2.9578	2.8150	2.6640	2.5848	2.5026	2.4170	2.3273	2.2325	2.1315
27	3.0618	2.9256	2.7827	2.6316	2.5522	2.4699	2.3840	2.2938	2.1985	2.0965
28	3.0320	2.8959	2.7530	2.6017	2.5223	2.4397	2.3535	2.2629	2.1670	2.0642
29	3.0045	2.8685	2.7256	2.5742	2.4946	2.4118	2.3253	2.2344	2.1379	2.0342
30	2.9791	2.8431	2.7002	2.5487	2.4689	2.3860	2.2992	2.2079	2.1108	2.0062
40	2.8005	2.6648	2.5216	2.3689	2.2880	2.2034	2.1142	2.0194	1.9172	1.8047
60	2.6318	2.4961	2.3523	2.1978	2.1154	2.0285	1.9360	1.8363	1.7263	1.6006
120	2.4721	2.3363	2.1915	2.0346	1.9500	1.8600	1.7628	1.6557	1.5330	1.3805
∞	2.3209	2.1847	2.0385	1.8783	1.7908	1.6964	1.5923	1.4730	1.3246	1.0000

Denominator degrees of freedom

(continued)

TABLE A-5 (continued)

F Distribution ($\alpha = 0.025$ in the right tail)

0.025

		Numerator degrees of freedom							
df_2 \ df_1	1	2	3	4	5	6	7	8	9
1	647.79	799.50	864.16	899.58	921.85	937.11	948.22	956.66	963.28
2	38.506	39.000	39.165	39.248	39.298	39.331	39.335	39.373	39.387
3	17.443	16.044	15.439	15.101	14.885	14.735	14.624	14.540	14.473
4	12.218	10.649	9.9792	9.6045	9.3645	9.1973	9.0741	8.9796	8.9047
5	10.007	8.4336	7.7636	7.3879	7.1464	6.9777	6.8531	6.7572	6.6811
6	8.8131	7.2599	6.5988	6.2272	5.9876	5.8198	5.6955	5.5996	5.5234
7	8.0727	6.5415	5.8898	5.5226	5.2852	5.1186	4.9949	4.8993	4.8232
8	7.5709	6.0595	5.4160	5.0526	4.8173	4.6517	4.5286	4.4333	4.3572
9	7.2093	5.7147	5.0781	4.7181	4.4844	4.3197	4.1970	4.1020	4.0260
10	6.9367	5.4564	4.8256	4.4683	4.2361	4.0721	3.9498	3.8549	3.7790
11	6.7241	5.2559	4.6300	4.2751	4.0440	3.8807	3.7586	3.6638	3.5879
12	6.5538	5.0959	4.4742	4.1212	3.8911	3.7283	3.6065	3.5118	3.4358
13	6.4143	4.9653	4.3472	3.9959	3.7667	3.6043	3.4827	3.3880	3.3120
14	6.2979	4.8567	4.2417	3.8919	3.6634	3.5014	3.3799	3.2853	3.2093
15	6.1995	4.7650	4.1528	3.8043	3.5764	3.4147	3.2934	3.1987	3.1227
16	6.1151	4.6867	4.0768	3.7294	3.5021	3.3406	3.2194	3.1248	3.0488
17	6.0420	4.6189	4.0112	3.6648	3.4379	3.2767	3.1556	3.0610	2.9849
18	5.9781	4.5597	3.9539	3.6083	3.3820	3.2209	3.0999	3.0053	2.9291
19	5.9216	4.5075	3.9034	3.5587	3.3327	3.1718	3.0509	2.9563	2.8801
20	5.8715	4.4613	3.8587	3.5147	3.2891	3.1283	3.0074	2.9128	2.8365
21	5.8266	4.4199	3.8188	3.4754	3.2501	3.0895	2.9686	2.8740	2.7977
22	5.7863	4.3828	3.7829	3.4401	3.2151	3.0546	2.9338	2.8392	2.7628
23	5.7498	4.3492	3.7505	3.4083	3.1835	3.0232	2.9023	2.8077	2.7313
24	5.7166	4.3187	3.7211	3.3794	3.1548	2.9946	2.8738	2.7791	2.7027
25	5.6864	4.2909	3.6943	3.3530	3.1287	2.9685	2.8478	2.7531	2.6766
26	5.6586	4.2655	3.6697	3.3289	3.1048	2.9447	2.8240	2.7293	2.6528
27	5.6331	4.2421	3.6472	3.3067	3.0828	2.9228	2.8021	2.7074	2.6309
28	5.6096	4.2205	3.6264	3.2863	3.0626	2.9027	2.7820	2.6872	2.6106
29	5.5878	4.2006	3.6072	3.2674	3.0438	2.8840	2.7633	2.6686	2.5919
30	5.5675	4.1821	3.5894	3.2499	3.0265	2.8667	2.7460	2.6513	2.5746
40	5.4239	4.0510	3.4633	3.1261	2.9037	2.7444	2.6238	2.5289	2.4519
60	5.2856	3.9253	3.3425	3.0077	2.7863	2.6274	2.5068	2.4117	2.3344
120	5.1523	3.8046	3.2269	2.8943	2.6740	2.5154	2.3948	2.2994	2.2217
∞	5.0239	3.6889	3.1161	2.7858	2.5665	2.4082	2.2875	2.1918	2.1136

Denominator degrees of freedom

(continued)

	Numerator degrees of freedom									
df_1 df_2	10	12	15	20	24	30	40	60	120	∞
1	968.63	976.71	984.87	993.10	997.25	1001.4	1005.6	1009.8	1014.0	1018.3
2	39.398	39.415	39.431	39.448	39.456	39.465	39.473	39.481	39.490	39.498
3	14.419	14.337	14.253	14.167	14.124	14.081	14.037	13.992	13.947	13.902
4	8.8439	8.7512	8.6565	8.5599	8.5109	8.4613	8.4111	8.3604	8.3092	8.2573
5	6.6192	6.5245	6.4277	6.3286	6.2780	6.2269	6.1750	6.1225	6.0693	6.0153
6	5.4613	5.3662	5.2687	5.1684	5.1172	5.0652	5.0125	4.9589	4.9044	4.8491
7	4.7611	4.6658	4.5678	4.4667	4.4150	4.3624	4.3089	4.2544	4.1989	4.1423
8	4.2951	4.1997	4.1012	3.9995	3.9472	3.8940	3.8398	3.7844	3.7279	3.6702
9	3.9639	3.8682	3.7694	3.6669	3.6142	3.5604	3.5055	3.4493	3.3918	3.3329
10	3.7168	3.6209	3.5217	3.4185	3.3654	3.3110	3.2554	3.1984	3.1399	3.0798
11	3.5257	3.4296	3.3299	3.2261	3.1725	3.1176	3.0613	3.0035	2.9441	2.8828
12	3.3736	3.2773	3.1772	3.0728	3.0187	2.9633	2.9063	2.8478	2.7874	2.7249
13	3.2497	3.1532	3.0527	2.9477	2.8932	2.8372	2.7797	2.7204	2.6590	2.5955
14	3.1469	3.0502	2.9493	2.8437	2.7888	2.7324	2.6742	2.6142	2.5519	2.4872
15	3.0602	2.9633	2.8621	2.7559	2.7006	2.6437	2.5850	2.5242	2.4611	2.3953
16	2.9862	2.8890	2.7875	2.6808	2.6252	2.5678	2.5085	2.4471	2.3831	2.3163
17	2.9222	2.8249	2.7230	2.6158	2.5598	2.5020	2.4422	2.3801	2.3153	2.2474
18	2.8664	2.7689	2.6667	2.5590	2.5027	2.4445	2.3842	2.3214	2.2558	2.1869
19	2.8172	2.7196	2.6171	2.5089	2.4523	2.3937	2.3329	2.2696	2.2032	2.1333
20	2.7737	2.6758	2.5731	2.4645	2.4076	2.3486	2.2873	2.2234	2.1562	2.0853
21	2.7348	2.6368	2.5338	2.4247	2.3675	2.3082	2.2465	2.1819	2.1141	2.0422
22	2.6998	2.6017	2.4984	2.3890	2.3315	2.2718	2.2097	2.1446	2.0760	2.0032
23	2.6682	2.5699	2.4665	2.3567	2.2989	2.2389	2.1763	2.1107	2.0415	1.9677
24	2.6396	2.5411	2.4374	2.3273	2.2693	2.2090	2.1460	2.0799	2.0099	1.9353
25	2.6135	2.5149	2.4110	2.3005	2.2422	2.1816	2.1183	2.0516	1.9811	1.9055
26	2.5896	2.4908	2.3867	2.2759	2.2174	2.1565	2.0928	2.0257	1.9545	1.8781
27	2.5676	2.4688	2.3644	2.2533	2.1946	2.1334	2.0693	2.0018	1.9299	1.8527
28	2.5473	2.4484	2.3438	2.2324	2.1735	2.1121	2.0477	1.9797	1.9072	1.8291
29	2.5286	2.4295	2.3248	2.2131	2.1540	2.0923	2.0276	1.9591	1.8861	1.8072
30	2.5112	2.4120	2.3072	2.1952	2.1359	2.0739	2.0089	1.9400	1.8664	1.7867
40	2.3882	2.2882	2.1819	2.0677	2.0069	1.9429	1.8752	1.8028	1.7242	1.6371
60	2.2702	2.1692	2.0613	1.9445	1.8817	1.8152	1.7440	1.6668	1.5810	1.4821
120	2.1570	2.0548	1.9450	1.8249	1.7597	1.6899	1.6141	1.5299	1.4327	1.3104
∞	2.0483	1.9447	1.8326	1.7085	1.6402	1.5660	1.4835	1.3883	1.2684	1.0000

Denominator degrees of freedom

(continued)

TABLE A-5 (continued)

F Distribution (α = 0.05 in the right tail)

Numerator degrees of freedom

df_2 \ df_1	1	2	3	4	5	6	7	8	9
1	161.45	199.50	215.71	224.58	230.16	233.99	236.77	238.88	240.54
2	18.513	19.000	19.164	19.247	19.296	19.330	19.353	19.371	19.385
3	10.128	9.5521	9.2766	9.1172	9.0135	8.9406	8.8867	8.8452	8.8123
4	7.7086	9.9443	6.5914	6.3882	6.2561	6.1631	6.0942	6.0410	6.9988
5	6.6079	5.7861	5.4095	5.1922	5.0503	4.9503	4.8759	4.8183	4.7725
6	5.9874	5.1433	4.7571	4.5337	4.3874	4.2839	4.2067	4.1468	4.0990
7	5.5914	4.7374	4.3468	4.1203	3.9715	3.8660	3.7870	3.7257	3.6767
8	5.3177	4.4590	4.0662	3.8379	3.6875	3.5806	3.5005	3.4381	3.3881
9	5.1174	4.2565	3.8625	3.6331	3.4817	3.3738	3.2927	3.2296	3.1789
10	4.9646	4.1028	3.7083	3.4780	3.3258	3.2172	3.1355	3.0717	3.0204
11	4.8443	3.9823	3.5874	3.3567	3.2039	3.0946	3.0123	2.9480	2.8962
12	4.7472	3.8853	3.4903	3.2592	3.1059	2.9961	2.9134	2.8486	2.7964
13	4.6672	3.8056	3.4105	3.1791	3.0254	2.9153	2.8321	2.7669	2.7144
14	4.6001	3.7389	3.3439	3.1122	2.9582	2.8477	2.7642	2.6987	2.6458
15	4.5431	3.6823	3.2874	3.0556	2.9013	2.7905	2.7066	2.6408	2.5876
16	4.4940	3.6337	3.2389	3.0069	2.8524	2.7413	2.6572	2.5911	2.5377
17	4.4513	3.5915	3.1968	2.9647	2.8100	2.6987	2.6143	2.5480	2.4943
18	4.4139	3.5546	3.1599	2.9277	2.7729	2.6613	2.5767	2.5102	2.4563
19	4.3807	3.5219	3.1274	2.8951	2.7401	2.6283	2.5435	2.4768	2.4227
20	4.3512	3.4928	3.0984	2.8661	2.7109	2.5990	2.5140	2.4471	2.3928
21	4.3248	3.4668	3.0725	2.8401	2.6848	2.5727	2.4876	2.4205	2.3660
22	4.3009	3.4434	3.0491	2.8167	2.6613	2.5491	2.4638	2.3965	2.3419
23	4.2793	3.4221	3.0280	2.7955	2.6400	2.5277	2.4422	2.3748	2.3201
24	4.2597	3.4028	3.0088	2.7763	2.6207	2.5082	2.4226	2.3551	2.3002
25	4.2417	3.3852	2.9912	2.7587	2.6030	2.4904	2.4047	2.3371	2.2821
26	4.2252	3.3690	2.9752	2.7426	2.5868	2.4741	2.3883	2.3205	2.2655
27	4.2100	3.3541	2.9604	2.7278	2.5719	2.4591	2.3732	2.3053	2.2501
28	4.1960	3.3404	2.9467	2.7141	2.5581	2.4453	2.3593	2.2913	2.2360
29	4.1830	3.3277	2.9340	2.7014	2.5454	2.4324	2.3463	2.2783	2.2229
30	4.1709	3.3158	2.9223	2.6896	2.5336	2.4205	2.3343	2.2662	2.2107
40	4.0847	3.2317	2.8387	2.6060	2.4495	2.3359	2.2490	2.1802	2.1240
60	4.0012	3.1504	2.7581	2.5252	2.3683	2.2541	2.1665	2.0970	2.0401
120	3.9201	3.0718	2.6802	2.4472	2.2899	2.1750	2.0868	2.0164	1.9588
∞	3.8415	2.9957	2.6049	2.3719	2.2141	2.0986	2.0096	1.9384	1.8799

Denominator degrees of freedom

(continued)

Numerator degrees of freedom

df_2 \ df_1	10	12	15	20	24	30	40	60	120	∞
1	241.88	243.91	245.95	248.01	249.05	250.10	251.14	252.20	253.25	254.31
2	19.396	19.413	19.429	19.446	19.454	19.462	19.471	19.479	19.487	19.496
3	8.7855	8.7446	8.7029	8.6602	8.6385	8.6166	8.5944	8.5720	8.5494	8.5264
4	5.9644	5.9117	5.8578	5.8025	5.7744	5.7459	5.7170	5.6877	5.6581	5.6281
5	4.7351	4.6777	4.6188	4.5581	4.5272	4.4957	4.4638	4.4314	4.3985	4.3650
6	4.0600	3.9999	3.9381	3.8742	3.8415	3.8082	3.7743	3.7398	3.7047	3.6689
7	3.6365	3.5747	3.5107	3.4445	3.4105	3.3758	3.3404	3.3043	3.2674	3.2298
8	3.3472	3.2839	3.2184	3.1503	3.1152	3.0794	3.0428	3.0053	2.9669	2.9276
9	3.1373	3.0729	3.0061	2.9365	2.9005	2.8637	2.8259	2.7872	2.7475	2.7067
10	2.9782	2.9130	2.8450	2.7740	2.7372	2.6996	2.6609	2.6211	2.5801	2.5379
11	2.8536	2.7876	2.7186	2.6464	2.6090	2.5705	2.5309	2.4901	2.4480	2.4045
12	2.7534	2.6866	2.6169	2.5436	2.5055	2.4663	2.4259	2.3842	2.3410	2.2962
13	2.6710	2.6037	2.5331	2.4589	2.4202	2.3803	2.3392	2.2966	2.2524	2.2064
14	2.6022	2.5342	2.4630	2.3879	2.3487	2.3082	2.2664	2.2229	2.1778	2.1307
15	2.5437	2.4753	2.4034	2.3275	2.2878	2.2468	2.2043	2.1601	2.1141	2.0658
16	2.4935	2.4247	2.3522	2.2756	2.2354	2.1938	2.1507	2.1058	2.0589	2.0096
17	2.4499	2.3807	2.3077	2.2304	2.1898	2.1477	2.1040	2.0584	2.0107	1.9604
18	2.4117	2.3421	2.2686	2.1906	2.1497	2.1071	2.0629	2.0166	1.9681	1.9168
19	2.3779	2.3080	2.2341	2.1555	2.1141	2.0712	2.0264	1.9795	1.9302	1.8780
20	2.3479	2.2776	2.2033	2.1242	2.0825	2.0391	1.9938	1.9464	1.8963	1.8432
21	2.3210	2.2504	2.1757	2.0960	2.0540	2.0102	1.9645	1.9165	1.8657	1.8117
22	2.2967	2.2258	2.1508	2.0707	2.0283	1.9842	1.9380	1.8894	1.8380	1.7831
23	2.2747	2.2036	2.1282	2.0476	2.0050	1.9605	1.9139	1.8648	1.8128	1.7570
24	2.2547	2.1834	2.1077	2.0267	1.9838	1.9390	1.8920	1.8424	1.7896	1.7330
25	2.2365	2.1649	2.0889	2.0075	1.9643	1.9192	1.8718	1.8217	1.7684	1.7110
26	2.2197	2.1479	2.0716	1.9898	1.9464	1.9010	1.8533	1.8027	1.7488	1.6906
27	2.2043	2.1323	2.0558	1.9736	1.9299	1.8842	1.8361	1.7851	1.7306	1.6717
28	2.1900	2.1179	2.0411	1.9586	1.9147	1.8687	1.8203	1.7689	1.7138	1.6541
29	2.1768	2.1045	2.0275	1.9446	1.9005	1.8543	1.8055	1.7537	1.6981	1.6376
30	2.1646	2.0921	2.0148	1.9317	1.8874	1.8409	1.7918	1.7396	1.6835	1.6223
40	2.0772	2.0035	1.9245	1.8389	1.7929	1.7444	1.6928	1.6373	1.5766	1.5089
60	1.9926	1.9174	1.8364	1.7480	1.7001	1.6491	1.5943	1.5343	1.4673	1.3893
120	1.9105	1.8337	1.7505	1.6587	1.6084	1.5543	1.4952	1.4290	1.3519	1.2539
∞	1.8307	1.7522	1.6664	1.5705	1.5173	1.4591	1.3940	1.3180	1.2214	1.0000

Denominator degrees of freedom

TABLE A-6

Critical Values of the Pearson Correlation Coefficient r

n	$\alpha = .05$	$\alpha = .01$
4	.950	.999
5	.878	.959
6	.811	.917
7	.754	.875
8	.707	.834
9	.666	.798
10	.632	.765
11	.602	.735
12	.576	.708
13	.553	.684
14	.532	.661
15	.514	.641
16	.497	.623
17	.482	.606
18	.468	.590
19	.456	.575
20	.444	.561
25	.396	.505
30	.361	.463
35	.335	.430
40	.312	.402
45	.294	.378
50	.279	.361
60	.254	.330
70	.236	.305
80	.220	.286
90	.207	.269
100	.196	.256

To test H_0: $\rho = 0$ against H_1: $\rho \neq 0$, reject H_0 if the absolute value of r is greater than the critical value in the table.

TABLE A-7

Critical Values for the Sign Test

n	.005 (one tail) .01 (two tails)	.01 (one tail) .02 (two tails)	.025 (one tail) .05 (two tails)	.05 (one tail) .10 (two tails)
1	*	*	*	*
2	*	*	*	*
3	*	*	*	*
4	*	*	*	*
5	*	*	*	0
6	*	*	0	0
7	*	0	0	0
8	0	0	0	1
9	0	0	1	1
10	0	0	1	1
11	0	1	1	2
12	1	1	2	2
13	1	1	2	3
14	1	2	2	3
15	2	2	3	3
16	2	2	3	4
17	2	3	4	4
18	3	3	4	5
19	3	4	4	5
20	3	4	5	5
21	4	4	5	6
22	4	5	5	6
23	4	5	6	7
24	5	5	6	7
25	5	6	7	7

NOTES:
1. * indicates that it is not possible to get a value in the critical region.
2. The null hypothesis is rejected if the number of the less frequent sign (x) is less than or equal to the value in the table.
3. For values of n greater than 25, a normal approximation is used with

$$z = \frac{(x + 0.5) - \left(\frac{n}{2}\right)}{\frac{\sqrt{n}}{2}}$$

TABLE A-8

Critical Values of T for the Wilcoxon Signed-Rank Test

n	.005 (one tail) .01 (two tails)	.01 (one tail) .02 (two tails)	.025 (one tail) .05 (two tails)	.05 (one tail) .10 (two tails)
			α	
5				1
6			1	2
7		0	2	4
8	0	2	4	6
9	2	3	6	8
10	3	5	8	11
11	5	7	11	14
12	7	10	14	17
13	10	13	17	21
14	13	16	21	26
15	16	20	25	30
16	19	24	30	36
17	23	28	35	41
18	28	33	40	47
19	32	38	46	54
20	37	43	52	60
21	43	49	59	68
22	49	56	66	75
23	55	62	73	83
24	61	69	81	92
25	68	77	90	101
26	76	85	98	110
27	84	93	107	120
28	92	102	117	130
29	100	111	127	141
30	109	120	137	152

Reject the null hypothesis if the test statistic T is less than or equal to the critical value found in this table. Fail to reject the null hypothesis if the test statistic T is greater than the critical value found in this table.

"Table of Critical Values of T for the Wilcoxon Signed-Rank Test" from *Some Rapid Approximate Statistical Procedures*, Copyright © 1949, 1964, Lederle Laboratories Division of American Cyanamid Company, All Rights Reserved, and Reprinted With Permission.

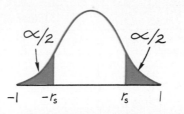

TABLE A-9

Critical Values of Spearman's Rank Correlation Coefficient r_s

n	$\alpha = 0.10$	$\alpha = 0.05$	$\alpha = 0.02$	$\alpha = 0.01$
5	.900	—	—	—
6	.829	.886	.943	—
7	.714	.786	.893	—
8	.643	.738	.833	.881
9	.600	.683	.783	.833
10	.564	.648	.745	.794
11	.523	.623	.736	.818
12	.497	.591	.703	.780
13	.475	.566	.673	.745
14	.457	.545	.646	.716
15	.441	.525	.623	.689
16	.425	.507	.601	.666
17	.412	.490	.582	.645
18	.399	.476	.564	.625
19	.388	.462	.549	.608
20	.377	.450	.534	.591
21	.368	.438	.521	.576
22	.359	.428	.508	.562
23	.351	.418	.496	.549
24	.343	.409	.485	.537
25	.336	.400	.475	.526
26	.329	.392	.465	.515
27	.323	.385	.456	.505
28	.317	.377	.448	.496
29	.311	.370	.440	.487
30	.305	.364	.432	.478

For $n > 30$ use $r_s = \pm z/\sqrt{n - 1}$, where z corresponds to the level of significance. For example, if $\alpha = 0.05$, then $z = 1.96$.

To test H_0: $\rho_s = 0$

against H_1: $\rho_s \neq 0$

E. G. Olds, "Distribution of sums of squares of rank differences to small numbers of individuals," *Annals of Statistics* 9 (1938): 133–148, and E. G. Olds, with amendment in *Annals of Statistics* 20 (1949): 117–118. Reprinted with permission.

TABLE A-10

Critical Values for Number of Runs G

	Value of n_2																		
	2	3	4	5	6	7	8	9	10	11	12	13	14	15	16	17	18	19	20
2	1	1	1	1	1	1	1	1	1	1	2	2	2	2	2	2	2	2	2
	6	6	6	6	6	6	6	6	6	6	6	6	6	6	6	6	6	6	6
3	1	1	1	1	2	2	2	2	2	2	2	2	2	3	3	3	3	3	3
	6	8	8	8	8	8	8	8	8	8	8	8	8	8	8	8	8	8	8
4	1	1	1	2	2	2	3	3	3	3	3	3	3	3	4	4	4	4	4
	6	8	9	9	9	10	10	10	10	10	10	10	10	10	10	10	10	10	10
5	1	1	2	2	3	3	3	3	3	4	4	4	4	4	4	4	5	5	5
	6	8	9	10	10	11	11	12	12	12	12	12	12	12	12	12	12	12	12
6	1	2	2	3	3	3	3	4	4	4	4	5	5	5	5	5	5	6	6
	6	8	9	10	11	12	12	13	13	13	13	14	14	14	14	14	14	14	14
7	1	2	2	3	3	3	4	4	5	5	5	5	5	6	6	6	6	6	6
	6	8	10	11	12	13	13	14	14	14	14	15	15	15	16	16	16	16	16
8	1	2	3	3	3	4	4	5	5	5	6	6	6	6	6	7	7	7	7
	6	8	10	11	12	13	14	14	15	15	16	16	16	16	17	17	17	17	17
9	1	2	3	3	4	4	5	5	5	6	6	6	7	7	7	7	8	8	8
	6	8	10	12	13	14	14	15	16	16	16	17	17	18	18	18	18	18	18
10	1	2	3	3	4	5	5	5	6	6	7	7	7	7	8	8	8	8	9
	6	8	10	12	13	14	15	16	16	17	17	18	18	18	19	19	19	20	20
11	1	2	3	4	4	5	5	6	6	7	7	7	8	8	8	9	9	9	9
	6	8	10	12	13	14	15	16	17	17	18	19	19	19	20	20	20	21	21
12	2	2	3	4	4	5	6	6	7	7	7	8	8	8	9	9	9	10	10
	6	8	10	12	13	14	16	16	17	18	19	19	20	20	21	21	21	22	22
13	2	2	3	4	5	5	6	6	7	7	8	8	9	9	9	10	10	10	10
	6	8	10	12	14	15	16	17	18	19	19	20	20	21	21	22	22	23	23
14	2	2	3	4	5	5	6	7	7	8	8	9	9	9	10	10	10	11	11
	6	8	10	12	14	15	16	17	18	19	20	20	21	22	22	23	23	23	24
15	2	3	3	4	5	6	6	7	7	8	8	9	9	10	10	11	11	11	12
	6	8	10	12	14	15	16	18	18	19	20	21	22	22	23	23	24	24	25
16	2	3	4	4	5	6	6	7	8	8	9	9	10	10	11	11	11	12	12
	6	8	10	12	14	16	17	18	19	20	21	21	22	23	23	24	25	25	25
17	2	3	4	4	5	6	7	7	8	9	9	10	10	11	11	11	12	12	13
	6	8	10	12	14	16	17	18	19	20	21	22	23	23	24	25	25	26	26
18	2	3	4	5	5	6	7	8	8	9	9	10	10	11	11	12	12	13	13
	6	8	10	12	14	16	17	18	19	20	21	22	23	24	25	25	26	26	27
19	2	3	4	5	6	6	7	8	8	9	10	10	11	11	12	12	13	13	13
	6	8	10	12	14	16	17	18·	20	21	22	23	23	24	25	26	26	27	27
20	2	3	4	5	6	6	7	8	9	9	10	10	11	12	12	13	13	13	14
	6	8	10	12	14	16	17	18	20	21	22	23	24	25	25	26	27	27	28

Value of n_1 (row labels, left axis)

The entries in this table are the critical G values assuming a two-tailed test with a significance level of $\alpha = 0.05$. The null hypothesis of randomness is rejected if the total number of runs G is less than or equal to the smaller entry or greater than or equal to the larger entry.

Adapted from C. Eisenhardt and F. Swed, "Tables for testing randomness of grouping in a sequence of alternatives," *Annals of Statistics* 14 (1943): 83–86. Reprinted with permission.

Appendix B

DATA SET I: REAL ESTATE DATA
(150 Randomly Selected Homes Recently Sold in Dutchess County, N.Y.)

Selling Price (dollars)	Living Area (sq. ft.)	Lot Size (acres)	Number of Rooms	Number of Baths
179,000	3,060	0.75	8	2
126,500	1,600	0.26	8	1.5
134,500	2,000	0.7	8	1
125,000	1,300	0.65	5	1
142,000	2,000	0.75	9	1.5
164,000	1,956	0.5	8	2.5
146,000	2,400	0.4	7	2.5
129,000	1,200	0.33	6	1
141,900	1,632	3	6	3
135,000	1,800	0.5	7	2
118,500	1,248	0.25	7	1
160,000	2,025	1.1	7	2
89,900	1,660	0.21	7	1
169,900	2,858	0.79	9	3
127,500	1,296	0.5	9	1
162,500	1,848	0.5	7	2.5
152,000	1,800	0.68	7	1.5
122,500	1,100	0.37	7	1
220,000	3,000	1.15	10	3.5
141,000	2,000	0.65	7	1
80,500	922	0.3	5	1
152,000	1,450	0.3	6	1.5
231,750	2,981	1.3	10	3.5
180,000	1,800	1.52	8	2.5

(continued)

Selling Price (dollars)	Living Area (sq. ft.)	Lot Size (acres)	Number of Rooms	Number of Baths
185,000	2,600	0.75	8	2
265,000	2,400	2	7	2
135,000	1,625	0.36	7	1.5
203,000	2,653	1.8	9	3
141,000	3,500	1	10	2.5
159,000	1,728	0.5	8	1.5
182,000	2,400	0.5	8	2.5
208,000	2,288	1.2	8	2.5
96,000	864	0.32	4	1
156,000	2,300	0.65	7	3
185,500	2,800	1.68	9	1.5
275,000	2,820	1	9	2.5
144,900	1,900	0.44	6	2
155,000	2,100	0.58	8	1.5
110,000	1,450	0.3	6	2
154,000	1,800	0.679	7	2
151,500	1,900	0.75	7	2
141,000	1,575	0.25	7	1.5
119,000	1,200	0.25	7	1
108,500	1,540	0.18	7	2
126,500	1,700	0.3037	8	2
302,000	2,130	11.91	8	1.5
130,000	1,800	0.3	7	1.5
140,000	1,650	0.5	8	2.5
123,500	1,362	0.4	7	2
153,500	3,700	1.1	10	3
194,900	2,080	1	8	2.5
165,000	2,320	0.4	8	2.5
179,900	2,790	0.75	13	2.5
194,500	2,544	0.28	9	2.5
127,500	1,850	0.26	9	2
170,000	2,277	0.8	8	3
160,000	1,900	1	8	2.5
135,000	1,400	0.35	6	2
117,000	1,248	0.3	6	1
235,000	3,150	0.3	11	4
223,000	1,680	14.37	8	2
163,500	2,276	1	8	2.5
78,000	821	2.3	4	1
187,000	2,080	1.23	8	2.5
133,000	1,100	0.33	6	1
125,000	1,200	0.33	5	1
116,000	1,100	1.1	6	1
135,000	1,800	1	8	2.5
194,500	2,300	0.91	8	2.5
99,500	1,000	0.49	4	1
152,500	1,786	0.3	8	2
141,900	1,950	0.75	8	2.5
139,900	1,839	2.6	7	1.5
117,500	1,300	0.29	6	1
150,000	1,564	0.3328	6	2

(continued)

Selling Price (dollars)	Living Area (sq. ft.)	Lot Size (acres)	Number of Rooms	Number of Baths
177,000	2,010	0.68	8	1.5
136,000	1,300	0.3	7	1
158,000	1,500	0.54	5	2.5
211,900	2,310	0.46	8	2.5
165,000	1,725	1.528	8	2.5
183,000	2,016	0.78	8	2.5
85,000	875	0.26	5	1
126,500	1,092	0.259	6	1
162,000	2,496	0.75	9	2.5
169,000	1,930	3	9	3
175,000	2,400	0.7	8	3
267,000	1,950	18.7	7	2.5
150,000	1,122	3.09	5	2
115,000	1,080	0.31	5	1
126,500	1,500	0.5	7	1.5
215,000	2,100	0.5	8	2.5
190,000	2,300	5.63	7	2.5
190,000	2,473	1.25	9	2.5
113,500	1,624	1.8	7	1.5
116,300	1,050	0.43	5	1.5
190,000	2,100	1.3	8	1.5
145,000	1,800	0.658	8	2.5
269,900	2,500	0.92	8	3
135,500	1,526	0.3	7	1.5
190,000	1,745	0.58	7	2.5
98,000	1,165	0.12	6	1
137,900	1,856	0.33	7	1.5
108,000	1,036	0.948	6	1
120,500	1,600	0.4	6	2
128,500	1,344	0.936	6	2
142,500	1,552	0.46	6	1.5
72,000	600	0.5	3	1
124,900	1,248	0.22	7	1
134,000	1,502	0.35	7	1.5
205,406	2,465	1.55	8	2.5
217,000	3,100	0.54	10	3.5
94,000	850	0.11	4	1
189,900	2,464	0.43	8	2.5
168,500	1,900	1.0636	7	2.5
133,000	2,000	0.5	8	2
180,000	2,272	0.41	9	2.5
139,500	1,610	0.45	8	1.5
210,000	2,100	0.5	8	2.5
126,500	1,050	1	5	1
285,000	2,516	8.1	7	2.5
195,000	2,265	0.85	8	2.5
97,000	1,300	0.37	5	1
117,000	1,008	0.5	6	1
150,000	1,600	1.84	7	2
180,500	2,000	0.6	9	2.5

(*continued*)

Selling Price (dollars)	Living Area (sq. ft.)	Lot Size (acres)	Number of Rooms	Number of Baths
160,000	1,760	0.05	7	2
181,500	2,250	0.33	9	2.5
124,000	1,783	0.22	8	1.5
125,900	1,118	0.56	7	1.5
165,000	2,680	0.5	9	3
122,000	1,950	0.5	7	1.5
132,000	2,000	0.108	8	2
145,900	1,680	0.5	6	1.5
156,000	3,000	0.5	11	2.5
136,000	1,750	0.5	7	2
142,000	1,500	0.41	7	1
140,000	1,403	0.5	6	2
144,900	1,450	0.3	7	1
133,000	1,908	0.46	7	2
196,800	1,960	1.33	8	2.5
121,900	1,300	0.78	6	1
126,000	1,232	0.314	6	2
164,900	1,980	0.7	8	2.5
172,000	2,100	1	8	2.5
100,000	1,338	0.12	6	1
129,900	1,070	1.69	5	1
110,000	1,289	0.25	6	1
131,000	1,066	0.33	5	1
107,000	1,100	0.17	5	1
165,900	1,840	1.162	8	2

DATA SET II: STUDENT DATA
(92 Students Taking Statistics)

Notes on Data

Sex: 0 represents female, 1 represents male.

Heights are in inches.

Weights are in pounds.

Pulse rates are in beats per minute.

Siblings: number of brothers plus the number of sisters.

Random: subjects were instructed to choose a three-digit random number.

SSN: subjects were instructed to write the last three digits of their social security numbers.

Hand: 0 represents left-handed, 1 represents right-handed.

Subject	Sex	Height	Weight	Pulse	Siblings	Random	SSN	Hand
1	0	66	115	34	2	213	001	1
2	1	66	128	82	5	169	224	1
3	1	70	150	74	2	812	212	1
4	0	65	107	65	4	125	097	1
5	1	71	160	77	3	749	900	1
6	1	69	140	62	5	137	327	1
7	0	64	110	70	2	202	406	1
8	0	64	128	52	2	344	551	1
9	0	63	130	72	2	496	126	1
10	0	66	104	80	3	348	091	1
11	1	73	163	78	0	714	523	1
12	1	70	155	58	2	765	103	1
13	0	65.5	120	90	1	831	541	1
14	1	69	175	85	5	491	066	1
15	1	67	145	74	2	169	174	1
16	0	68	125	67	2	312	366	0
17	1	73	145	66	3	263	508	1
18	1	62	180	71	3	192	392	1
19	0	62	125	62	2	584	361	1
20	0	63	140	85	3	968	045	1
21	0	63	189	80	9	377	621	1
22	1	71	170	58	4	403	436	1
23	0	64	112	99	3	372	854	1
24	1	68	154	80	2	123	032	1
25	1	71	185	65	4	493	002	1
26	0	71.75	150	76	3	894	281	1
27	0	51.5	120	75	3	016	452	1
28	1	72	180	60	0	682	762	1
29	0	62	110	88	1	390	776	1
30	1	71	155	54	1	123	669	1
31	0	66	140	92	1	325	327	1
32	0	64	115	76	3	734	635	1
33	0	62	104	80	3	316	506	1
34	1	67	147	75	1	357	028	0
35	1	66	190	71	6	945	739	1
36	0	66	140	92	2	208	459	1
37	0	63	114	40	1	115	721	1
38	0	68	135	69	2	776	675	1
39	0	67	130	66	4	143	939	1
40	1	72	155	74	1	628	867	0
41	1	70	170	73	1	479	348	1
42	1	71	152	82	2	316	585	1
43	0	63	138	85	2	229	177	1
44	1	70	150	84	1	781	105	1
45	0	66.5	116	78	2	628	829	1
46	0	66	120	87	2	356	772	1
47	0	66	138	92	4	195	894	0
48	1	73	150	78	3	199	635	1
49	1	71	190	92	3	223	783	1
50	1	72	115	73	1	114	965	0
51	0	66.5	120	72	3	264	072	1
52	0	66	140	48	1	308	198	1

(continued)

Subject	Sex	Height	Weight	Pulse	Siblings	Random	SSN	Hand
53	0	67	110	75	8	105	605	1
54	0	61.75	98	72	4	357	205	1
55	1	66	155	57	6	333	632	1
56	1	70	140	100	0.5	421	975	1
57	1	72	200	71	1	107	971	1
58	0	70	145	68	1	311	609	1
59	0	64	127	79	1	458	106	1
60	1	72	145	80	1	007	077	1
61	1	70	165	60	1	323	683	1
62	1	68	200	60	3	487	201	1
63	1	73	185	60	0	598	010	1
64	0	66	117	10	3	475	207	1
65	1	70	148	55	2	145	409	1
66	0	61	118	72	8	999	128	1
67	0	63	115	75	5	635	914	1
68	1	70	170	60	1	350	742	1
69	1	72	158	78	5	732	965	1
70	1	72	170	61	2	328	768	1
71	1	71	145	73	3	433	916	1
72	0	62	125	60	2	969	507	1
73	1	71	150	68	3	452	688	1
74	0	62	130	50	3	333	365	1
75	1	73	180	68	5	312	863	0
76	1	73.5	135	72	1	392	397	1
77	0	61	125	70	2	968	666	1
78	1	71	190	68	2	888	071	1
79	0	63	115	90	2	357	394	1
80	1	73	190	70	2	397	244	1
81	0	61	98	90	1	033	923	1
82	0	64	120	90	1	123	008	1
83	1	73	160	70	2	735	818	1
84	1	67	135	75	0	777	231	1
85	1	68	135	70	1	984	345	0
86	0	61	110	65	1	222	890	1
87	1	72	164	65	3	285	251	1
88	1	70	180	80	2	729	757	0
89	0	67.5	140	96	1	276	057	0
90	1	70	145	65	2	696	586	1
91	1	69	190	76	4	323	381	1
92	0	64	100	84	2	357	416	1

Appendix C: Minitab

There are two different levels of software recommended for use with this text. STATDISK, developed specifically for this text, is provided free to colleges that adopt this book. It is an easy-to-use software package designed for students with little or no prior computer experience. It is available for the IBM PC, Apple IIe, and compatible models. A separate *STATDISK Manual/Workbook* is available. We highly recommend STATDISK for those who can spare little or no class time for discussion of computer use. It is menu-driven, and students can easily use it on their own.

For those who incorporate computer usage as a major component of the course, we recommend Minitab, although other popular packages (such as BMDP, SAS, SPSS, or Microstat) can also be used with this text. An inexpensive version of Minitab is available from Benjamin/ Cummings Publishing Company, 390 Bridge Parkway, Redwood City, CA 94065.

Minitab is a command-driven language. This means that there are special words (commands) that are used to accomplish various tasks. We first list the important *general* commands that are used throughout the course, then we list *specific* commands that apply to each particular chapter. In the commands given below, we use CAPITAL letters to highlight the necessary part of the expression and we use lowercase letters to represent parts that may be omitted. When running Minitab, either capital letters or lowercase letters may be used. For example, the command "RANK C1 and put ranks in C2" may be entered in an abbreviated form as "RANK C1 C2" or "rank c1 c2."

General Commands

Minitab Command	Use

SET C1	Enter data in a column. For example,
	
	    SET C1
	    66 65 64 64 63
	
	causes the five scores to be stored in the column identified as C1.

READ C1 C2	Enter data in two different columns. Use READ C1 C2 C3 for three sets of scores, and so on. (Note: Both SET and READ allow you to enter data, but SET is generally better with *one* set of data, while READ tends to be better with two or more sets of data. With SET, you enter all of the data in one operation; with READ you enter data one row at a time.)
ENDOFDATA	Signals the end of the entry of data. Can be abbreviated as simply END.
NAME C1 'HEIGHT'	Gives the column of data C1 the more meaningful name of 'HEIGHT'. When such a name is used, it must always be enclosed within single quotes.
PRINT C1	Displays the data stored in the column C1. This is especially useful for verifying that data have been entered correctly. Also, PRINT 'HEIGHT' will display the scores in the column named 'HEIGHT'.
SAVE 'GRADES'	Saves the data in a computer file. To save on a disk in drive B, type: SAVE 'B:GRADES'. File names and column names should be different.
RETRIEVE 'GRADES'	Retrieves data previously saved in a computer file. To retrieve data from a disk in drive B, type: RETRIEVE 'B:GRADES'.
ERASE C1	Erases the data stored in column C1.
STOP	Signals the end of the use of Minitab.

In the sample run shown below and on the next page, five scores are entered in column C1 and printed. Then five pairs of data are entered in columns C1 and C2. They are named and printed. Everything immediately to the right of the symbol > is entered by the user; everything else is displayed by Minitab.

```
MTB > SET C1
DATA> 66 65 64 64 63
DATA> ENDOFDATA
MTB > PRINT C1

C1
    66    65    64    64    63
```

```
MTB > READ C1 C2
DATA> 66 115
DATA> 65 107
DATA> 64 110
DATA> 64 128
DATA> 63 130
DATA> ENDOFDATA
       5 ROWS READ
MTB > NAME C1 'HEIGHT'
MTB > NAME C2 'WEIGHT'
MTB > PRINT C1 C2

  ROW  HEIGHT  WEIGHT
    1      66     115
    2      65     107
    3      64     110
    4      64     128
    5      63     130
```

Miscellaneous notes:

1. Don't use commas in numbers. For example, enter 734527.80 instead of 734,527.80.

2. To correct an error of entering a wrong number:
   a. If you haven't yet hit the RETURN key, backspace and type over.
   b. If you have already hit the RETURN key, you can reenter the correct data set. If you prefer to replace, delete, or insert a score, use the formats suggested by the following examples.

LET C3(7) = 9   *Replaces* the 7th entry of column C3 with the number 9

DELETE 3 C5   *Deletes* the entry in the 3rd row of column C5

INSERT 5 6 C1 9   *Inserts* a 9 in column C1 between rows 5 and 6

3. Use LET to do arithmetic with data. For example,

LET C3 = C1 + C2   Column C3 is created by adding the column C1 values to those in column C2.

LET C2 = C2/2   Each entry of column C2 is divided by 2.

LET C5 = C2 - C1   Column C5 is created by subtracting the column C1 values from column C2.

4. Specific numbers have been used in many of the commands that follow, but other values may also be used. For example, BINOMIAL $n=5$ $p=.3$ can be replaced by BINOMIAL $n=32$ $p=.75$. The particular numbers used have been included only to illustrate typical cases.

Specific Minitab Commands

Chapter 2   (See page 104.)

`MEAN C1`	Gives the mean of the values in column C1
`MEDIAN C1`	Gives the median of the values in column C1
`STDEV C1`	Gives the standard deviation of the values in column C1
`SUM C1`	Gives the sum of the values in column C1
`SSQ C1`	Gives the sum of the squares of the values in column C1
`MAXIMUM C1`	Gives the maximum value found in column C1
`MINIMUM C1`	Gives the minimum value found in column C1
`HISTOGRAM C1`	Gives a histogram of the values found in column C1
`STEM-AND-LEAF C1`	Gives a stem-and-leaf plot of the values in column C1
`DESCRIBE C1`	Gives the mean, median, standard deviation, minimum, maximum, first and third quartiles, and the number of scores.

Chapter 3   (See example below.)

```
RANDOM 50 C1;
INTEGERS randomly selected from 1 to 6.
```
Randomly generates 50 integers, where each is between 1 and 6 inclusive. (INTEGERS provides a distribution in which the numbers are equally likely. Some other distributions are also available.)

```
RANDOM 6 C1;
UNIFORM 1 5.
```
Randomly generates 6 numbers between 1 and 5. Numbers come from a population with a uniform distribution.

```
MTB > RANDOM 50 C1;
SUBC> INTEGERS 1 6.
MTB > PRINT C1

C1
    4   4   2   5   2   3   1   3   4   5   2   2   2   4   5
    2   5   5   2   3   2   4   4   4   5   6   5   1   1   4
    6   1   3   3   1   2   2   3   1   2   6   4   6   1   2
    4   2   2   2   5

MTB > RANDOM 6 C1;
SUBC> UNIFORM 1 5.
MTB > PRINT C1

C1
    3.33638   1.04711   2.88852   1.06512   1.14015   2.51822
```

Chapter 4 (See pages 229 and 249.)

```
PDF;                      Gives probabilities for a binomial experiment in which n = 5
BINOMIAL n=5 p=.3.        and p = 0.3. (PDF represents a "probability density function.")

BOXPLOT C1               Gives a boxplot of the data in column C1
```

Chapter 5 (See example below.)

```
RANDOM 36 C1;
NORMAL mu=100 sigma=15.
```

Randomly generates 36 numbers from a normally distributed population with a mean of 100 and a standard deviation of 15.

```
MTB > RANDOM 36 C1;
SUBC> NORMAL mu=100 sigma=15.
MTB > PRINT C1
```

```
C1
    102.789     94.374    115.105    115.908     93.704    102.303    107.026
     89.323    108.569     98.781     84.783     74.635     81.170     90.567
     97.233    111.450    138.779     86.234     85.722     82.985    119.258
     97.943     84.047    100.943     73.494    103.590     98.168     96.171
     98.488    102.445     96.174    104.018     97.192     95.563     87.498
    114.200
```

Chapter 6   (See example below.)

```
ZTEST mean=100 sigma=15 with data in C1
```

Two-tailed test of the null hypothesis that the mean equals 100 where the population standard deviation is known to be 15. Sample data are in C1.

```
ZTEST mean=100 sigma=15;
ALTERNATIVE=+1.
```

Right-tailed test

```
ZTEST mean=100 sigma=15;
ALTERNATIVE=-1.
```

Left-tailed test

```
TTEST mean=100 with data in C1
```

Two-tailed t test of the null hypothesis that the mean equals 100. The sample data are in C1.

```
TTEST mean=100 with data in C1;
ALTERNATIVE=+1.
```

Right-tailed test

```
TTEST mean=100 with data in C1;
ALTERNATIVE=-1.
```

Left-tailed test

(*continued*)

```
MTB > SET C1
DATA> 106 103 102 101 99 103
DATA> 102 105 100 104 97 107
DATA> ENDOFDATA
MTB > TTEST mean=100 with data in C1

TEST OF mu=100.000 VS mu N.E. 100.000

          N       MEAN    STDEV    SE MEAN        T     P VALUE
C1       12    102.417    2.906      0.839     2.88       0.015
```

Chapter 7 (See page 408.)

`ZINTERVAL 95 percent and sigma = 15 for the data in C1`

> Constructs the 95% confidence interval estimate of the mean using the data in column C1. Assumes that the population standard deviation is known to be 15.

`TINTERVAL with 95 percent for data in C1`

> Construct the 95% confidence interval estimate of the mean using the data from column C1.

Chapter 8 (See page 460.)

`TWOSAMPLE 95 percent C1 C2`

> Construct the confidence interval and test the claim that two populations have the same mean. The first set of sample data is in C1 and the second set is in C2.

```
TWOSAMPLE 95 percent C1 C2;
POOLED.
```

> If the two population variances are believed to be approximately equal, we use this format to get the confidence interval and to test the claim that the two means are equal.

```
TWOSAMPLE 95 percent C1 C2;
POOLED;
ALTERNATIVE=+1.
```

> Same as above, except that the test is right-tailed. Use -1 for a left-tailed test.

Chapter 9 (See pages 532.)

`PLOT C1 VS C2`   Gives a scatter diagram.

`CORRELATION C1 and C2`

> Gives the value of the linear correlation coefficient r.

746

```
REGRESSION C2 1 C1
```
Gives the equation of the regression line. The second variable is expressed in terms of the first.

```
REGRESSION C3 2 C2 C1
```
Gives the multiple regression equation where the third variable is expressed in terms of the first two variables.

Chapter 10 (See pages 586 and 603.)

```
CHISQUARE analysis for C1 C2 C3
```
Gives the chi-square test statistic for a contingency table with columns C1, C2, C3.

```
AOVONEWAY C1 C2 C3
```
Uses analysis of variance (one-way) to test the claim that the three sets of data (in columns C1, C2, C3) have the same mean. Can also be used with more than three sets of data.

Chapter 11

```
LET C3=C2-C1
STEST difference in median=0 for data in C3,
```
Sign test for paired data in columns C1 and C2.

```
LET C3=C2-C1
WTEST for data in C3
```
Wilcoxon signed-ranks test for paired data in columns C1 and C2.

```
MANN-WHITNEY C1 C2
```
Wilcoxon rank-sum test (equivalent to the Mann-Whitney test) for the data in columns C1 and C2.

```
KRUSKAL-WALLIS C1 C2
```
Kruskal-Wallis test for the data in columns C1 and C2. Column C1 contains all sample data while column C2 contains identifying numbers.

```
RANK C1 and put ranks in C3
RANK C2 and put ranks in C4
CORRELATION C3 and C4
```
Gives the rank correlation coefficient.

```
RUNS test above and below 1179 for data in C1
```
Does runs test for randomness above and below the median of 1179 for the data in C1. (The median can be found by entering MEDIAN C1.)

Appendix D: Glossary

Addition rule. Rule for determining the probability that, on a single trial, either event A occurs, or event B occurs, or they both occur.

Alternative hypothesis. Denoted by H_1, the statement that is equivalent to the negation of the null hypothesis.

Analysis of variance. A method analyzing population variance in order to make inferences about the population.

ANOVA. See analysis of variance.

Arithmetic mean. The sum of a set of scores divided by the number of scores.

Average. Any one of several measures designed to reveal the central tendency of a collection of data.

Bimodal. Having two modes.

Binomial experiment. An experiment with a fixed number of independent trials. Each outcome falls into exactly one of two categories.

Binomial probability formula. See Formula 4-6 in Section 4-4.

Bivariate data. Data arranged in pairs.

Box-and-whisker diagram. Graphic method of showing the spread of a set of data.

Central limit theorem. Theorem stating that sample means tend to be normally distributed.

Centroid. The point $(\bar{x}, \bar{y})$ determined from a collection of bivariate data.

Chebyshev's theorem. Uses standard deviation to provide information about the distribution of data. See Section 2-5.

Chi-square distribution. Continuous probability distribution with selected values (Table A-4).

Class boundaries. Values obtained from a frequency table by increasing the upper class limits and decreasing the lower class limits by the same amount so that there are no gaps between consecutive classes.

Classical approach to probability. Determining the probability of an event by dividing the number of ways the event can occur by the total number of possible outcomes.

Class marks. The midpoints of the classes in a frequency table.

Class width. The difference between two consecutive lower class limits in a frequency table.

Cluster sampling. Population is divided into sections and a sample of sections is randomly selected.

Coefficient of determination. The amount of the variation in y that is explained by the regression line.

Combinations rule. Rule for determining the number of different combinations of selected items.

Complement of an event. All outcomes in which the original event does not occur.

Completely randomized design. In analysis of variance, each element is given the same chance of belonging to the different categories or treatments.

Compound event. A combination of simple events.

Conditional probability. The conditional probability of event B, given that event A has already occurred, is $P(A \text{ and } B)/P(A)$.

Confidence interval. A range of values used to estimate some population parameter with a specific level of confidence.

Confidence interval limits. The two numbers that are used as the high and low boundaries of a confidence interval.

Contingency table. A table of observed frequencies where the rows correspond to one variable of classification and the columns correspond to another variable of classification.

Continuity correction. An adjustment made when a discrete random variable is being approximated by a continuous random variable (see Section 5-5).

Continuous random variable. A random variable with infinite values that can be associated with points on a continuous line interval.

Convenience sampling. Sample data are selected because they are readily available.

Correlation. Statistical association between two variables.

Correlation coefficient. A measurement of the strength of the relationship between two variables.

Countable set. A set with either a finite number of values or values that can be made to correspond to the positive integers.

Critical region. The area under a curve containing the values that lead to rejection of the null hypothesis.

Critical value. Value separating the critical region from the values of the test statistic that would not lead to rejection of the null hypothesis.

Cumulative frequency table. Frequency table in which each class and frequency represents cumulative data up to and including that class.

Data. The numbers or information collected in an experiment.

Decile. The 9 deciles divide the ranked data into 10 groups with 10% of the scores in each group.

Degree of confidence. Probability that a population parameter is contained within a particular confidence interval.

Degrees of freedom. The number of values that are free to vary after certain restrictions have been imposed on all values.

Denominator degrees of freedom. The degrees of freedom corresponding to the denominator of the F test statistic.

Dependent events. Events that are not independent. (See independent events.)

Dependent samples. The values in one sample are related to the values in another sample.

Descriptive statistics. The methods used to summarize the key characteristics of known population data.

Discrete random variable. A random variable with either a finite number of values or a countable number of values.

Efficiency. Measure of the sensitivity of a nonparametric test in comparison to a corresponding parametric test.

Empirical approximation of probability. Estimated value of probability based on actual observations.

Empirical rule. Uses standard deviation to provide information about data with a bell-shaped distribution. See Section 2-5.

Event. A result or outcome of an experiment.

Expected frequency. Theoretical frequency for a cell of a contingency table or multinomial table.

Expected value. For a discrete random variable, the sum of the products obtained by multiplying each value of the random variable by the corresponding probability.

Experiment. Process that allows observations to be made.

Explained deviation. For one pair of values in a collection of bivariate data, the difference between the predicted y value and the mean of the y values.

Explained variation. The sum of the squares of the explained deviations for all pairs of bivariate data in a sample.

Exploratory data analysis (EDA). Branch of statistics emphasizing the investigation of data.

Factorial rule. n different items can be arranged $n!$ different ways.

F distribution. A continuous probability distribution with selected values given in Table A-5.

Finite population correction factor. Factor for correcting the standard error of the mean when a sample size exceeds 5% of the size of a finite population.

Frequency polygon. Graphical method for representing the distribution of data using connected straight-line segments.

Frequency table. A list of categories of scores along with their corresponding frequencies.

Fundamental counting rule. For a sequence of two events in which the first event can occur m ways and the second can occur n ways, the events together can occur a total of $m \cdot n$ ways.

Hinge. The median value of the bottom (or top) half of a set of ranked data.

Histogram. A graph of vertical bars representing the frequency distribution of a set of data.

H test. See the Kruskal-Wallis test.

Hypothesis. A statement or claim that some population characteristic is true.

Hypothesis test. A method for testing claims made about populations. Also called test of significance.

Independent events. The case when the occurrence of any one of the events does not affect the probabilities of the occurrences of the other events.

Independent samples. The values in one sample are not related to the values in another sample.

Inferential statistics. The methods of using sample data to make generalizations or inferences about a population.

Interquartile range. The difference between the first and third quartiles.

Interval. Level of measurement of data: data can be arranged in order, and differences between data values are meaningful.

Interval estimate. (See confidence interval.)

Kruskal-Wallis test. A nonparametric hypothesis test used to compare three or more independent samples.

Least-squares property. For a regression line, the sum of the squares of the vertical deviations of the sample points from the regression line is the smallest sum possible.

Left-tailed test. Hypothesis test in which the critical region is located in the extreme left area of the probability distribution.

Linear correlation coefficient. Measure of strength of relationship between two variables.

Lower class limits. The smallest numbers that can actually belong to the different classes in a frequency table.

Maximum error of estimate. The largest difference between a point estimate and the true value of a population parameter.

Mean. The sum of a set of scores divided by the number of scores.

Mean deviation. The measure of dispersion equal to the sum of the deviations of each score from the mean, divided by the number of scores.

Measure of central tendency. Value intended to indicate the center of the scores in a collection of data.

Measure of dispersion. Any of several measures designed to reflect the amount of variability among a set of scores.

Median. The middle value of a set of scores arranged in order of magnitude.

Midquartile. One-half of the sum of the first and third quartiles.

Midrange. One-half the sum of the highest and lowest scores.

Mode. The score that occurs most frequently.

Multimodal. Having more than two modes.

Multinomial experiment. An experiment with a fixed number of independent trials and each outcome falls into exactly one of several categories.

Multiple coefficient of determination. Measure of how well a multiple regression equation fits the sample data.

Multiple regression. Study of linear relationships among three or more variables.

Multiplication rule. Rule for determining the probability that event A will occur on one trial while event B occurs on a second trial.

Mutually exclusive events. Events that cannot occur simultaneously.

Nominal. Level of measurement of data: data consist of names, labels, or categories only.

Nonparametric methods. Statistical procedures for testing hypotheses or estimating parameters; they are not based on population parameters and do not require many of the restrictions of parametric tests.

Nonsampling errors. Errors from external factors not related to sampling.

Normal distribution. A bell-shaped probability distribution described algebraically by Equation 5-1 in Section 5-1.

Null hypothesis. Denoted by H_0, it is the claim made about some population characteristic. It usually involves the case of no difference.

Numerator degrees of freedom. The degrees of freedom corresponding to the numerator of the F test statistic.

Observed frequency. The actual frequency count recorded in one cell of a contingency table or multinomial table.

Odds against. The odds against event A are obtained by finding $P(\bar{A})/P(A)$, usually expressed in the form of $a{:}b$ where a and b are integers having no common factors.

Odds in favor. The odds in favor of event A are obtained by finding $P(A)/P(\bar{A})$, usually expressed as the ratio of two integers with no common factors.

Ogive. Graphical method of representing a cumulative frequency table.

One-way analysis of variance. Analysis of variance involving data classified into groups according to a single criterion only.

Ordinal. Level of measurement of data: data may be arranged in order, but differences between data values either cannot be determined or they are meaningless.

Parameter. A measured characteristic of a population.

Parametric methods. Statistical procedures for testing hypotheses or estimating parameters; based on population parameters.

Pearson's product moment. (See linear correlation coefficient.)

Percentile. The 99 percentiles divide the ranked data into 100 groups with 1% of the scores in each group.

Permutations rule. Rule for determining the number of different arrangements of selected items.

Pie chart. Graphical method for representing data in the form of a circle containing wedges.

Point estimate. A single value that serves as an estimate of a population parameter.

Pooled estimate of p_1 and p_2. The probability obtained by combining the data from two sample proportions and dividing the total number of successes by the total number of observations.

Population. The complete and entire collection of elements to be studied.

Predicted value. Using a regression equation, the value of one variable given a value for the other variable.

Probability. A measure of the likelihood that a given event will occur. Mathematical probabilities are expressed as numbers between 0 and 1.

Probability distribution. Collection of values of a random variable along with their corresponding probabilities.

P-value. The probability that a test statistic in a hypothesis test is at least as extreme as the one actually obtained.

Quartile. The three quartiles divide the ranked data into four groups with 25% of the scores in each group.

Random sample. A sample selected in a way that allows every member of the population to have the same chance of being chosen.

Random selection. Sample elements are selected in such a way that all elements available for selection have the same chance of being selected.

Random variable. The values that correspond to the numbers associated with events in a sample space.

Range. The measure of dispersion that is the difference between the highest and lowest scores.

Rank. The numerical position of an item in a sample set arranged in order.

Rank correlation coefficient. Measure of the strength of the relationship between two variables; based on the ranks of the values.

Ratio. Level of measurement of data: data can be arranged in order, differences between data values are meaningful, and there is an inherent zero starting point.

Regression line. A straight line that summarizes the relationship between two variables.

Right-tailed test. Hypothesis test in which the critical region is located in the extreme right area of the probability distribution.

Rigorously controlled design. In analysis of variance, all factors are forced to be constant so that effects of extraneous factors are eliminated.

Run. Used in the runs test for randomness, a sequence of data exhibiting the same characteristic.

Runs test. Nonparametric method used to test for randomness.

Sample. A subset of a population.

Sample space. In an experiment, the set of all possible outcomes or events that cannot be further broken down.

Sampling errors. Errors resulting from the sampling process itself.

Scatter diagram. Graphical method for displaying bivariate data.

Semi-interquartile range. One-half of the difference between the first and third quartiles.

Significance level. The probability that serves as a cutoff between results attributed to chance and results attributed to significant differences.

Sign test. A nonparametric hypothesis test used to compare samples from two populations.

Simple event. An experimental outcome that cannot be further broken down.

Single factor analysis of variance. (See one-way analysis of variance.)

Slope. Measure of steepness of a straight line.

Spearman's rank correlation coefficient. (See rank correlation coefficient.)

Standard deviation. The measure of dispersion equal to the square root of the variance.

Standard error of estimate. Measure of spread of sample points about the regression line.

Standard error of the mean. The standard deviation of all possible sample means $\bar{x}$.

Standard normal distribution. A normal distribution with a mean of 0 and a standard deviation equal to 1.

Standard score. Also called z score, it is the number of standard deviations that a given value is above or below the mean.

Statistic. A measured characteristic of a sample.

Statistics. The collection, organization, description, and analysis of data.

Stem-and-leaf plot. Method of sorting and arranging data to reveal the distribution.

Stratified sampling. Samples are drawn from each stratum (class).

Student t distribution. (See t distribution.)

Systematic sampling. Every kth element is selected for a sample.

t distribution. A bell-shaped distribution usually associated with small sample experiments. Also called the Student t distribution.

10–90 percentile range. The difference between the 10th and 90th percentiles.

Test of significance. (See hypothesis test.)

Test statistic. Used in hypothesis testing, it is the sample statistic based on the sample data.

Total deviation. The sum of the explained deviation and unexplained deviation for a given pair of values in a collection of bivariate data.

Total variation. The sum of the squares of the total deviation for all pairs of bivariate data in a sample.

Tree diagram. Graphical depiction of the different possible outcomes in a compound event.

Two-tailed test. Hypothesis test: the critical region is divided between the left and right extreme areas of the probability distribution.

Two-way table. (See contingency table.)

Type I error. The mistake of rejecting the null hypothesis when it is true.

Type II error. The mistake of failing to reject the null hypothesis when it is false.

Unexplained deviation. For one pair of values in a collection of bivariate data: difference between y coordinate and predicted value.

Unexplained variation. The sum of the squares of the unexplained deviations for all pairs of bivariate data in a sample.

Uniform distribution. A distribution of values evenly distributed over the range of possibilities.

Upper class limits. Largest numbers that can belong to the different classes in a frequency table.

Variance. The measure of dispersion found by using Formula 2-5 in Section 2-5.

Variance between samples. In analysis of variance, the variation among the different samples.

Variation within samples. In analysis of variance, the variation that is due to chance.

Variation due to error. In analysis of variance, the variation within samples that is due to chance.

Variation due to treatment. (See variance between samples.)

Weighted mean. Mean of a collection of scores that have been assigned different degrees of importance.

Wilcoxon rank-sum test. A nonparametric hypothesis test used to compare two independent samples.

Wilcoxon signed-ranks test. A nonparametric hypothesis test used to compare two dependent samples.

y-intercept. Point at which a straight line crosses the y-axis.

z score. (See standard score.)

Appendix E: Bibliography

Adler, I. 1966. *Probability and Statistics for Everyman*. New York: New American Library.

Anderson, R., and T. Bancroft. 1952. *Statistical Theory in Research*. New York: McGraw-Hill.

Armore, S. 1975. *Statistics: A Conceptual Approach*. Columbus, Ohio: Charles E. Merrill.

Bacheller, M. 1978. *The Hammond Almanac*. Maplewood, N.J.: Hammond Almanac.

Barnett, V. 1973. *Comparative Statistical Inference*. New York: Wiley.

Bashaw, W. 1969. *Mathematics for Statistics*. New York: Wiley.

Brook, R., and others, eds. 1986. *The Fascination of Statistics*. New York: Dekker.

Byrkit, D. 1987. *Statistics Today: A Comprehensive Introduction*. Menlo Park, Calif.: Benjamin/ Cummings.

Chou, Y. 1975. *Statistical Analysis*. 2nd ed. New York: Holt, Rinehart & Winston.

Christensen, H. 1977. *Statistics Step by Step*. Boston: Houghton Mifflin.

Cochran, W. 1982. *Contributions to Statistics*. New York: Wiley.

Conover, W. 1980. *Practical Nonparametric Statistics*. 2nd ed. New York: Wiley.

Devore, J., and R. Peck. 1986. *Statistics: The Exploration and Analysis of Data*. St. Paul, Minn.: West Publishing.

Elzey, F. 1966. *A Programmed Introduction to Statistics*. Belmont, Calif.: Brooks/Cole.

Fairley, W., and F. Mosteller. 1977. *Statistics and Public Policy*. Reading, Mass.: Addison-Wesley.

Fisher, R. 1966. *The Design of Experiments*. 8th ed. New York: Hafner.

Freedman, D., R. Pisani, and R. Purves. 1978. *Statistics*. New York: Norton.

Freund, J. 1988. *Modern Elementary Statistics*. 7th ed. Englewood Cliffs, N.J.: Prentice-Hall.

Freund, J., and E. Walpole. 1987. *Mathematical Statistics*. 4th ed. Englewood Cliffs, N.J.: Prentice-Hall.

Freund, J., and R. Smith. 1986. *Statistics: A First Course*. 4th ed. Englewood Cliffs, N.J.: Prentice-Hall.

Goldman, R., and J. Weinberg. 1985. *Statistics: An Introduction*. Englewood Cliffs, N.J.: Prentice-Hall.

Grant, E. 1964. *Statistical Quality Control*. 3rd ed. New York: McGraw-Hill.

Guenther, W. 1973. *Concepts of Statistical Inference*. 2nd ed. New York: McGraw-Hill.

Haber, A., and R. Runyon. 1973. *General Statistics*. 2nd ed. Reading, Mass.: Addison-Wesley.

Hamburg, M. 1977. *Statistical Analysis for Decision Making*. 2nd ed. New York: Harcourt Brace Jovanovich.

Hauser, P. 1975. *Social Statistics in Use*. New York: Russell Sage Foundation.

Heerman, E., and L. Braskam. 1970. *Readings in Statistics for the Behavioral Sciences*. Englewood Cliffs, N.J.: Prentice-Hall.

Hoaglin, D., F. Mosteller, and J. Tukey, eds. 1983. *Understanding Robust and Exploratory Data Analysis*. New York: Wiley.

Hoel, P. 1983. *Elementary Statistics*. 4th ed. New York: Wiley.

Hollander, M., and D. Wolfe. 1973. *Nonparametric Statistical Methods*. New York: Wiley.

Hooke, R. 1983. *How to Tell the Liars from the Statisticians*. New York: Dekker.

Huff, D. 1954. *How to Lie with Statistics*. New York: Norton.

Johnson, R. 1988. *Elementary Statistics*. Boston: Duxbury Press.

Johnson, R., and G. Bhattacharyya. 1985. *Statistics: Principles and Methods*. New York: Wiley.

Kimble, G. 1978. *How to Use (and Misuse) Statistics*. Englewood Cliffs, N.J.: Prentice-Hall.

King, R., and B. Julstrom. 1982. *Applied Statistics Using the Computer*. Sherman Oaks, Calif.: Alfred.

Kirk, R., ed. 1972. *Statistical Issues: A Reader for the Behavioral Sciences*. Belmont, Calif.: Brooks/Cole.

Kotz, S., and D. Stroup. 1983. *Educated Guessing—How to Cope in an Uncertain World*. New York: Dekker.

Langley, R. 1970. *Practical Statistics Simply Explained*. New York: Dover.

Lapin, L. 1975. *Statistics: Meaning and Method*. New York: Harcourt Brace Jovanovich.

Lindley, D. 1971. *Making Decisions*. New York: Wiley.

McClave, J., and F. Dietrich. 1985. *Statistics*. 3rd ed. Riverside, N.J.: Dellen/Macmillan.

McClave, J., and P. Benson. 1982. *Statistics for Business and Economics*. San Francisco: Dellen.

McGhee, J. 1985. *Introductory Statistics*. St. Paul, Minn.: West Publishing.

Mendenhall, W. 1987. *Introduction to Probability and Statistics*. 7th ed. Boston: Duxbury Press.

Mood, A., and others. 1974. *Introduction to the Theory of Statistics*. 3rd ed. New York: McGraw-Hill.

Moore, D. 1979. *Statistics: Concepts and Controversies*. San Francisco: Freeman.

Mosteller, F., and R. Rourke. 1973. *Sturdy Statistics, Nonparametrics and Order Statistics*. Reading, Mass.: Addison-Wesley.

Mosteller, F., R. Rourke, and G. Thomas, Jr. 1970. *Probability with Statistical Applications*. 2nd ed. Reading, Mass.: Addison-Wesley.

Neter, J., W. Wasserman, and M. Kutner. 1985. *Applied Linear Statistical Models*. Homewood, Ill.: Irwin.

Nobile, P., and J. Deedy, eds. 1972. *The Complete Ecology Fact Book*. Garden City, N.Y.: Doubleday.

Noether, G. 1976. *Elements of Nonparametric Statistics*. 2nd ed. New York: Wiley.

Ott, L., and W. Mendenhall. 1985. *Understanding Statistics*. 4th ed. Boston: Duxbury Press.

Owen, D. 1962. *Handbook of Statistical Tables*. Reading, Mass.: Addison-Wesley.

Raiffa, H. 1968. *Decision Analysis: Introductory Lectures on Choices Under Uncertainty*. Reading, Mass.: Addison-Wesley.

Reichard, R. 1974. *The Figure Finaglers*. New York: McGraw-Hill.

Reichmann, W. 1962. *Use and Abuse of Statistics*. New York: Oxford University Press.

Roscoe, J. 1975. *Fundamental Research Statistics for the Behavioral Sciences*. 2nd ed. New York: Holt, Rinehart & Winston.

Ryan, T., B. Joiner, and B. Ryan. 1985. *Minitab Student Handbook*. 2nd ed. Boston: Duxbury.

Schmid, C. 1983. *Statistical Graphics*. New York: Wiley.

Siegal, S. 1956. *Nonparametric Statistics for the Behavioral Sciences*. New York: McGraw-Hill.

Snedecor, G., and W. Cochran. 1980. *Statistical Methods*. 7th ed. Ames, Iowa: Iowa State University Press.

Spear, M. 1969. *Practical Charting Techniques*. New York: McGraw-Hill.

Steger, J., ed. 1971. *Readings in Statistics for the Behavioral Sciences*. New York: Holt, Rinehart & Winston.

Tanur, J., ed. 1978. *Statistics: A Guide to the Unknown*. 2nd ed. San Francisco: Holden-Day.

Tukey, J. 1977. *Exploratory Data Analysis*. Reading, Mass.: Addison-Wesley.

Walker, H., and J. Lev. 1969. *Elementary Statistical Methods*. 3rd ed. New York: Holt, Rinehart & Winston.

Wayne, D. 1978. *Applied Nonparametric Statistics*. Boston: Houghton Mifflin.

Weisberg, S. 1985. *Applied Linear Regression*. 2nd ed. New York: Wiley.

Winer, B. 1971. *Statistical Principles of Experimental Design*. 2nd ed. New York: McGraw-Hill.

Winkler, R., and W. Hays. 1975. *Statistics: Probability, Inference and Decision*. 2nd ed. New York: Holt, Rinehart & Winston.

Witte, R. 1985. *Statistics*. 2nd ed. New York: Holt, Rinehart & Winston.

Wonnacott, R. and T. Wonnacott. 1985. *Introductory Statistics*. 4th ed. New York: Wiley.

Yamane, T. 1973. *Statistics: An Introductory Analysis*. 3rd ed. New York: Harper & Row.

Zeisel, H. 1968. *Say It with Figures*. 5th ed. New York: Harper & Row.

Appendix F: Answers to Odd-Numbered Exercises

Chapter 1

1-1. The 800 respondents are a sample. The population consists of all divorced people.

1-3. One answer: In recent years a large proportion of women with no prior experience have entered the job market for the first time.

1-5. Unlisted numbers would be excluded and your sample could be biased and not representative.

1-7. (a) Ordinal   (d) Interval
     (b) Ratio   (e) Nominal
     (c) Nominal

1-9. Alumni with lower salaries would be less inclined to respond, so the reported salaries will tend to be disproportionately higher. Also, those who cheat on their income taxes might not want to reveal their true income. Others may want to retain privacy.

1-11. 50%. Not necessarily. No matter how good the operating levels are, about 50% will be below average.

1-13. a, c, d

1-15. Respondents sometimes tend to round off to a nice even number like 50.

1-17. The figure is very precise, but it is probably not very accurate. The use of a precise number may incorrectly suggest that it is also accurate.

1-19. A family is a group of two or more people related by birth, marriage, or adoption who are living together in a household. A household consists of one or more people sharing a housing unit (house, apartment, etc.). A person living alone would count as a household but not a family.

1-21. (a) ratio   (b) interval

1-23. Fahrenheit temperatures are at the interval level of measurement, so that ratios are not meaningful. Three times 300° F is not the same as 900° F.

1-25. The bars are not drawn in their proper proportions.

1-27. Sample: the 703 business owners surveyed
Population: all owners of businesses with fewer than 500 employees

1-29. Since there are groups of 20 animals each, all percentages of success should be multiples of 5. The given percentages cannot be correct.

1-31. According to the *New York Times*, "it would have to remove all the plaque, remove it again, and then remove it for a third time plus some more still."

Chapter 2

2-1. 5, 8, 11, 14, 17

2-3. 2.4, 4.9, 7.4, 9.9, 12.4

2-5. 30, 50, 70, 90

2-7. 10.0, 12.5, 15.0, 17.5, 20.0, 22.5, 25.0

2-9. 5   2-11. 0.25

2-13. 8   2-15. 5.0

2-17. 83.5, 91.5, 99.5, 107.5, 115.5

2-19. 18.65, 23.65, 28.65, 33.65, 38.65

2-21. Lower: 79.5, 87.5, 95.5, 103.5, 111.5
Upper: 87.5 95.5, 103.5, 111.5, 119.5

2-23. Lower: 16.15, 21.15, 26.15, 31.15, 36.15
Upper: 21.15, 26.15, 31.15, 36.15, 41.15

2-25.

I.Q.	Freq.
Less than 88	16
Less than 96	53
Less than 104	103
Less than 112	132
Less than 120	146

2-27.

Height	Cum. Freq.
Less than 21.2	16
Less than 26.2	31
Less than 31.2	43
Less than 36.2	51
Less than 41.2	54

2-29. 80–84, 85–89, 90–94, 95–99, 100–104, 105–109, 110–114, 115–119

2-31. 16.0–19.9, 20.0–23.9, 24.0–27.9, 28.0–31.9, 32.0–35.9, 36.0–39.9, 40.0–43.9

2-33. 110–129

2-35. 17.3–19.4

2-37.

Miles	Frequency
1– 5	11
6–10	9
11–15	6
16–20	5
21–25	6
26–30	0
31–35	2
36–40	1

2-39.

Kwh	Frequency
714–733	4
734–753	3
754–773	1
774–793	6
794–813	0
814–833	5
834–853	5
854–873	4
874–893	2
894–913	1

2-41. The distributions appear to be very different. The distributions are very difficult to observe from the raw data.

2-43. The third guideline is clearly violated since the class width varies. Although the class limits don't appear to be convenient numbers (5th guideline), they do correspond to special classes, including preschool (under 5), elementary school (5–13), and high school (14–17).

2-45. (a) 0.5, 8.5, 16.5, 24.5, 32.5, 40.5
(b) 4.5, 12.5, 20.5, 28.5, 36.5
(c) 0.5, 8.5, 16.5, 24.5, 32.5, 40.5

2-47.

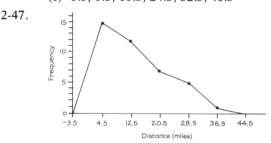

2-49. (a) −0.5, 3999.5, 7999.5, 11999.5, 15999.5, 19999.5, 23999.5, 27999.5
(b) 1999.5, 5999.5, 9999.5, 13999.5, 17999.5, 21999.5, 25999.5
(c) −0.5, 3999.5, 7999.5, 11999.5, 15999.5, 19999.5, 23999.5, 27999.5

2-51.

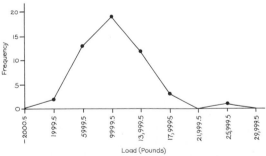

2-53. (a) 699.5, 719.5, 739.5, 759.5, 779.5, 799.5, 819.5, 839.5, 859.5, 879.5, 899.5, 919.5
(b) 709.5, 729.5, 749.5, 769.5, 789.5, 809.5, 829.5, 849.5, 869.5, 889.5, 909.5
(c) 699.5, 719.5, 739.5, 759.5, 779.5, 799.5, 819.5, 839.5, 859.5, 879.5, 899.5, 919.5

2-55.

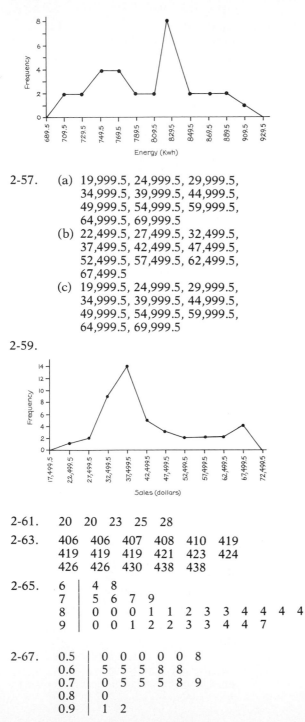

2-57. (a) 19,999.5, 24,999.5, 29,999.5,
34,999.5, 39,999.5, 44,999.5,
49,999.5, 54,999.5, 59,999.5,
64,999.5, 69,999.5
(b) 22,499.5, 27,499.5, 32,499.5,
37,499.5, 42,499.5, 47,499.5,
52,499.5, 57,499.5, 62,499.5,
67,499.5
(c) 19,999.5, 24,999.5, 29,999.5,
34,999.5, 39,999.5, 44,999.5,
49,999.5, 54,999.5, 59,999.5,
64,999.5, 69,999.5

2-59.

2-61. 20 20 23 25 28

2-63. 406 406 407 408 410 419
419 419 419 421 423 424
426 426 430 438 438

2-65.
6	4 8
7	5 6 7 9
8	0 0 0 1 1 2 3 3 4 4 4 4 5 6 9
9	0 0 1 2 2 3 3 4 4 7

2-67.
0.5	0 0 0 0 0 8
0.6	5 5 5 8 8
0.7	0 5 5 5 8 9
0.8	0
0.9	1 2

2-69.

x	f
0– 9	28
10– 19	22
20– 29	20
30– 39	22
40– 49	21
50– 59	11
60– 69	8
70– 79	0
80– 89	4
90– 99	1
100–109	2

2-71. (a)

x	f
15	1
16	3
17	2
18	7
19	6
20	9
21	8
22	3
23	5
24	2
25	1

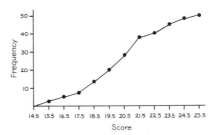

(c) 28   (d) 8.5%

2-73. (a) The histogram for π is more
uniform.
(b) π is irrational, whereas 22/7 is
rational. The decimal form of π is
nonrepeating, whereas the decimal
form of 22/7 has the digits 142857
repeating.

2-75. The heights of the bars will be approximately halved. Also, depending on the
data, the general shape may be changed.

2-77. Mean 3.1; median 3.0; mode 3; midrange 3.5

2-79. Mean 71.5; median 72.0; mode 77; midrange 71.0

2-81. Mean 9.3; median 9.0; mode
9; midrange 11.0

2-83. Mean 8.33; median 8.40; mode 8.40; midrange 8.40

2-85. Mean 1.3339; median 1.2780; mode
None; midrange 1.4185

2-87. Mean 0.709; median 0.725; mode 0.57,
0.71, 0.74, 0.85; midrange 0.7000

2-89. Mean 13.3; median 11.0; mode 1, 5; midrange 20.5

2-91. Mean 806.9; median 818.0; mode 752, 774; midrange 811.0

2-93. (a) 19.5, 59.5, 99.5, 139.5, 179.5 (b) 95.7

2-95. (a) 2, 3, 4, 5, 6, 7, 8 (b) 3.2

2-97. (a) 0.105, 0.125, 0.145, 0.165, 0.185 (b) 0.131

2-99. (a) 1249.5, 4999.5, 9999.5, 13,749.5, 17,499.5, 22,499.5, 29,999.5, 42,499.5
    (b) 25,032.8

2-101. (a) Mean: 1029.66; median: 1026.70; mode: none; midrange: 1031.95.
    (b) Mean: 29.66; median: 26.70; mode: none; midrange: 31.95.
    (c) The part (b) answers are 1000 less.
    (d) They change by k.
    (e) Mean: 0.02966; median: 0.02670; mode: none; midrange: 0.03195.
    (f) Mean: 296.6; median: 267.0; mode: none; midrange: 319.5.
    (g) The part (f) answers are 10,000 times the answers from part (e).
    (h) They are also multiplied by k.
    (i) Some unmanageable data sets can be made manageable by adding or subtracting a constant or by scaling the numbers up or down through multiplication or division.

2-103. The averages are the same for both sets and do not reveal any differences, but the lists are different in the degree of variation among the scores.

2-105. Mean is 503.6 instead of 506.6. The result from Exercise 2-100 is likely to be distorted less.

2-107. 1.092

2-109. Skewed right: 2-77, 2-78; skewed left: 2-79, 2-80.

2-111. $147,606

2-113. Range 3.0; variance 0.8; standard deviation 0.9

2-115. Range 12.0; variance 22.7; standard deviation 4.8

2-117. Range 8.0; variance 4.7; standard deviation 2.2

2-119. Range 6.40; variance 3.83; standard deviation 1.96

2-121. Range 0.8110; variance 0.0814; standard deviation 0.2854

2-123. Range 0.580; variance 0.026; standard deviation 0.161

2-125. Range 39.0; variance 92.4; standard deviation 9.6

2-127. Range 194.0; variance 2763.1; standard deviation 52.6

2-129. Range 8.0; variance 4.7; standard deviation 2.2 (same results)

2-131. Range 80.0; variance 467.8; standard deviation* 21.6

2-133. Variance 1451.0; standard deviation 38.1

2-135. Variance 1.6; standard deviation 1.3

2-137. Variance 0.000385; standard deviation 0.0196

2-139. Variance 162,750,559.3; standard deviation 12,757.4

2-141. The statistics students are a more homogeneous group and should therefore have a smaller standard deviation.

2-143. If all scores are the same, the standard deviation is zero, but it cannot be negative.

2-145. Group C: range is 19.0, standard deviation is 6.0
    Group D: range is 16.0, standard deviation is 6.5

2-147. (a) 68% (b) 95% (c) 44.0, 68.0

2-149. $\sigma = 2.05$, $R = 8$, and $2.05 \leq 8/2$

2-151. Mean is 100.0 and standard deviation is 47.6

2-153. $\bar{x} = 4.56$, $s = 1.67$, $s^2 = 2.78$

2-155. (a) 15% (b) 20% (c) 95%

2-157. (a) 2.00 (b) -2.00 (c) 0.50

2-159. 0.55   2-161. -1.62 2-163. -1.50

2-165. Test b since $z = 0.60$ is greater than $z = 0.30$

―――――――――

*Range and standard deviation are multiplied by 10, but the variance is multiplied by 100.

2-167. Test b since $z = 3.00$ is greater than 2.00 or 2.67

2-169. 9   2-171. 73   2-173. $117,500

2-175. $184,000   2-177. $126,500

2-179. $179,000   2-181. 8

2-183. 72   2-185. 6400

2-187. 13,450   2-189. 7470   2-191. 7930

2-193 (a) $52,500   (d) Yes; yes
(b) $152,750   (e) No
(c) $96,703

2-195. Answer varies.

2-197. Inferential, since it is an inference about a population based on sample data.

2-199. (a) 16.0   (e) 17.0
(b) 17.0   (f) 39.1
(c) 8   (g) 6.3
(d) 16.5

2-201.

x	f
0– 9	2
10–19	4
20–29	10
30–39	5
40–49	7
50–59	2
60–69	6
70–79	2
80–89	7
90–99	5

2-203. (a) 27   (b) 44   (c) 32

2-205. (a) 1851.8   (e) 1860.0
(b) 1878.0   (f) 277, 577.5
(c) 2000   (g) 526.9
(d) 2130.0

2-207. (a) −1.24   (b) 1.04

2-209.

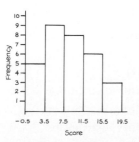

2-211.

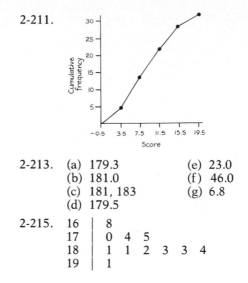

2-213. (a) 179.3   (e) 23.0
(b) 181.0   (f) 46.0
(c) 181, 183   (g) 6.8
(d) 179.5

2-215.

16	8
17	0 4 5
18	1 1 2 3 3 4
19	1

Chapter 3

3-1. 1.2, 77/75, −1/2, 5, 1.001, $\sqrt{2}$

3-3. 1/30   3-5. 1/5   3-7. 0.340

3-9. 3/8   3-11. 7/25   3-13. 0.0900

3-15. 11/19   3-17. 11/20

3-19. 47/198   3-21. 1/20

3-23. 0.0183

3-25. (a) bb, bg, gb, gg   (b) 1/4   (c) 1/2

3-27. (a) ttt, ttf, tft, tff, ftt, ftf, fft, fff
(b) 1/8   (c) 1/8   (d) 1/2

3-29. (a) rrr, rrr, rrw, rwr, rwr, rww, rwr, rwr, rww, wrr, wrr, wrw, wwr, wwr, www, wwr, wwr, www, brr, brr, brw, bwr, bwr, bww, bwr, bwr, bww
(b) 2/27   (c) 11/27
(d) 5/27   (e) 16/27

3-31. 1   3-33. Mutually exclusive: b only

3-35. 0.199 3-37. 61/366

3-39. 30/43   3-41. 0.256

3-43. 0.761   3-45. 10/29

3-47. 24/29   3-49. 69/758

3-51. 453/758   3-53. 0.500

3-55. 0.490   3-57. 0.550

3-59. 0.530   3-61. 17/60

3-63. 1/8

3-65. (a) $P(A \text{ or } B) = 0.9$
(b) $P(A \text{ or } B) < 0.9$
(c) $0.4 \leq P(B) \leq 0.8$, and they may or may not be mutually exclusive

3-67. $P(A \text{ or } B) = P(A) + P(B) - 2P(A \text{ and } B)$

3-69. Independent: a, b, e

3-71. 0.384   3-73. 0.106

3-75. 0.00583   3-77. 1/15

3-79. 0.0000000206   3-81. 55/96

3-83. (a) 3/10   (b) 9/25

3-85. 1/1024   3-87. 0.000000250

3-89. (a) 0.995   (b) 0.995

3-91. 0.787

3-93. (a) 0.00346   (b) 0.00826

3-95. 0.00000256   3-97. 0.431

3-99. (a) 4/5   (e) 8/25
(b) 1/5   (f) 16
(c) 2/5   (g) Yes: 24, 4, 6
(d) 3/5

3-101. The system works successfully.

3-103. All answers are "true."

3-105. 1/10; 9/10   3-107. 1/5; 4/5

3-109. 1/2; 1/2   3-111. 3/5; 2/5

3-113. 5:2; 2:5   3-115. 3:2; 2:3

3-117. 1/9   3-119. 3/10

3-121. 0.190   3-123. 0.986

3-125. 0.401   3-127. 0.999957

3-129. 9:1   3-131. 49:1

3-133. 35:1   3-135. 31:23

3-137. 0.779   3-139. 9:1; 1/10

3-141. 0.999999179   3-143. 1:31

3-145. (a) $P(B) = 0$ (b) $P(B) = 1$
(c) $P(B) \leq 0.3$

3-147. 973:27   3-149. No

3-151. 0.970   3-153. 5040

3-155. 4830   3-157. 720

3-159. 30   3-161. 120

3-163. 1326   3-165. $n!$

3-167. 1   3-169. 6

3-171. 5040   3-173. 646,646

3-175. $10! = 3,628,800$

3-177. (a) 5040   (b) 1/5040

3-179. 1,000,000,000   3-181. 43,680

3-183. 125,000 permutations

3-185. 18,720

3-187. (a) 2002   (b) 0.675

3-189. (a) 3,268,760   (b) 1/3,268,760

3-191. 6.20×10^{23}

3-193. (a) 1/5,461,512
(b) $\dfrac{1}{{}_N C_n}$ or $\dfrac{(N-n)!n!}{N!}$

3-195. 120

3-197. Calculator: 3.0414093×10^{64}
Approximation: 3.0363452×10^{64}

3-199. 10

3-201. (a) 0.08   (d) 4:1
(b) 0.6   (e) 1/13
(c) 0.8

3-203. 0.442   3-205. (a) 1:3   (b) 9/16

3-207. 1/5040

3-209. (a) 7/8   (b) 7/16   (c) 1

3-211. 1/4   3-213. 49/60

3-215. 6,135,755:1

3-217. (a) 40,320   (c) 45
(b) 20,160   (d) 3160

3-219. 2/5   3-221. 1/120

3-223. 336   3-225. 119/354

3-227. 0.994   3-229. 768

3-231. 11,880   3-233. 63:1

3-235. 24

Chapter 4

4-1. a, c, d are continuous.

4-3. Probability distribution

4-5. No, $\Sigma P(x) \neq 1$.

4-7. Probability distribution

4-9. Probability distribution

4-11. Probability distribution

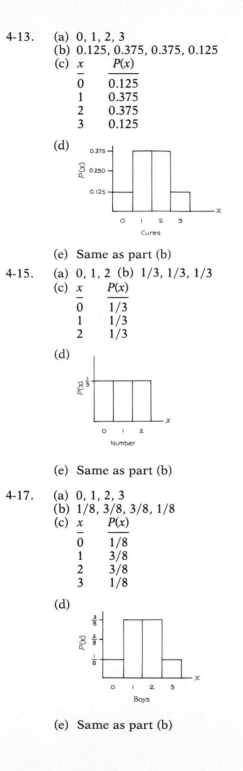

4-13. (a) 0, 1, 2, 3
(b) 0.125, 0.375, 0.375, 0.125
(c)

x	$P(x)$
0	0.125
1	0.375
2	0.375
3	0.125

(d)

(e) Same as part (b)

4-15. (a) 0, 1, 2 (b) 1/3, 1/3, 1/3
(c)

x	$P(x)$
0	1/3
1	1/3
2	1/3

(d)

(e) Same as part (b)

4-17. (a) 0, 1, 2, 3
(b) 1/8, 3/8, 3/8, 1/8
(c)

x	$P(x)$
0	1/8
1	3/8
2	3/8
3	1/8

(d)

(e) Same as part (b)

4-19. (a) 2, 3, 4, 5, 6, 7, 8, 9, 10, 11, 12
(b) 1/36, 2/36, 3/36, 4/36, 5/36, 6/36, 5/36, 4/36, 3/36, 2/36, 1/36
(c)

x	$P(x)$
2	1/36
3	2/36
4	3/36
5	4/36
6	5/36
7	6/36
8	5/36
9	4/36
10	3/36
11	2/36
12	1/36

(d)

(e) Same as part (b)

4-21. No, $\Sigma P(x) > 1$.

4-23. 0.179, 0.111, 0.015

4-25. Mean 1.1; variance 0.5; standard deviation 0.7

4-27. Mean 2.3; variance 0.6; standard deviation 0.8

4-29. Mean 21.3; variance 304.7; standard deviation 17.5

4-31. Mean 6.0; variance 5.9; standard deviation 2.4

4-33. Mean 2.4; variance 1.0; standard deviation 1.0

4-35. Mean 3.9; variance 1.4; standard deviation 1.2

4-37. Mean 0.4; variance 0.3; standard deviation 0.6

4-39. Mean 5.9; variance 1.1; standard deviation 1.1

4-41. $275

4-43. −$70

4-45. $122,500

4-47. $60 per day

4-49. (a) bbbbb bbbbg bbbgb bbbgg bbgbb
bbgbg bbggb bbggg bgbbb bgbbg
bgbgb bgbgg bggbb bggbg bgggb
bgggg gbbbb gbbbg gbbgb gbbgg
gbgbb gbgbg gbggb gbggg ggbbb
ggbbg ggbgb ggbgg gggbb gggbg
ggggb ggggg
(b) 1/32, 5/32, 10/32, 10/32, 5/32, 1/32
(c) see figure below
(d) 2.5   (e) 1.25
(f) 1.12

4-51.
$$\mu = \Sigma x \cdot P(x) = \left(1 \cdot \frac{1}{n}\right) + \left(2 \cdot \frac{1}{n}\right) + \cdots + \left(n \cdot \frac{1}{n}\right)$$

$$= \frac{1}{n}(1 + 2 + \cdots + n)$$

$$= \frac{1}{n} \cdot \frac{n(n + 1)}{2} = \frac{n + 1}{2}$$

$$\sigma^2 = \Sigma x^2 \cdot P(x) - \mu^2$$

$$= \left(1^2 \cdot \frac{1}{n}\right) + \left(2^2 \cdot \frac{1}{n}\right) + \cdots + \left(n^2 \cdot \frac{1}{n}\right) - \left(\frac{n + 1}{2}\right)^2$$

$$= \frac{1}{n}(1^2 + 2^2 + \cdots + n^2) - \left(\frac{n + 1}{2}\right)^2$$

$$= \frac{1}{n} \cdot \frac{n(n + 1)(2n + 1)}{6} - \frac{n^2 + 2n + 1}{4}$$

$$= \frac{2n^2 + 3n + 1}{6} - \frac{n^2 + 2n + 1}{4}$$

$$= \frac{4n^2 + 6n + 2 - (3n^2 + 6n + 3)}{12} = \frac{n^2 - 1}{12}$$

4-53. a, c, d, e

4-55. (a) 0.117   (c) 0.478   (e) 0+
(b) 0.215   (d) 0+

4-57. (a) 10   (c) 8   (e) 1
(b) 1   (d) 56

4-59. (a) 45/512   (b) 8/27   (c) 10/243

4-61. 0.205   4-63. 0.346   4-65. 0.0133

4-67. (a) 0.047   (b) 0.276   (c) 0.456

4-69. 0.200   4-71. 0.0743

4-73. (a) 0.105   (b) 0.243   (c) 0.274

4-75. 0.000297

4-77.

x	P(x)
0	0.296
1	0.444
2	0.222
3	0.037

4-79. 0.336   4-81. 0.00274

4-83. 0.00581   4-85. 0.052

4-87. 0.417

4-89. Mean 32.0; variance 16.0; standard deviation 4.0

4-91. Mean 4.8; variance 1.9; standard deviation 1.4

4-93. Mean 9.0; variance 6.8; standard deviation 2.6

4-95. Mean 92.4; variance 76.4; standard deviation 8.7

4-97. Mean 3.2; variance 2.6; standard deviation 1.6

4-99. Mean 168.7; variance 56.2; standard deviation 7.5

4-101. Mean 25.0; variance 12.5; standard deviation 3.5

4-103. Mean 2.0; variance 1.0; standard deviation 1.0

4-105. Mean 30.7; variance 16.0; standard deviation 4.0

4-107. Mean 8.8; variance 4.9; standard deviation 2.2

4-109. Mean 94.0; variance 49.8; standard deviation 7.1

4-111. Mean 4.4; standard deviation 1.7

4-113. Mean 338.0; standard deviation 17.1

4-115. Mean 88.8; standard deviation 7.5

4-117. Mean 6.9; standard deviation 2.4

4-119. Mean 127.6; standard deviation 6.8

4-121. Yes   4-123. Yes

4-125. (a) 0.2   (c) 0.5   (e) 0
(b) 0.8   (d) 0.65

4-127. (a) 0.333   (d) 0.667
(b) 0.733   (e) 0.500
(c) 0.267

4-129. (a) 0.5   (c) 0.6   (e) 0.5
(b) 0.8   (d) 0.5

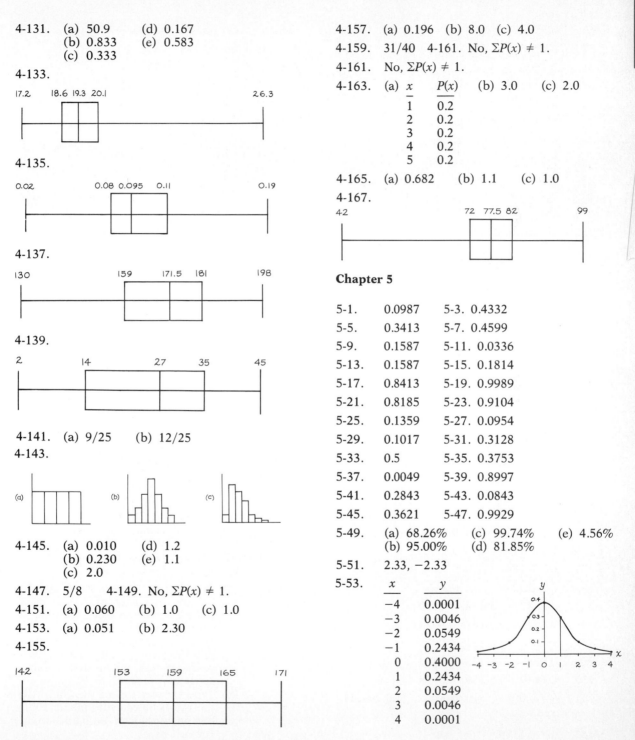

4-131. (a) 50.9   (d) 0.167
       (b) 0.833   (e) 0.583
       (c) 0.333

4-133.

4-135.

4-137.

4-139.

4-141. (a) 9/25   (b) 12/25
4-143.

4-145. (a) 0.010   (d) 1.2
       (b) 0.230   (e) 1.1
       (c) 2.0

4-147. 5/8   4-149. No, $\Sigma P(x) \neq 1$.

4-151. (a) 0.060   (b) 1.0   (c) 1.0

4-153. (a) 0.051   (b) 2.30

4-155.

4-157. (a) 0.196   (b) 8.0   (c) 4.0
4-159. 31/40   4-161. No, $\Sigma P(x) \neq 1$.
4-161. No, $\Sigma P(x) \neq 1$.
4-163. (a)

x	$P(x)$
1	0.2
2	0.2
3	0.2
4	0.2
5	0.2

(b) 3.0   (c) 2.0

4-165. (a) 0.682   (b) 1.1   (c) 1.0
4-167.

Chapter 5

5-1. 0.0987   5-3. 0.4332
5-5. 0.3413   5-7. 0.4599
5-9. 0.1587   5-11. 0.0336
5-13. 0.1587   5-15. 0.1814
5-17. 0.8413   5-19. 0.9989
5-21. 0.8185   5-23. 0.9104
5-25. 0.1359   5-27. 0.0954
5-29. 0.1017   5-31. 0.3128
5-33. 0.5   5-35. 0.3753
5-37. 0.0049   5-39. 0.8997
5-41. 0.2843   5-43. 0.0843
5-45. 0.3621   5-47. 0.9929
5-49. (a) 68.26%   (c) 99.74%   (e) 4.56%
       (b) 95.00%   (d) 81.85%
5-51. 2.33, −2.33
5-53.

x	y
−4	0.0001
−3	0.0046
−2	0.0549
−1	0.2434
0	0.4000
1	0.2434
2	0.0549
3	0.0046
4	0.0001

5-55.	0.1587	
5-57.	0.4987	5-59. 0.0038
5-61.	0.1596	5-63. 0.5
5-65.	0.4332	5-67. 0.2881
5-69.	41.68%	5-71. 6.68%
5-73.	0.6780	5-75. 0.3190
5-77.	0.4743	5-79. 94.84%
5-81.	0.3202	5-83. 0.2186
5-85.	34.04	5-87. 5.5

5-89. (a) 50.5 (c) 46.67%
(b) 2.2 (d) 35.31%

5-91. Uniform: 0.5769; normal: 0.6568

5-93. 1.645 5-95. −1.645

5-97. −1.645, 1.645 5-99. −1.645, 1.28

5-101. 0.25 5-103. 0.67

5-105. 3.957 5-107. 79.8 seconds

5-109. 64.4 in. 5-111. 4.35

5-113. 75.3, 124.7 5-115. 2.432

5-117. 24.87 mm, 25.40 mm

5-119. 3.470, 5.756

5-121. 29.6 minutes

5-123. (a) 68.1 (b) 10.0 (c) 82 (d) 84.6

Is approx. suitable?	μ	σ
5-125. Yes	6.25	2.17
5-127. No	4.32	2.03
5-129. No	2.50	1.57
5-131. No	79.8	2.00

5-133.	(a) 0.121	(b) 0.1173	
5-135.	(a) 0.114	(b) 0.1215	
5-137.	(a) 0.194	(b) 0.1922	
5-139.	(a) 0.227	(b) 0.2327	
5-141.	0.0287	5-143. 0.0019	
5-145.	0.9985	5-147. 0.0023	
5-149.	0.1034	5-151. 0.0099	
5-153.	0.0869	5-155. 0.0222	
5-157.	0.6119	5-159. 0.0198	
5-161.	(a) 0.001	(b) 0.0012	(c) 0.0011
5-163.	262		
5-165.	(a) 0.1179	(b) 0.4641	

5-167. (a) 0.6844 (b) 0.9987

5-169. (a) 0.1293 (b) 0.4772

5-171. 0.9332 5-173. 0.7734

5-175. 0.5636 5-177. 0.2005

5-179. 0.2766 5-181. It is halved.

5-183. (a) 1.06 (c) 0.628 (e) 2.12
(b) 0.454 (d) 0.217

5-185. 0.9898 5-187. 0.8805

5-189. (a) $\mu = 8.0$, $\sigma = 5.4$
(b) 2,3 2,6 2,8 2,11 2,18
3,6 3,8 3,11 3,18 6,8 6,11
6,18 8,11 8,18 11,18
(c) 2.5, 4.0, 5.0, 6.5, 10.0, 4.5, 5.5, 7.0,
10.5, 7.0, 8.5, 12.0, 9.5, 13.0, 14.5
(d) $\mu_{\bar{x}} = 8.0$, $\sigma_{\bar{x}} = 3.4$

5-191. $\mu = 28.4$; $\sigma = 9.4$; 0.1151

5-193. (a) 0.6293 (b) 0.9706 (c) 0.9798

5-195. (a) 0.4222 (b) 0.8531 (c) 0.2417

5-197. (a) 0.0398 (d) 18.05 years
(b) 0.6591 (e) 0.9920
(c) 11.22 years

5-199. 0.7357

5-201. (a) 0.6568 (c) 0.0336 (e) 84.4
(b) 0.3707 (d) 124.7

5-203. 0.9656

Chapter 6

6-1. (a) $H_0: \mu \leq 30$; $H_1: \mu > 30$
(b) $H_0: \mu \leq 100$; $H_1: \mu > 100$
(c) $H_0: \mu \geq 100$; $H_1: \mu < 100$
(d) $H_0: \mu \geq 12{,}300$; $H_1: \mu < 12{,}300$
(e) $H_0: \mu = 3271$; $H_1: \mu \neq 3271$

6-3. (a) Type I error: Reject the claim that
the mean age of professors is 30
years or less when their mean age is
actually 30 years or less.
Type II error: Fail to reject the claim
that the mean age of professors is 30
years or less when that mean is
actually greater than 30 years.
(b) Type I error: Reject the claim that
the mean IQ of criminals is 100 or
less when their mean IQ is actually
100 or less.

Type II error: Fail to reject the claim that the mean IQ of criminals is 100 or less when that mean is actually greater than 100.

(c) Type I error: Reject the claim that the mean IQ of college students is at least 100 when it really is at least 100.
Type II error: Fail to reject the claim that the mean IQ of college students is at least 100 when it is really less than 100.

(d) Type I error: Reject the claim that the mean annual household income is at least $12,300 when it really is at least $12,300.
Type II error: Fail to reject the claim that the mean annual household income is at least $12,300 when it is actually less than $12,300.

(e) Type I error: Reject the claim that the mean monthly cost equals $3271 when it does equal that amount.
Type II error: Fail to reject the claim that the mean monthly cost equals $3271 when it does not equal that amount.

6-5. Right-tailed: a, b
Left-tailed: c, d
Two-tailed: e

6-7. (a) 1.645   (d) −2.575, 2.575
(b) 2.33   (e) −1.645
(c) −1.96, 1.96

6-9. Test statistic: $z = 1.44$. Critical value: $z = 2.33$. Fail to reject H_0: $\mu \le 100$.
There is not sufficient evidence to warrant rejection of the claim that the mean is less than or equal to 100.

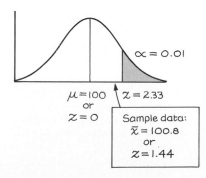

6-11. Test statistic: $z = -4.33$. Critical value: $z = -1.645$. Reject H_0: $\mu \ge 20$.
There is sufficient evidence to warrant rejection of the claim that the mean is at least 20.

6-13. Test statistic: $z = 1.73$. Critical values: $z = -1.645$ and $z = 1.645$. Reject H_0: $\mu = 500$.
There is sufficient evidence to warrant rejection of the claim that the mean is equal to 500.

6-15. Test statistic: $z = 1.77$. Critical value: $z = 1.645$. Reject H_0: $\mu \le 40$.
There is sufficient evidence to support the claim that the mean is greater than 40.

6-17. Test statistic: $z = 2.47$. Critical value: $z = 1.645$. Reject H_0: $\mu \le 90,000$.
There is sufficient evidence to support the claim that the mean is more than 90,000 miles.

6-19. Test statistic: $z = -1.31$. Critical value: $z = -2.33$. Fail to reject H_0: $\mu \ge 7.5$.
There is not sufficient evidence to support the claim that the mean is less than 7.5 years.

6-21. Test statistic: $z = 1.76$. Critical values: $z = 1.96$ and $z = -1.96$. Fail to reject H_0: $\mu = 13.20$.
There is not sufficient evidence to warrant rejection of the claim that the mean for the region is the same as the mean of the population.

6-23. Test statistic: $z = -8.63$. Critical value: $z = 1.645$. Fail to reject H_0: $\mu \le 420$.
The sample data do not support the claim of improved reliability. In fact, it appears that reliability actually deteriorated.

6-25. Test statistic: $z = -4.22$. Critical values: $z = -1.96$ and $z = 1.96$. Reject H_0: $\mu = 22.6$.
There is sufficient evidence to warrant rejection of the claim that the new spark plug does not change fuel consumption. (The new spark plug seems to consume more fuel since the mpg rating appears to be significantly lower.)

6-27. Test statistic: $z = 0.26$. The critical value depends on the significance level chosen, but the test statistic will not be in the critical region for any reasonable choice. Fail to reject H_0: $\mu = 180$.
There is not sufficient evidence to warrant rejection of the claim that the population mean is 180.

6-29. $\bar{x} = 104.3$ and $s = 13.5$. Test statistic: $z = 1.86$. Critical value: $z = 1.88$. Fail to reject H_0: $\mu \leq 100$.
There is not sufficient evidence to support the claim that the mean is above 100.

6-31. $\bar{x} = 28.4$, $s = 9.5$. Test statistic: $z = -1.57$. Critical values: $z = -2.05$ and $z = 2.05$. Fail to reject H_0: $\mu = 30.0$.
There is not sufficient evidence to warrant rejection of the claim that the mean equals 30.0. The sample mean does not differ significantly from the claimed mean of 30.0.

6-33. Reject the null hypothesis.

6-35. Fail to reject the null hypothesis.

6-37. *P*-value: 0.0808. Fail to reject the null hypothesis.

6-39. *P*-value: 0.0262. Reject the null hypothesis.

6-41. *P*-value: 0.0970. Fail to reject the null hypothesis.

6-43. *P*-value: 0.0524. Fail to reject the null hypothesis.

6-45. Test statistic: $z = -2.24$. *P*-value: 0.0125. Reject H_0: $\mu \geq 100$.
There is sufficient evidence to warrant rejection of the claim that the mean is greater than or equal to 100.

6-47. Test statistic: $z = 2.40$. *P*-value: 0.0164. Fail to reject H_0: $\mu = 75.6$.
There is not sufficient evidence to warrant rejection of the claim that the mean is 75.6.

6-49. Test statistic: $z = -0.96$. *P*-value: 0.1685. Fail to reject H_0: $\mu \geq 10.00$.
There is not sufficient evidence to warrant rejection of the claim that the mean is at least 10.00 minutes.

6-51. Test statistic: $z = 2.59$. *P*-value: 0.0096. Reject H_0: $\mu = 25$.
There is sufficient evidence to warrant rejection of the claim that the mean equals 25 years. (It appears to be greater than 25 years.)

6-53. The null hypothesis is the claim that the mean is equal to some value.

6-55. 13.52

6-57. (a) -2.056, 2.056   (d) -4.032, 4.032
(b) 1.337   (e) -2.467
(c) -3.365

6-59. Test statistic: $t = 1.500$. Critical value: $t = 1.860$. Fail to reject H_0: $\mu \leq 10$.
There is not sufficient evidence to warrant rejection of the claim that the mean is 10 or less.

6-61. Test statistic: $t = -2.121$. Critical value: $t = -2.110$. Reject H_0: $\mu \geq 98.6$.
There is sufficient evidence to warrant rejection of the claim that the mean is 98.6 or more.

6-63. Test statistic: $t = 2.014$. Critical values: $t = 2.145$ and $t = -2.145$. Fail to reject H_0: $\mu = 75$.
There is not sufficient evidence to warrant rejection of the claim that the mean equals 75.

6-65. Test statistic: $t = -4.845$. Critical value: $t = -2.861$. Reject H_0: $\mu \geq 154$.
The new tires will be purchased.

6-67. Test statistic: $t = 3.615$. Critical value: $t = 1.796$. Reject H_0: $\mu \leq 500$.
There is sufficient evidence to support the claim that the mean is more than 500 meters.

6-69. Test statistic: $t = -2.549$. Critical values: $t = 2.776$ and $t = -2.776$. Fail to reject H_0: $\mu = 17,850$.
There is not sufficient evidence to warrant rejection of the claim that the mean equals 17,850 pounds.

6-71. Test statistic: $t = 1.518$. Critical value: $t = 2.064$. Fail to reject H_0: $\mu \leq 1500$.
There is not sufficient evidence to support the claim that the mean amount charged is greater than $1500.

6-73. Test statistic: $t = 2.000$. Critical values: $t = 2.132$ and $t = -2.132$. Fail to reject H_0: $\mu = 5.4$.
There is not sufficient evidence to warrant rejection of the claim that the mean equals 5.4 mV.

6-75. Test statistic: $t = 2.598$. Critical value: $t = 1.315$. Reject H_0: $\mu \le 32$.
There is sufficient evidence to support the brewery's claim that the mean is greater than 32 ounces.

6-77. Test statistic: $t = 0.472$. Critical value: $t = 1.314$. Fail to reject H_0: $\mu \le 5$.
There is not sufficient evidence to support the claim that the mean is greater than 5 years.

6-79. Test statistic: $t = -4.021$. Critical values: $t = -2.977$ and $t = 2.977$. Reject H_0: $\mu = 41.9$.
There is sufficient evidence to conclude that the population mean for third-graders is different than 41.9.

6-81. Test statistic: $t = 5.541$. Critical values: $t = -2.093$ and $t = 2.093$. Reject H_0: $\mu = 1.000$.
There is sufficient evidence to warrant rejection of the claim that the mean residual lung volume equals 1.000 liter.

6-83. The sample yields a mean of 78.1 and a standard deviation of 5.6. Test statistic: $t = 2.476$. Critical value: $t = 1.729$. Reject H_0: $\mu \le 75$.
There is sufficient evidence to support the claim that the class is above average.

6-85. The critical z score from Table A-2 will be less than the corresponding t score from Table A-3 so that the critical region will be larger than it should be, and you are more likely to reject the null hypothesis.

6-87. Using $z = 1.645$, the table and the approximation both result in $t = 1.833$.

6-89. Test statistic: $z = 3.27$. Critical values: $z = -1.96$; $z = 1.96$. Reject H_0: $p = 0.3$.
There is sufficient evidence to warrant rejection of the claim that the proportion of defects is equal to 0.3. (It seems to be significantly higher.)

6-91. Test statistic: $z = -1.23$. Critical value: $z = -2.33$. Fail to reject H_0: $p \ge 0.7$.
There is not sufficient evidence to support the claim that the proportion of voters who favor nuclear disarmament is less than 0.7.

6-93. Test statistic: $z = 6.91$. Critical value: $z = 2.05$. Reject H_0: $p \le 1/4$.
There is sufficient evidence to support the claim that more than 1/4 of all white-collar criminals have attended college.

6-95. Test statistic: $z = 1.48$. Critical values: $z = -1.645$ and $z = 1.645$. Fail to reject H_0: $p = 0.71$.
There is not sufficient evidence to warrant rejection of the claim that the actual percentage is 71%.

6-97. Test statistic: $z = -4.37$. Critical value: $z = -2.33$. Reject H_0: $p \ge 0.10$.
There is sufficient evidence to support the claim that less than 10% prefer pediatrics.

6-99. Test statistic: $z = 3.73$. Critical value: $z = 2.33$. Reject H_0: $p \le 0.784$.
There is sufficient evidence to support the claim that the on-time arrival rate is higher than 78.4%.

6-101. Test statistic: $z = -1.75$. Critical value: $z = -1.645$. Reject H_0: $p \ge 0.07$.
There is sufficient evidence to support the claim that the no show rate is lower than 7%.

6-103. Test statistic: $z = 1.41$. Critical value: $z = 2.33$. Fail to reject H_0: $p \le 1/3$.
There is not sufficient evidence to support the claim that more than 1/3 of all adults smoke.

6-105. Test statistic: $z = 1.84$. Critical value: $z = 1.645$. Reject H_0: $p \le 0.03$.
There is sufficient evidence to warrant rejection of the manager's claim that production is not out of control. It appears that production is out of control.

6-107. Test statistic: $z = -4.4$. Critical value: $z = -2.33$. Reject H_0: $p \geq 1/2$.
There is sufficient evidence to support the claim that fewer than 1/2 of San Francisco residential telephones have unlisted numbers.

6-109. 1.977%   6-111. 93.55%

6-113. (a) 8.907, 32.852   (d) 40.289
(b) 10.117   (e) 4.075
(c) 8.643, 42.796

6-115. Test statistic: $\chi^2 = 50.44$. Critical value: $\chi^2 = 38.885$. Reject H_0: $\sigma^2 \leq 100$.
There is sufficient evidence to support the claim that the variance is greater than 100.

6-117. Test statistic: $\chi^2 = 9.500$. Critical value: $\chi^2 = 7.962$. Fail to reject H_0: $\sigma^2 \geq 416$.
There is not sufficient evidence to warrant rejection of the claim that the variance is at least 416.

6-119. Test statistic: $\chi^2 = 23.109$. Critical values: $\chi^2 = 2.603, 26.757$. Fail to reject H_0: $\sigma^2 = 2.38$.
There is not sufficient evidence to warrant rejection of the claim that the variance is equal to 2.38.

6-121. Test statistic: $\chi^2 = 35.743$. Critical value: $\chi^2 = 5.697, 35.718$. Reject H_0: $\sigma = 10.0$.
There is sufficient evidence to warrant rejection of the claim that the standard deviation is equal to 10.0.

6-123. Test statistic: $\chi^2 = 108.889$. Critical value: $\chi^2 = 101.879$. Reject H_0: $\sigma^2 \leq 9.00$.
There is sufficient evidence to warrant rejection of the claim that the variance is equal to or less than 9.00.

6-125. Test statistic: $\chi^2 = 22.337$. Critical value: $\chi^2 = 21.920$. Reject H_0: $\sigma \leq 2.00$.
There is sufficient evidence to support the claim that the standard deviation is greater than 2.00 hours.

6-127. Test statistic: $\chi^2 = 44.8$. Critical value: $\chi^2 = 51.739$. Reject H_0: $\sigma^2 \geq 0.0225$.
There is sufficient evidence to support the claim that the new machine produces less variance.

6-129. Test statistic: $\chi^2 = 69.135$. Critical value: $\chi^2 = 67.505$. Reject H_0: $\sigma^2 \leq 19.7$.
There is sufficient evidence to support the claim that women have a larger standard deviation.

6-131. Test statistic: $\chi^2 = 35.490$. Critical value: $\chi^2 = 35.479$. Reject H_0: $\sigma \leq 50$.
There is sufficient evidence to support the claim that the standard deviation of the hardness indices is greater than 50.0.

6-133. Test statistic: $\chi^2 = 92.364$. Critical values: $\chi^2 = 51.739, 90.531$. Reject H_0: $\sigma = 97.23$.
There is sufficient evidence to warrant rejection of the claim that the standard deviation of women is equal to the 97.23 standard deviation for men.

6-135. $n = 12, \bar{x} = 33.05, s = 1.13$. Test statistic: $\chi^2 = 3.511$. Critical value: $\chi^2 = 3.816$. Reject H_0: $\sigma \geq 2.0$.
There is sufficient evidence to support the claim that the standard deviation is less than 2.0 mg.

6-137. (a) Estimated values: 73.772, 129.070. Table A-4 values: 74.222, 129.561.
(b) 116.643, 184.199

6-139. (The sample standard deviation is $s = 22.9$.) Test statistic: $\chi^2 = 77.350$. Critical value: $\chi^2 = 83.298$. (Interpolated critical value is 82.110.) Fail to reject H_0: $\sigma \leq 20$.
There is not sufficient evidence to support the claim that the standard deviation is greater than 20 kwh.

6-141. (a) $z = -1.645$ (d) $\chi^2 = 10.856$
(b) $z = 2.33$ (e) $\chi^2 = 16.047, 45.72$
(c) $t = -3.106, 3.106$

6-143. (a) H_0: $\mu \geq 20.00$ (b) Left-tailed
(c) Rejecting the claim that the mean is at least 20.0 minutes when it really is at least 20.0 minutes.
(d) Failing to reject the claim that the mean is at least 20.0 minutes when it is really less than 20.0 minutes.
(e) 0.01

6-145. Test statistic: $z = 1.22$. Critical values: $z = -1.96$ and $z = 1.96$. Fail to reject H_0: $p = 0.20$. (*continued*)

There is not sufficient evidence to warrant rejection of the claim that the percentage who can't read is equal to 20%.

6-147. Test statistic: $\chi^2 = 9.310$. Critical values: $\chi^2 = 8.907, 32.852$. Fail to reject H_0: $\sigma = 4.0$.
There is not sufficient evidence to warrant rejection of the claim that the standard deviation equals 4.0 grams.

6-149. Test statistic: $z = -3.00$. Critical values: $z = -1.96$ and $z = 1.96$. Reject H_0: $\mu = 10.0$.
There is sufficient evidence to warrant rejection of the claim that the mean equals 10.0 seconds.

6-151. Test statistic: $z = 2.47$. Critical value: $z = 1.645$. Reject H_0: $p \leq 1/5$.
There is sufficient evidence to support the claim that more than 1/5 believe that birth control pills prevent venereal disease.

6-153. Test statistic: $z = -2.73$. Critical value: $z = -2.33$. Reject H_0: $\mu \geq 5.00$
There is sufficient evidence to support the claim that the mean radiation dosage is below 5.00 milliroentgens.

6-155. Test statistic: $\chi^2 = 77.906$. Critical value: $\chi^2 = 74.397$. (Interpolated critical value is 73.274.) Reject H_0: $\sigma^2 \leq 6410$.
There is sufficient evidence to support the counselor's claim that the current group has more varied aptitudes.

6-157. Test statistic: $z = 3.35$. Critical value: $z = 1.645$. Reject H_0: $p \leq 0.5$.
There is sufficient evidence to support the claim that the majority are opposed to the increase in taxes.

6-159. $\bar{x} = 98.0$, $s = 8.4$, $n = 15$. Test statistic: $t = -0.922$. Critical values: $t = 2.977$ and $t = -2.977$. Fail to reject H_0: $\mu = 100$.
There is not sufficient evidence to warrant rejection of the claim that the mean is equal to 100.

Chapter 7

7-1. (a) 1.96   (d) 2.093
(b) 2.33   (e) 2.977
(c) 2.05

7-3.   2.94   7-5. 20.6
7-7.   3.8625   7-9. $68.8 < \mu < 72.0$
7-11. $98.1 < \mu < 99.1$
7-13. $3.84 < \mu < 4.04$
7-15.   5.08 years $< \mu <$ 5.22 years
7-17. $20.12 < \mu < 33.56$
7-19. $0.994 < \mu < 0.998$
7-21.   110   7-23. 5968
7-25. $27.8 < \mu < 29.8$
7-27. $12.8 < \mu < 16.6$
7-29. $66.2 < \mu < 70.8$
7-31. $13.9 < \mu < 18.5$
7-33.   208   7-35. 61
7-37. $\bar{x} = 3.558$, $s = 1.233$; $2.676 < \mu < 4.440$
7-39. $\bar{x} = \$52,490$, $s = \$29,100$; $\$42,980 < \mu < \$62,000$ (assuming 95% confidence)
7-41.   90%   7-43. 52
7-45.   Multiply both sides by $\sqrt{n}$, then divide both sides by E, then square both sides.
7-47.   3254
7-49. $\hat{p} = 0.2$; $\hat{q} = 0.8$; point estimate 0.2; $E = 0.0351$
7-51. $\hat{p} = 0.304$; $\hat{q} = 0.696$; point estimate 0.304; $E = 0.0276$
7-53. $0.208 < p < 0.292$
7-55. $0.553 < p < 0.654$
7-57.   423
7-59. $23.9\% < p < 26.0\%$
7-61. $0.243 < p < 0.311$
7-63. $70.3\% < p < 73.7\%$
7-65.   553   7-67. 4145
7-69. $0.00181 < p < 0.0159$
7-71. $78.1\% < p < 83.9\%$
7-73. $0.828 < p < 0.852$; yes
7-75. $49.9\% < p < 52.3\%$
7-77. $p = 0.5$, $q = 0.5$
7-79.   89%
7-81.   (a) 12   (b) 97   (c) 9.886, 45.559

7-83. (a) 3.247, 20.483 (b) 1335 (c) 1.44

7-85. $4.0 < \sigma^2 < 11.3$

7-87. $58.5 < \sigma^2 < 208.5$

7-89. $14.01\ h < \sigma < 23.88\ h$

7-91. $7.3\ h < \sigma < 12.3\ h$

7-93. $6.73 < \sigma < 12.09$

7-95. 19.5 years $< \sigma < 24.6$ years

7-97. $0.064 < \sigma < 0.406$ $(s^2 = 0.0136)$

7-99. $0.211 < \sigma^2 < 1.486$

7-101. 98% 7-103. $2.7 < \sigma < 2.9$

7-105. (a) 83.2 (b) $82.16 < \mu < 84.24$

7-107. 354 7-109. 404

7-111. $7.1 < \sigma < 19.0$

7-113. $0.128 < p < 0.212$

7-115. (a) 1.645 (c) 2.262
(b) 2.700, 19.023 (d) 0.25

7-117. (a) 16.81 (b) $3.03 < \sigma < 6.35$

7-119. $0.215 < p < 0.265$ 7-121. 271

7-123. $15.082 < \mu < 17.718$ 7-125. 8687

7-127. $5.570 < \mu < 8.264$

Chapter 8

8-1. Test statistic: $F = 2.0000$. Critical value: $F = 4.0260$. Fail to reject H_0: $\sigma_1^2 = \sigma_2^2$. There is not sufficient evidence to warrant rejection of the claim that the variances are equal.

8-3. Test statistic: $F = 7.7931$. The critical value of F is between 8.6565 and 8.5599. Fail to reject H_0: $\sigma_1^2 = \sigma_2^2$. There is not sufficient evidence to warrant rejection of the claim that the variances are equal.

8-5. Test statistic: $F = 2.4091$. Critical value: $F = 2.4523$. Fail to reject H_0: $\sigma_1^2 = \sigma_2^2$. There is not sufficient evidence to warrant rejection of the claim that the variances are equal.

8-7. Test statistic: $F = 169.0000$. Critical value: $F = 3.8919$. Reject H_0: $\sigma_1^2 = \sigma_2^2$. There is sufficient evidence to warrant rejection of the claim that the variances are equal.

8-9. Test statistic: $F = 4.0000$. Critical value: $F = 3.1789$. Reject H_0: $\sigma_1^2 \leq \sigma_2^2$. There is sufficient evidence to warrant rejection of the claim that the variance of population A exceeds that of population B.

8-11. Test statistic: $F = 1.4603$. Critical value: $F = 1.6664$. Fail to reject H_0: $\sigma_1^2 \leq \sigma_2^2$. There is not sufficient evidence to support the claim that the variance of population A exceeds that of population B.

8-13. Test statistic: $F = 1.7333$. Critical value: $F = 2.5848$. Fail to reject H_0: $\sigma_1^2 = \sigma_2^2$. There is not sufficient evidence to support the claim of unequal variances.

8-15. Test statistic: $F = 1.8526$. Critical value: $F = 1.8055$. Reject H_0: $\sigma_1^2 = \sigma_2^2$. There is sufficient evidence to support the claim that the second scale produces greater variance.

8-17. Test statistic: $F = 1.7923$. Critical value: $F = 1.6373$. Reject H_0: $\sigma_1 = \sigma_2$. There is sufficient evidence to warrant rejection of the claim that the two sample groups come from populations with the same standard deviation.

8-19. Test statistic: $F = 1.3786$. Critical value: $F = 1.8363$. Fail to reject H_0: $\sigma_1 = \sigma_2$. There is not sufficient evidence to warrant rejection of the claim that the two groups come from populations with the same standard deviation.

8-21. Test statistic: $F = 1.3478$. Critical value: $F = 2.6171$. Fail to reject H_0: $\sigma_1^2 = \sigma_2^2$. There is not sufficient evidence to warrant rejection of the claim that both groups come from populations with the same variance.

8-23. Test statistic: $F = 1.2478$. Critical value: $F = 3.0502$. Fail to reject H_0: $\sigma_1 = \sigma_2$. There is not sufficient evidence to warrant rejection of the claim that both groups come from populations with the same standard deviation.

8-25. (a) $F_L = 0.2484$, $F_R = 4.0260$.
(b) $F_L = 0.2315$, $F_R = 5.5234$.
(c) $F_L = 0.1810$, $F_R = 4.3197$.
(d) $F_L = 0.3071$, $F_R = 4.7290$.
(e) $F_L = 0.2115$, $F_R = 3.2560$.

8-27. (a) No (b) No (c) No

8-29. Test statistic: $z = -1.67$. Critical values: $z = -1.96$ and $z = 1.96$. Fail to reject H_0: $\mu_1 = \mu_2$.
There is not sufficient evidence to warrant rejection of the claim that the two population means are equal.

8-31. F test results: Test statistic is $F = 4.1927$. Critical value: $F = 2.8621$. Reject H_0: $\sigma_1^2 = \sigma_2^2$.
Test of means: Test statistic is $t = 0.334$. Critical values: $t = -2.132, 2.132$. Fail reject H_0: $\mu_1 = \mu_2$.
There is not sufficient evidence to warrant rejection of the claim that the two population means are equal.

8-33. $\bar{d} = 0.141$, $s_d = 0.376$. Test statistic: $t = 1.185$. Critical values: $t = -2.262, 2.262$. Fail to reject H_0: $\mu_1 = \mu_2$.
There is not sufficient evidence to warrant rejection of the claim that both positions have the same mean.

8-35. F test results: Test statistic is $F = 1.6531$. Critical F value is between 2.2090 and 2.1460. Fail to reject H_0: $\sigma_1^2 = \sigma_2^2$. Test of means: Test statistic is $t = -3.94$. Critical values: $t = -1.96, 1.96$. Reject H_0: $\mu_1 = \mu_2$. There is sufficient evidence to warrant rejection of the claim that the mean amount of acetaminophen is the same for each brand.

8-37. Test statistic: $z = -1.45$. Critical values: $z = -1.645$, $z = 1.645$. Fail to reject H_0: $\mu_1 = \mu_2$.
There is not sufficient evidence to warrant rejection of the claim that the two populations have the same mean.

8-39. F test results: Test statistic is $F = 3.7539$. Critical value of F is between 3.2497 and 3.1532. Fail to reject H_0: $\sigma_1^2 = \sigma_2^2$. Test of means: Test statistic is $t = -1.916$. Critical values: $t = -2.201, 2.201$. Fail to reject H_0: $\mu_1 = \mu_2$.
There is not sufficient evidence to warrant rejection of the claim that the mean down times are equal.

8-41. $\bar{d} = -6.575$, $s_d = 2.018$. Test statistic: $t = -9.214$. Critical values: $t = -2.365$ and $t = 2.365$. Reject H_0: $\mu_1 = \mu_2$.
There is sufficient evidence to warrant rejection of the claim that the before and after responses have the same mean.

8-43. F test results: Test statistic is $F = 1.6198$. Critical F value is close to 2.0677. Fail to reject H_0: $\sigma_1^2 = \sigma_2^2$. Test of means: Test statistic is $t = -1.412$. Critical values: $t = -1.960, 1.960$. Fail to reject H_0: $\mu_1 = \mu_2$.
There is not sufficient evidence to warrant rejection of the claim that the means are equal. Both brands appear to have the same mean nicotine content.

8-45. Test statistic: $z = 0.78$. Critical values: $z = -2.33$ and $z = 2.33$. Fail to reject H_0: $\mu_1 = \mu_2$.
There is not sufficient evidence to warrant rejection of the claim that the two populations have the same mean.

8-47. F test results: Test statistic is $F = 2.6406$. Critical F value is between 2.2989 and 2.3567. Reject H_0: $\sigma_1^2 = \sigma_2^2$. Test of means: Test statistic is $t = 2.327$. Critical value: $t = -1.714$. Fail to reject H_0: $\mu_1 \geq \mu_2$.
There is not sufficient evidence to support the claim that System 2 has a greater mean than System 1. (With $\bar{x}_1 > \bar{x}_2$ we can never conclude that $\mu_1 < \mu_2$.)

8-49. F test results: Test statistic is $F = 1.2478$. Critical value: $F = 3.0502$. Fail to reject H_0: $\sigma_1^2 = \sigma_2^2$. Test of means: Test statistic is $t = -0.405$. Critical values: $t = -2.056, 2.056$. Fail to reject H_0: $\mu_1 = \mu_2$.
There is not sufficient evidence to warrant rejection of the claim that the placebo and calcium populations have the same mean.

8-51. $\bar{d} = -17.6$, $s_d = 20.0$. Test statistic: $t = -2.783$. Critical value: $t = -1.833$. Reject H_0: $\mu_1 \geq \mu_2$.
There is sufficient evidence to support the claim that the mean of the after scores is significantly greater than the mean of the before scores. The course does appear to be effective.

8-53. $\bar{d} = -11.2$, $s_d = 16.1$. Test statistic: $t = -2.200$. Critical values: $t = -2.262$ and $t = 2.262$. Fail to reject H_0: $\mu_1 = \mu_2$.
There is not sufficient evidence to warrant rejection of the claim that the course has no effect on the mean; that is, the differences are not significant and the course appears to have no effect.

8-55. F test results: Test statistic is $F = 1.0271$. Critical value: $F = 3.8049$. Fail to reject $H_0: \sigma_1^2 = \sigma_2^2$. Test of means: Test statistic is $t = 1.468$. Critical values: $t = -2.485$, 2.485. Fail to reject $H_0: \mu_1 = \mu_2$.
There is not sufficient evidence to warrant rejection of the claim that the male and female populations have the same mean.

8-57. Test statistic: $z = 8.56$. Critical value: $z = 2.33$. Reject $H_0: \mu_1 \geq \mu_2$.
There is sufficient evidence to support the claim that older men come from a population with a mean less than that for men in the 25–34 age bracket.

8-59. $\bar{d} = -4.5$, $s_d = 3.6$. Test statistic: $t = -3.953$. Critical values: $t = -2.821$ and $t = 2.821$. Reject $H_0: \mu_1 = \mu_2$.
There is sufficient evidence to warrant rejection of the claim that both test versions have the same mean.

8-61. $9.3 < (\mu_1 - \mu_2) < 14.7$

8-63. $-7.1 < \mu_d < -1.9$

8-65. $\hat{p}_1 = 0.335$; $\hat{p}_2 = 0.370$; $\bar{p} = 0.358$; $\bar{q} = 0.642$

8-67. $\hat{p}_1 = 0.200$; $\hat{p}_2 = 0.240$; $\bar{p} = 0.220$; $\bar{q} = 0.780$

8-69. (a) -2.05 (c) Reject $H_0: p_1 = p_2$
(b) -1.96, 1.96

8-71. (a) -2.15
(b) -2.33
(c) There is not sufficient evidence to support the manager's claim of a lower rate of defects.

8-73. Test statistic: $z = 0.71$. Critical values (assuming $\alpha = 0.05$): $z = -1.96$, 1.96. Fail to reject $H_0: p_1 = p_2$.
There is not sufficient evidence to warrant rejection of the claim that there is no significant difference between the two sample proportions (that is, the sample proportions don't differ by a significant amount).

8-75. Test statistic: $z = -1.50$. Critical values: $z = -1.645$ and $z = 1.645$. Fail to reject $H_0: p_1 = p_2$.
There is not sufficient evidence to warrant rejection of the claim that the proportion of speeding convictions is the same for both counties.

8-77. Test statistic: $z = -5.46$. Critical value: $z = -1.645$. Reject $H_0: p_1 \geq p_2$.
There is sufficient evidence to support the claim that the percentage of male surgeons is greater than the percentage of female surgeons.

8-79. Test statistic: $z = -1.19$. Critical values: $z = -1.96$ and $z = 1.96$. Fail to reject $H_0: p_1 = p_2$.
There is not sufficient evidence to warrant rejection of the claim that the two professors have the same failure rate.

8-81. Test statistic: $z = 2.26$. Critical value: $z = 2.33$. Fail to reject $H_0: p_1 \leq p_2$.
There is not sufficient evidence to support the claim that the more recent rate is greater than the rate in 1960.

8-83. Test statistic: $z = 1.62$. Critical values: $z = -1.96$ and $z = 1.96$. Fail to reject $H_0: p_1 = p_2$.
There is not sufficient evidence to warrant rejection of the claim that four-year public and private colleges have the same freshmen dropout rate.

8-85. Test statistic: $z = 23.67$. Critical value: $z = 2.33$. Reject $H_0: p_1 \leq p_2$.
There is sufficient evidence to support the claim that Queens County has a lower DWI conviction rate.

8-87. Test statistic: $z = 7.05$. Critical value: $z = 2.05$. Reject $H_0: p_1 \leq p_2$. There is sufficient evidence to support the claim that region A has a greater proportion of listeners who prefer country music.

8-89. Test statistic: $z = -0.75$. Critical values: $z = -1.96$ and $z = 1.96$. Fail to reject $H_0: p_1 = p_2$. There is not sufficient evidence to support the claim that there is a difference between the proportion of males who voted for the Democrat and the proportion of females who voted for the Democrat.

8-91. Test statistic: $z = -2.40$. Critical values: $z = -1.96$ and $z = 1.96$. Reject $H_0: p_1 - p_2 = 0.25$.
There is sufficient evidence to warrant rejection of the claim that the California percentage exceeds the New York percentage by an amount equal to 25%.

8-93. Test statistic: $z = -1.23$. Critical values: $z = -2.33$ and $z = 2.33$. Fail to reject H_0: $p_1 = p_2$.
There is not sufficient evidence to warrant rejection of the claim that both plants have the same rate of defects.

8-95. Test statistic: $F = 1.8225$. Critical F value is between 3.9639 and 3.8682. Fail to reject H_0: $\sigma_1 = \sigma_2$.
There is not sufficient evidence to warrant rejection of the claim that the standard deviations are equal.

8-97. Test statistic: $z = 3.94$. Critical values: $z = -1.96$ and $z = 1.96$. Reject H_0: $\mu_1 = \mu_2$.
There is sufficient evidence to warrant rejection of the claim of no difference between the means of the two groups. They appear to be different.

8-99. $\bar{d} = -1.21$, $s_d = 1.23$. Test statistic: $t = -2.95$. Critical values: $t = -2.306$ and $t = 2.306$. Reject H_0: $\mu_1 = \mu_2$.
There is sufficient evidence to warrant rejection of the claim that both means are equal.

8-101. F test results: Test statistic is $F = 27.5625$. Critical value of F is close to 2.1952. Reject H_0: $\sigma_1^2 = \sigma_2^2$. Test of means: Test statistic is $t = -4.843$. Critical values: $t = -2.093$, 2.093. Reject H_0: $\mu_1 = \mu_2$.
There is sufficient evidence to warrant rejection of the claim that the means are equal.

8-103. Test statistic: $z = -2.45$. Critical values: $z = -1.645$ and $z = 1.645$. Reject H_0: $p_1 = p_2$.
There is sufficient evidence to support the claim that the sample proportions are significantly different.

8-105. Test statistic: $F = 3.4490$. The critical F value is between 2.9222 and 2.8249. Reject H_0: $\sigma_1 = \sigma_2$.
There is sufficient evidence to warrant rejection of the claim that the two production methods yield batteries whose lives have equal standard deviations.

8-107. F test results: Test statistic is $F = 1.0374$. The critical F value is close to 2.0677. Fail to reject H_0: $\sigma_1^2 = \sigma_2^2$. Test of means: Test statistic is $t = -1.008$. Critical values: $t = -1.96$, 1.96. Fail to reject H_0: $\mu_1 = \mu_2$.
There is not sufficient evidence to warrant rejection of the claim that both means are the same.

8-109. $\bar{d} = -234.0$, $s_d = 199.5$. Test statistic: $t = -3.709$. Critical value: $t = -1.833$. Reject H_0: $\mu_1 = \mu_2$.
There is sufficient evidence to support the claim that the program is effective.

8-111. Test statistic: $z = -8.06$. Critical values: $z = -2.24$ and $z = 2.24$. Reject H_0: $\mu_1 = \mu_2$.
There is sufficient evidence to warrant rejection of the claim that there is no difference between the two population means. There does appear to be a difference.

Chapter 9

9-1. (a) Positive correlation
(b) Positive correlation
(c) Positive correlation
(d) Negative correlation
(e) No correlation

9-3. (a) Significant positive linear correlation
(b) Significant negative linear correlation
(c) No significant linear correlation
(d) No significant linear correlation
(e) No significant linear correlation

9-5. (b) 4   (e) 49
(c) 7   (f) 20
(d) 15   (g) -0.191

9-7. (b) 5   (e) 81
(c) 9   (f) 47
(d) 31   (g) 0.917

9-9. (b) 0.997   (c) 0.878
(d) Significant positive linear correlation

9-11. (b) 0.967 (c) 0.811
(d) Significant positive linear correlation

9-13. (b) 0.287 (c) 0.707
(d) No significant linear correlation

9-15. (b) 0.897 (c) 0.707
(d) Significant positive linear correlation

9-17. (b) 0.885 (c) 0.707
(d) Significant positive linear correlation

9-19. (b) 0.845 (c) 0.707
(d) Significant positive linear correlation

9-21. (b) -0.341 (c) 0.632
(d) No significant linear correlation

9-23. (b) 0.883 (c) 0.707
(d) Significant positive linear correlation

9-25. (b) 0.310 (c) 0.514
(d) No significant linear correlation

9-27. (b) 0.600 (c) 0.707
(d) No significant linear correlation

9-29. 0.900, 0.805, 0.549, 0.378, 0.306, 0.258, 0.231, 0.164

9-31. In attempting to calculate r we get a denominator of zero, so a real value of r does not exist. However, it should be obvious that the value of x is not at all related to the value of y.

9-33. r decreases to 0.912. The effect on an extreme value can be minimal or severe, depending on the other data.

9-35. With $r = 0.963$ and $n = 11$, it is reasonable to conclude that there is a significant positive linear correlation. This section of the parabola can be approximated by a straight line.

9-37. $y' = x + 2$

9-39. $y' = -0.36x + 3.64$

9-41. $y' = 0.027x - 0.00048$

9-43. $y' = 1.49x - 14.10$

9-45. $y' = 0.0693x + 0.0464$

9-47. $y' = 1.14x - 16.50$

9-49. $y' = 0.85x + 4.05$

9-51. $y' = 30.116x - 8.217$

9-53. $y' = -0.11x + 22.61$

9-55. $y' = 49.88x + 58.26$

9-57. $y' = 0.034x + 3.880$

9-59. $y' = 0.000124x - 0.210$

9-61. (a) 110.0 (d) 25.0
(b) 25.0 (e) 110.0
(c) 110.0

9-63. (a) 6 (d) 6
(b) 17 (e) 6
(c) -13

9-65. $y' = 0.133x + 108$. The effect can be substantial.

9-67. They are equal.

9-69. Note that s_x and s_y are never negative.

9-71. The sum of the squares using the regression line ($y' = -2.00x + 7.25$) is 0.75. The sum of the squares using the line $y' = -x + 6$ is 5.0.

9-73. 0.111; 11.1%

9-75. 0.640; 64.0%

9-77. (a) 1 (c) 1 (e) 0
(b) 0 (d) 1

9-79. (a) 140.4973 (d) 0.9950
(b) 0.7027 (e) 0.4840
(c) 141.2000

9-81. (a) 14
(b) Since the standard error of estimate is zero, $E = 0$ and there is no "interval" estimate.

9-83. (a) 18.9 (b) $17.1 < y < 20.8$

9-85. (a) $(n - 2)s_e^2$
(b) $\dfrac{r^2 \cdot \text{(unexplained variation)}}{1 - r^2}$
(c) -0.949

9-87. The equation of the regression line is the same, but s_e changes from 0.466 to 0.403.

9-89. 85 9-91. 75

9-93. $y' = -128 + 1.03x_1 + 12.4x_2 + 2.30x_3 - 1.06x_4$

9-95. 255

9-97. $y' = 4.73 + 1.07x_1 + 0.869x_2$ and $R^2 = 0.230$

9-99. $y' = 3.72 + 1.27x_2 + 0.297x_3$ and $R^2 = 0.164$

9-101. (a) Taxes $= -0.399 + 0.0172(\text{SP}) - 0.0143(\text{LA})$
(b) 0.787

9-103. (a) Taxes $= 2.78 + 0.320(\text{LA}) - 0.939(\text{acreage}) - 0.756(\text{rooms})$
(b) 0.619

9-105. $y = 2.17 + 2.44x + 0.464x^2$; since $R^2 = 1$, the parabola fits perfectly.

9-107. With no variable x_2 the equations become $\Sigma y = bn + m\Sigma x_1$; $\Sigma x_1 y = b\Sigma x_1 + m\Sigma x_1^2$. Treating these as a system of two equations with the variables m, b and solving, we get Formulas 9-2 and 9-3.

9-109. (a) Correlation   (d) Correlation
(b) Regression   (e) Correlation
(c) Regression

9-111. (a) 0.919   (b) 0.878
(c) Significant positive linear correlation
(d) $y' = 0.70x + 32.33$

9-113. (a) 0.687   (b) 0.811
(c) No significant linear correlation
(d) $y' = 0.22x + 5.41$

9-115. (a) 0.988   (b) 0.707
(c) Significant positive linear correlation
(d) $y' = 1.80x + 1.16$

9-117. (a) 0.887   (b) 0.707
(c) Significant positive linear correlation
(d) $y' = 0.95x + 7.01$

9-119. (a) 0.884   (b) 0.707
(c) Significant positive linear correlation
(d) $x_2' = 0.177x_1 - 0.184$

9-121. (a) 0.286   (b) 0.707
(c) No significant linear correlation
(d) $x_3' = 2.51x_2 + 13.2$

9-123. (a) $y' = 958 + 0.171x_1 - 11.6x_2 - 0.134x_3$
(b) 0.430   (c) Not very well

9-125. (a) 4.0720   (c) 14.7500   (e) 2.3106
(b) 10.6780   (d) 0.2761

9-127. (a) 1   (c) 1   (e) 0
(b) 0   (d) 1

Chapter 10

10-1. The expected frequencies are 20, 20, 20, 20, 20.
(a) 5.900   (b) 13.277
(c) Fail to reject the claim that absences occur on the five days with equal frequency.

10-3. Test statistic: $\chi^2 = 4.200$. Critical value: $\chi^2 = 16.919$. Fail to reject the claim that the digits are uniformly distributed.

10-5. Test statistic: $\chi^2 = 8.733$. Critical value: $\chi^2 = 13.277$. Fail to reject the claim that the proportions of smokers in the different age brackets are all equal.

10-7. Test statistic: $\chi^2 = 4.800$. Critical value: $\chi^2 = 14.067$. Fail to reject the claim that the aspirins are equally effective.

10-9. Test statistic: $\chi^2 = 1.954$. Critical value: $\chi^2 = 12.592$. Fail to reject the claim that the given percentages are correct.

10-11. Test statistic: $\chi^2 = 22.600$. Critical value: $\chi^2 = 19.675$. Reject the claim that months were selected with equal frequencies.

10-13. Test statistic: $\chi^2 = 9.013$. Critical value: $\chi^2 = 9.488$. Fail to reject the claim that the given percentages are correct.

10-15. Test statistic: $\chi^2 = 9.561$. Critical value: $\chi^2 = 14.067$. Fail to reject the claim that the figures agree.

10-17. Test statistic: $\chi^2 = 253.450$. Critical value: $\chi^2 = 9.210$. Reject the claim that the observed values are compatible with the dean's expectation.

10-19. 180 rolls; reject the claim of equal frequencies.

10-21. (a) Critical value: $\chi^2 = 3.841$

$$\chi^2 = \frac{\left(f_1 - \frac{f_1 + f_2}{2}\right)^2}{\frac{f_1 + f_2}{2}} + \frac{\left(f_2 - \frac{f_1 + f_2}{2}\right)^2}{\frac{f_1 + f_2}{2}}$$

$$= \frac{(f_1 - f_2)^2}{f_1 + f_2}$$

(continued)

(b) Critical value: $z^2 = 1.96^2 = 3.842$

$$z^2 = \frac{\left(\dfrac{f_1}{f_1 + f_2} - 0.5\right)^2}{\dfrac{1/4}{f_1 + f_2}} = \frac{(f_1 - f_2)^2}{f_1 + f_2}$$

10-23. Combining time slots 1 and 2, 5 and 6, 7 and 8, and 9 and 10, we get a test statistic of $\chi^2 = 4.118$ and a critical value of $\chi^2 = 11.071$ (with $\alpha = 0.05$). Fail to reject the claim that observed and expected frequencies are compatible.

10-25. (a) 27.778, 22.222, 22.222, 17.778
(b) 11.025   (c) 3.841
(d) Reject the claim that voting is independent of party.

10-27. Test statistic: $\chi^2 = 12.321$. Critical value: $\chi^2 = 3.841$. Reject the claim that success and group are independent.

10-29. Test statistic: $\chi^2 = 0.202$. Critical value: $\chi^2 = 5.991$. Fail to reject the claim that payer and group are independent.

10-31. Test statistic: $\chi^2 = 1.505$. Critical value: $\chi^2 = 3.841$. Fail to reject the claim that the accident rate is independent of the use of cellular phones.

10-33. Test statistic: $\chi^2 = 11.825$. The critical value depends on the significance level. With $\alpha = 0.05$, the critical value is $\chi^2 = 9.488$ and we reject the claim that day of the week is independent of the number of defects.

10-35. Test statistic: $\chi^2 = 3.271$. Critical value: $\chi^2 = 5.991$. Fail to reject the claim that the cause of death is independent of the determining source.

10-37. Test statistic: $\chi^2 = 0.615$. Critical value: $\chi^2 = 7.815$. Fail to reject the claim that smoking is independent of age group.

10-39. Test statistic: $\chi^2 = 17.344$. Critical value: $\chi^2 = 13.362$. Reject the claim that the subject and grade are independent.

10-41. (Use categories of East, Central, West.) Test statistic: $\chi^2 = 10.362$. Critical value: $\chi^2 = 9.488$. Reject the claim that region and opinion are independent.

10-43. It is multiplied by the same constant.

10-45. Test statistic: $F = 1.6373$. Critical value: $F = 3.2317$. Fail to reject the claim of equal means.

10-47. Test statistic: $F = 1.5430$. Critical value: $F = 2.6060$. Fail to reject the claim of equal means.

10-49. Test statistic: $F = 1.6500$. Critical value: $F = 3.0556$. Fail to reject the claim of equal means.

10-51. Test statistic: $F = 8.2086$. Critical value: $F = 3.8853$. Reject the claim of equal means.

10-53. Test statistic: $F = 15.8140$. Critical value: $F = 3.6823$. Reject the claim of equal means.

10-55. Test statistic: $F = 9.8683$. The critical value depends on the significance level. With a 0.05 level of significance, the critical value is $F = 3.6823$ and we fail to reject the claim of equal means.

10-57. Test statistic: $F = 3.7239$. Critical value: $F = 6.2262$. Fail to reject the claim of equal means.

10-59. Test statistic: $F = 11.6744$. Critical value: $F = 4.6755$. Reject the claim of equal means.

10-61. Test statistic: $F = 0.1587$. Critical value: $F = 3.3541$. Fail to reject the claim of equal means.

10-63. Test statistic: $F = 5.0793$. Critical value: $F = 3.3541$. Reject the claim of equal means.

10-65. Test statistic: $F = 1.8318$. Critical value: $F = 3.1504$. Fail to reject the claim of equal means.

10-67. Test statistic: $F = 4.8601$. Critical value: $F = 2.4472$. Reject the claim of equal means.

10-69. (a) With test statistic $z = -8.41$ and critical values $z = -1.96, 1.96$, reject the claim that $\mu_1 = \mu_2$.
(b) With test statistic $z = 8.58$ and critical values $z = -1.96, 1.96$, reject the claim that $\mu_2 = \mu_3$.
(c) With test statistic $z = 1.21$ and critical values $z = -1.96, 1.96$, fail to reject the claim that $\mu_1 = \mu_3$.

(d) With test statistic $F = 48.2442$ and critical value $F = 3.0000$ (approx.), reject the claim of equal means.

10-71. (a) The test statistic does not change.
(b) The test statistic is multiplied by the square of the constant.

10-73. Test statistic: $\chi^2 = 0.633$. Critical value: $\chi^2 = 5.991$. Fail to reject the claim that the candidate's popularity is independent of geographic region.

10-75. Test statistic: $F = 10.8266$. Critical value: $F = 3.0718$. Reject the claim of equal means.

10-77. Test statistic: $\chi^2 = 27.325$. Critical value: $\chi^2 = 12.592$. Reject the claim that customers arrive on the different days with equal frequencies.

10-79. Test statistic: $\chi^2 = 2.723$. Critical value: $\chi^2 = 9.488$. Fail to reject the claim that grade distribution is independent of the subject.

10-81. Test statistic: $F = 4.3124$. Critical value: $F = 4.8740$. Fail to reject the claim of equal means.

10-83. Test statistic: $\chi^2 = 10.000$. Critical value: $\chi^2 = 11.345$. Fail to reject the claim that the characteristics are equally likely.

Chapter 11

11-1 $x = 2$ (the smaller of 2 and 8); the critical value is 1. Since $x = 2$ is not less than or equal to the critical value, fail to reject the claim of no difference.

11-3. $x = 4$ (the smaller of 4 and 8) and the critical value is 2. Since $x = 4$ is not less than or equal to the critical value, fail to reject the null hypothesis that the x and y samples come from the same population. They appear to come from the same population.

11-5. $x = 1$ (the smaller of 1 and 8) and the critical value is 1. Since $x = 1$ is less than or equal to the critical value, reject the null hypothesis of no effect. The pill appears to lower blood pressure.

11-7. $x = 2$ (the smaller of 2 and 8), $n = 10$, and the critical value is 1. Since $x = 2$ is not less than or equal to the critical value, fail to reject the null hypothesis that the diet has no effect. The diet appears to be ineffective.

11-9. The test statistic $x = 4$ is not less than or equal to the critical value of 0. Fail to reject the null hypothesis of no difference. There appears to be no difference between the ages of husbands and wives.

11-11. The test statistic $x = 0$ is less than or equal to the critical value of 0. Reject the claim that the before and after responses are the same.

11-13. The test statistic $x = 5$ is less than or equal to the critical value of 5. Reject the null hypothesis that half (or fewer) of the students feel that they understand the purpose of their student senate. The data support the claim that most understand.

11-15. The test statistic $x = 7$ is not less than or equal to the critical value of 5. Fail to reject the null hypothesis that both parties are preferred equally. There is not a significant difference.

11-17. The test statistic $x = 32$ is converted to $z = -1.15$. The critical value is $z = -1.645$. Since the test statistic is not less than or equal to the critical value, fail to reject the null hypothesis that the median life is at least 40 hours.

11-19. The statistic $x = 18$ is converted to the test statistic $z = -1.84$. The critical value is $z = -1.645$. Since the test statistic is less than or equal to the critical value, reject the null hypothesis that production was unchanged or lowered by the new fertilizer. It appears that production was increased.

11-21. The statistic $x = 6$ is converted to the test statistic $z = -2.83$. The critical values are $z = -2.575$ and $z = 2.575$. Since the test statistic is less than or equal to the critical values, reject the null hypothesis that the drink had no effect.

11-23. The statistic $x = 22$ is converted to the test statistic $z = -0.71$. The critical value is $z = -1.28$. Since the test statistic is not less than or equal to the critical value, fail to reject the null hypothesis that at most half of the voters favor the bill. There is not sufficient evidence to support the claim that the majority favor the bill.

11-25. With $k = 3$ we get $6 < M < 17$.

11-27. 78

11-29. The given entries 5, 8, 10, 12, 15 correspond to ranks 1, 2, 3, 4, 5.

11-31. The given entries 50, 100, 150, 200, 400, 600 correspond to ranks 1, 2, 3, 4, 5, 6.

11-33. The given entries 6, 8, 8, 9, 12, 20 correspond to ranks 1, 2.5, 2.5, 4, 5, 6.

11-35. The given entries 12, 13, 13, 13, 14, 15, 16, 16, 18 correspond to ranks 1, 3, 3, 3, 5, 6, 7.5, 7.5, 9.

11-37. (a) 3, -7, -2, -4, -1, 0 (discard), -10, -15
(b) 3, 5, 2, 4, 1, 6, 7
(c) $+3$, -5, -2, -4, -1, -6, -7
(d) $T = 3$ (the smaller of 3 and 25)

11-39. (a) 1, -1, 2, -3, -6, 5, -10, -20, 0 (discard)
(b) 1.5, 1.5, 3, 4, 6, 5, 7, 8
(c) $+1.5$, -1.5, $+3$, -4, -6, $+5$, -7, -8
(d) $T = 9.5$ (the smaller of 9.5 and 26.5)

11-41. (a) 25; reject H_0
(b) 25; reject H_0

11-43. (a) 35; reject H_0
(b) 8; reject H_0

11-45. $T = 6.5$, $n = 9$. The test statistic $T = 6.5$ is greater than the critical value of 6, so fail to reject the null hypothesis of equal results.

11-47. $T = 15$, $n = 10$. The test statistic $T = 15$ is greater than the critical value of 8, so fail to reject the null hypothesis that both positions have the same distance.

11-49. $T = 4.5$, $n = 8$. The test statistic $T = 4.5$ is greater than the critical value of 4, so fail to reject the null hypothesis of equal perceptions of depth.

11-51. $T = 11.5$, $n = 14$. The test statistic $T = 11.5$ is less than or equal to the critical value of 21, so reject the null hypothesis of equal anxiety levels.

11-53. $T = 3$, $n = 10$. The test statistic $T = 3$ is less than or equal to the critical value of 8, so reject the null hypothesis of equal test results.

11-55. $T = 0$, $n = 8$. The test statistic $T = 0$ is less than or equal to the critical value of 4, so reject the null hypothesis that the before and after responses have the same distribution.

11-57. (a) $T = 51.5$, $n = 36$. Test statistic $z = -4.42$ is less than or equal to the critical value of $z = -2.575$, so reject the null hypothesis that both systems require the same times. (It appears that the scanner system is faster.)
(b) $T = 380.5$, $n = 39$. Test statistic $z = -0.13$ and the critical value is $z = -2.575$, so fail to reject the null hypothesis that both systems require the same times. After subtracting 20, the results are radically affected.

11-59. 1954

11-61. $\mu_R = 150$, $\sigma_R = 17.32$, $R = 166$, $z = 0.92$. The test statistic $z = 0.92$ is not in the critical region bounded by $z = -1.96$, 1.96, so fail to reject the null hypothesis of equal performances.

11-63. $\mu_R = 144$, $\sigma_R = 16.25$, $R = 163.5$, $z = 1.20$. The test statistic of $z = 1.20$ is not in the critical region bounded by the critical values $z = -1.96$ and $z = 1.96$, so fail to reject the null hypothesis of equal ratings.

11-65. $\mu_R = 203$, $\sigma_R = 21.76$, $R = 184$, $z = -0.87$. The test statistic $z = -0.87$ is not in the critical region bounded by the critical values of $z = -1.96$, $z = 1.96$, so fail to reject the null hypothesis of equal times.

11-67. $\mu_R = 201.5$, $\sigma_R = 23.89$, $R = 163$, $z = -1.61$. The test statistic $z = -1.61$ is not in the critical region bounded by the critical values of $z = -2.575$ and $z = 2.575$, so fail to reject the null hypothesis of equality.

11-69. $\mu_R = 224$, $\sigma_R = 20.26$, $R = 270.5$, $z = 2.29$. The test statistic $z = 2.29$ is in the critical region bounded by the critical values $z = -1.96$, 1.96, so reject the null hypothesis of equal attitudes.

11-71. $\mu_R = 162$, $\sigma_R = 19.44$, $R = 86$, $z = -3.91$. The test statistic $z = -3.91$ is in the critical region bounded by the critical values $z = -1.96$, 1.96, so reject the null hypothesis of no difference between the two treatments.

11-73. $\mu_R = 370$, $\sigma_R = 31.411$, $R = 254$, $z = -3.69$. The test statistic $z = -3.69$ is in the critical region bounded by the critical values $z = -1.96$, 1.96, so reject the null hypothesis of equal scores.

11-75. $\mu_R = 224$, $\sigma_R = 20.265$, $R = 254$, $z = 1.48$. The test statistic $z = 1.48$ is not in the critical region bounded by the critical values $z = -1.96$, 1.96, so fail to reject the null hypothesis that both populations are the same.

11-77. (a) $\mu_R = 232.5$, $\sigma_R = 24.11$, $R = 120$, $z = -4.67$. The test statistic $z = -4.67$ is in the critical region bounded by the critical values $z = -1.96$, 1.96, so reject the null hypothesis that both groups come from the same population.

(b) $\mu_R = 232.5$, $\sigma_R = 24.11$, $R = 225$, $z = -0.31$. The test statistic $z = -0.31$ is not in the critical region bounded by the critical values of $z = -1.96$, 1.96, so fail to reject the null hypothesis that both groups come from the same population.

11-79. (a)

ABAB	4
ABBA	5
BBAA	7
BAAB	5
BABA	6

(b)

R	p
3	1/6
4	1/6
5	2/6
6	1/6
7	1/6

(c) No, the most extreme rank distribution has a probability of at least 1/6, and we can never get into a critical region with a probability of 0.10 or less.

11-81. 0

11-83. The test statistic $H = 2.435$ is less than the critical value of 5.991 so fail to reject the null hypothesis that the samples come from identical populations.

11-85. The test statistic of $H = 12.167$ exceeds the critical value of 5.991, so reject the null hypothesis that the samples come from identical populations.

11-87. The test statistic of $H = 0.775$ is less than the critical value of 9.210, so fail to reject the null hypothesis of equally effective methods.

11-89. The test statistic of $H = 1.589$ does not exceed the critical value of 5.991, so fail to reject the null hypothesis that the living areas are the same in all three zones.

11-91. The test statistic of $H = 1.732$ does not exceed the critical value of 5.991, so fail to reject the null hypothesis that the taxes are the same in all three zones.

11-93. The test statistic is $H = 8.756$. The critical value depends on the significance level chosen. With $\alpha = 0.05$, the critical value is 9.488, and we fail to reject the null hypothesis that the methods are equally effective.

11-95. The test statistic $H = 31.086$ exceeds the critical value of 9.488, so reject the null hypothesis that the samples come from identical populations.

11-97. $H = \dfrac{1}{1176}(R_1^2 + R_2^2 + \cdots + R_8^2) - 147$

11-99. Dividing the unrounded value of H by the correction factor results in $1.1375604 \div 0.99923077$, which is rounded off to 1.138, the same value obtained without the correction factor.

11-101. (a) ± 0.450 (d) ± 0.475
(b) ± 0.280 (e) ± 0.342
(c) ± 0.373

11-103.

x	1	2	4	3	5
y	5	4	3	2	1

11-105. $d = 3, 0, 0, 3$; $\Sigma d^2 = 18$; $n = 4$; $r_s = -0.8$

11-107. $d = 3, 3, 1, 2, 2, 5$; $\Sigma d^2 = 52$; $n = 6$; $r_s = -0.486$

11-109. $r_s = 0.571$. Critical values: $r_s = \pm 0.738$. No significant correlation.

11-111. $r_s = 0.476$. Critical values: $r_s = \pm0.738$. No significant correlation.

11-113. $r_s = 0.814$. Critical values: $r_s = \pm0.886$. No significant correlation.

11-115. $r_s = 1.000$. Critical values: $r_s = \pm0.738$. Significant positive correlation.

11-117. $r_s = 0.855$. Critical values: $r_s = \pm0.648$. Significant positive correlation.

11-119. $r_s = 0.818$. Critical values: $r_s = \pm0.525$. Significant positive correlation.

11-121. $r_s = -0.905$. Critical values: $r_s = \pm0.738$. Significant negative correlation.

11-123. $r_s = 0.458$. Critical values: $r_s = \pm0.738$. No significant correlation.

11-125. (a) 123, 132, 213, 231, 312, 321
(b) 1, 0.5, 0.5, −0.5, −0.5, −1
(c) 1/6 or 0.167

11-127. (a) The rank correlation coefficient, because the trend is not linear
(b) Sign changes

11-129. $n_1 = 5$, $n_2 = 3$, $G = 3$, critical values: 1, 8

11-131. $n_1 = 4$, $n_2 = 6$, $G = 2$, critical values: 2, 9

11-133. $n_1 = 7$, $n_2 = 2$, $G = 2$, critical values: 1, 6

11-135. $n_1 = 10$, $n_2 = 10$, $G = 9$, critical values: 6, 16

11-137. $\bar{x} = 11.0$; BBBBBABAAA; $n_1 = 6$, $n_2 = 4$, $G = 4$. The critical values are 2, 9. Fail to reject randomness.

11-139. $\bar{x} = 9.2$; $n_1 = 6$, $n_2 = 8$, $G = 2$. Critical values are 3, 12. Reject randomness.

11-141. Median is 9.5, $n_1 = 5$, $n_2 = 5$, $G = 2$. Critical values are 2, 10. Reject randomness.

11-143. Median is 9, so delete the three 9s to get BBBBBBBAAAAAAA. $n_1 = 7$, $n_2 = 7$, $G = 2$. Critical values are 3, 13. Reject randomness.

11-145. $n_1 = 9$, $n_2 = 18$, $G = 4$. Critical values are 8, 18. Reject randomness.

11-147. $n_1 = 30$, $n_2 = 15$, $G = 14$, $\mu_G = 21$, $\sigma_G = 2.94$. Test statistic is $z = -2.38$. Critical values are $z = -1.96$, 1.96. Reject randomness.

11-149. Median is 834. $n_1 = 8$, $n_2 = 8$, $G = 2$. Critical values are 4, 14. Reject randomness. (There appears to be a downward trend.)

11-151. $n_1 = 17$, $n_2 = 28$, $G = 12$, $\mu_G = 22.16$, $\sigma_G = 3.11$. Test statistic is $z = -3.27$. Critical values are $z = -1.96$, 1.96. Reject randomness.

11-153. $n_1 = 26$, $n_2 = 14$, $G = 21$, $\mu_G = 19.2$, $\sigma_G = 2.833$. Test statistic is $z = 0.64$. Critical values are $z = -1.96$, 1.96. Fail to reject randomness.

11-155. $n_1 = 49$, $n_2 = 51$, $G = 43$, $\mu_G = 50.98$. $\sigma_G = 4.9727$. Test statistic is $z = -1.60$. Critical values are $z = -1.96$, 1.96. Fail to reject randomness.

11-157. $n_1 = 30$, $n_2 = 30$, $G = 8$, $\mu_G = 31$, $\sigma_G = 3.84$. Test statistic is $z = -5.99$. Critical values are $z = -1.96$, 1.96. Reject randomness.

11-159. $n_1 = 40$, $n_2 = 40$, $G = 6$, $\mu_G = 41$, $\sigma_G = 4.44$. Test statistic is $z = -7.88$. Critical values are $z = -1.96$, 1.96. Reject randomness.

11-161. Minimum is 2, maximum is 4. Critical values of 1 and 6 can never be realized so that the null hypothesis of randomness can never be rejected.

11-163. The 84 sequences yield two runs of 2, seven runs of 3, twenty runs of 4, twenty-five runs of 5, twenty runs of 6, and ten runs of 7, so that $P(2\text{ runs}) = 2/84$, $P(3\text{ runs}) = 7/84$, $P(4\text{ runs}) = 20/84$, $P(5\text{ runs}) = 25/84$, $P(6\text{ runs}) = 20/84$, and $P(7\text{ runs}) = 10/84$. From this we conclude that the G values of 3, 4, 5, 6, 7 can easily occur by chance while $G = 2$ is unlikely since $P(2\text{ runs})$ is less than 0.025. The lower critical G value is therefore 2, and this agrees with Table A-10. The table lists 8 as the upper critical value, but it is impossible to get 8 runs using the given elements.

11-165. Differences: −10, 1, −5, −4, −6, −3, −1, −2, −11, and 0 (which is discarded). Ranks: 8, 1.5, 6, 5, 7, 4, 1.5, 3, 9. Signed ranks: −8, 1.5, −6, −5, −7, −4, −1.5, −3, −9. $T = 1.5$ (the smaller of 1.5 and 43.5). The critical value is 6 from Table A-8 with $n = 9$. The test statistic $T = 1.5$ is less than or equal to the critical value of 6, so we reject the null hypothesis of equal indices before and after the course.

11-167. Using the Wilcoxon rank-sum test for two independent samples and assuming $\alpha = 0.05$ we get: $\mu_R = 137.5$, $\sigma_R = 17.260$, $R = 177.5$, $z = 2.32$. The test statistic $z = 2.32$ is in the critical region bounded by the critical values of $z = -1.96, 1.96$, so reject the null hypothesis of no difference between Kings County and Queens County. (They appear to be different.)

11-169. $r_s = -0.911$. Critical values: $r_s = \pm 0.738$. There is a negative correlation between age and extra service income.

11-171. $\mu_R = 137.50$, $\sigma_R = 17.26$, $R = 177.5$, $z = 2.32$. The test statistic $z = 2.32$ is in the critical region bounded by $z = -1.96$, 1.96. Reject the null hypothesis of no difference.

11-173. $T = 12$, $n = 7$. The test statistic $T = 12$ is greater than the critical value of 2, so fail to reject the null hypothesis that there is no difference between the scores of Judge A and Judge B.

11-175. $r_s = 0.619$. Critical values: $r_s = \pm 0.738$. No significant correlation.

11-177. Using the sign test, discard the zero to get 10 negative signs and 9 positive signs, so that $x = 9$, which is not less than or equal to the critical value of 4 found from Table A-7. Fail to reject the null hypothesis that the median is 38.

11-179. Using the Kruskal-Wallis test, the test statistic is $H = 10.500$ and the critical value is 5.991. Reject the null hypothesis that the machines produce the same amount.

11-181. $n_1 = 20$, $n_2 = 5$, $G = 5$. Critical values: 5, 12. Reject randomness.

11-183. The test statistic of $H = 13.501$ exceeds the critical value of 5.991, so reject the null hypothesis that the three groups come from identical populations.

11-185. $n_1 = 20$, $n_2 = 10$, $G = 5$, and the critical values are 9 and 20. Since $G = 5$ is less than the lower critical value of 9, reject the null hypothesis of randomness.

11-187. Discard the two zeros to get 11 negative signs and 1 positive sign, so that $x = 1$, which is less than the critical value of 2 found from Table A-7. Reject the null hypothesis of no effect.

Index

abuse of statistics, 8
accuracy, 13
Adams, John, 117
addition rule, 132–140
airlines, 7, 376
alternative hypothesis, 332–333
analysis of variance, 593–605
area, 203
arithmetic mean, 58
aspirin, 441, 483
Associated Press, 14
average, 9, 57–69

bar graph, 11
baseball, 598
Bayes' Theorem, 155
Bennett, William, 136
Berman, Ronald, 238
Bills of Mortality, 5
bimodal distribution, 62, 246
binomial approximated by
    normal, 293–302
binomial distribution, 218–239
binomial distribution mean, 235
binomial distribution standard
    deviation, 235
binomial distribution variance,
    235
binomial experiment, 218
binomial probability formula, 222
binomial probability table,
    702–709
bivariate data, 503
bound on error estimate, 400
box-and-whisker, 247
boxplot, 247
Brooks, Juanita, 145
Bryson, Maurice, 13
Bureau of Census, 11, 23, 40, 79
Bureau of Labor Statistics, 11, 69

Campbell, John, 337, 352
Cannell, Dr. John, 10
cancer, 375, 470
capture-recapture, 8
Census, 79
central limit theorem, 305–317
central tendency, measures of,
    57–69
centroid, 514
Chebyshev's theorem, 83–84, 94
chi-square distribution, 381, 426,
    654

chi-square table, 712
cigarettes, 565, 567, 579, 627
class, 32
class attendance, 646
class boundaries, 31
classical approach to probability,
    120
class marks, 31
class size paradox, 65
class width, 31–32
clusters, 245
cluster sampling, 705
coefficient of determination, 539
coefficient of variation, 96
coincidences, 117
coins, 120
Collins case, 145
combinations, 177
complementary events, 161–164
completely randomized design,
    604
compound event, 132
composite sampling, 225
conditional probability, 146–156
confidence coefficient, 400
confidence interval, 400
confidence interval for mean, 400
confidence interval for proportion,
    417
confidence interval for standard
    deviation, 427, 428
confidence interval for variance,
    427
confidence interval limits, 402
confidential survey, 707
consistency, 425
consumer price index, 98, 314
Continental Airlines, 14
contingency table, 579–586
continuity correction, 297
continuous random variable, 201
control charts, 285
convenience sampling, 705
correlation, 503–516, 662–671
correlation coefficient tables, 720,
    723
countable, 200
counting, 172–181
critical region, 333
critical value, 333
cumulative frequency polygon, 46
cumulative frequency table, 36
cycles, 686

Darwin, Charles, 522
deciles, 98
decision theory, 213
deductive, 6
degree of confidence, 400
degrees of freedom, 360, 381, 382,
    444, 456, 508, 542, 570, 584,
    596, 599, 655
dependent, 148, 454
descriptive statistics, 28
discrete random variable, 200
disobedience, 18
dispersion, 77–78
Disraeli, Benjamin, 9
distance, 17
distribution, 29
distribution-free tests, 620
distribution shapes, 243
draft, 680
drug approval, 331

economic indicators, 538
economics, 686
efficiency, 621
empirical approximation of
    probability, 120
empirical rule, 83–84, 94–95
estimates, 398
ethics, 708–709
event, 119
exclusive or, 132, 144
exit poll, 492
expectation, 212–213
expected frequency, 569, 581
expected value, 213
experiment, 118
explained deviation, 537–538
explained variation, 539
exploratory data analysis, 47

factorial, 174
factorial rule, 174
failure, 219
F distribution, 443–444
F distribution tables, 714–719
Federal Aviation Administration,
    197, 229
Federal Bureau of Investigation,
    41
Federalist papers, 30
finite population correction factor,
    313
Fisher, R. A., 569

frequency, 30
frequency polygon, 45
frequency table, 30
fundamental counting rule, 173

Gallup, George, 13, 706
Galton, Sir Francis, 522, 548
Gardner, Martin, 117
gender gap, 3, 4, 11
geometric distribution, 234
geometric mean, 75–76
geometry, 7
goodness-of-fit, 567
Gosset, William, 359
Graunt, John, 5
guessing, 122

Hamilton, Alexander, 30
Hamlet, 136
harmonic mean, 75
Harris, Lou, 270
heights, 548
hemline index, 538
Hertz, 300
hinge, 247
histogram, 42
Hite, Shere, 420, 704
H test, 654
Huff, Darrell, 9, 14
hypergeometric distribution, 234
hypothesis, 330
hypothesis test, 328–395

inclusive or, 132, 144
independent, 148, 454
index numbers, 98
inductive, 6
inference, 29
inferential statistics, 29
infinity, 200
inflation, 314
interquartile range, 103
interval, 17
interval estimate, 400
interval estimate of mean, 400
interval estimate of predicted
    value, 542
interval estimate of proportion,
    417
interval estimate of variance, 427

Jay, John, 30
Jefferson, Thomas, 117

Kay, Alan, 14
Kruskal-Wallis test, 593, 654-657

Landon, Alfred, 13
Law of Large Numbers, 121

least squares property, 530
left-tailed test, 339–340
leukemia, 245
level of confidence, 400
levels of measurement, 15
Likert scale, 20
linear correlation coefficient,
    504–506
linear correlation coefficient table,
    720
Literary Digest, 13
lottery, 162, 181, 203, 213
lower class limits, 31

Madison, James, 30
magazine survey, 297
Mann-Whitney U test, 645, 653
maximum error of estimate of
    mean, 400
maximum error of estimate of
    proportion, 417
mean, 58, 66, 68, 685
mean deviation, 81
mean for binomial distribution,
    235
mean for discrete random
    variable, 210
measures of central tendency, 57
measures of dispersion, 77–88
measures of position, 96–104
median, 60, 68, 623, 635, 685
Mendel, Gregor, 569
Miami Herald, 13
midquartile, 103
midrange, 63, 68
Milgram, Stanley, 18
Minitab, 64, 104, 229, 249, 408,
    460, 486, 532, 549, 586, 603,
    741–747
mode, 62, 68
multimodal distribution, 62
multinomial distribution, 235
multinomial experiments,
    567–573
multiple coefficient of determi-
    nation, 550
multiple regression, 547–552
multiplication rule, 144–156
mutually exclusive, 35, 136

National Safety Council, 23
negative correlation, 503, 508, 666
Nielsen television ratings, 86
nominal, 15, 627
nonparametric, 620–699
nonsampling error, 706
nonstandard normal distribution, 276
normal approximation to bino-
    mial, 293–302

normal distribution, 263–327
nuclear power, 153
null hypothesis, 332–333

observed frequency, 569
odds, 164–168
odds against, 165
odds in favor, 165
ogive, 46
one-way analysis of variance, 604
ordinal, 17
outlier, 94

parachuting, 199
parameter, 6, 620
Pearson, Karl, 514
Pearson's index of skewness, 96
Pearson's product moment, 514
percentiles, 98–102
permutations, 176
phone surveys, 703
pi, 56
pie chart, 41–42
point estimate, 399, 415, 425
point estimate of mean, 399
point estimate of proportion, 415
point estimate of variance, 425
Poisson distribution, 234
police, 42, 485
polio, 482
polls, 705
pooled estimate of proportion, 481
Pope, 13
population, 6
population size, 313
position, measures of, 96–104
positive correlation, 503, 508, 666
precision, 13
predicted value, 524
prediction interval, 542
probability:
    addition rule, 135
    complementary events, 161
    defined, 120
    distributions, 197–261
    multiplication rule, 149
    nature of, 116, 118
    value, 350
product testing, 367
proportions, tests of, 372–376
P-value, 350–355, 365, 376, 461,
    486

quadratic mean, 76
quartiles, 97

random, 703
random sample, 703
random selection, 122

random variable, 198–199
range, 78–79
rank correlation, 662–671
rank correlation coefficient table, 723
ranked data, 622
ratio, 18
redundancy, 150
regression, 522–532
regression analysis, 522
regression line, 522
Reichard, Robert, 14
relative frequency, 120
reliability, 265
report, 707
right-tailed test, 339–340
rigorously controlled design, 604
robust, 425
Roosevelt, Franklin D., 13
root mean square, 76
rounding, 64, 83, 97, 126, 406
run, 677
runs test, 677-686
runs test table, 724

Salk vaccine, 482
sample, 6–7
sample proportion, 414–415
sample size for estimating mean, 405
sample size for estimating proportion, 418
sample size for estimating standard deviation, 428–429
sample size for estimating variance, 428–429
sample space, 119
sampling error, 706
S.A.T. scores, 122, 529, 619
scatter diagram, 604, 505, 666
Schwartz, Noel, 367
seat belts, 624
semi-interquartile range, 103
serial numbers, 247
Shakespeare, 136, 238, 294
Shickin, Julius, 69
sigma, 58
significance level, 333
sign test, 623–630
sign test table, 721
simple event, 119
simulation, 194
single factor analysis of variance, 604

skewed distribution, 76, 246
slope, 523
smoking, 565, 567, 579, 627
Spearman, Charles, 664
Spock, Dr. Benjamin, 678
Stamp, Josiah, 9
standard deviation:
    calculation of, 80, 85–87
    definition, 80
    for binomial distribution, 235
    for discrete random variable, 211
    for frequency table, 87
    units, 81
standard error of estimate, 541–542
standard error of the mean, 309
standard normal distribution, 267–273
standard normal distribution table, 710
standard score, 97, 276
STATDISK, 64, 103, 228, 298, 301, 366, 408, 460, 467, 468, 532, 549, 585, 603
statistic, 6
statistics, 4, 118
stem-and-leaf plot, 47–50
stratified sampling, 704
Student t distribution, 359
Student t distribution table, 711
success, 219
summarizing data, 30–36
Super Bowl omen, 538
survey medium, 481, 703
syphilis, 225, 708
systematic sampling, 704

tables, 701–724
tally marks, 32
t distribution, 359
t distribution table, 711
teacher ratings, 508
telephone numbers, 172
telephone surveys, 417, 703
television, 86, 640, 663
testing hypotheses, 328–395
tests of proportions, 372–376
tests of significance, 330
tests of variances, 380–386
test statistic, 332
Titanic, 117
total deviation, 537–538
total variation, 539

tree diagram, 147, 156
trends, 686
triangle, 7
trimmed mean, 77
t test, 357–367
Tukey, John, 47
two-tailed test, 339–340
two-way tables, 580
type I error, 332, 336
type II error, 332, 336, 349–350

unbiased estimator, 398
unemployment, 69
unexplained deviation, 537–538
unexplained variation, 539
uniform distribution, 243
upper class limits, 31
USA Today, 14, 20, 40

validity, 265
variability, 78
variable, 59
variance:
    calculation of, 81
    definition, 81
    for binomial distribution, 235
    for discrete random variable, 211
    notation, 81
    tests of, 380–386, 442
    units, 81
variance between samples, 595, 598
variance within samples, 595, 598
variation, 536–544
Voltaire, 181
volume, 12
voting power, 445

Warner, Stanley, 120
weather, 165, 267
weighted mean, 65–66
whales, 8
Wickramashinghe, N. C., 138
Wilcoxon rank-sum test, 645–649
Wilcoxon signed-ranks table, 722
Wilcoxon signed-ranks test, 636–640

Yankelovich, Daniel, 704
y-intercept, 523

ZIP codes, 597
z score, 97, 276